Social Bioarcl

MW01201260

BLACKWELL STUDIES IN GLOBAL ARCHAEOLOGY

Series Editors: Lynn Meskell and Rosemary A. Joyce

Blackwell Studies in Global Archaeology is a series of contemporary texts, each carefully designed to meet the needs of archaeology instructors and students seeking volumes that treat key regional and thematic areas of archaeological study. Each volume in the series, compiled by its own editor, includes 12–15 newly commissioned articles by top scholars within the volume's thematic, regional, or temporal area of focus.

What sets the *Blackwell Studies in Global Archaeology* apart from other available texts is that their approach is accessible, yet does not sacrifice theoretical sophistication. The series editors are committed to the idea that useable teaching texts need not lack ambition. To the contrary, the *Blackwell Studies in Global Archaeology* aim to immerse readers in fundamental archaeological ideas and concepts, but also to illuminate more advanced concepts, thereby exposing readers to some of the most exciting contemporary developments in the field. Inasmuch, these volumes are designed not only as classic texts, but as guides to the vital and exciting nature of archaeology as a discipline.

1 Mesoamerican Archaeology: Theory and Practice
Edited by Julia A. Hendon and Rosemary A. Joyce
2 Andean Archaeology
Edited by Helaine Silverman
3 African Archaeology: A Critical Introduction
Edited by Ann Brower Stahl
4 Archaeologies of the Middle East: Critical Perspectives
Edited by Susan Pollock and Reinhard Bernbeck
5 North American Archaeology
Edited by Timothy R. Pauketat and Diana DiPaolo Loren
6 The Archaeology of Mediterranean Prehistory
Edited by Emma Blake and A. Bernard Knapp
7 Archaeology of Asia
Edited by Miriam T. Stark
8 Archaeology of Oceania: Australia and the Pacific Islands
Edited by Ian Lilley
9 Historical Archaeology
Edited by Martin Hall and Stephen W. Silliman
10 Classical Archaeology
Edited by Susan E. Alcock and Robin G. Osborne
11 Prehistoric Britain
Edited by Joshua Pollard
12 Prehistoric Europe
Edited by Andrew Jones
13 Egyptian Archaeology
Edited by Willeke Wendrich
14 Social Bioarchaeology
Edited by Sabrina C. Agarwal and Bonnie A. Glencross

Social Bioarchaeology

Edited by

Sabrina C. Agarwal and
Bonnie A. Glencross

A John Wiley & Sons, Ltd., Publication

This edition first published 2011
© 2011 Blackwell Publishing Ltd except for editorial material and organization © 2011 Sabrina C. Agarwal and Bonnie A. Glencross

Blackwell Publishing was acquired by John Wiley & Sons in February 2007. Blackwell's publishing program has been merged with Wiley's global Scientific, Technical, and Medical business to form Wiley-Blackwell.

Registered Office
John Wiley & Sons Ltd, The Atrium, Southern Gate, Chichester, West Sussex, PO19 8SQ, United Kingdom

Editorial Offices
350 Main Street, Malden, MA 02148-5020, USA
9600 Garsington Road, Oxford, OX4 2DQ, UK
The Atrium, Southern Gate, Chichester, West Sussex, PO19 8SQ, UK

For details of our global editorial offices, for customer services, and for information about how to apply for permission to reuse the copyright material in this book please see our website at www.wiley.com/wiley-blackwell.

The right of Sabrina C. Agarwal and Bonnie A. Glencross to be identified as the author of the editorial material in this work has been asserted in accordance with the UK Copyright, Designs and Patents Act 1988.

All rights reserved. No part of this publication may be reproduced, stored in a retrieval system, or transmitted, in any form or by any means, electronic, mechanical, photocopying, recording or otherwise, except as permitted by the UK Copyright, Designs and Patents Act 1988, without the prior permission of the publisher.

Wiley also publishes its books in a variety of electronic formats. Some content that appears in print may not be available in electronic books.

Designations used by companies to distinguish their products are often claimed as trademarks. All brand names and product names used in this book are trade names, service marks, trademarks or registered trademarks of their respective owners. The publisher is not associated with any product or vendor mentioned in this book. This publication is designed to provide accurate and authoritative information in regard to the subject matter covered. It is sold on the understanding that the publisher is not engaged in rendering professional services. If professional advice or other expert assistance is required, the services of a competent professional should be sought.

Library of Congress Cataloging-in-Publication Data

Social bioarchaeology / edited by Sabrina C. Agarwal and Bonnie A. Glencross.
p. cm. – (Blackwell studies in global archaeology)
Includes bibliographical references and index.
ISBN 978-1-4051-9187-6 (hardback); ISBN 978-1-4443-3767-9 (paperback)
1. Human remains (Archaeology) 2. Social archaeology.
3. Adaptation (Biology)–History. 4. Human growth–History. 5. Aging–History.
6. Life spans (Biology)–History. I. Agarwal, Sabrina C. II. Glencross, Bonnie A.
CC79.5.H85S634 2010
930.1–dc22

2010038142

A catalogue record for this book is available from the British Library.

Set in 10/12.5pt Plantin by Toppan Best-set Premedia Ltd
Printed and bound in Malaysia by Vivar Printing Sdn Bhd

01 2011

Contents

List of Tables and Figures

Tables

Figures

Notes on Contributors

Sabrina C. Agarwal is an Assistant Professor in the Department of Anthropology at the University of California Berkeley. She received her B.A. and M.Sc. from the University of Toronto, and Ph.D. from the same institution, working jointly in the Samuel Lunenfeld Research Institute of Mount Sinai Hospital, Toronto, and subsequently enjoyed two years as a Social Sciences and Humanities Research Council (SSHRC) Postdoctoral Fellow in the Department of Anthropology at McMaster University. Her research interests are focused broadly upon the age- and sex-related changes in bone quantity and quality, and particularly in the application of biocultural and evolutionary approaches to the study of bone fragility. More recently, she is particularly interested in the application of research in bone maintenance to dialogues of social identity and embodiment in bioarchaeology. She has examined age-related changes in cortical bone microstructure, trabecular architecture, and mineral density in several British archaeological populations, and is currently examining the long-term effect of growth and reproduction on the human and non-human primate maternal skeleton, studying samples from Turkey and Japan. She has recent publications in the *American Journal of Physical Anthropology*, and is co-editor of the volume *Bone Loss and Osteoporosis: An Anthropological Perspective* (with Sam Stout).

Valerie A. Andrushko is an Assistant Professor in the Department of Anthropology at Southern Connecticut State University. She received her Ph.D. in 2007 from the University of California, Santa Barbara. Since 1999, she has led a regional research program on prehistoric human burials from Cuzco, Peru, the ancient capital of the Inca Empire. Her recent articles include "Prehistoric Trepanation in the Cuzco Region of Peru: A View into an Ancient Andean Practice" (with John Verano; 2008) and "Strontium Isotope Evidence for Prehistoric Migration at Chokepukio, Valley of Cuzco, Peru" (with Michele Buzon et al.; 2009). Along with her long-standing research program in Cuzco, she has also investigated trophy-taking among prehistoric groups in California ("Trophy-Taking and Dismemberment as Warfare Strategies in Prehistoric Central California" with Al Schwitalla et al.; 2010).

George J. Armelagos is the Goodrich C. White Professor of Anthropology at Emory University in Atlanta, Georgia. His research has focused on diet and disease in human adaptation. He has co-authored *Demographic Anthropology* (with Alan Swedlund) and *Consuming Passions: The Anthropology of Eating* (with Peter Farb), and has co-edited *Paleopathology at the Origins of Agriculture* (with Mark Cohen) and *Disease in Populations in Transition: Anthropological and Epidemiological Perspective* (with Alan Swedlund). He has authored or co-authored over 260 articles and has had four of his papers reprinted 60 times. He has been President of American Association of Physical Anthropologists and Chair of the Anthropology Section of the American Association for the Advancement of Science. He is the recipient of the Viking Fund Medal for 2005, one of anthropology's highest honor. He was the Distinguished Lecturer for the Biological Anthropology Section of the American Anthropological Association for 2005. In 2008, the American Anthropological Association awarded him the Franz Boas Award for Exemplary Service. In 2009, he received the Charles Darwin Award for Lifetime Achievement from the American Association of Physical Anthropologists.

Autumn R. Barrett is Assistant Director of the Institute for Historical Biology at the College of William and Mary, and Adjunct Assistant Professor at the College of William and Mary and Virginia Commonwealth University. She received her M.A. from the College of William and Mary, where she is currently a Ph.D. candidate. She specializes in historical anthropology of the African Diaspora, combining bioarchaeological, ethnographic, and documentary analyses. She was a contributor to the skeletal biology component of the New York African Burial Ground Research Project. Her research interests include childhood health within the context of life histories, the role of childhood and child labor in colonialism and nation-building, and the political economy of historical narratives. Her dissertation compares representations of African enslavement and revolt in Brazil and the United States, focusing on dialogue and socio-political contest surrounding the commemoration of two historic African cemeteries.

Patrick Beauchesne is currently a Ph.D. candidate at University of California Berkeley. He earned his master's degree from the University of Western Ontario and his bachelor's from McMaster University. His research interests began with histological investigations of bone pathology in archaeological skeletal remains and now include skeletal biology, growth and development, applications of computed tomography in biological anthropology, life history and evolutionary theory, and bioarchaeology. He has bioarchaeological field and laboratory experience in various countries including Peru, Italy, and Turkey. His most recent publication, Beauchesne and Saunders (2006), dealt with applying a simple hand-grinding method to the production of bone histological slides using archaeological skeletons. His current dissertation work involves exploring growth and development, physiological stress and life history theory in a Roman archaeological assemblage.

Michael L. Blakey is the National Endowment for the Humanities Professor of Anthropology, Professor of American Studies, and Founding Director of the Institute for Historical Biology at the College of William and Mary. He is a key advisor on prominent museum projects including the Race Exhibition of the American Anthropological Association and the National Museum of African American History and Culture of the Smithsonian Institution, where he held several prior research positions at the Natural History Museum. He was Scientific Director of New York City's colonial African Burial Ground archaeological site that has become a National Monument. He has numerous publications in flagship journals on bioarchaeology, race, racism, ethics, and the history of anthropology. He has held professorships at Spelman College, Columbia, Brown, Rome, and Howard University where he founded the W. Montague Cobb Biological Anthropology Laboratory.

Bonnie A. Glencross is an Assistant Professor in the Department of Archaeology and Classical Studies at Wilfrid Laurier University in Ontario, Canada. Bonnie studied at the University of Toronto where she received her B.Sc., M.A. and Ph.D. (2003) in anthropology, and produced her doctoral thesis on skeletal injury patterns and lifetime fracture risk in prehistoric hunter-gatherers from Indian Knoll, Kentucky. Beginning in 2006, she held a two-year Social Sciences and Humanities Research Council (SSHRC) Postdoctoral Fellowship in the Department of Anthropology, University of California Berkeley. Her postdoctoral research remains a focus and is contributing to a long-term collaborative and interdisciplinary investigation of biocultural adaptations in the Neolithic community of Çatalhöyük, Turkey. In addition to her research in Turkey, she is also the primary investigator for a project on work-related injury and fatalities during the Victorian era in Ontario. She has published in the *International Journal of Osteoarchaeology* on the identification of childhood skeletal trauma (Glencross, B. & Stuart-Macadam, P. (2000). Childhood trauma in the archaeological record. *International Journal of Osteoarchaeology* 10:198–209), and on quantitative methods in bioarchaeology (Glencross, B. & Sawchuk, L. (2003). The person-years construct: Ageing and the prevalence of health-related phenomena from skeletal samples. *International Journal of Osteoarchaeology*, 13:369–74).

Siân E. Halcrow was awarded her Ph.D. in 2007 from the University of Otago, New Zealand, and is now a postdoctoral fellow at the University. She has research expertise in the analysis of health and disease of infants and children from prehistoric Southeast Asia and has worked as a bioarchaeologist on projects in Thailand, Laos, Cambodia, and New Zealand. She has published papers on dental formation, dental pathology, paleodemography, diet and childhood in the past.

Sandra E. Hollimon is an Instructor of Anthropology in the Behavioral Sciences Department of Santa Rosa Junior College. She is also the Director of the campus museum, which specializes in Native American art. She received her Ph.D. in anthropology from the University of California, Santa Barbara. Her research concerns the antiquity of gender systems in Native North American cultures. In 2006, she published "The Archaeology of Non-binary Genders in Native North American Societies," in the volume *Handbook of Gender in Archaeology*, edited by Sarah M. Nelson.

Mary Jackes studied social anthropology in Australia for her bachelor's and master's degrees and moved to Canada to do her Ph.D. at the University of Toronto. She became a fossil hunter at Olduvai and Laetoli, but was always frustrated by the necessity of making sweeping statements based on tiny paleontological samples, so her Ph.D. made sweeping statements about 12,000 disarticulated vertebrae from Kleinburg, a proto-historic Iroquoian site. Her research since 1984 has focused on Iberia and Algeria. Methodological problems are of major interest to her: comparison of proto- and post-contact sites in Ontario and of sites bracketing the transition to agriculture in central Portugal allow her to test hypotheses on, for example, demographic change and dental pathology. Her recent publications include "The Mid-Seventeenth Century Collapse of Iroquoian Ontario: Examining the Last Great Burial Place of the Neutral Nation" in L. Buchet, C. Rigeade, I. Séguy, M. Signoli (eds.), 9e Journées Anthropologiques de Valbonne, *Vers une anthropologie des catastrophes* Editions APDCA/INED, Valbonne: CÉPAM, pp. 347–373 (2009); "Teeth and the Past in Portugal: Pathology and the Mesolithic-Neolithic Transition" in T. Koppe, G. Meyer and K. W. Alt (eds.), *Comparative Dental Morphology, Frontiers of Oral Biology* 13:167–172 (2009).

Judith Littleton received her Ph.D. from the Australian National University in 1993, and she is currently Associate Professor in Biological Anthropology at the University of Auckland, New Zealand. Her main research interests are with biocultural models of morbidity and mortality among skeletal, historic, and contemporary populations particularly exploring the concepts of local biologies and syndemics. Current projects include analysis of human remains from Bahrain and Australia as well as a large interdisciplinary project on tuberculosis among the diverse populations of New Zealand. Recent publications include a co-edited volume with J. Park, A. Herring, and T. Farmer titled *Multiplying and Dividing: Tuberculosis in Canada and Aotearoa/New Zealand* (RAL-e Auckland: University of Auckland 2008), "From the Perspective of Time: Hunter-Gatherer Burials in Southeastern Australia," *Antiquity* (2007), and with J. Park "Tuberculosis and Syndemics: Implications for Pacific Health in New Zealand," *Social Science and Medicine* (2009).

Tracy L. Prowse is an Assistant Professor in the Department of Anthropology at McMaster University. Her research explores diet and health in past populations using paleopathological and isotopic analyses of human bones and teeth. Her current project is the bioarchaeological investigation of a rural Roman cemetery on an Imperial estate at Vagnari, south Italy. Her interdisciplinary research combines skeletal, isotopic, and archaeological evidence embedded within the historical context of the populations she studies. She has published on paleodiet and migration in Roman Italy in the *American Journal of Physical Anthropology* and the *Journal of Archaeological Science*.

Charlotte Roberts is a full Professor of Archaeology at Durham University, England. She is a bioarchaeologist of 25 years, with a background in nursing, archaeology, environmental archaeology, and human bioarchaeology. She is specifically interested in the interaction of people with their environments utilizing multiple lines of evidence in the past and exploring patterns of health and disease, especially health problems common today. She utilizes a holistic approach to understanding why and how people and communities today experience health problems, especially infectious diseases, including a consideration of the impact of age, sex, gender, ethnicity, religion, and social, economic and political status on disease occurrence. Her research has so far generated over 150 publications, including papers in a range of journals, book chapters, and authored books: *The Archaeology of Disease* (2005, with Keith Manchester), *Health and Disease in Britain: from Prehistory to the Present Day* (2003, with Margaret Cox), *The Bioarchaeology of Tuberculosis: A Global View on a Remerging Disease* (2003, with Jane Buikstra), *Human Remains in Archaeology. A Handbook* (2009), and co-edited texts: *The Past and Present of Leprosy* (2002, with Mary Lewis and Keith Manchester). She currently holds UK research council grants exploring the biomolecular archaeology of tuberculosis, and the skeletons of the people buried in Anglo-Saxon Bamburgh, and is heavily involved in the Global History of Health project based at Ohio State University. She has been elected President for the Paleopatholgoy Association for 2011–2013.

Joanna Sofaer is a Senior Lecturer in Archaeology at the University of Southampton. She has published widely on archaeological theory, osteoarchaeology and European prehistory. She is the author of *The Body as Material Culture* (2006), editor of *Children and Material Culture* (2000), *Material Identities* (2007), and *Biographies and Space* (with Dana Arnold) (2008).

Nancy Tayles is a Senior Lecturer at the University of Otago, Dunedin, New Zealand. She has worked as a bioarchaeologist in Southeast Asia and the Pacific for over 20 years with a research focus on quality of life of prehistoric people and communities. She is particularly interested in the biological and demographic responses to changes in subsistence, ecological and climatic variation, development of technology, and socio-political variation. She is a member of a multidisciplinary, international project researching the late prehistoric site of Ban Non Wat, Thailand, and has the pleasure of working with a large sample ($n = 637$) of burials spanning around 2,000 years of occupation at the site, starting with a small farming community through to the end of the prehistoric period. She is also interested in issues of skeletal repatriation. She has published a co-edited book *Bioarchaeology in Southeast Asia* (with Marc Oxenham) and papers on dental and skeletal paleopathology, repatriation, isotopes and migration and field anthropology.

Bethany L. Turner is an Assistant Professor in the Department of Anthropology at Georgia State University. She specializes in multi-isotopic analysis of ancient human skeletal remains in order to assess diet and residential mobility, and their relationships to disease, well-being, and cultural variables such as gender, status, and political economic processes. She has excavated and/or studied Peruvian skeletal populations from Machu Picchu and the Cuzco region since 2004; her current research is based in Cuzco, once the capital of the Inca imperial state. Her recent publications can be found in *Journal of Archaeological Science* and *International Journal of Osteoarchaeology*. In addition to her work in Peru, she has also published and presented collaborative isotopic analyses of populations from Christian-era Sudanese Nubia, pre-Mississippian North Florida and Imperial Mongolia.

Estella Weiss-Krejci teaches at the Department of Social and Cultural Anthropology of the University of Vienna, Austria. Her research focuses on mortuary behavior of the ancient Maya, medieval and post-medieval Central Europe and Neolithic and Copper Age Iberia. Additionally, she regularly conducts archaeological fieldwork in Belize where she investigates ancient Maya water storage features. Her research results have been published in archaeological journals such as *Antiquity*, *Latin American Antiquity*, and *Journal of Social Archaeology* and in various edited books.

Sonia Zakrzewski is a Senior Lecturer in Archaeology at the University of Southampton, England. She received her Ph.D. in biological anthropology from the University of Cambridge. Her research interests focus upon the recognition of social identity, migration, mobility, and religion within human osteology, particularly in ancient Egypt and medieval Spain. Her previous research has focused upon the bioarchaeological and osteoarchaeological changes associated with state formation in ancient Egypt, and she is currently writing a monograph that focuses upon Egyptian skeletal identity. Her publications have appeared in the journals *American Journal of Physical Anthropology*, *International Journal of Osteoarchaeology*, and *Journal of Archaeological Science*.

Molly K. Zuckerman is currently a Postdoctoral Fellow at the South Carolina Institute of Archaeology and Anthropology (SCIAA), University of South Carolina. She pursued her undergraduate studies in Anthropology at Pennsylvania State University, complemented with a second B.A. in Women's Studies, and engaged in post-baccalaureate studies at the Smithsonian Institution's National Museum of Natural History, where she received training in paleopathology and osteology. Her primary research interests lie in the biosocial determinants of health disparities in the past, with a special focus on the evolution and ecology of infectious disease. Her dissertation research investigates the social and ecological history of syphilis in Early Modern England, and her current research is based in England, but she has also participated in interdisciplinary fieldwork in Kenya, Scotland, and the American West. Her work has been published in several edited volumes and she is also currently engaged in collaborative isotopic and trace element analyses of a sample of Imperial-era Mongolian mummies and of skeletal samples from the English Industrial Revolution.

Series Editors' Preface

This series was conceived as a collection of books designed to cover central areas of undergraduate archaeological teaching. Each volume in the series, edited by experts in the area, includes newly commissioned articles written by archaeologists actively engaged in research. By commissioning new articles, the series combines one of the best features for readers, the presentation of multiple approaches to archaeology, with the virtues of a text conceived from the beginning as intended for a specific audience. While the model reader for the series is conceived of as an upper-division undergraduate, the inclusion in the volumes of researchers actively engaged in work today will also make these volumes valuable for more advanced researchers who want a rapid introduction to contemporary issues in specific subfields of global archaeology.

Each volume in the series will include an extensive introduction by the volume editor that will set the scene in terms of thematic or geographic focus. Individual volumes, and the series as a whole, exemplify a wide range of approaches in contemporary archaeology. The volumes uniformly engage with issues of contemporary interest, interweaving social, political, and ethical themes. We contend that it is no longer tenable to teach the archaeology of vast swaths of the globe without acknowledging the political implications of working in foreign countries and the responsibilities archaeologists incur by writing and presenting other people's pasts. The volumes in this series will not sacrifice theoretical sophistication for accessibility. We are committed to the idea that usable teaching texts need not lack ambition.

Blackwell Studies in Global Archaeology aims to immerse readers in fundamental archaeological ideas and concepts, but also to illuminate more advanced concepts, exposing readers to some of the most exciting contemporary developments in the field.

Lynn Meskell and Rosemary A. Joyce

1

Building a Social Bioarchaeology

Sabrina C. Agarwal and
Bonnie A. Glencross

> "Every man is the builder of a temple, called his body, to the god he worships, after a style purely his own, nor can he get off by hammering marble instead. We are all sculptors and painters, and our material is our own flesh and blood and bones" (Henry David Thoreau).

Perhaps nothing so easily invokes the mind to reflect on life and death as the human skeleton. While often regarded as the dry and inanimate remainder of the corpse, during life the skeleton forms the building blocks of our personal and social experience with the world. As such, for the archaeologist, human skeletal remains provide an assemblage like no other. Skeletal remains offer not only corporeal evidence of human existence, but also a biological material that has been crafted and shaped through the cultural experiences of life and death.

The duality of skeletal remains as both a biological and cultural entity has formed the basis of bioarchaeological theoretical inquiry. Bioarchaeology has developed and long used various biocultural models that emphasize the synergistic relationship of social, cultural, and physical forces in shaping the skeletal body (Armelagos et al. 1982; Armelagos and Van Gerven 2003; Schutkowski 2008). The use of biocultural approaches are best exemplified in population-based bioarchaeological studies that strive to interpret indicators of health and disease as adaptive responses of the skeleton to large-scale change, such as shifts in subsistence (Cohen and Armelagos 1984; Cohen and Crane-Kramer 2007), political or economic change (Goodman et al. 1992), or periods of contact (Larsen and Milner 1993). This population-based perspective and emphasis on the adaptive response of the skeleton to environmental (cultural) forces can be considered the founding and first wave of theoretical

Social Bioarchaeology. Edited by Sabrina C. Agarwal and Bonnie A. Glencross
© 2011 Blackwell Publishing Ltd

engagement in bioarchaeology. The commitment of the field to move away from the historical legacy of typological classification in skeletal biology, and more recently to also go beyond mere paleopathological description (Armelagos 2003), has continued to fuel studies of skeletal stress and adaptation in past populations in a growing number of areas such as the examination of growth and development (Cardosa 2007; Hoppa and Fitzgerald 1999; Saunders 2008), metabolic stress (Brickley and Ives 2008), postcranial biomechanical adaptation (Ruff 2008; Ruff and Larsen 2001; Stock 2006), activity-related markers and degeneration (Kennedy et al. 1999; Weiss 2007; Weiss and Jurmain 2007), and trauma (Grauer and Roberts 1996, Milner et al. 1991; Torres-Rouff and Costa Junqueria 2006; Walker 2001) (For comprehensive overviews of the field at large see Cox and Mays 2000; Larsen 1997.)

The second wave of engagement in the field of bioarchaeology has focused on two somewhat divergent areas of research. First, in recent decades there has been a growing interest in the application of state-of-the-art technologies to investigate health and lifestyle in past populations (Iscan and Kennedy 1989; Katzenberg and Saunders 2008; Saunders and Katzenberg 1992). For example, there has been ongoing advancement in the field in the use of isotopic methods for reconstructing diet and migration patterns (see, for example, Dupras and Tocheri 2007; Eckhardt et al. 2009; Katzenberg 2008; Sealy 2001; White et al. 2004), ancient DNA analysis of pathological conditions (Roberts and Ingram 2008), and the use of noninvasive micro-imaging technologies (Cooper et al. 2004; Rühli et al. 2002). While the use of state-of-the-art technologies to examine archaeological material is considered groundbreaking, this work has been criticized as primarily descriptive and lacking in the analytical and population-based focus of the field at large (Armelagos and Van Gerven 2003, Hens and Godde 2008). However, these studies have been instrumental in advancing the scientific arsenal of methods available to bioarchaeologists and opening new avenues of inquiry to be pursued in studies with larger-scale questions. The second area of research in recent decades has been the critical examination of the nature of archaeological skeletal samples themselves. While the early work of 20th-century skeletal biologists and even modern popular representations of the field, particularly in relation to forensic anthropology, give the impression of skeletal remains as unbiased and unambiguous sources of clear biological evidence, bioarchaeologists have long known the reality and challenges of their data set. The publication of the seminal paper by Wood et al. (1992), entitled "The Osteological Paradox," brought to the forefront the nature of the skeletal record, specifically the role of selective mortality and hidden heterogeneity in susceptibility to illness (or frailty), their influence on the formation of skeletal samples, and how these affect our interpretation of health and disease in past populations. The challenges first outlined by Wood et al. (1992) are now expected and routine considerations in all bioarchaeological investigations of health in the past, and in fact have served to push our understanding of the biological and cultural component of individual and population frailty in disease processes further (Byers 1994; Cohen et al. 1994; Goodman 1993; Jackes 1993; Milner et al. 2008; Wright and Yoder 2003). Similarly, although early assessments of the challenges in the construction of demographic profiles in past populations were considered to be insurmountable roadblocks (Bocquet-Appel and Masset 1982), there has been much research on ways to better estimate and deal with the effects of demographic change on the composition of archaeological samples (Hoppa and Vaupel 1992; Jackes 1992; Konigsberg and Frankenberg 2002; Milner et al. 2008; Paine 2000).

The third wave of engagement within the field of bioarchaeology is anchored in the greater contextualization of archaeological skeletal remains. While the incorporation of archaeological contextual information has been central in the study of mortuary practice for some time (for example, see Beck 1995; Chapman et al. 2009; Parker Pearson 2001), only recent studies have emphasized the deeper understanding of past life ways gained through the close and simultaneous consideration of archaeological, historical, and ethnographic sources of data along with skeletal analyses (Blakey and Rankin-Hill 2004; Buikstra and Beck 2006). Attention to the biocultural adaptation of the skeleton and the use of state-of-the-art methodology is still maintained in the field; however, current research seeks to integrate elements from biological, behavioral, ecological, and social research. The goal of this new bioarchaeological practice is to transcend the skeletal body into the realm of lived experience and to make a significant contribution to our understanding of social processes and life in the past. Simultaneously, contemporary bioarchaeologists are more keenly aware of the present social world in which their work is practiced, and in which the skeletal remains they study are still situated in. Careful thought and attention is now paid to ethical considerations and the role of multiple stakeholders in bioarchaeological research. While early studies in human osteology emphasized biological and evolutionary change, contemporary bioarchaeology is now clearly a discipline poised to engage with social theory. In building a social bioarchaeology, scientists are engaged in the construction of the biological and social essence of individuals from the ground, or if you will skeleton up. Ultimately the interest is in reconstructing the biological footings of the skeletal body and cultural framework that has together created the social spaces and the social creatures that inhabit them. The collection of papers here considers the constituent parts in the construction of social bioarchaeological analyses; an emphasis is placed on skeletons and skeletal samples as basic building materials that are contextually situated, engaged in a biocultural framework, and that support and reflect social representations of identity in health and disease.

The opening chapters in *Part I, Materials and Meaning: The Nature of Skeletal Samples,* present the potential interpretive meaning and practical nature of skeletal samples. The focus in these chapters is on how the skeletal materials we study came to be, and how people in the past can affect and transform people today. In Chapter 2, Zuckerman and Armelagos begin with a recount of the origins of bioarchaeology in North America, from its initial focus on racial typology in the 19th and early 20th centuries, toward the population- and evolutionary-based research of the "New Physical Anthropology." Building from earlier critiques of the field (Armelagos 2003, Armelagos and Van Gerven 2003; Armelagos et al. 1982; Goodman 1998), the authors also expound the practical challenges in recent decades in grounding the field in a biocultural-oriented anthropology. They go beyond, to further outline how theoretical sophistication in the field has led to the greater historical and archaeological contextualization of skeletal samples, and growing concern for uncovering aspects of the social past. Most importantly, Zuckerman and Armelagos emphasize how the strength of the biocultural approach lies in its power to keep bioarchaeology socially relevant with the exploration of the links between biology and sociopolitical processes affecting identity, health, and disease.

In Chapter 3, Turner and Andrushko explore the meaning and importance of human remains to both archaeologists and culturally affiliated descent groups. In recent years there has been much written on the ethical practice of excavation, curation, and study of indigenous human skeletal remains by bioarchaeologists (Turner et al. 2004; Walker 2008; Walker and Larsen 2004), and particularly in regards to the U.S. federal NAGPRA law (Bowman 1989; Rose et al. 1996; Thomas 2000; Watkins 2004). However, there has been little discussion of the role of descendent populations and the ethical practice of bioarchaeology internationally. With their presentation of the practical and ethical concerns of studying human skeletal remains in Peru, Turner and Andrushko begin the dialogue on how bioarchaeologists trained in the post-NAGPRA age can carry out ethical bioarchaeology outside North America. They describe the very different ethical landscapes in researching indigenous Peruvian remains in comparison to those in North America, which are created by the unique relationships between indigenous Peruvians and the central government, tourism, and foreign archaeologists. In doing so the authors are able to delineate the potential pitfalls and successes possible in the practice of ethical bioarchaeology internationally. The emphasis in their paper is one of how descent groups form different views and agendas for research involving human remains, and the critical importance of developing a deep understanding of the contextual influences of descendant perspectives and beliefs that is widely applicable to bioarchaeological research in any geographical area.

The remaining two chapters in Part I deal with the realities of the skeletal record itself, specifically the processes of taphonomy that create the burial record and the biased construction of skeletal populations. Although there has been much recent discussion in mortuary archaeology on the social formation of burials (Beck 1995; Rakita et al. 2005), there have been few systematic and detailed bioarchaeological studies on the formation of prehistoric skeletal samples. In Chapter 4, Weiss-Krejci provides a detailed discussion of the formation processes of mortuary skeletal samples using a graphic model exemplified with historic and ethnographic examples. She then goes on to provide a retrospective analysis of the formation of mortuary deposits from the prehistoric Mayan site of Tikal. She illustrates how preconceived notions of "burials" and what constitutes a grave, can color our perceptions of mortuary behavior in the past. Further, this work reminds us that we must remain vigilant of the fact that the data we collect and how we categorize it is in fact part of our interpretation, and to use our own cultural norms as the basis for these categorizations can render the results removed from the social realities of the past. In Chapter 5, Jackes critically examines human skeletal samples as the underpinnings of social bioarchaeological analyses. She prudently crafts the groundwork noting inherent properties of skeletal samples and their implications for reconstructing the social past. This work begins with a demonstration of the influence of changing social and historical context from immigration, emigration, and war to religious affiliation, occupation, social status, and personal choice, on the composition of the historic cemetery of St. John's Anglican Church in Sydney, Australia. Jackes correctly reminds us that even the basic materials of bioarchaeology are designed through the influence of numerous social and physical factors, and that this has far-reaching implications for our interpretations of the past. Above all, this paper emphasizes the need to view both the skeleton and its context as integrated and inseparable, requiring the incorporation of both for the effect of a stronger foundation to social bioarchaeology.

In *Part II, Social Identity: Bioarchaeology of Sex, Gender, Ethnicity, and Disability*, chapter authors address the complex structure of social identity. Identities are composed of multiple features that are connected together in an edifice of individual personhood. While this includes sex and gender, constructs of identity also include aspects such as age, religion, ethnicity, and disability. Hollimon (Chapter 6) begins with the universal variable of biological sex, which underlies not only socially constructed gendered identities but research of any type involving human remains. She starts with an examination of theoretical issues relating to the sex/gender duality touching upon discussions of embodiment and nonbinary genders in bioarchaeology. Hollimon then goes on to discuss the study of gender identity through various sources in bioarchaeology including mortuary analysis, activity reconstruction, body modification, health and disease, diet, violence and warfare. She points out that since bioarchaeologists investigate human biocultural interactions framed within their historical and cultural context, they are well equipped to supply a comprehensive picture of sex and gender in the archaeological record. In outlining future avenues of research in the field, Hollimon points to the use of queer theory and life course perspectives to better understand the construction of gender identity in the past.

In Chapter 7, Zakrzewski emphasizes the need to address complexity and fluidity of identity through holistic analyses. In her chapter, she builds up evidence for the often intangible feature of ethnicity as expressed in a medieval skeletal assemblage from Iberia. Metric and nonmetric skeletal traits have long been considered important in the application of biological distance studies of inter-group variation among large temporal or geographically separated groups (for reviews see Pietrusewsky 2008; Saunders and Rainey 2008). Using these traits, Zakrzewski demonstrates how observed bioarchaeological identity within the Islamic Écija is composed of a variety of divergent yet superimposed aspects, including religion and ethnicity, and even age and gender. This work enables new insights on interpretation of intra-group variation and the identification of small subgroups of individuals with unique social and ethnic identities. Most importantly, Zakrzewski illustrates how ethnic identity is formed within a dynamic system constructed of social processes and personal choices, and how in the Iberian example migration and religious conversion are driving forces.

The chapter by Barrett and Blakey (Chapter 8) is also concerned with the reconstruction of identity, along with social inequality, as articulated through the skeletal and historical life histories of enslaved Africans in colonial New York. Employing multiple lines of evidence, including the skeletal, archaeological, and documentary records, the authors reconstruct the lives of the men and women of the New York African Burial Ground. Contextualizing the New York African Burial Ground project within the greater field of African American bioarchaeology, Barrett and Blakey utilize skeletal stress markers to make distinctions between the quality of life and health shared by enslaved individuals born free in Africa, and those born into slavery in New York City. This work shows the great promise of combining detailed biocultural analyses of paleopathological data with known historic contexts in order to more fully elucidate the history of African Diaspora communities. Finally, in Chapter 9 Roberts explores identity and embodiment through disease experiences in the past. Adding to the recent use of disability theory to examine the physical and social costs of disease in the past (Hawkey 1998; Roberts 2000), Roberts considers quality of life, for those stricken with leprosy and tuberculosis in late medieval England in the context of perceptions and stigma associated with these diseases. While individuals with either disease are stigmatized today, in late medieval England leprosy was commonplace in special hospitals as well as social constructions of health. Interestingly, the bioarchaeological record provides little evidence to support the idea of nonacceptance of people with leprosy or tuberculosis by the communities in which they resided. Roberts notes that bioarchaeologists, in conducting disease research, have long dealt with aspects of identity although they have failed to be explicit in stating so, and this may be due to a preoccupation with material rather than conceptual understandings of disease. Since bioarchaeologists routinely rely on syntheses of clinical, archaeological, and historical evidence, she demonstrates how they can potentially contribute to our understanding of not only the physical, but also the mental and social experience of disease.

In the third and final part of the volume, *Part III, Growth and Aging: The Life Course of Health and Disease,* six papers address aspects of biological and social age in the past, and the application of life course theory to research on growth, aging, diet, and injury in bioarchaeology. In Chapter 10, Sofaer examines the theoretical foundations for the social understanding of age and aging in relation to the study of human remains. Determination of age or age structure is a core component of all skeletal-based investigations. Bioarchaeologists and paleopathologists, in dealing with biological materials, have necessarily relied on biological age estimations as the basis for interpretation. Through the exploration and comparison of alternate approaches to the study of age in philosophy, psychology, and sociology, Sofaer provides us with provocative and novel theoretical directions for the social understanding of age and the aging process in bioarchaeology. She also highlights the challenges faced by bioarchaeologists to reconcile their interpretative focus on biology with a wider understanding of age as more than a category, but rather a process situated within human development, including life experiences and attitudes toward age and aging. Agarwal and Beauchesne (Chapter 11) further critically examine aging in bioarchaeology in their discussion of the plasticity and development of the skeleton. The authors first provide an overview of the history and limitations of the concept of plasticity and adaptation in human biology and bioarchaeology. They then explore alternative perspectives on human morphology in development and plasticity, particularly drawing from theoretical approaches of developmental systems theory (DST) and life course theory. Agarwal and Beauchesne outline new directions in the study of bone maintenance and aging of the skeleton that are possible with the integration of ideas in both biological and social theory. More importantly, this work calls for a focus on the pivotal role of ontogenetic processes and embodied lived experience in the construction of the healthy and aged skeleton.

Chapters 12 and 13 consider the bioarchaeology of childhood. Halcrow and Tayles focus specifically on the practical and theoretical issues surrounding social bioarchaeological investigations of children and childhood. They first discuss the issues of terminology in subadult research, and provide a historical overview of the primary methods of assessing subadult health in the skeletal record such as the examination of nonspecific stress indicators, diet, and trauma. The authors then present a detailed examination of the construction of childhood and the social child. Echoing earlier chapters in this volume, they argue that biosocial approaches to age and childhood need to integrate the (bio)archaeological data with childhood social theory, and not view biological and social aspects of the body as mutually exclusive. Chapter 13 provides us with a bioarchaeological example of modeling childhood from the skeletal record. Using multiple indicators of skeletal stress, Littleton presents a comprehensive picture of growing up in the Hellenistic period in Bahrain through an exploration of the biological, environmental, and sociocultural influences on the risk of survival. She seeks to identify what were particularly "risky" periods of childhood at the local or community level and touches upon the challenges of tackling the issue of individual heterogeneity in health and disease. Further, this work highlights how localized ecological conditions place children at risk or may in fact protect them from the external environment.

The final two chapters in Part III explore the use of life course theory as a developmental framework in bioarchaeology, and emphasize the powerful connections between people's lives and the changing social and historical context in which they unfold. In Chapter 14, Glencross argues that paleopathology, and particularly skeletal injury, are unique sources of biological data that when combined with other contextual information have the ability to make significant contributions to the exploration of social identity, cultural age, and social agency across the life course. She emphasizes the strong relationship between growth, development, chronological age, fracture patterns, and associated behaviors as forming the basis for identifying age-related patterns of skeletal injury. As an example, Glencross integrates evidence from mortuary data and age-centered patterns of skeletal injury at Indian Knoll within the context of traditional value systems. She demonstrates for the group, peak fracture frequencies at adolescence, middle age, and in old adults, with distinct patterns between males and females. Glencross shows that when interpreted in the context of traditional value systems, the patterns of fracture risk highlight how communities shape and guide individual behavior in social relations and responsibilities across the life course. This work also clearly shows how the added dimension of viewing skeletal fractures as accumulated pathology boosts our ability to understand skeletal injury in the context of lifelong processes. In the final chapter (Chapter 15), Prowse uses life course theory to explore diet and dental health in the Roman Imperial (first–third centuries A.D.) skeletal sample from Isola Sacra, Italy. Like Glencross, her approach is interdisciplinary, integrating the analysis of stable isotopes and dental pathology data within the historical context of Roman Italy. She specifically integrates isotopic and dental evidence with literary and archaeological evidence of food choice in the Roman diet. Prowse clearly shows that men and women were eating different diets and points to the historical evidence from the Roman period that suggests tighter control over the bodies and behaviors of women, and the disparate status of Roman women as compared to men. The combined data not only provide us with the opportunity to explore diet in this Roman period sample but provides insight into the variability between sexes and among different age cohorts which can be explained within the framework of social relations and status.

The overarching theme of all the chapters of this volume and social bioarchaeology more broadly, is that of the contextualization of human skeletal remains. The inclusion and integration of mortuary, archaeological, archival, and/or ethnographic data along with what is found in the skeletal record, is now becoming routinely part of the bioarchaeologist's toolkit. Current bioarchaeology has also never been more noticeably distinct from the descriptive and reductionist skeletal biology of the past. Contemporary bioarchaeologists are much more engaged with social theory as they strive to better connect the biology and social construction of the skeleton. Easily stemming from this and ethics in archaeology, is the growing interest in the practice of a bioarchaeology that involves community outreach and consideration of multiple stakeholders (Swidler 1997). Further, as evinced in these chapters, the success of the field is in the use of multidisciplinary and multiscalar research that can skillfully glide between varying levels of analysis and builds on collaborative perspectives. Perhaps most exciting is the realization that the temporal and regional boundaries of bioarchaeology can be pushed open. It is becoming increasingly apparent that methods and lines of inquiry are shared interests; no longer is the work of researchers limited to an audience of a specific geographical expertise, or tied to the historic/prehistoric divide. It is with these new directions and hopes that social bioarchaeology is poised as a truly holistic field to rebuild the lives and people of the past.

REFERENCES

Armelagos, G. J. 2003 Bioarchaeology as Anthropology. Archaeological Papers of the American Anthropological Association 13(1):27–40.

Armelagos, G.J., and D. P. Van Gerven 2003 A Century of Skeletal Biology and Paleopathology: Contrasts, Contradictions, and Conflicts. American Anthropologist 105(1):51–62.

Armelagos, G. J., D. S. Carlson, and D. P. Van Gerven 1982 The Theoretical Foundations and Development of Skeletal Biology. *In* A History of American Physical Anthropology, 1930–1980. F. Spencer, ed. Pp. 305–329. New York: Academic Press.

Beck, L. ed. 1995 Regional Approaches to Mortuary Analysis. New York: Plenum Press.

Blakey, M. L., and L. M. Rankin-Hill 2004 The African Burial Ground Project. Skeletal Biology Final Report. Volumes 1 and 2. Washington, DC: The African Burial Ground Project, Howard University for the United States General Services Administration, Northeast and Caribbean Region.

Bocquet-Appel, J. P., and C. Masset 1982 Farewell to Palaeodemography. Journal of Human Evolution 11:321–333.

Bowman M. B. 1989 The Reburial of Native American Skeletal Remains: Approaches to the Resolution of a Conflict. Harvard Environmental Law Review 13:147–208.

Brickley, M., and R. Ives 2008 The Bioarchaeology of Metabolic Bone Disease. New York: Academic Press.

Buikstra, J., and L. Beck 2006 Bioarchaeology: The Contextual Analysis of Human Remains. New York: Academic Press.

Byers, S. N. 1994 On Stress and Stature in the "Osteological Paradox." Current Anthropology 35(3):282.

Cardosa, H. F. V. 2007 Environmental Effects on Skeletal Versus Dental Development: Using Documented Subadult Skeletal to Test a Basic Assumption in Human Osteological Research. American Journal of Physical Anthropology 132:223–233.

Chapman, R., I. Kinnes, and K. Randsborg eds. 2009 The Archaeology of Death (New Directions in Archaeology). Cambridge: Cambridge University Press.

Cohen, M. N., and G. J. Armelagos eds. 1984 Paleopathology at the Origins of Agriculture. New York: Academic Press.

Cohen, M. N., and G. M. M. Crane-Kramer eds. 2007 Ancient Health: Skeletal Indicators of Agricultural and Economic Intensification (Bioarchaeological Interpretations of the Human Past: Local, Regional, and Global Perspectives). Gainesville: University Press of Florida.

Cohen, M. N., J. W. Wood, and G. R. Milner 1994 The Osteological Paradox Reconsidered. Current Anthropology 35(5):629–637.

Cooper, D. M. L., J. R. Matyas, M. A. Katzenberg, and B. Hallgrimsson 2004 Comparison of Microcomputed Tomographic and Microradiographic Measurements of Cortical Bone Porosity. Calcified Tissue International 74:437–447.

Cox, M., and S. Mays (2000) Human Osteology: In Archaeology and Forensic Science. London: Greenwich Medical Media.

Dupras, T. L., and M. W. Tocheri 2007 Reconstructing Infant Weaning Histories at Roman Period Kellis, Egypt Using Stable Isotope Analysis of Dentition. American Journal of Physical Anthropology 134:63–74.

Eckardt, H., C. Chenery, P. Booth, J. A. Evans, A. Lamb, and G. Muldner, 2009 Oxygen and Strontium Isotope Evidence for Mobility in Roman Winchester. Journal of Archaeological Science 36:2816–2825.

Grauer, A. L., and C. A. Roberts 1996 Paleoepidemiology, Healing and Possible Treatment of Trauma in the Medieval Cemetery Population of St. Helen-on-the-Walls, York, England. American Journal of Physical Anthropology 130:60–70.

Goodman, A. H. 1993 On the Interpretation of Health from Skeletal Remains. Current Anthropology 34(3):281.

Goodman, A. H. 1998 The Biological Consequences of Inequality in Antiquity. *In* Building a New Biocultural Synthesis: Political-Economic Perspectives on Human Biology. A. Goodman, and T. Leatherman, eds. Pp. 141–169. Ann Arbor, MI: University of Michigan Press.

Goodman, A. H., D. L. Martin, and G. J. Armelagos 1992 Health, Economic Change, and Regional Political Economic Relations: Examples from Prehistory. MASCA 9:51–59.

Hawkey, D. 1998 Disability, Compassion and the Skeletal Record: Using Musculoskeletal Stress Markers (MSM) to Construct an Osteobiography from Early New Mexico. International Journal of Osteoarchaeology 8:326–340.

Hens, S. M., and K. Godde 2008 Brief Communication: Skeletal Biology Past and Present: Are We Moving in the Right Direction? American Journal of Physical Anthropology 137(2):234–239.

Hoppa, R. D., and C. M. Fitzgerald 1999 Human Growth in the Past: Studies from Bones and Teeth. Cambridge: Cambridge University Press.

Hoppa, R. D., and J. W. Vaupel 2002 Paleodemography: Age Distributions from Skeletal Samples. Cambridge: Cambridge University Press.

Iscan, M. Y., and K. A. R. Kennedy 1989 Reconstruction of Life from the Skeleton. New York: Wiley.

Jackes, M. 1992 Paleodemography: Problems and Techniques. *In* Skeletal Biology of Past Peoples: Research Methods. S. R. Saunders and M. A. Katzenberg, eds. Pp. 189–224. New York: Wiley.

Jackes, M. (1993). On Paradox and Osteology. Current Anthropology 34(4):434–439.

Katzenberg, M. A. 2008 Stable Isotope Analysis: A Tool for Studying Past Diet, Demography, and Life History. *In* Biological Anthropology of the Human Skeleton. 2nd Edition. M. A. Katzenberg, and S. R. Saunders, eds. Pp. 413–442. Hoboken, NJ: Wiley.

Katzenberg, A., and S. R. Saunders, eds. 2008 Biological Anthropology of the Human Skeleton. 2nd Edition. New York: Wiley.

Kennedy, K. A. R., L. Capasso, and C. A. Wilczak 1999 Atlas of Occupational Markers on Human Remains. Edigrafica S.P.A., Teramo, Italy. Journal of Paleontology Monograph Publications 3.

Konigsberg, L. W., and S. R. Frankenberg 2002 Deconstructing Death in Paleodemography. American Journal of Physical Anthropology 117:297–307.

Larsen, C. S. 1997 Bioarchaeology: Interpreting Behavior from the Human Skeleton. Cambridge, MA: Cambridge University Press.

Larsen, C. S., and G. R. Milner eds. 1993 In the Wake of Contact: Biological Responses to Conquest. New York: Wiley.

Milner, G. R., E. Anderson and V. G. Smith 1991 Warfare in Late Prehistoric West-Central Illinois. American Antiquity 56:581–603.

Milner, G. R., J. W. Wood, and J. Boldsen 2008 Advances in Palaeodemography. *In* Biological Anthropology of the Human Skeleton. 2nd Edition. M. A. Katzenberg, and S. R. Saunders, eds. Pp. 561–600. New York: Wiley.

Paine, R. R. 2000 If a Population Crashes in Prehistory and There is No Paleodemographer There to Hear It, Does It Make a Sound? American Journal of Physical Anthropology 112(2):181–190.

Parker Pearson, M. 2001 The Archaeology of Death and Burial. Texas: Texas A&M University Press.

Pietrusewsky, M. 2008 Metric Analysis of Skeletal Remains: Methods and Applications. *In* Biological Anthropology of the Human Skeleton. 2nd Edition. M. A. Katzenberg, and S. R. Saunders, eds. Pp. 487–532. New York: Wiley.

Rakita, G. F.M., J. E. Buikstra, L. A. Beck, and S. R. Williams eds. 2005 Interacting With the Dead: Perspectives on Mortuary Archaeology for the New Millennium. Gainesville: University Press of Florida.

Roberts, C. 2000 Did They Take Sugar? The Use of Skeletal Evidence in the Study of Disability in Past Populations. *In* Madness, Disability and Social Exclusion. The Archeology and Anthropology of Difference. J. Hubert, ed. Pp. 46–59. London: Routlege.

Roberts, C., and S. Ingham 2008 Using Ancient DNA Analysis in Palaeopathology: A Critical Analysis of Published Papers with Recommendations for Future Work. International Journal of Osteoarcheology 18(6):600–613.

Rose, J. C., T. J. Green, and V. D. Green 1996 NAGPRA is Forever: Osteology and the Repatriation of Skeletons. Annual Review of Anthropology 25:81–103.

Ruff, C. 2008 Biomechanical Analyses of Archaeological Human Skeletons. *In* Biological Anthropology of the Human Skeleton. 2nd Edition. M. A. Katzenberg, and S. R. Saunders, eds. Pp. 183–206. New York: Wiley.

Ruff, C., and C. S. Larsen 2001 Reconstructing Behavior in Spanish Florida: The Biomechanical Evidence. *In* Bioarchaeology of Spanish Florida: The Impact of Colonialism. C. S. Larsen, ed. Pp. 113–145. Gainesville: University Press of Florida.

Rühli, F. J., C. Lanz, S. Ulrich-Bochsler, and K. W. Alt 2002 State-of-the-Art Imaging in Palaeopathology: The Value of Multislice Computed Tomography in Visualizing Doubtful Cranial Lesions. International Journal of Osteoarchaeology 12(5):372–379.

Saunders, S. R. 2008 The Juvenile Skeletons and Growth Related Studies. *In* Biological Anthropology of the Human Skeleton. 2nd Edition. M. A. Katzenberg, and S. R. Saunders, eds. Pp. 117–148. New York: Wiley.

Saunders, S. R., and M. A. Katzenberg 1992 Skeletal Biology of Past Peoples: Research Methods. New York: Wiley.

Saunders, S. R., and D. L. Rainey 2008 Nonmetric Trait Variation in the Skeleton: Abnormalities, Anomalies, and Atavisms. *In* Biological Anthropology of the Human Skeleton. 2nd Edition. M. A. Katzenberg, and S. R. Saunders, eds. Pp. 533–560. New York: Wiley.

Schutkowski, H. ed. 2008 Between Biology and Culture. Cambridge: Cambridge University Press.

Sealy, J. 2001 Body Tissue Chemistry and Palaeodiet. *In* Handbook of Archaeological Science. D. R. Brothwell, and A. M. Pollard, eds. Pp. 269–279. Chichester: Wiley.

Stock J. T. 2006 Hunter-Gatherer Postcranial Robusticity Relative to Patterns of Mobility, Climatic Adaptation, and Selection for Tissue Economy. American Journal of Physical Anthropology 131(2):194–204.

Swidler, N. ed. 1997 Native Americans and Archaeologists: Stepping Stones to Common Ground. Walnut Creek, CA: Alta Mira.

Thomas, D. H. 2000 Skull Wars: Kennewick Man, Archaeology and the Battle for Native Identity. New York: Basic Books.

Torres-Rouff, C., and M. A. Costa Junqueira 2006 Interpersonal Violence in Prehistoric San Pedro de Atacama, Chile: Behavioral Implications of Environmental Stress. American Journal of Physical Anthropology 130:60–70.

Turner, B. L., D. S. Toebbe, and G. J. Armelagos 2004 To the Science, to the Living, to the Dead: Ethics and Bioarchaeology. *In* The Nature of Difference: Science, Society, and Human Biology. G. T. H. Ellison, and A. H. Goodman, eds. Pp. 203–225. Boca Raton: Taylor & Francis.

Walker, P. 2001 A Bioarchaeological Perspective on the History of Violence. Annual Review of Anthropology 30:573–596.

Walker, P. 2008 Bioarchaeological Ethics: A Historical Perspective on the Value of Human Remains. *In* Biological Anthropology of the Human Skeleton. 2nd edition. M. A. Katzenberg, and S. R. Saunders, eds. Pp. 3–39. New York: Wiley.

Walker, P., and C. S. Larsen 2004 The Ethics of Bioarchaeology. *In* Ethical Issues in Biological Anthropology. Trudy Turner, ed. Albany: State University of New York Press.

Watkins, J. 2004 Becoming American or Becoming Indian? Journal of Social Archaeology 4(1):60–80.

Weiss, E. 2007 Muscle Markers Revisited: Activity Pattern Reconstruction with Controls in a Central California Amerind Population. American Journal of Physical Anthropology 133(3):931–940.

Weiss, E., and R. Jurmain 2007 Osteoarthritis Revisited: A Contemporary Review of Aetiology. International Journal of Osteoarchaeology 17(5):437–450.

White, C. D., R. Storey, F. J. Longstaffe, and M. W. Spence 2004 Immigration, Assimilation, and Status in the Ancient City of Teotihuacan: Stable Isotope Evidence from Tlajinga 33. Latin American Antiquity 15:176–197.

Wood, J. W., G. R. Milner, H. C. Harpending, and K. M. Weiss 1992 The Osteological Paradox: Problems of Inferring Prehistoric Health from Skeletal Samples. Current Anthropology 33:343–370.

Wright, L. E., and C. J. Yoder 2003 Recent Progress in Bioarcheology: Approaches to the Osteological Paradox. Journal of Archaeological Research 11:43–70.

Part I

Materials and Meaning: The Nature of Skeletal Samples

2

The Origins of Biocultural Dimensions in Bioarchaeology

Molly K. Zuckerman and
George J. Armelagos

Introduction

Over the past half century, the biocultural approach has emerged as an integrative intellectual force in biological anthropology (Goodman and Leatherman 1998; Goodman et al. 1988). In the face of myriad challenges to a holistic four-field anthropology and proliferating specialization within the discipline (Blanchard 2006; Borofsky 2002; Brown and Yoffee 1992; Holden 1993; Nichols et al. 2003; Peacock 1995; Weiner 1995), it provides a unique avenue for synthetic research that extends across the subdisciplines.

Despite this potential, challenges have been raised against this perspective. For example, Segal and Yanagisako's (2005) edited volume, *Unwrapping the Sacred Bundle: Reflections on the Disciplining of Anthropology*, challenges the wisdom of the biocultural approach. In their introduction and Yanagisako's chapter, they advocate a new era in anthropology which eschews the holistic approach that has been central to the discipline since its inception (Boas 1904; Wolf 1974). Instead, they suggest increased specialization and separation between the biological, archaeological, and sociocultural subdisciplines. Over a century ago, Boas (1904:523) had conceptualized anthropology as "... the biological history of mankind in all its varieties; linguistics applied to people without written languages; the ethnology of people without historic records; and prehistoric archeology." Instead, Segal and Yanagisako (2005), following Stocking (1988:346), contest this holistic vision. They posit that Boas actually viewed four-field anthropology as a "historically contingent" phenomenon already blemished by "indications of breaking up" (Boas 1904:523). With this interpretation, it is curious that the four-field approach has persisted – albeit with change – for over a century.

Social Bioarchaeology. Edited by Sabrina C. Agarwal and Bonnie A. Glencross
© 2011 Blackwell Publishing Ltd

In their text, Segal and Yanagisako (2005) characterize the biocultural approach as reductive, deterministic, and preferential to biological and adaptationist interpretations rather than more sociocultural approaches to the detriment of both. They also argue that it bears the stain of evolutionary perspectives developed during the earliest and most racist decades of the discipline. This, however, misinterprets the fundamental tenets of the biocultural approach. It endows contemporary practitioners with the original sins of our discipline, and most importantly, misrepresents the utility, significance, and potential of the biocultural approach for modern biological anthropology and by extension, bioarchaeology.

In this chapter, we trace the development and maturation of bioarchaeology from its origins in the biocultural approach of biological anthropology. Multiple different "bioarchaeologies" have been proposed throughout the 20th century, each differing in its foci, usage, and applications (e.g., Angel 1946; Blakely 1977a; Goodman 1998; Krogman 1935; Larsen 1997; Saul 1972). Amidst these, a bioculturally-oriented bioarchaeology emerged in the 1970s, arising from a combination of methods, data, and theoretical orientations from across anthropology, a devotion to a rigorous scientific approach (e.g., Platt 1964), and an embrace of the biocultural approach in the larger field of biological anthropology (Armelagos 2003). Bioculturally-oriented bioarchaeology provides a method for cultural comparison with an evolutionary approach suitable for application to both archaeological skeletal samples and human populations throughout time and place (Armelagos 2008).

Origins of the Biocultural Approach

Biological anthropology and its subfields have traditionally been distinct from the other more humanistic fields of anthropology. This is largely because biological anthropology and skeletal biology have had little if any concern for culture or history for the majority of their existence (Armelagos and Goodman 1998; Blakey 1987; Gould 1981; Smedley 1993). Throughout the 19th and early 20th centuries, this predisposition was evident in an obsession with race; relying primarily on the use of cranial morphometrics, many practitioners strove to establish racial typologies for various regions and cultural contexts (e.g., Morton 1839; 1844). Overall, this reflected the contemporary fascination with hereditarianism active in contemporary anthropology and the other social sciences (Dufour 2006). In biological anthropology in particular, Armelagos et al. (1982) attributes this esoteric interest to the field's origins in biology and its Linnaean and Darwinian emphasis on biological discreteness and categorization. Buikstra (2006a) deems it a result of the predominance of biomedically trained researchers in biological anthropology at the time; these had little interest in the archaeological context of the skeletal remains they studied. This approach to categorizing human variation persisted as a tenet of skeletal biology and dominant research focus well into the 20th century (e.g., Bernhard and Kandler 1974) despite growing doubts about the value of such classifications (Livingstone 1962; Montagu 1952). This thematic devotion even overshadowed innovative features of some analyses (e.g., Hooton 1930) that seemed to anticipate the modern biocultural approach (Armelagos 2003, 2008).

Emergence of the foundations of the biocultural approach in the 1950s finally bridged the gap between biological anthropology and the larger discipline. This shift was largely due to a declining interest in pure typological description and racial classification and a turn toward population-based analyses in the 1950s. In part, the rise of Nazism and institutionalized, genocidal racism in Germany in the 1940s contributed by undermining an interest in racial studies among researchers and encouraging them to find new ways to apply their methods to societal issues (Armelagos and Goodman 1998; Blakey 1987). More significantly, the "Origins and Evolution of Man" symposium, held at the Cold Springs Harbor Biological Laboratory, and its corresponding proceedings (Warren 1951) effectively introduced the populational approach to biological anthropology.[1] During the symposium, proponents of the existing typological paradigm clashed with those invested in the populational approach, which was a recent adoption from the emergent fields of population and evolutionary biology (Armelagos et al. 1982; Smocovits 1992). The symposium's typological studies laid emphasis on partitioning human variation into static racial categories; the populational analyses instead prioritized the genetic characteristics of breeding populations. In doing so, these created the first opportunity for examining the mechanics and effects of evolution in biological anthropology. Ironically, however, half of the presented papers – both typological and populational in approach – were focused on the genetics of race or on racial traits; by today's standards all of them would be classed as typological (Armelagos 2008).

This conflict was reflected in the reception to Washburn's (1951; 1953) seminal publication, "New Physical Anthropology," among biological anthropologists. These papers proposed a revolutionary, strategic redirection from the existing mode, which Washburn considered to be driven by anthropometry and speculation. Instead, the New Anthropology would be characterized by synthetic theory-driven research and motivated by hypothesis testing premised on models of evolution and adaptation. Some anthropologists eagerly and successfully adopted this new paradigm in the 1950s (see Haraway 1990). Others unsuccessfully attempted to integrate the new method of blood group analysis into this evolutionary paradigm (Boyd 1950), ultimately propagating only typological reconstructions of cultural and genetic relationships between human populations (Armelagos et al. 1982). But for the most part, like the more revolutionary components of Hooton's (1930) study, scholars largely neglected Washburn's contributions for decades (Armelagos 2003).

Nearly simultaneously with Washburn, Neumann (1952) published "Archeology and Race in the American Indian" which was heralded as a breakthrough in American archaeology and biological anthropology. Neumann was at the forefront of a number of biological anthropologists using racial models to define the morphology of Native American groups in order to reconstruct their cultural history. The paper provided methods for reconstructing the culture history and genetic relationships among Native Americans via cranial morphometrics. Ultimately, it inspired a generation of similar research on racial histories (e.g., Coon 1962; 1965; Howells 1973; 1989). Though they represented wholly oppositional paradigms and directions for the development of biological anthropology, both Washburn and Neumann's contributions were recognized as benchmarks in the field and selected for reprinting in the *Yearbook of Physical Anthropology* (Neumann 1954; Washburn 1953). The paradox of these selections did not appear to challenge notions of consistency in the minds of biological anthropologists (Armelagos 2008).

Finally, at the end of the 1950s Livingstone published a transformative paper that stands as a landmark in biocultural studies (Armelagos 2008). Working from research that established that the sickle cell trait, a genetic hemoglobin variant, had a protective effect against malaria (Allison 1954; Haldane 1949), Livingstone (1958) unraveled the linkages between population growth, subsistence strategy, the natural history of mosquitoes as disease vectors, and distribution of sickle cell trait in West Africa (Dufour 2006). Specifically, Livingstone demonstrated that selection for the trait – and frequencies of the gene – was highest in regions heavily populated by *Anopheles* mosquitoes, which carry the plasmodium parasite that causes malaria. Swidden agriculture had created environments that enabled the mosquito's proliferation, making many areas uninhabitable by human populations. In response, as the heterozygous form is protective against malaria, selection for the gene had allowed swidden agriculturalists to re-penetrate and utilize these areas. Importantly, this was one of the first studies in anthropology to conceptualize the "environment" as more than the external physical conditions of a human population (Dufour 2006). As the gene had been previously characterized as a racial marker of Negroid populations (Tapper 1995; 1998) this interpretation also struck at the core of traditional racial interpretations in anthropology. Whereas these had inhibited evolutionary investigations, Livingstone's work has also indirectly inspired a long-running series of genomic and biocultural studies on the adaptive relationship between humans, the *Anopheles* mosquito, and the *Plasmodium* parasite throughout history (e.g., Armelagos and Harper 2005; Escalante and Ayala 1994; Joy et al. 2003; Tishkoff et al. 2001).

Between the 1960s and 1980s, the biocultural approach matured through contributions from the ecological approach in anthropology (Goodman and Leatherman 1998); this same development also played a critical role in laying the foundations of bioculturally-oriented bioarchaeology. Livingstone's (1958) study promptly inspired extensive research in biological anthropology on "human adaptability": genetic adaptation, nongenetic acclimatization responses, and phenotypic plasticity to various environmental and socially influenced challenges in human populations (e.g., Frisancho 1993; Frisancho et al. 1995; Gross and Underwood 1971; Lee 1968). This also coincided with the rise of ecological concerns in popular culture and in response, the social and biological sciences (Goodman and Martin 2002). In anthropology, these studies were framed in a general "ecological approach" which conceptualized the cultural, biological, and physical components of humans' environments as an integrated whole that influenced human biology and behavior (Moran 1982; Thomas 1997; Thomas et al. 1979). In doing so, a materialist, holistic approach was generated that bridged across the subdisciplines (e.g., Bennett 1966; Rappoport 1967) and lent biological anthropology an unprecedented four-field orientation (Baker 1996; Harris 1968; Little 1982; Thomas et al. 1979).

Not all of the subdiscipline was in bloom, however. Many researchers had rejected the use of ecological, evolutionary, and processual perspectives in their research. This left much of skeletal biology in particular still mired in typology and descriptive historicism (see Armelagos 2003; Armelagos et al. 1982). While contemporary researchers may have spoken of a "processual biological anthropology" in the 1960s, Lasker (1970:2) instead characterized it as merely a "handmaiden to history." By the 1970s and 1980s the ecological approach had also become highly controversial. The challenges arose primarily from within the new perspectives of processual ecology and political-economy in sociocultural anthropology (Goodman and Leatherman 1998). They were largely directed at the approach's "adaptationist program" and tendency to naturalize social processes, a weakness that commonly arises in functionalist, evolutionary perspectives (Gould and Lewontin 1979; Haraway 1989; Lewontin 1978; Orlove 1980; Ortner 1994; Singer 1996).

Processual ecology and political economy have been critical for developing a political-economic perspective in biocultural research in biological anthropology and bioarchaeology. There are a variety of political-economic paradigms in anthropology. Overall though, they focus on the intersection of global and local systems through history and the sculpting effect that they have on the social relations and institutions that control access to fundamental resources; power and its distribution are persistent central foci. They are also explicitly concerned with class, ethnicity, gender, culture, politics, and the unequal relations of cultural institutions (Roseberry 1988). Processual ecology developed in the 1980s and places greater emphasis on mechanisms of change, actor-based models, and on conceptualizing adaptive strategies as being constrained by scarce resources and hierarchical goals (e.g., Bennett 1966; Orlove 1980). Political-economic influences on the field also encouraged greater attention to the role of power and hierarchy in resource distribution and the source of constraints acting on humans (e.g., DeWalt 1998; DeWalt and Pelto 1985).

In the 1980s and 1990s, the tenets of processual ecology, human adaptability, and political economy began to be interjected into biological anthropology. This was spurred by concerns among some researchers that physiological responses to stress and malnutrition (e.g., stunted growth) (Seckler 1982) among contemporary populations were being superficially interpreted as cost-free adaptations rather than the consequences of larger social and political processes (see Goodman 1994; Goodman and Leatherman 1998; Martorell 1989; Pelto and Pelto 1989; Thomas 1998). Since then, these perspectives have been used to investigate the effects of social inequality on human biology in past and present societies (e.g., Blakey 1998a; Crooks 1996; Leatherman 1996; Leatherman 2005; Leatherman et al. 1986; Thomas 1998). In bioarchaeology, this development was complemented by the rise of post-processual archaeology (Goodman and Leatherman 1998). This paradigm stresses attention to social inequality and the role of social relations in shifting modes of production in the past and in doing so, acts as a corrective to the "stage" analyses common to contemporary neoevolutionary approaches (Cobb 1993; Maguire 1992; 1993; Paynter 1989; Saitta 1988; 1998). From it, bioarchaeology gained political-economic perspectives and an emphasis on framing hypotheses within political, social, and economic contexts and issues, including social inequality, gender, and violence (Armelagos 2003; Goodman and Armelagos 1985; Goodman et al. 1988).

The Biocultural Approach within Biological Anthropology and Bioarchaeology

Definitions of both the biocultural approach and bioculturally-oriented bioarchaeology have shifted over the past three decades, following changes in research agendas and theoretical orientations in biological anthropology. Overall, the biocultural approach explicitly emphasizes the dynamic interaction between humans and their larger social, cultural, and physical environments. Human variability is viewed as a function of responsiveness to factors within this larger environment that both mediate and produce each other (Blakely 1977a; Dufour 2006; Van Gerven et al. 1974); effectively, biology and culture are held as dialectically intertwined (Levins and Lewontin 1985). Perhaps the most important distinction between this paradigm and others in biological anthropology is that it explicitly considers social and cultural components of the environment, as well as physical, in regards to human adaptation.

With the interjection of political economy in the 1980s and 1990s, the biocultural approach became more engaged and action-oriented than the earlier adaptationist paradigm (Buikstra 2006a). It also became firmly grounded in a consideration of the effects of social relations, particularly power relations, on human biologies (Leatherman and Goodman 1997). Researchers began to investigate the effects of sociocultural, historical, ideological, and political-economic processes, such as the control, production, and distribution of material resources, on human biologies in given cultural systems and in turn, the reciprocal influence of compromised biologies on the social fabric (Blakey 2001; Goodman and Leatherman 1998). This revised approach draws attention to the influences of both microenvironmental, proximate conditions as well as complex, ultimate political, social, and economic realities on processes of biological and cultural adaptation (Goodman and Leatherman 1998). While a descriptive bioarchaeology (or paleopathology) places the emphasis on questions about the presence, absence, or degree of a given pathology in a given temporal, geographical, or cultural context, a bioculturally-oriented modern bioarchaeology (or paleopathology) focuses instead on examining the pattern of a pathology in order to elucidate the effects of social, ecological, and political processes on health within and between populations (Goodman 1998).

As detailed by Goodman and Leatherman (1998: 19-20), a political-economic biocultural approach emphasizes several interrelated issues in biological anthropology. First, it emphasizes the importance of analyzing the social processes underlying biological variation, such as in labor and access to basic resources, rather than just indicators such as socioeconomic status or gender. It highlights the interconnections between global and local (macro–micro) conditions, particularly in the linkages between regional and even international processes. The approach also draws attention to the importance of historical specificity and contingency in understanding social change and biological consequences as well as to the active, agential role of humans in constructing their environments. Lastly, this approach emphasizes the role of ideology and knowledge in comprehending human social relations and their biological consequences. Overall, this perspective seeks to investigate the effects of localized, proximate conditions on human biologies and the linkage between these contexts and larger historical political-economic processes.

A bioculturally-oriented bioarchaeology can significantly contribute to this enterprise. Through use of this approach, bioarchaeology can go beyond simply augmenting archaeological or historical interpretations and instead illuminate experiences that affect human biology but remain invisible or altered in archaeological or historical information (Joyce 2005; Sofaer 2006; Swedlund and Herring 2003). This is because individuals cannot easily mask their biological responses to disease or malnutrition (Krieger 2005). Biocultural approaches can also facilitate bioarchaeology and biological anthropology's integration into the larger discipline of anthropology and unite practitioners across it (Blakely 1977a). More importantly, political-economic perspectives also render both anthropology in general and bioarchaeological studies of the past more relevant to contemporary societies (Martin 1998). For example, biocultural research can deconstruct and in doing so, denaturalize human suffering. Careful analysis of the social relations that structure access to material resources can reveal that unequal distributions of disease, undernutrition, trauma, and biological adaptation are the products of human action and interest rather than natural and inevitable stressors equally borne by all members of a society (Leatherman and Goodman 1997). Revealing these underlying social contexts and processes can also indicate solutions for coping with and combating them in the modern era (Goodman 1998).

The Emergence and Development of a Bioculturally-Oriented Bioarchaeology

Bioarchaeology emerged in the 1970s from the nexus of the new physical anthropology, the development of ecological approaches in both biological anthropology and the discipline as a whole, the emergence of the new archaeology, and finally, the anthropologization of paleopathology (Armelagos 2003; Armelagos and Van Gerven 2003). Bioarchaeology is premised on three primary components. These include the application of a population perspective; the recognition that culture is an adaptive force within human environments that is inextricably linked to biological adaptation; and the existence of methods for testing alternative hypotheses on the interaction between biological and cultural dimensions of the adaptive process. Bioculturally-oriented bioarchaeology in particular embraces the concept that cultural systems, such as technology, social organization, and ideology, can inhibit or encourage biological processes such as undernutrition and disease (Armelagos and Van Gerven 2003). This linkage grants bioarchaeology its creative and interpretive power for answering significant questions on the adaptive experiences of past populations on regional and broader levels (Armelagos 2003).

Much like the New Physical Anthropology, the processual or New Archaeology transformed archaeological studies. Unlike previous paradigms that were rooted in description and studies of cultural diffusion, the new archaeology adopted an empirical and ecological approach to investigating the adaptive relationship between cultural systems and their environments (Binford 1962; 1964; Binford and Binford 1968). This paradigm gave an emphasis on ecological perspectives, process, hypothesis testing, and regional-level analysis to bioarchaeology (Armelagos 2003). This regional focus inspired several of the first bioculturally-oriented analyses in bioarchaeology. Hooton (1930) pioneered[2] regional-level integrations of archaeological and biological data, but the uniform practice emerged only in the 1970s from a blend of data, methods, and theory from the two new paradigms (Larsen 1997). This began with contributions to Blakely's (1977a) influential edited volume, which demonstrated the strategy and benefits of this holistic approach (Blakely 1977b; Buikstra 1977). Subsequent researchers employed the same approach to investigate population-level responses to regionally specific environmental factors, such as synchronic versus diachronic disease patterns, particularly in relation to agricultural intensification (Buikstra 1991; Cook 1979). The majority of these studies incorporated holism, interdisciplinary research, complex systems based approaches, and new methodologies, such as isotopic dietary reconstructions, drawn from other disciplines (e.g., DeNiro and Schoeninger 1983; Gilbert and Mielke 1985; Huss-Ashmore et al. 1982; van der Merwe and Vogel 1978); these same themes persist in modern biocultural analyses (Buikstra 2006b).

The emergence of a biocultural and anthropological perspective in paleopathology and this field's shift to an ecologically-oriented paleoepidemiology in the 1960s were also critical to the development of biocultural bioarchaeology (Armelagos 2003; Goodman 1998). Before the late 1960s, paleopathological research was largely descriptive, atheoretical, and restricted to case studies presenting a particularistic diagnosis of singular lesions and disease episodes in solitary skeletons (Larsen 1997). The majority paid little attention to the role of culture in mediating disease occurrences and were more biomedically than anthropologically-oriented (see Angel 1981; Armelagos et al. 1982; Buikstra and Cook 1980; Ubelaker 1982). Spurred in part by harsh critiques of the field (Brothwell 1967; Jarcho 1966) researchers in the late 1960s and 1970s began to generate populational research that paid unprecedented attention to cultural context (Cook and Powell 2006; Ubelaker 1982). While many researchers continued (and continue) to produce case study-based research (Lovejoy et al. 1982; Stojanowksi and Buikstra 2005), there was a dramatic increase in the number of studies that integrated archaeological, ethnographic, historical, and skeletal data to explain patterns of nutrition, trauma, and disease in skeletal samples, particularly those from the New World and Sudanese Nubia (e.g., Armelagos 1968a; Cassidy 1972; Hoyme and Bass 1962; Lallo 1973; Rose 1973).

Much of this new population-level bioculturally-oriented research in paleopathology – and by the 1970s in bioarchaeology – was facilitated by the recognition and analysis of suites of skeletal stress indicators (Buikstra 1991). Skeletal stress indicators can be used to elucidate both a given cultural system's success in protectively buffering environmental stressors as well as that of individuals within (Larsen 1987; 1997). Following their recognition in the 1970s (Steinbock 1976), they were integrated into a systemic "stress concept" gleaned from human biology (Selye 1950), and from there into an increasingly sophisticated set of models for estimating interactions between host resistance, cultural systems, and environmental and sociopolitical stressors (Goodman 1991; Goodman and Armelagos 1989; Goodman et al. 1984; Goodman and Martin 2002; Goodman et al. 1988). Since then they have been continuously used to gage the success of biocultural adaptations in past populations (e.g., Angel 1966a; 1966b; Brothwell 1967; Jannsens 1970; Kerley and Bass 1967; Mittler and Van Gerven 1994; Van Gerven et al. 1974). By the 1970s and 1980s stress indicators were commonly employed in large-scale populational analyses on adaptive responses to agricultural intensification in prehistory, especially during the Neolithic Revolution (e.g., Cohen and Armelagos 1984). These fell to the wayside in the 1990s but have been revived by Steckel and Rose's (2002) *The Backbone of History*, a paleoepidemiological analysis of health in the Western hemisphere (Cook and Powell 2006). Contributions to this volume employ health indices, which are primarily derived from stress indicators. Among others, Cook and Powell (2006) have argued that this emphasis on stress was responsible for shifting paleopathology toward an anthropologically-oriented diachronic attention to population processes. In bioarchaeology, it introduced attention to wide-scale patterns of disease at the populational-level and analytical questions of biocultural adaptation and *in situ* evolution (Larsen 1997).

The acceleration of research on functional (adaptive) anatomy and morphology and biomechanics in the 1970s also bolstered the development of the biocultural approach in biological anthropology and bioarchaeology (Armelagos et al. 1982). Since then, reconstructing the behavior and lifestyles of past individuals from skeletal and archaeological data in a biocultural context has become a primary goal in bioarchaeology (Bridges 1992; 1994; Hawkey and Merbs 1995; Pearson and Buikstra 2006; Ruff 2000). Drawing upon 19th- and early 20th-century European anatomical work (Pearson and Buikstra 2006), statistical analyses (van der Klaauw 1945; 1952), and early functional craniology (Moss 1972; Moss and Young 1960), several early analyses (e.g., Hylander 1975) followed Livingstone (1958) by deconstructing the black box of skeletal features attributed to racial traits and opening them to evolutionary interpretations (Adams and Van Gerven 1978). For example, Carlson, Van Gerven, and colleagues' (Carlson and Van Gerven 1977; 1979; Van Gerven et al. 1974) functional craniometric analyses of ancient Nubian skulls demonstrated that changes in facial reduction originally attributed to migration patterns of different racial groups (Elliott-Smith and Wood Jones 1910) were instead related to dietary shifts and the cultural evolution of food production and preparation technologies.

Bioculturally-oriented functional analyses of postcranial remains gained impetus in the late 20th century. Investigations have centered on topics such as mobility patterns, the biocultural consequences of the shift to agriculture (Bridges 1989; Larsen 1981, 1982, 1995, 2006; Ruff et al. 1984), the sexual and gendered division of labor (Larsen 1998; Mays 1999; Sofaer Derevenski 2000), and the effects of undernutrition on skeletal morphology (Sibley et al. 1992) across temporal, cultural, and geographical contexts. These analyses have been premised on archaeologically, historically, and ethnographically contextualized analyses of long bone cross-sectional geometry (Ruff 2000), arthritis (Angel 1966a; Bridges 1990; Jurmain 1975, 1977a, 1977b; Merbs 1969, 1980, 1983), trauma (Lambert 1997, 2002; Lovejoy and Heiple 1981; Walker 2001), and musculoskeletal markers (Kelley and Angel 1987; Merbs 1983; Ruff and Hayes 1983). In many of these studies behavioral interpretations of various pathologies were explicitly generated from culturally specific ethnohistorical accounts. Important linkages have also been made between skeletal morphology and climatic shifts, changes in subsistence strategies, and even shifts in social relations and the control and distribution of resources (e.g., Spielmann et al. 2009). For example, though Larsen and colleagues (2001) do not employ a political-economic approach they have documented significant shifts in the biomechanical costs of coerced labor and dietary inadequacy between pre-contact and post-contact indigenous populations involved in the Spanish colonial mission system in Florida.

However, as Armelagos and others (Armelagos et al. 1982; Armelagos and Van Gerven 2003; Blackith and Reyment 1971) have noted, like other methodological and analytical advances, complex statistical and multivariate analyses of skeletal morphology do not inherently produce nontypological research. They also do not necessarily generate nondescriptive or even novel research questions. On one hand, techniques such as multivariate statistics, isotopic dietary and provenience reconstructions, histological analyses, and biomolecular assays have made significant contributions to paleopathological and bioarchaeological research. They have been used to document past diets and mobility patterns (see Katzenberg and Harrison 1997; Schwarcz et al. 1991; Schwarcz and Schoeninger 1991; White et al. 2000; White et al. 1998), detect evidence of biochemical prevention and treatment for pathologies (Bassett et al. 1980; Rasmussen et al. 2008), and augment paleopathological data (Katzenberg and Harrison 1997; Katzenberg and Pfeiffer 1995). More controversially, these techniques have also been used to improve diagnostic capabilities for several diseases, including anemia, infection, and treponemal disease (Schultz 2001; von Hunnius et al. 2006; Wapler et al. 2004). More bioculturally-oriented analyses have used various biochemical techniques to explore human host–pathogen co-evolution in response to human cultural, economic, and demographic shifts (Donoghue et al. 2005), assess the respective roles of environmental stressors and/or dietary inadequacy in causing nonspecific pathologies (Turner 2008; Turner et al. 2008; White and Armelagos 1997), and examine the effects of seasonality and cultural food production technologies on diet (Kohn et al. 1998; Turner et al. 2007; White 1993; White et al. 2004).

However, rather than generating new research questions, many of these techniques have been used primarily to circumnavigate the limitations of skeletal evidence, such as the relative insensitivity of bone to most diseases (Ortner 1992). In doing so, these studies are often undertaken to address longstanding questions, such as the evolutionary interaction of tuberculosis and leprosy in the medieval period (Manchester 1991), the origins and antiquity of syphilis (Bouwman and Brown 2005; von Hunnius et al. 2007), and the pre-Columbian presence of tuberculosis in the New World (Salo et al. 1994). In some cases, particularly in paleopathology, when a technique is novel or newly introduced from another discipline its mere application to skeletal analyses seems to have sufficed as the primary motivating factor for a study (Armelagos et al. 1982; Armelagos and Martin 1990). These techniques have an enormous potential to generate new bioculturally-oriented research and complicate questions of *how* social relations and the differential distribution of resources, exposure to stressors, power, and knowledge affected health in the past, but have been infrequently applied to these issues.

By the 1980s and 1990s, this wide range of influences had produced a bioarchaeology that was inherently anthropological, evolutionary, and ecological in orientation. It centered on populational analyses of the processes driving past health disparities and was driven by biocultural concerns (Goodman 1998). Within the larger field of biological anthropology, bioculturally-oriented research was becoming increasingly focused on documenting patterns of biological dysfunction in the form of undernutrition, disease, and reduced resilience to biological insults. These parameters were interpreted in relation to proximate social indicators, such as socioeconomic status, in impoverished environments and the biological consequences of ubiquitous social change (e.g., Mazess 1975). However, in most bioarchaeological analyses investigation halted at these proximate indicators; very few researchers pursued their underlying root causes or the ultimate reasons for the differential distribution of resources, stressors, power, and knowledge across past societies (Goodman and Leatherman 1998).

This restraint can be attributed to several mutually reinforcing issues. Among other concerns, the biocultural approach can be very challenging in practice (Dressler 1995; Dufour 2006; McElroy 1990). For example, in a biocultural analysis, researchers are typically out to assess the effects of a culturally defined variable – an independent variable – on some aspect of human biology. According to Dufour, these variables are often very difficult to operationalize, especially when they are composed of multiple, intersecting social, ecological, and economic variables. Successfully operationalizing them in ways that are ethnographically or historically valid and replicable requires location-specific ethnographic knowledge (or archaeological and historical) (Dressler 1995). For many skeletal samples this degree of contextual information may simply be non-existent (e.g., Djurić-Srejić and Roberts 2001). Practitioners must also grapple with understanding the complex mechanics and effects of major constructs such as health or poverty (Dufour 2006). Poverty is a multidimensional social, economic, material, and even psychological phenomena and specifics of the context – ethnographic, epidemiological, nutritional, etc., as well as characteristics of the human-built and physical environments – can lead to a multiplicity of research questions and approaches (see Narayan 2000). Lastly, understanding complex interactions between biology and culture necessitates defining and measuring multiple causal pathways, which can be extremely challenging (Dufour 2006).

In bioarchaeology and paleopathology, material and theoretical factors can also prevent researchers from fully utilizing the biocultural approach in their research (Goodman 1998). Material limitations include sampling biases, the differential preservation of various skeletal elements, and persistent difficulties in accurately aging skeletons (e.g., Binford 1971; Bocquet-Appel and Masset 1982; Henderson 1987; Jackes 2000; Waldron 1994). The primary theoretical concern for bioarchaeology and related fields premised on skeletal analysis remains the variable representativeness of archaeological skeletal samples to their original living populations (Ortner 1991; Wood et al. 1992). This remains a source of active and continuous investigation (Wright and Yoder 2003).

Goodman (1998) and Armelagos and Van Gerven (2003) also highlight the fact that many researchers simply do not ask bioculturally-oriented research questions. According to Goodman (1998), the majority of studies are steered away from either pursuing or fully realizing a biocultural approach by a persistent view of societies as integrated and functional wholes and by a focus on evolutionary questions within a narrow ecological framework. These, Goodman argues, are legacies of the ecological bias from processual archaeology and human adaptability studies. In general, these limitations prematurely prevent researchers from achieving a fully fleshed out consideration of the relationships between sociopolitical factors, such as governance, access to ideology and political power, and patterns of disease in past populations.

Perhaps the most important limitation is that bioculturally-oriented research also reflects just one approach for divining the nature and implications of health and disease in the past. Since the 1950s, skeletal biology, bioarchaeology, and paleopathology have been characterized by a distinct directional schism: one side tends toward processual, analytical, and biocultural approaches; while the other tends toward descriptive, method-driven research (Armelagos 2003; Larsen 2002). In paleopathology in particular, researchers can be divided into those who ally the field more closely with biomedicine and those who adhere to a more anthropological, bioculturally-oriented direction. Both paradigms have been equally ardently recommended during the ensuing decades (e.g., Armelagos 1994; 2003; 2008; Armelagos et al. 1982; Armelagos and Van Gerven 2003; Garn 1962; Jarcho 1966; Ortner 1991; 1992; 1994; Washburn 1951; 1953); their actual impact on the field is best determined through content analyses (Stojanowksi and Buikstra 2005). For example, Lovejoy et al.'s (1982) survey of publications in skeletal biology from the *American Journal of Physical Anthropology* (*AJPA*) (1930–1980) documented an increase in "analytical" papers but an overall excess of "descriptive" and insufficiently theoretical publications. Armelagos and Van Gerven's (2003) re-visitation reached a similar conclusion. In a survey of later *AJPA* papers (1980–1984; 1996–2000) they found that descriptive approaches remained dominant, methodological innovations seemed to compel descriptive rather than analytical research and bioculturally-oriented analyses had stalled. In a rejoinder, Stojanowski and Buikstra's (2005) evaluation of bioarchaeological papers from the same period instead demonstrated little change in the visibility of analytical research since the 1980s and that it packs a significantly greater impact than descriptive. Like Hens and Godde (2008), who replicated Lovejoy et al.'s approach for *AJPA* and *International Journal of Osteoarchaeology* publications (1980–2004), Stojanowski and Buikstra argue that bioarchaeology has found a healthy, vital balance. It resides fruitfully between the descriptive foundations of a material-based science and the analytical and occasionally biocultural approach required to drive meaningful, explanatory, and theory-building research.

The Current State of Bioculturally-Oriented Bioarchaeological Research

Since the late 1990s, use of the biocultural approach has increased significantly in bioarchaeology and to a lesser degree, in paleopathology (Armelagos 2003; Buikstra 2006b; Goodman 1998; Larsen and Creekmore 2002). As the field matures beyond its roots as descriptive osteology and continues to address salient critiques of its basic operational assumptions (e.g., Bocquet-Appel and Masset 1982; Cadien et al. 1974; Wood et al. 1992), bioarchaeologists are engaging in research that unites social and evolutionary theory with an enhanced concern for historically and archaeologically contextualizing human remains (Knudson and Stojanowksi 2008). Bioculturally-oriented research continues on established foci, such as the evolution of human diet and cuisine (Atkins et al. 2007; Schutkowski 2008; Trevathan et al. 2008; Ungar 2007; Ungar and Teaford 2002), violence (Lambert 1997, 2002; Walker 2001) and gender-based violence (Martin 1998; Martin and Frayer 1997), studies of the African Diaspora (Blakey 1998b; 2001), and the emergence, ecology, and evolution of disease (Armelagos 2004; Armelagos et al. 1996; Armelagos and Harper 2005; Barnes et al. 1999; Barrett et al. 1998; Ortner and Schutkowski 2008) and its relationship to social inequality (Armelagos et al. 2005). But researchers are also addressing wholly new topics and perspectives. In paleopathology, Knudson and Stojanowski (2008) have identified a shift from using data from individuals to look at broad-scale processes toward using similar data to look at individual experiences of health and disease in the context of larger social issues. For example, Fay (2006) used a combination of paleopathological, bioarchaeological, and historical evidence to argue that variation in the burial context of individuals with leprosy represents variable and potentially hierarchical social and economic roles during life. This deconstructs and complicates monolithic views that posit a dichotomy between those who suffer from chronic illness and those who do not (Knudson and Stojanowksi 2008). Some paleopathologists also incorporate disability theory, which is derived from a variety of disciplines, into their analyses in order to divine the social and functional costs of severe trauma and debilitating disease in the past (Hawkey 1998; Roberts 1999; Shakespeare 1999). These interpretations have profound limits, however; DeGusta (2003), Keenleyside (2003), and most forcefully Dettwyler (1991), advise that cultural variability wholly precludes finding evidence for care giving or compassion in the record. Paleopathology, however, can certainly demonstrate evidence of cultural support (e.g., Trinkaus and Zimmerman 1982; Walker et al. 1982).

According to Knudson and Stojanowski (2008), an increased concern with contextualizing skeletal remains and integrating social theory into current research has also made bioarchaeology uniquely well suited to the study of social identity. Social identity encompasses gender, age, social and socioeconomic status, ethnic affiliation, and religion, as well as their associated roles and behavioral expectations. In addition to work on health, individual-level disease experiences, and disability, researchers are currently exploring gender identity and its relationship to biological sex in cross-cultural contexts (Geller 2008; Joyce 2005; Perry and Joyce 2001; Sofaer 2006), gender-based patterns of diet and mobility (Bentley et al. 2007; White 2005), queer theory (Geller 2008), and the health, behavior, and sexuality of women in the past (Grauer 2003; Grauer and Stuart-Macadam 1998; Hamilton et al. 2007; Hollimon 2000; Nelson 2004, 2007). The established area of biodistance analyses is being revisited with the intent of recovering evidence of social phenomena like kinship structures (Corruccini and Shimada 2002; Geller 2006; Shimada et al. 2004), ethnogenesis (Klaus 2008), and the relationship between social and biological identities (Nystrom 2006; Stojanowski and Schillaci 2006). The bioarchaeology of childhood and old age, which have been sorely neglected in the field (Lucy 2005), are also receiving increased attention (Finlay 2000; Lewis 2007; Perry 2005; Stoodley 2000). Studies of status, bodily modification, and the political and economic factors underlying systematic access to resources and exposure to stress are also being revisited in a more theoretically sophisticated biocultural context (Blom 2005; Buzon 2006; Buzon and Richman 2007; Geller 2006). Researchers have also begun to incorporate theoretical work on the concept of embodiment (e.g., Blom 2005; Sofaer 2006), which has been long established in social epidemiology and archaeology of the body (Fowler 2004; Joyce 2005; Krieger 2005).

As mentioned above, the overall scarcity of contextual information for many skeletal samples seriously limits or entirely precludes biocultural analyses and attempts at reconstructing social identities. Researchers should also exercise significant caution in this process, given the varying but omnipresent degrees of incompleteness and biases of historical, archaeological, and skeletal data (Cox 1995). When possible, however, these reconstructions can also make bioarchaeology increasingly resonant for contemporary societies. For example, the examination of age and gender identities in the past can denaturalize them – reveal the temporal and cultural contingency of these categories – and in doing so undermine the facile, unexamined foundations for misperceptions and discrimination based on age, gender, and sexual orientation (Knudson and Stojanowksi 2008).

Bioculturally-oriented bioarchaeological and paleopathological analyses can also act dialectically. In other words, investigations of the linkage between sociopolitical processes and their biological effects on past populations can be used to reveal the causes of health disparities in contemporary societies and vice versa. Influenced by the experimental traditions of New Archaeology, researchers have investigated the causes and social impacts of various pathologies frequently encountered in the archaeological record, such as enamel hypoplasias and dental caries, in contemporary populations (e.g., Goodman et al. 1992; May et al. 1993; Stuart-Macadam 1989, 1992; but see Holland and O'Brien 1997; Walker and Hewlett 1990). Bioarchaeological investigations can also be performed dialogically in order to engage and benefit contemporary societies and descendant communities (Martin 1998). For example, local African American communities helped to set the research agenda for the *African Burial Ground Project*, a bioculturally-oriented analysis of a historic African American cemetery (Blakey 1998b). The agenda included issues ranging from investigations of direct ancestry in the descent community to reconstructions of past health, diet, and behavior. Likewise, collaborative and multidisciplinary research projects between contemporary North American indigenous groups, bioarchaeologists, paleopathologists, and researchers from other scientific fields have been used to investigate the historical origins of health problems currently endemic in these groups (Brand et al. 1990; Smith et al. 1991). Many of these conditions, such as diabetes, are archaeologically invisible, but Martin (1998) states that they can be inferentially tied to historical changes in dietary quality and activity levels and have led to pragmatic solutions for preventing health disparities (e.g., Cowen 1990).

Conclusion

Biocultural approaches in bioarchaeology have revolutionized the field, facilitating its transition from a descriptive enterprise to socially, culturally, and politically informed dynamic force in biological anthropology. Just as the biocultural approach has enabled bioarchaeology's ascendance throughout the late 20th century, it also promises to extend between the sociocultural and biological subdisciplines in anthropology and provide yet needed synthesis. Ultimately, the biocultural approach may also enable researchers to understand challenges to human health in the past as well as in the present and future. Many scholars (e.g., Armelagos and Van Gerven 2003; Goodman 1998; Tylor 1881; White 1965) have argued that to constitute a socially and scientifically valid endeavor, anthropological research must be relevant to contemporary societies. This requires a synthetic and holistic approach that recognizes human variation as a response to environmental variation, and disease as a response to environmental inadequacy. In turn, this requires an interdisciplinary and intradisciplinary attention to all of the factors involved in human plasticity and in the human environmental milieu across time and place. Biological anthropologists must also be attentive to on-the-ground applications for their research and the potential of their results to prevent or accommodate disease and lessen and denaturalize human suffering. Only by illuminating the social, cultural, and environmental origins of biological stress can biological anthropology and bioarchaeology be at the forefront of a relevant, practical, and synthetic anthropology (Goodman and Leatherman 1998).

ACKNOWLEDGMENTS

Thanks are extended to Bethany Turner. The lead author also completed this publication as a National Science Foundation Fellow.

NOTES

1 See http://library.cshl.edu/symposia/1950/index.html.
2 Hooton's (1930) publication contributed what many researchers have described as an epidemiological or populational approach to biological anthropology (i.e., Angel 1981; Buikstra and Cook 1980; Ubelaker 1982) as well as the first example of an integration of archaeological and skeletal information from a single skeletal sample in the published literature. Overall, however, his research was characterized by a racial and typological approach to skeletal morphology and human variation (Armelagos 1968b).

REFERENCES

Adams, W. Y., and D. P. Van Gerven 1978 The Retreat from Migrationism. Annual Review of Anthropology 7(1):483–532.

Allison, A. C. 1954 The Distribution of the Sickle-Cell Trait in East Africa and Elsewhere, and its Apparent Relationship to the Incidence of Subtertian Malaria. Royal Society of Tropical Medicine and Hygiene 48:312–318.

Angel, J. L. 1946 Social Biology of Greek Culture Growth. American Anthropologist 48:493–553.

Angel, J. L. 1966a Early Skeletons from Tranquility, California Volume 2. Washington, DC: Smithsonian Press.

Angel, J. L. 1966b Porotic Hyperostosis, Anemias, Malarias, and Marshes in the Prehistoric Eastern Mediterranean. Science 153:760–763.

Angel, J. L. 1981 History and Development of Paleopathology. American Journal of Physical Anthropology 56(4):509–515.

Armelagos, G. J. 1968a Paleopathology of Three Archaeological Populations from Sudanese Nubia. Dissertation, University of Colorado.

Armelagos, G. J. 1968b Aiken's Fremont Hypothesis and the Use of Skeletal Material in Archaeological Interpretation. American Antiquity 33(3):385–386.

Armelagos, G. J. 1994 Review: Human Paleopathology: Current Syntheses and Future Options. Journal of Field Archaeology 21(2):239–243.

Armelagos, G. J. 2003 Bioarchaeology as Anthropology. Archaeological Papers of the American Anthropological Association 13(1):27–40.

Armelagos, G. J. 2004 Emerging Disease in the Third Epidemiological Transition. *In* The Changing Face of Disease: Implications for Society. N. Mascie-Taylor, J. Peters, and S. McGarvey, eds. Pp. 7–23. Society for the Study of Human Biology, Vol. 43. Boca Raton, FL: CRC.

Armelagos, G. J. 2008 Biocultural Anthropology at its Origins. *In* The Tao of Anthropology. J. Kelso, ed. Pp. 269–282. Gainesville: University Press of Florida.

Armelagos, G. J., K. C. Barnes, and J. Lin 1996 Disease in Human Evolution: The Re-emergence of Infectious Disease in the Third Epidemiological Transition. AnthroNotes 18(3):1–7.

Armelagos, G. J., P. J. Brown, and B. L. Turner 2005 Evolutionary, Historical and Political Economic Perspectives on Health and Disease. Social Science and Medicine 61:755–765.

Armelagos, G. J., D. S. Carlson, and D. P. Van Gerven 1982 The Theoretical Foundations and Development of Skeletal Biology. *In* A History of American Physical Anthropology, 1930–1980. F. Spencer, ed. Pp. 305–329. New York: Academic Press.

Armelagos, G. J., and A. H. Goodman 1998 Race, Racism and Anthropology. *In* Building a New Biocultural Synthesis: Political-Economic Perspectives on Human Biology. A. Goodman, and T. Leatherman, eds. Ann Arbor, MI: University of Michigan Press.

Armelagos, G. J., and K. N. Harper 2005 Genomics at the Origins of Agriculture, Part Two. Evolutionary Anthropology 14:109–121.

Armelagos, G. J., and D. L. Martin 1990 Advances in Paleopathology: A Biocultural Approach to Understanding Osteoporosis in Prehistoric Populations. *In* Para Conocer el Hombre. S. Genovés, ed. Pp. 281–291: Instituto de Investigaciones Antropológicas, Universidad Autónoma de México.

Armelagos, G. J., and D. P. Van Gerven 2003 A Century of Skeletal Biology and Paleopathology: Contrasts, Contradictions, and Conflicts. American Anthropologist 105(1):51–62.

Atkins, P. J., P. Lummel, and D. J. Oddy 2007 Food and the City in Europe Since 1800. Aldershot: Ashgate.

Baker, P. 1996 Adventures in Human Population Biology. Annual Review of Anthropology 25:1–18.

Barnes, K. C., G. J. Armelagos, and S. C. Morreale 1999 Darwinian Medicine and the Emergence of Allergy. *In* Evolutionary Medicine. W. Trevathan, J. McKenna, and E. Smith, eds. New York: Oxford University Press.

Barrett, R. et al. 1998 Emerging and Re-Emerging Infectious Diseases: The Third Epidemiologic Transition. Annual Review of Anthropology 27:247–271.

Bassett, E. J. et al. 1980 Tetracycline-Labeled Human Bone from Ancient Sudanese Nubia. Science 209:1532–1534.

Bennett, J. W. 1966 Northern Plainsmen: Adaptive Strategies and Agrarian Life. Chicago: Aldine.

Bentley, R. A. et al. 2007 Shifting Gender Relations at Khok Phanom Di, Thailand: Isotopic Evidence from the Skeletons. Current Anthropology 48:301–314.

Bernhard, W., and A. Kandler 1974 Bevölkerungsbiologie. Stuttgart: Fischer.

Binford, L. R. 1962 Archaeology as Anthropology. American Antiquity 28:217–225.
Binford, L. R. 1964 A Consideration of Archaeological Research Design. American Antiquity 29:425–441.
Binford, S. R., and L. R. Binford 1968 New Perspectives in Archaeology. Chicago: Aldine.
Blackith, R. E., and R. A. Reyment 1971 Multivariate Morphometries. New York: Academic Press.
Blakely, R. L., ed. 1977a Biocultural Adaptation in Prehistoric America, Southern Anthropological Society Proceedings, No. 11. Athens: The University of Georgia Press.
Blakely, R. L. 1977b Sociocultural Implications of Demographic Data from Etowah, Georgia. *In* Biocultural Adaptation in Prehistoric America, Southern Anthropological Society Proceedings, No. 11. R. Blakely, ed. Pp. 45–66. Athens: University of Georgia Press.
Blakey, M. L. 1987 Skull Doctors Revisited: Intrinsic Social and Political Bias in the History of American Physical Anthropology, with Special Reference to the Work of Ales Hrdlička. Critique of Anthropology 2:7–35.
Blakey, M. L. 1998a Beyond European Enlightenment: Toward a Critical and Humanistic Human Biology. *In* Building a New Biocultural Synthesis: Political-Economic Perspectives on Human Biology. A. Goodman, and T. Leatherman, eds. Pp. 379–406. Ann Arbor, MI: University of Michigan Press.
Blakey, M. L. 1998b The New York African Burial Ground Project: An Examination of Enslaved Lives, A Construction of Ancestral Ties. Transforming Anthropology 7(1):53–58.
Blakey, M. L. 2001 Bioarchaeology of the African Diaspora in the Americas: Its Origins and Scope. Annual Review of Anthropology 30:387–422.
Blanchard, S. 2006 Obscurantist Holism Versus Clear-Cut Analysis: Will Anthropology Obviate the Biology-Culture Divide? Dialectical Anthropology 30(1–2):1–25.
Blom, D. E. 2005 Embodying Borders: Human Body Modification and Diversity in Tiwanaku society. Journal of Anthropological Archaeology 24:1–24.
Boas, F. 1904 The History of Anthropology. Science 20(512):513–524.
Bocquet-Appel, J.-P., and C. Masset 1982 Farewell to Paleodemography. Journal of Human Evolution 11:321–333.
Borofsky, R. 2002 The Four Subfields: Anthropologists as Mythmakers. American Anthropologist 104(2):463–480.
Bouwman, A. S., and T. A. Brown 2005 The Limits of Biomolecular Palaeopathology: Ancient DNA Cannot be Used to Study Venereal Syphilis. Journal of Archaeological Science 32(5):703–713.
Boyd, W. C. 1950 Genetics and the Races of Man. Boston: Heath.
Brand, J. C. et al. 1990 Plasma Glucose and Insulin Responses to Traditional Pima Indian Meals. American Journal of Clinical Nutrition 51:416–420.
Bridges, P. S. 1989 Changes in Activities with the Shift to Agriculture in the Southeastern United States. Current Anthropology 30:385–394.
Bridges, P. S. 1990 Osteological Correlates of Weapon Use. Scientific Papers Number 6. *In* A Life in Science: Papers in Honor of Lawrence J Angel. J. Buikstra, ed. Pp. 87–98. Kampsville, IL: American Anthropological Association/Center for American Archaeology.
Bridges, P. S. 1992 Prehistoric Arthritis in the Americas. Annual Review of Anthropology 21:67–91.
Bridges, P. S. 1994 Vertebral Arthritis and Physical Activities in the Prehistoric Southeastern United States. American Journal of Physical Anthropology 93:83–93.
Brothwell, D. 1967 The Biocultural Background to Disease. *In* Diseases in Antiquity: A Survey of the Diseases, Injuries, and Surgery of Early Populations. D. Brothwell, and A. Sandison, eds. Springfield, IL: C. C. Thomas.

Brown, P.J., and N. Yoffee 1992 Is Fission the Future of Anthropology. Anthropology Newsletter 33(7):1–21.

Buikstra, J. E. 1977 Biocultural Dimensions of Archaeological Study: A Regional Perspective. *In* Biocultural Adaptation in Prehistoric America, Southern Anthropological Society Proceedings, No. 11. R. Blakely, ed. Pp. 67–84. Athens: University of Georgia Press.

Buikstra, J. E. 1991 Out of the Appendix and into the Dirt: Comments on Thirteen Years of Bioarchaeological Research. *In* What Mean These Bones? Studies in Southeastern Bioarchaeology. M. Powell, P. Bridges, and A. Mires, eds. Pp. 172–189. Tuscaloosa: University of Alabama Press.

Buikstra, J. E. 2006a A Historical Introduction. *In* Bioarchaeology: The Contextual Analysis of Human Remains. J. Buikstra, and L. Beck, eds. Pp. 7–27. Amsterdam: Academic Press.

Buikstra, J. E. 2006b On the 21st Century. *In* Bioarchaeology: The Contextual Analysis of Human Remains. L. Beck, and J. Buikstra, eds. Pp. 347–357. New York: Academic Press.

Buikstra, J. E., and D. C. Cook 1980 Paleopathology: An American Account. Annual Review of Anthropology 9:433–470.

Buzon, M. R. 2006 Life in New Kingdom Nubia: A Bioarchaeological Analysis of Ethnicity, Biological Affinities, and Health. Acta Nubica: Proceedings of the X Conference for Nubian Studies, Rome, 2006, pp. 213–218. Instituto Poligrafico e Zecca dello Stato Spa.

Buzon, M. R., and R. Richman 2007 Traumatic Injuries and Imperialism: The Effects of Egyptian Colonial Strategies in Upper Nubia. American Journal of Physical Anthropology 133:783–791.

Cadien, J. D. et al. 1974 Biological Lineages, Skeletal Populations, and Microevolution. Yearbook of Physical Anthropology 18:194–201.

Carlson, D. S., and D. P. Van Gerven 1977 Masticatory Function and Post-Pleistocene Evolution in Nubia. American Journal of Physical Anthropology 46(3):495–506.

Carlson, D. S., and D. P. Van Gerven 1979 Diffusion, Biological Determinism and Biocultural Adaptation in the Nubian Corridor. American Anthropologist 81(3):561–580.

Cassidy, C. M. 1972 A Comparison of Nutrition and Health in Pre-Agricultural and Agricultural Amerindian Skeletal Populations. Unpublished Ph.D. Dissertation, University of Wisconsin.

Cobb, C. R. 1993 Archaeological Approaches to the Political Economy of Nonstratified Societies. *In* Archaeological Method and Theory. M. Schiffer, ed. Pp. 43–100. Tucson: University of Arizona Press.

Cohen, M. N., and G. J. Armelagos, eds. 1984 Paleopathology at the Origins of Agriculture. Orlando, FL: Academic Press.

Cook, D. C. 1979 Subsistence Base and Health in Prehistoric Illinois Valley: Evidence from the Human Skeleton. Medical Anthropology 3(1):109–124.

Cook, D. C., and M. L. Powell 2006 The Evolution of American Paleopathology. *In* Bioarchaeology: The Contextual Analysis of Human Remains. J. Buikstra, and L. Beck, eds. Pp. 281–323. Amsterdam: Academic Press.

Coon, C. S. 1962 The Origin of Races. New York: Alfred A. Knopf.

Coon, C. S. 1965 The Living Races of Man. New York: Alfred A. Knopf.

Corruccini, R., and I. Shimada 2002 Dental Relatedness Corresponding to Mortuary Patterning at Huaca Loro, Peru. American Journal of Physical Anthropology 117:113–121.

Cowen, R. 1990 Seeds of Protection: Ancestral Menus May Hold a Message for Diabetes-Prone Descendants. Science News 137:350–351.

Cox, M. 1995 A Dangerous Assumption: Anyone Can Be a Historian! The Lessons from Christ Church Spitalfields. *In* Grave Reflections: Portraying the Past Through Skeletal Studies. S. Saunders, and A. Herring, eds. Toronto: Canadian Scholars' Press.

Crooks, D. L. 1996 American Children at Risk: Poverty and its Consequences for Children's Health, Growth, and School Performance. Yearbook of Physical Anthropology 38: 57–86.

DeGusta, D. 2003 Aubesier 11 is Not Evidence of Neanderthal Conspecific Care. Journal of Human Evolution 45:91–94.

DeNiro, M. J., and M. J. Schoeninger 1983 Stable Carbon and Nitrogen Isotope Ratios of Bone Collagen: Variations Within Individuals, Between Sexes, and Within Populations Raised on Monotonous Diets. Journal of Archaeological Science 10(3):199–203.

Dettwyler, K. A. 1991 Can Paleopathology Provide Evidence for "Compassion"? American Journal of Physical Anthropology 84:375–384.

DeWalt, B. R. 1998 The Political Ecology of Population Increase and Malnutrition in Southern Honduras. *In* Building a New Biocultural Synthesis: Political-Economic Perspectives on Human Biology. A. Goodman, and T. Leatherman, eds. Pp. 295–317. Ann Arbor, MI: University of Michigan Press.

DeWalt, B. R., and P. J. Pelto, eds. 1985 Micro and Macro Levels of Anlaysis in Anthropology: Issues in Theory and Research. Boulder, CO: Westview.

Djurić-Srejić, M., and C. Roberts 2001 Palaeopathological Evidence of Infectious Disease in Skeletal Populations from Later Medieval Serbia. International Journal of Historical Archaeology 11(5):311–320.

Donoghue, H. D. et al. 2005 Co-infection of *Mycobacterium tuberculosis* and *Mycobacterium leprae* in Human Archaeological Samples: A Possible Explanation for the Historical Decline of Leprosy. Proceedings of the Royal Society B-Biological Sciences 272(1561): 389–394.

Dressler, W. M. 1995 Modeling Biocultural Interactions: Examples from Studies of Stress And Cardiovascular Disease. Yearbook of Physical Anthropology 38:27–56.

Dufour, D. L. 2006 Biocultural Approaches in Human Biology. American Journal of Human Biology 18(1):1–9.

Elliott-Smith, G., and F. Wood Jones 1910 Reports of the Human Remains, the Archaeological Series of Nubia, Report for 1907–1908, II. Cairo.

Escalante, A. A., and F. J. Ayala 1994 Phylogeny of the Malaria Genus *Plasmodium* Derived from rRNA Gene Sequences. Proceedings of the National Academy of Sciences 91:11373–11377.

Fay, I. 2006 Text, Space and the Evidence Of Human Remains in English Late Medieval and Tudor Disease Culture: Some Problems and Possibilities. *In* Social Archaeology of Funerary Remains. R. Gowland and C. Knüsel, eds. Pp. 190–209. Oxford: Oxbow Books.

Finlay, N. 2000 Outside of Life: Traditions of Infant Burial from Cillin to Cist. World Archaeology 31(3):407–422.

Fowler, C. 2004 The Archaeology of Personhood: An Anthropological Approach. London: Routledge.

Frisancho, A. R. 1993 Human Adaptation and Accommodation. Anne Arbor, MI: University of Michigan Press.

Frisancho, A. R. et al. 1995 Developmental, Genetic, and Environmental Components of Aerobic Capacity at High Altitude. American Journal of Physical Anthropology 96(4):431–442.

Garn, S. M. 1962 The Newer Physical Anthroplogy. American Anthropologist 64(5):917–918.

Geller, P. L. 2006 Altering Identities: Body Modifications and the Pre-Columbian Maya. *In* Social Archaeology of Funerary Remains. R. Gowland, and C. Knüsel, eds. Pp. 279–291. Oxford: Alden Press.

Geller, P. L. 2008 Conceiving Sex: Fomenting a Feminist Bioarchaeology. Journal of Social Archaeology 8(1):113–138.

Gilbert, R. I., and J. H. Mielke 1985 The Analysis of Prehistoric Diets. Orlando, FL: Academic Press.

Goodman, A. H. 1991 Health, Adaptation, and Maladaptation in Past Societies. *In* Health in Past Societies: Biocultural Interpetations of Human Skeletal Remains in Archaeological Contexts. H. Bush, and M. Zvelebil, eds. Pp. 31–38: British Archaeological Reports, International Series.

Goodman, A. H. 1994 Cartesian Reductionism and Vulgar Adaptationism: Issues in the Interpretation of Nutritional Status in Prehistory. *In* Paleonutrition: The Diet and Health of Prehistoric Americans. K. Sobolik, ed. Pp. 163–177. Carbondale, IL: Occassional Paper No. 2: Center for Archaeological Investigations.

Goodman, A. H. 1998 The Biological Consequences of Inequality in Antiquity. *In* Building a New Biocultural Synthesis: Political-Economic Perspectives on Human Biology. A. Goodman, and T. Leatherman, eds. Pp. 141–169. Ann Arbor, MI: The University of Michigan Press.

Goodman, A. H., and G. J. Armelagos 1985 Death and Disease at Dr. Dickson's Mounds. Natural History Magazine 94(9):12–18.

Goodman, A. H., and G. J. Armelagos 1989 Infant and Childhood Morbidity and Mortality Risks in Archaeological Populations. World Archaeology 21:227–242.

Goodman, A. H. et al. 1984 Health Changes at Dickson Mounds, Illinois (AD 950–1300). *In* Paleopathology at the Origins of Agriculture. M. Cohen, and G. Armelagos, eds. Pp. 271–306. New York: Academic Press.

Goodman, A.H., and T. L. Leatherman 1998 Traversing the Chasm Between Biology and Culture: An Introduction. *In* Building a New Biocultural Synthesis: Political-Economic Perspectives on Human Biology. A. Goodman and T. Leatherman, eds. Pp. 3–43. Ann Arbor, MI: University of Michigan Press.

Goodman, A. H., and D. L. Martin 2002 Reconstructing Health Profiles from Skeletal Remains. *In* The Backbone of History: Health and Nutrition in the Western Hemisphere. R. Steckel and J. Rose, eds. Pp. 11–61. Cambridge: Cambridge University Press.

Goodman, A. H., G. J. Armelagos, and J. C. Rose 1992 Socioeconomic and Anthropometric Correlates of Linear Enamel Hypoplasia in Children from Solis, Mexico. Journal of Paleopathology 2:373–380.

Goodman, A. H., R. B. Thomas, A. Swedlund, and G. J. Armelagos 1988 Biocultural Perspectives on Stress in Prehistoric, Historical, and Contemporary Population Research. American Journal of Physical Anthropology 31(S9):169–202.

Gould, S. 1981 The Mismeasure of Man. New York: Norton.

Gould, S. J., and R. Lewontin 1979 The Spandrels of San Marcos and the Panglossian Paradigm: A Critique of the Adaptationist Programme. Proceedings of the Royal Society of London, Series B 205:581–598.

Grauer A. 2003 Where Were the Women? *In* Human Biologists in the Archives: Demography, Health, Nutrition, and Genetics in Historical Populations. D. Herring, and A. Swedlund, eds. Pp. 266–288. Cambridge: Cambridge University Press.

Grauer A., and P. Stuart-Macadam eds. 1998 Sex and Gender in Paleopathological Perspective. Cambridge: University of Cambridge Press.

Gross, D., and J. Underwood 1971 Technological Change and Caloric Costs: Sisal Agriculture in Northeastern Brazil. American Anthropologist 73(3):725–740.

Haldane, J. B. S. 1949 Disease and Evolution. Supplement to La Ricerca Scientifica 19:68–76.

Haraway, D. 1989 Primate Visions: Gender, Race, and Nature in the World of Modern Science. New York: Routledge.

Hamilton, S., R. Whitehouse, and K. Wright 2007 Archaeology and Women: Ancient and Modern Issues. Walnut Creek, CA: Left Coast Press.

Haraway, D. 1990 Remodelling the Human Way of Life: Sherwood Washburn and the New Physical Anthropology, 1950–1980. *In* Bones, Bodies, Behavior: Essays on Biological Anthropology. G. J. Stocking, ed. Pp. 206–260. History of Anthropology, Volume 5. Madison: University of Wisconsin Press.

Harris, M. 1968 The Rise of Anthropological Theory. New York: TY Crowell.

Hawkey, D. 1998 Disability, Compassion and the Skeletal Record: Using Musculoskeletal Stress Markers (MSM) to Construct an Osteobiography from Early New Mexico. International Journal of Osteoarchaeology 8:326–340.

Hawkey, D., and C. F. Merbs 1995 Activity-Induced Muskuloskeletal Stress Markers (MSM) and Subsistence Strategy Changes Among Hudson Bay Eskimos. International Journal of Osteoarchaeology 5:324–338.

Henderson, J. 1987 Factors Determining the State of Preservation of Human Remains. *In* Death, Decay, and Reconstruction: Approaches to Archaeology and Forensic Science. A. Boddington, A. Garland, and R. Janaway, eds. Manchester: Manchester University Press.

Hens, S. M., and K. Godde 2008 Brief Communication: Skeletal Biology Past and Present: Are We Moving in the Right Direction? American Journal of Physical Anthropology 137:234–239.

Holden, C. 1993 Failing to Cross the Biology-Culture Gap. Science 262(5139):1641–1642.

Holland, T. D., and M. J. O'Brien 1997 Parasites, Porotic Hyperostosis, and the Implications of Changing Perspectives. American Antiquity 62(2):183–193.

Hollimon, S. 2000 Archaeology of the 'Aqi: Gender and Sexuality in Prehistoric Chumash Society. *In* Archaeologies of Sexuality. R. Schmidt, and B. Voss, eds. Pp. 179–196. New York: Routledge.

Hooton, E. A. 1930 The Indians of Pecos Pueblo: A Study of Their Skeletal Remains. New Haven, CT: Yale University Press.

Howells, W. W. 1973 Cranial Variation in Man: A Study by Multivariate Analysis of Patterns of Difference Among Recent Human Populations. *In* Papers of the Peabody Museum of Archaeology and Ethnology, Harvard University, 67. Cambridge, MA: Peabody Museum of Archaeology and Ethnology, Harvard University.

Howells, W. W. 1989 Skull Shapes and the Map: Craniometric Analyses in the Dispersion of Modern Homo. *In* Papers of the Peabody Museum of Archaeology and Ethnology, Harvard University, 79. Cambridge, MA: Peabody Museum of Archaeology and Ethnology, Harvard University.

Hoyme, L. E., and W. M. Bass 1962 Human Skeletal Remains from the Tolifero (Ha6) and Clarksville (Mc14) Sites, John H. Kerr Reservoir Basin, Virginia. Bulletin of the Bureau of American Ethnology 182:239–400.

Huss-Ashmore, R., A. H. Goodman, and G. J. Armelagos 1982 Nutritional Inference from Paleopathology. Advances in Archaeological Method and Theory 5:395–474.

Hylander, W. C. 1975 The Adaptive Significance of Eskimo Craniofacial Morphology. *In* Oro-Facial Growth and Development. A. Dahlberg, and T. Graber, eds. Pp. 129–169. The Hague: Mouton.

Jackes, M. 2000 Building the Bases for Paleodemographic Analysis: Adult Age Determination. *In* Biological Anthropology of the Human Skeleton. M. Katzenberg, and S. Saunders, eds. Pp. 417–466. New York: Wiley-Liss.

Jannsens, P. A. 1970 Paleopathology: Diseases and Injuries of Prehistoric Man. London: Paul Baker.

Jarcho, S. 1966 The Development and Present Condition of Human Paleopathology in the United States. *In* Human Paleopathology. S. Jarcho, ed. Pp. 3–30. New Haven, CT: Yale University Press.

Joy, D.A. et al. 2003 Early Origin and Recent Expansion of *Plasmodium falciparum*. Science 300(5617):318–321.

Joyce, R. A. 2005 Archaeology of the Body. Annual Review of Anthropology 34:139–158.

Jurmain, R. D. 1975 Distribution of Degenerative Joint Disease in Skeletal Populations. Unpublished Ph.D. Dissertation, Harvard University.

Jurmain, R. D. 1977a Paleoepidemiology of Degenerative Knee Disease. Medical Anthropology 1:1–14.

Jurmain, R. D. 1977b Stress and the Etiology of Osteoarthritis. American Journal of Physical Anthropology 46:353–366.

Katzenberg, M. A., and A. Harrison 1997 What's in a Bone? Recent Advances in Archaeological Bone Chemistry. Journal of Archaeological Research 5:265–293.

Katzenberg, M. A., and S. Pfeiffer 1995 Nitrogen Isotope Evidence for Weaning Age in a Nineteenth Century Skeletal Sample. *In* Bodies of Evidence. A. Grauer, ed. Pp. 221–235. New York: Wiley.

Keenleyside, A. 2003 An Unreduced Dislocated Mandible in an Alaskan Eskimo: A Case of Altruism or Adaptation? International Journal of Osteoarchaeology 13:384–389.

Kelley, J. O., and J. L. Angel 1987 Life Stress of Slavery. American Journal of Physical Anthropology 74:199–211.

Kerley, E. R., and W. M. Bass 1967 Paleopathology: Meeting Ground for Many Disciplines. Science 157:638–644.

Klaus, H. D. 2008 Out of Light Came Darkness: Bioarchaeology of Mortuary Ritual, Health, and Ethnogenesis in the Lambayeque Valley Complex, North Coast Peru (AD 900–1750). Unpublished Ph.D. Dissertation, Ohio State University.

Knudson, K. J., and C. M. Stojanowksi 2008 New Directions in Bioarchaeology: Recent Contributions to the Study of Human Social Identities. Journal of Archaeological Research 16(4):397–432.

Kohn, M. J., M. J. Schoeninger, and J. W. Valley 1998 Variability in Oxygen Isotope Compositions of Herbivore Teeth: Reflections of Seasonality or Developmental Physiology? Chemical Geology 152(1–2):97–112.

Krieger, N. 2005 Embodiment: A Conceptual Glossary for Epidemiology. Journal of Epidemiology & Community Health 59:350–355.

Krogman, W. M. 1935 Life Histories Recorded in Skeletons. American Anthropologist 37(1):92–103.

Lallo, J. 1973 The Skeletal Biology of Three Prehistoric American Indian Societies from Dickson Mounds. Unpublished Ph.D. Dissertation, University of Massachusetts.

Lambert, P. M. 1997 Patterns of Violence in Prehistoric Hunter Gatherer Societies of Coastal Southern California. *In* Troubled Times: Violence and Warfare in the Past. D. Martin, and D. Frayer, eds. Amsterdam: Gordon and Breach.

Lambert, P. M. 2002 The Archaeology of War: A North American Perspective. Journal of Archaeological Research 10:207–241.

Larsen, C.S. 1981 Functional Implications of Postcranial Size Reduction on the Prehistoric Georgia Coast, USA. Human Evolution 10:489–502.

Larsen, C.S. 1982 The Anthropology of St. Catherine's Island. 3. Prehistoric Human Biological Adaptation Anthropological Papers, American Museum of Natural History 57 (Part 3):155–270.

Larsen, C.S. 1987 Bioarchaeological Interpretation of Subsistence Economy and Behavior from Human Skeletal Remains. Advances in Archaeolological Method and Theory 10:27–56.

Larsen, C.S. 1995 Biological Changes in Human Populations with Agriculture. Annual Review of Anthropology 24:185–213.

Larsen, C.S. 1997 Bioarchaeology: Interpreting Behavior from the Human Skeleton. Cambridge: Cambridge University Press.

Larsen, C.S. 1998 Gender, Health, and Activity in Foragers and Farmers in the American Southeast: Implications for Social Organization in the Georgia Bight. *In* Sex and Gender in Paleopathological Perspective. A. Grauer, and P. Stuart-Macadam, eds. Pp. 165–189. Cambridge: Cambridge University Press.

Larsen, C.S. 2002 Bioarchaeology: The Lives and Lifestyles of Past People. Journal of Archaeological Research 10(2):119–166.

Larsen, C.S. 2006 The Agricultural Revolution as Environmental Catastrophe: Implications for Health and Lifestyle in the Holocene. Quaternary International 150(1):12–20.

Larsen, C. S., and A. Creekmore 2002 Bioarchaeology of the Late Prehistoric Guale: South End Mound I, St. Catherines Island, Georgia. New York: American Museum of Natural History.

Larsen, C. S. et al. 2001 Frontiers of Contact: Bioarchaeology of Spanish Florida. Journal of World Prehistory 15(1):69–123.

Lasker, G. W. 1970 Physical Anthropology: Search for General Processes and Principles. American Anthropologist 72:1–8.

Leatherman, T. 1996 A Biocultural Perspective on Health and Household Economy in Southern Peru. Medical Anthropology Quarterly 10(4):476–495.

Leatherman, T. 2005 A Space of Vulnerability in Poverty and Health: Political-Ecology and the Biocultural Analysis. ETHOS 33(1):46–70.

Leatherman, T. et al. 1986 Illness and Political Economy: An Andean Dialectic. Cultural Survival Quarterly 10(3):19–21.

Leatherman, T., and A. H. Goodman 1997 Expanding the Biocultural Synthesis: Toward a Biology of Poverty. American Journal of Physical Anthropology 102:1–3.

Lee, R. 1968 What Hunters Do for a Living or How to Make Out on Scarce Resources. *In* Man the Hunter. R. Lee and I. DeVore, eds. Pp. 30–48. Chicago: Aldine.

Levins, R., and R. Lewontin 1985 The Dialectical Biologist. Cambridge, MA: Havard University Press.

Lewis, M. E. 2007 The Bioarchaeology of Children: Perspectives from Biological and Forensic Anthropology. Cambridge: Cambridge University Press.

Lewontin, R. 1978 Adaptation. Scientific American 239(3):156–169.

Little, M. 1982 Development of Ideas on Human Ecology and Adaptation. *In* A History of American Physical Anthropology, 1930–1980. F. Spencer, ed. New York: Academic Press.

Livingstone, F. 1958 Anthropological Implications of Sickle Cell Gene Distribution in West Africa. American Anthropologist 60:533–562.

Livingstone, F. 1962 On the Non-Existence of Human Races. Current Anthropology 3:279–281.

Lovejoy, C. O., and K. G. Heiple 1981 Analysis of Fractures in Skeletal Populations with an Example from the Libben Site, Ottawa County. American Journal of Physical Anthropology 55:529–541.

Lovejoy, C. O., R. P. Mensforth, and G. J. Armelagos 1982 Five Decades of Skeletal Biology as Reflected in the American Journal of Physical Anthropology. *In* A History of American Physical Anthropology 1930–1980. F. Spencer, ed. Pp. 329–336. New York: Academic Press.

Lucy, S. 2005 The Archaeology of Age. *In* The Archaeology of Identity: Approaches to Gender, Age, Status, Ethnicity, and Religion. M. Diaz-Andreu, S. Lucy, S. Babic, and D. Edwards, eds. Pp. 43–66. London: Routledge.

Maguire, R. H. 1992 A Marxist Archaeology. Orlando: Academic Press.

Maguire, R. H. 1993 Archaeology and Marxism. *In* Archaeological Method and Theory. M. Schiffer, ed. Pp. 101–157. Tucson: University of Arizona Press.

Manchester, K. 1991 Tuberculosis and Leprosy: Evidence for Interaction of Disease. *In* Human Paleopathology: Current Syntheses and Future Options. D. Ortner, and A. Aufderheide, eds. Pp. 23–35. Washington, DC: Smithsonian Institution Press.

Martin, D. L. 1998 Owning the Sins of the Past: Historical Trends, Misssed Opportunities and New Directions in the Study of Human Remains. *In* Building a New Biocultural Synthesis: Political-Economic Perspectives on Human Biology. A. Goodman, and T. Leatherman, eds. Pp. 171–190. Ann Arbor, MI: University of Michigan Press.

Martin, D. L., and D. W. Frayer, eds. 1997 Troubled Times: Violence and Warfare in the Past. New York: Routledge.

Martorell, R. 1989 Body Size, Adaptation, and Function. Human Organization 48(1): 15–20.

May, R. L., A. H. Goodman, and R. S. Meindl 1993 Response of Bone and Enamel Formation to Nutritional Supplementation and Morbidity Among Malnourished Guatemalan Children. American Journal of Physical Anthropology 92(1):37–51.

Mays, S. 1999 A Biomechanical Study of Activity Patterns in a Medieval Skeletal Assemblage. International Journal of Osteoarchaeology 9:68–73.

Mazess, R. B. 1975 Biological Adaptation: Aptitudes and Acclimatization. *In* Biosocial Interactions in Population Adaptation. E. Watts, F. Johnston, and G. Lasker, eds. Pp. 9–18. The Hague: Mouton.

McElroy, A. 1990 Biocultural Models in Studies of Human Health and Adaptation. Medical Anthropology Quarterly 4:243–265.

Merbs, C. F. 1969 Patterns of Activity-Induced Pathology in a Canadian Eskimo Isolate. Unpublished Ph.D. Dissertation, University of Wisconsin, Madison.

Merbs, C. F. 1980 The Pathology of a La Jollan Skeleton from Punta Minitas, Bajas California. Pacific Coast Archaeological Society Quarterly 16(4):37–43.

Merbs, C. F. 1983 Patterns of Activity-Induced Pathology in a Canadian Inuit Population. Ottawa: National Museum of Canada.

Mittler, D. M., and D. P. Van Gerven 1994 Developmental, Diachronic, and Demographic Analysis of Cribra Orbitalia in the Medieval Christian Populations of Kulubnarti. American Journal of Physical Anthropology 93(3):287–297.

Montagu, M. A. 1952 Man's Most Dangerous Myth: The Fallacy of Race. New York: Harper.

Moran, E. F. 1982 Human Adaptability: An Introduction to Ecological Anthropology. Boulder, CO: Westview Press.

Morton, S. G. 1839 Crania Americana. London: Simpkin, Marshall & Co.

Morton, S. G. 1844 An Inquiry into the Distinctive Characteristics of the Aboriginal Race of America. Philadelphia: Dobson.

Moss, M. L., and R. W. Young 1960 A Functional Approach to Craniology. American Journal of Physical Anthropology 18:281–291.

Moss, M. O. 1972 Twenty Years of Functional Cranial Analysis. American Journal of Orthodontics 61:478–485.

Narayan, D. 2000 Voices of the Poor: Can Anyone Hear Us? New York: Oxford University Press.

Nelson, S. 2004 Gender in Archaeology: Analyzing Power and Prestige. Walnut Creek, CA: AltaMira Press.

Nelson, S. 2007 Women in Antiquity: Theoretical Approaches to Gender and Archaeology. Lanham, CA: AltaMira Press.

Neumann, G. K. 1952 Archaeology and Race in the Amerindian. *In* Archaeology of the Eastern United States. J. Griffin, ed. Pp. 13–34. Chicago: University of Chicago Press.

Neumann, G. K. 1954 Archeology and Race in the American Indian. Yearbook of Physical Anthropology 8:213–242.

Nichols, D. L., R. A. Joyce, and S. D. Gillespie 2003 Is Archaeology Anthropology? Archaeological Papers of the American Anthropological Association 13(1):3–13.

Nystrom, K. C. 2006 Late Chachapoya Population Structure Prior to Inka Conquest. American Journal of Physical Anthropology 131:334–342.

Orlove, B. 1980 Ecological Anthropology. Annual Review of Anthropology 9:235–273.

Ortner, D. J. 1991 Theoretical and Methodological Issues in Paleopathology. *In* Human Paleopathology: Current Syntheses and Future Options. D. Ortner. and A. Aufderheide, eds. Pp. 5–11. Washington, DC: Smithsonian University Press.

Ortner, D. J. 1992 Skeletal Pathology: Probabilities, Possibilities, and Impossibilities. *In* Disease and Demography in the Americas: Changing Patterns Before and After 1492. J. Verano, and D. Ubelaker, eds. Pp. 5–13. Washington, DC: Smithsonian Institution Press.

Ortner, D. J. 1994 Descriptive Methodology in Paleopathology. *In* Skeletal Biology in the Great Plains: Migration, Warfare, Health, and Subsistence. D. Owsley, and R. Jantz, eds. Pp. 73–80. Washington, DC: Smithsonian Institution Press.

Ortner, D. J., and H. Schutkowski 2008 Ecology, Culture and Disease in Past Populations. *In* Between Biology and Culture. H. Schutkowski, ed. Pp. 105–129. Cambridge: Cambridge University Press.

Ortner, S. B. 1984 Theory in Anthropology Since the Sixties. Comparative Studies in Social History 26:126–166.

Paynter, R. 1989 The Archaeology of Equality and Inequality. Annual Review of Anthropology 18:369–399.

Peacock, J. 1995 Claiming Common Ground. Anthropology Newsletter 4:1–3.

Pearson, O. M., and J. E. Buikstra 2006 Behavior and the Bones. *In* Bioarchaeology: The Contextual Analysis of Human Remains. L. Beck, and J. Buikstra, eds. Pp. 207–227. New York: Academic Press.

Pelto, G. H., and P. J. Pelto 1989 Small But Healthy? An Anthropological Perspective. Human Organization 48(1):11–15.

Perry, E., and R. Joyce 2001 Providing a Past for "Bodies that Matter": Judith Butler's Impact on the Archaeology of Gender. International Journal of Sexuailty and Gender Studies 6(1/2):63–76.

Perry, M. A. 2005 Redefining Childhood Through Bioarchaeology: Toward an Archaeological and Biological Understanding of Children in Antiquity. Archeological Papers of the American Anthropological Association 15(1):89–111.

Platt, J. R. 1964 Strong Inference. Science 146(3642):347–353.

Rappoport, R. A. 1967 Pigs for the Ancestors. New Haven, CT: Yale University Press.

Rasmussen, K. L. et al. 2008 Mercury Levels in Danish Medieval Human Bones. Journal of Archaeological Science 35(8):2295–2306.

Roberts, C. 1999 Disability in the Skeletal Record: Assumptions, Problems, and Some Examples. Archaeological Review from Cambridge 15:79–98.

Rose, J. C. 1973 Analysis of Dental Microdefects of Prehistoric Populations from Illinois. Unpublished Ph.D. Dissertation, University of Massachusetts.

Roseberry, W. 1988 Political Economy. Annual Review of Anthropology 17:161–185.

Ruff, C. B. 2000 Biomechanical Analyses of Archaeological Human Skeletal Samples. *In* Skeletal Biology of Past Peoples: Research Methods. S. Saunders, and M. Katzenberg, eds. Pp. 71–102. New York: Wiley.

Ruff, C. B., and W. C. Hayes 1983 Cross-Sectional Geometry of Pecos Pueblo Femora and Tibiae: A Biomechanical Investigation. II. Sex, Age, and Side Differences. American Journal of Physical Anthropology 60(3):383–400.

Ruff, C. B., C. S. Larsen, and W. C. Hayes 1984 Structural Changes in the Femur with the Transition to Agriculture on the Georgia Coast. American Journal of Physical Anthropology 64(2):125–136.

Saitta, D. J. 1988 Marxism, Prehistory and Primitive Communism. Rethinking Marxism 1:145–168.

Saitta, D. J. 1998 Linking Political Economy and Human Biology: Lessons from North American Archaeology. *In* Building a New Biocultural Synthesis: Political-Economic Perspectives on Human Biology. A. Goodman, and T. Leatherman, eds. Pp. 127–147. Ann Arbor, MI: University of Michigan Press.

Salo, W. L. et al. 1994 Identification of Mycobacterium-Tuberculosis DNA in a Pre-Columbian Peruvian Mummy. Proceedings of the National Academy of Sciences 91(6):2091–2094.

Saul, F. P. 1972 The Human Skeletal Remains of the Altar de Sacrificios: An Osteobiographic Analysis. Vol. 63 (2), Papers of the Peabody Museum of Archaeology and Ethnology. Cambridge, MA: Peabody Museum of Archaeology and Ethnology, Harvard University.

Schultz, M. 2001 Paleohistopathology of Bone: A New Approach to the Study of Ancient Diseases. American Journal of Physical Anthropology 116(S33):106–147.

Schutkowski, H. 2008 Thoughts for Food: Evidence and Meaning of Past Dietary Habits. *In* Between Biology and Culture. H. Schutkowski, ed. Pp. 141–165. Cambridge: Cambridge University Press.

Schwarcz, H., L. Gibbs, and M. Knyf 1991 Oxygen Isotope Analysis as an Indicator of Place of Origin. *In* Snake Hill: An Investigation of a Military Cemetery from the War of 1812. S. Pfeiffer, and R. Williamson, eds. Pp. 263–268. Toronto: Dundurn Press.

Schwarcz, H., and M. Schoeninger 1991 Stable Isotope Analyses in Human Nutritional Ecology. Yearbook of Physical Anthropology 34:283–321.

Seckler, D. 1982 Small But Healthy: A Basic Hypothesis in the Theory, Measurment, and Policy of Malnutrition. *In* Newer Concepts in Nutrition and Their Implications for Policy. P. Sukhatme, ed. Pune, India: Maharashtra Association for the Cultivation of Science.

Segal, D. A., and S. J. Yanagisako 2005 Unwrapping the Sacred Bundle: Reflections on the Disciplining of Anthropology. Durham: Duke University Press.

Selye, H. 1950 Stress. Montreal: Medical Publishers.

Shakespeare, T. 1999 Commentary: Observations on Disability and Archaeology. Archaeological Review from Cambridge 15(2):99–101.

Shimada, I., K-I. Shinoda, J. Farnum, R. Corruccini, and H. Watanabe 2004 An Integrated Analysis of Pre-Hispanic Mortuary Practices: A Middle Sican Case Study. Current Anthropology 34:369–402.

Sibley, L., G. J. Armelagos, and D. P. Van Gerven 1992 Obstetric Dimensions of the True Pelvis in a Medieval Population from Sudanese Nubia. American Journal of Physical Anthropology 89(4):421–430.

Singer, M. 1996 Farewell to Adaptationism: Unnatural Selection and the Politics of Biology. Medical Anthropology Quarterly 10(4):496–515.

Smedley, A. 1993 Race in North America: Origins and Evolution of a World View. Boulder, CO: Westview.

Smith, C. J., S. F. Schakel, and R. G. Nelson 1991 Selected Traditional and Contemporary Foods Currently Used by the Pima Indians. Journal of the American Dietetic Association 91:338–341.

Smocovits, V. B. 1992 Unifying Biology: The Evolutionary Synthesis and Evolutionary Biology. Journal of the History of Biology 25(1):1–65.

Sofaer Derevenski, J. R. 2000 Sex Differences in Activity-Related Osseous Change in the Spine and the Gendered Division of Labor at Ensay and Wharram Percy, UK. American Journal of Physical Anthropology 111:333–354.

Sofaer, J. R. 2006 The Body as Material Culture: A Theoretical Osteoarchaeology. Cambridge: Cambridge University Press.

Spielmann, K. A. et al. 2009 "… being weary, they had rebelled": Pueblo Subsistence and Labor Under Spanish Colonialism. Journal of Anthropological Archaeology 28(1): 102–125.

Steckel, R. H., and J. C. Rose, eds. 2002 The Backbone of History: Health and Nutrition in the Western Hemisphere. Cambridge: Cambridge University Press.

Steinbock, R. T. 1976 Paleopathological Diagnosis and Interpretation. Springfield, IL: CC Thomas.

Stocking, G.W. Jr. 1988 Guardians of the Sacred Bundle: The American Anthropological Association and the Representation of Holistic Anthropology. *In* Learned Societies and the Evolution of the Disciplines, American Council of Learned Societies, Occassional Paper No. 5. S. Cohen, D. Bromwich, and G. J. Stocking, eds. New York: American Council of Learned Societies.

Stojanowksi, C. M., and J. E. Buikstra 2005 Research Trends in Human Osteology: A Content Analysis of Papers Published in the *American Journal of Physical Anthropology*. American Journal of Physical Anthropology 128:98–109.

Stojanowski, C. M., and M. A. Schillaci 2006 Phenotypic Approaches for Understanding Patterns of Intracemetery Biological Variation. Yearbook of Physical Anthropology 131:49–88.

Stoodley, N. 2000 From the Cradle to the Grave: Age Organization and the Early Anglo-Saxon Burial Rite. World Archaeology 31:456–472.

Stuart-Macadam, P. 1989 Nutritional Deficiency Disease: A Survey of Scurvy, Rickets, and Iron Deficiency Anemia. *In* Reconstruction of Life from the Skeleton. M. Iscan, and K. Kennedy, eds. Pp. 201–222. New York: Wiley.

Stuart-Macadam, P. 1992 Porotic Hyperostosis: A New Perspective. American Journal of Physical Anthropology 87:39–47.

Swedlund, A. C., and D. H. Herring 2003 Human Biologists in the Archives: Demography, Health, Nutrition, and Genetics in Historical Populations. *In* Human Biologists in the Archives: Demography, Health, Nutrition, and Genetics in Historical Populations. D. Herring, and A. Swedlund, eds. Pp. 1–10. Cambridge: Cambridge University Press.

Tapper, M. 1995 Interrogating Bodies: Medico-Racial Knowledge, Politics and the Study of a Disease. Comparative Studies in Society and History 37(1):76–93.

Tapper, M. 1998 In the Blood: Sickle Cell Anemia and the Politics of Race. Philadelphia: University of Pennsylvania Press.

Thomas, R. B. 1997 Wandering Toward the Edge of Adaptability: Adjustments of Andean People to Change. *In* Human Adaptability Past, Present, and Future. R. Huss-Ashmore and R. Ulijaszek, eds. Pp. 183–232. New York: Oxford University Press.

Thomas, R. B. 1998 The Evolution of Human Adaptability Paradigms: Toward a Biology of Poverty. *In* Building a New Biocultural Synthesis: Political-Economic Perspectives on Human Biology. A. Goodman, and T. Leatherman, eds. Pp. 43–74. Ann Arbor, MI: University of Michigan Press.

Thomas, R. B., T. B. Gage, and M. A. Little 1979 Reflections on Adaptive and Ecological Models. *In* Human Population Biology: A Transdisciplinary Perspective. M. Little, and J. Haas, eds. Pp. 296–319. Oxford: Oxford University Press.

Tishkoff, S. A. et al. 2001 Haplotype Diversity and Linkage Disequilibrium at Human G6PD: Recent Origin of Alleles that Confer Malarial Resistance. Science 293(5529): 455–462.

Trevathan, W., E. O. Smith, and J. J. McKenna 2008 Evolutionary Medicine and Health: New Perspectives. New York: Oxford University Press.

Trinkaus, E., and M. R. Zimmerman 1982 Trauma Among the Shanidar Neanderthals. American Journal of Physical Anthropology 57(1):61–76.

Turner, B. L. 2008 The Servants of Machu Picchu: Life Histories and Population Dynamics in Late Horizon Peru. Unpublished Ph.D. Dissertation, Emory University.

Turner, B. L. et al. 2007 Age-Related Variation in Isotopic Indicators of Diet at Medieval Kulubnarti, Sudanese Nubia. International Journal of Osteoarchaeology 17(1):1–25.

Turner, B. L. et al. 2008 Insights into Immigration and Social Class at Machu Picchu, Peru Based on Oxygen, Strontium, and Lead Isotopic Analysis. Journal of Archaeological Science 36(2):317–332.

Tylor, E. B. 1881 Anthropology: An Introduction to the Study of Man and Civilization. Volume 1 and 2. London: Watts.

Ubelaker, D. H. 1982 The Development of Human Paleopathology. *In* A History of American Physical Anthropology 1930–1980. F. Spencer, ed. Pp. 337–356. New York: Academic Press.

Ungar, P. S. 2007 Evolution of the Human Diet: The Known, the Unknown, and the Unknowable. Oxford: Oxford University Press.

Ungar, P. S., and M. F. Teaford 2002 Human Diet: Its Origin and Evolution. Westport, CT: Bergin & Garvey.

van der Klaauw, C. J. 1945 Cerebral Skull and Facial Skull. Archives Neerlandaises de Zoologie 7:16–38.

van der Klaauw, C. J. 1952 Functional Components of the Skull. Long Island, NJ: EJ Brill.

van der Merwe, N. J., and J. C. Vogel 1978 13C Content of Human Collagen as a Measure of Prehistoric Diet in Woodland North America. Nature 276:815–816.

Van Gerven, D. P., D. S. Carlson, and G. J. Armelagos 1974 Racial History and Biocultural Adaptation of Nubian Archaeological Populations. Journal of African History 24(4):555–564.

von Hunnius, T. E. et al. 2006 Histological Identification of Syphilis in Pre-Columbian England. American Journal of Physical Anthropology 129(4):566–599.

von Hunnius, T. E. et al. 2007 Digging Deeper into the Limits of Ancient DNA Research on Syphilis. Journal of Archaeological Science 34:2091–2100.

Waldron, T. 1994 Counting the Dead: The Epidemiology of Skeketal Populations. Chichester: Wiley.

Walker, A., M. R. Zimmerman, and R. E. F. Leakey 1982 A Possible Case of Hypervitaminosis A in *Homo erectus*. Nature 296:248–250.

Walker, P. L. 2001 A Bioarchaeological Perspective on the History of Violence. Annual Review of Anthropology 30:573–596.

Walker, P. L., and B. S. Hewlett 1990 Dental Health, Diet, and Social Status Among Central African Foragers and Farmers. American Anthropologist 92:383–398.

Wapler, U., E. Crubezy, and M. Schultz 2004 Is Cribra Orbitalia Synonymous with Anemia? Analysis and Interpretation of Cranial Pathology in Sudan. American Journal of Physical Anthropology 123:333–339.

Warren, K. B. ed. 1951 Origin and Evolution of Man. New York: Long Island Biological Association.

Washburn, S. L. 1951 The New Physical Anthropology. Transactions of the New York Academy of Sciences 13(2):258–304.

Washburn, S. L. 1953 The Strategy of Physical Anthropology. *In* Anthropology Today. A. Kroeber, ed. Pp. 714–727. Chicago: University of Chicago Press.

Weiner, A.B. 1995 Culture and Our Discontents. American Anthropologist 97:14–21.

White, C. 1993 Isotopic Determination of Seasonality in Diet and Death from Nubian Mummy Hair. Journal of Archaeological Science 20(6):657–666.

White, C., F. J. Longstaffe, and K. R. Law 2004 Exploring the Effects of Environment, Physiology and Diet on Oxygen Isotope Ratios in Ancient Nubian Bones and Teeth. Journal of Archaeological Science 31(2):233–250.

White, C. D. 2005 Gendered Food Behavior Among the Maya: Time, Place, Status and Ritual. Journal of Social Archaeology 5:356–382.

White, C. D., and G. J. Armelagos 1997 Osteopenia and Stable Isotope Ratios in Bone Collagen of Nubian Female Mummies. American Journal of Physical Anthropology 103(2):185–199.

White, L. A. 1965 Anthropology 1964: Retrospect and Prospect. American Anthropologist 67:629–637.

Wolf, E.R. 1974 Anthropology. New York: Norton.

Wood, J. et al. 1992 The Osteological Paradox: Problems of Inferring Prehistoric Health from Skeletal Samples. Current Anthropology 33:343–370.

Wright, L. E., and C. J. Yoder 2003 Recent Progress in Bioarchaeology: Approaches to the Osteological Paradox. Journal of Archaeological Research 11(1):44–70.

Yanagisako, S. J. 2005 Flexible Disciplinarity: Beyond the American Tradition. *In* Unwrapping the Sacred Bundle: Reflections on the Discipline of Anthropology. D. Segal, and S. Yanagisako, eds. Pp. 78–98. Durham: Duke University Press.

3

Partnerships, Pitfalls, and Ethical Concerns in International Bioarchaeology

Bethany L. Turner and
Valerie A. Andrushko

Introduction

Analyses of human remains serve as valuable bridges between the past and present. The information unlocked through bioarchaeology provides insights into diet, disease, behavior, and less-tangible elements of cultural practice in antiquity, all of which provide beneficial knowledge to the living. As Buikstra (1995:232) succinctly states, "the challenge for the archaeologist of death is to establish the nature of the coded message left us by the living." Culturally affiliated descent communities and bioarchaeologists are therefore similar in that both hold strong connections to human remains. However, these two groups are positioned differently in the ways that they view human remains, and often diverge in their agendas regarding the excavation, storage, analysis, and reburial of human skeletal populations. Moreover, descent communities themselves vary in the ways they regard the remains and cultural objects of their ancestors, with the consequence that there is no single strategy through which bioarchaeologists can align their analytical goals with those of descendants.

Over the last two decades, much attention has been paid to finding common ground between bioarchaeologists and indigenous descent communities in the United States (Ubelaker and Grant 1989; Walker 2000), Canada (Cybulski 2001; Katzenberg 2001), and Australia (Smith 2004). In accordance, increased attention to ethical issues in studying human remains has become an integral part of training students and scholars in bioarchaeology in the United States. However, many of these bioarchaeologists conduct their research in countries where fewer codified

Social Bioarchaeology. Edited by Sabrina C. Agarwal and Bonnie A. Glencross
© 2011 Blackwell Publishing Ltd

standards exist pertaining to human remains, but where ethical issues surrounding human remains nonetheless abound. In contexts such as these, the standards pertaining to ethical practice that inform training in North American bioarchaeology may fall short in negotiating the different local actors, unspoken rules, and competing perspectives that frame the ethical landscapes outside of North America. It then becomes necessary to fine-tune the ethical stance imparted by one's training to the particulars of a foreign field location. This in turn necessitates a deeper understanding of the contextual factors shaping the ways in which local populations view research on human remains, particularly those of indigenous groups.

In this chapter, we seek to explore some of these issues using one such country, Peru, as a case study. Despite unclear and regionally varying policies for excavating and accessing human remains, and the problems that ensue, the social and cultural backdrop of Peruvian bioarchaeology is strongly imbued with ethical concerns. Moreover, bioarchaeologists trained in the United States find different ethical landscapes in researching indigenous Peruvian remains than those involving indigenous North Americans. These differences are tied to historically different relationships between indigenous Peruvians and the central government, the intersection of archaeology and Peru's booming tourism economy, and relationships between Peruvian and foreign archaeologists. Arguably, the experience of these differences between North American training and foreign research contexts is not unique to Peru. In fact, we aim to use this case study to argue that all bioarchaeologists need to familiarize themselves with the multifaceted social and historical contexts in which their research is conducted, whether codified legal frameworks exist or not.

Given both the tremendous opportunities and critical uncertainties surrounding bioarchaeological research in Peru, we feel that it serves as an important case study through which to examine the ways in which bioarchaeologists navigate them. We begin by tracing the historical and current ethical issues that have been central in North American bioarchaeology – particularly the United States – and which form the core of ethics training among North American bioarchaeologists. We then briefly trace the development of the field in Peru, and then identify what we feel are key themes in recent, ongoing, and future research from the perspective of North American bioarchaeology. The following sections are not meant to serve as comprehensive treatments of the field of Peruvian or Andean bioarchaeology, but rather to highlight studies and researchers engaged in the ongoing negotiation of ethical praxis. Our discussion focuses heavily on the social, political, historical, and economic contexts surrounding bioarchaeological research in Peru, and the different groups involved. However, we hope that our focus on key themes, including collaborative partnerships, dissemination of findings, training of local students, and the rise of novel forms of analysis, will help to frame similar discourse among bioarchaeologists working in other areas of the world as well.

Ethics and Bioarchaeology in the United States

Issues of ethical practice in the excavation, curation, and study of indigenous human remains have been central to bioarchaeology in the United States for decades. This is, in large part, because the treatment of human remains is tied inextricably to a sordid history of ill treatment of Native Americans by the United States government, and by society at large (Swidler 1997; Thomas 2000). Stemming from broken government treaties, forced relocation to reservations in the 19th century, and laws designed to deny rights well into the 20th century, Native American groups have experienced systematic, widespread marginalization (Peterson II 1990–1). The effects of this are manifold: numerous studies have documented significant disparities in health related to socioeconomic status (Cantwell et al. 1998; Flores et al. 1999), nutritional status (Dillinger et al. 1999), environmental quality (Harris and Harper 2001; McGinnis and Davis 2001), mental health (Chester et al. 1999; De Coteau et al. 2003; Grossman et al. 1991), and violent encounters (Lester 1995) in Native American populations relative to other American subpopulations. This marginalization has extended beyond the living to the deceased as well. Throughout American history (Thomas 2000) and as recently as the 1970s (Bowman 1989; Dongoske 1996), Native American skeletal remains and associated artifacts were treated differently than remains from Euro or African Americans. Often, remains were stored as specimens in laboratories and museums with little attention to the wishes or concerns of descent communities, and no guarantee that the remains would even be analyzed (Bowman 1989; Rose et al. 1996). Those bioarchaeologists who forged meaningful partnerships with descent populations (e.g., Anyon et al. 1997; Baker et al. 2001; Martin 1997; Walker 2004; Walker 2000:30) did so by choice rather than necessity, with the result that many descent groups were excluded from decision-making processes regarding how their ancestors were stored, treated, and studied.

The passage of NAGPRA , short for the Native American Graves Protection and Repatriation Act of 1990 (Public Law 101-601), made what had previously been an ethical choice on the part of archaeologists into a federal mandate. Decisions regarding the fate of Native American remains and artifacts uncovered on federal lands, or accidentally uncovered during excavations or development projects, were placed in the hands of culturally affiliated descent groups (Lannan 1998; Rose et al. 1996). Moreover, both public and private museums and universities, and any other institutions receiving federal funding, were required to inventory all of their Native American collections and present those inventories to government officials, subject to review by culturally affiliated descent groups (Rose et al. 1996). In the 19 years since the passage of NAGPRA, there have been complex, often unforeseen issues in the ways in which NAGPRA policies are implemented. For example, the standards by which cultural affiliation is established between living and ancient Native Americans has proved periodically problematic (Anyon et al. 1997; Cantwell 2000; Johnson 2002; Zimmerman 1997). The controversy surrounding the more than 9,000-year-old skeleton known as Kennewick Man is perhaps the most high-profile example of the difficulty associated with affiliating ancient remains and/or those with no associated artifacts to their living descendants (Holden 2004; Lannan 1998; Powell and Rose 1999; Thomas 2000). In addition, Native American descent groups must belong to federally recognized tribes or nations in order to exert authority over affiliated ancestral remains, a policy which has impacted unrecognized tribes (Larsen and Walker 2005: 115). Finally, a long history of forced migration or relocation of Native American nations has resulted in unaffiliated tribes maintaining control over land, and the remains therein, that originally belonged to other tribes (Peterson II 1990–1).

Another set of unforeseen outcomes came from within bioarchaeology. Initial reactions to the passage of NAGPRA by many bioarchaeologists were negative, given the uncertainty of future research on Native American archaeological populations. However, within a few years it became clear that NAGPRA had in fact revitalized the discipline (Dongoske 1996; Fine-Dare 2002; Rose et al. 1996). Skeletal collections that had long sat in boxes on museum shelves were examined for the first time, while others were reexamined using new equipment and analytical techniques. The need to inventory collections on such a large scale resulted in a long-needed standardization of data collection (Buikstra and Ubelaker 1994) that now permits greater comparison between skeletal populations. In negotiating with descent groups for permission to study human remains, participating in reburial ceremonies (Baker et al. 2001), and including Native American students and consultants (Lippert 1997; Rice 1997), archaeologists and bioarchaeologists have also forged new relationships, or redefined existing relationships, with Native American communities (Dongoske 1996; Ferguson 1996; Ravesloot 1997; Zimmerman 1989). The result is that the majority of North American bioarchaeologists are open supporters of NAGPRA, and more than 19 years after its passage, NAGPRA and a strong attention to ethical practice are central aspects of training for undergraduate and graduate students. It is therefore reasonable to assume that the majority of active researchers in bioarchaeology carry this ethical imperative with them into field and laboratory settings, whether their research is conducted on Native American remains or on remains from outside of the United States.

Those bioarchaeologists whose research foci lie outside of the United States, but similarly center on indigenous remains, must therefore translate these ethical standards to their specific research areas. Some countries, such as Canada, the United Kingdom, and Australia, have their own federal or provincial guidelines pertaining to indigenous sites, artifacts, and skeletal populations (Jones and Harris 1998; Smith 2004), but others do not. However, the absence of codified laws and policies should not be interpreted as representing an absence of ethical standards. In the Peruvian context, the absence of federal laws instead translates to situational, dynamic, and sometimes unclear routes for ethical practices in excavating, storing, and analyzing human remains.

Bioarchaeological Praxis in Peru: A Brief Contextual Overview

The practice of bioarchaeology in Peru is influenced by cultural and political economic processes, as it is in the United States. However, many of the processes operating in Peru differ qualitatively from the U.S. examples described above, owing to substantially different historical and modern lived experiences among Peruvians.

Prior to the Spanish conquest of western South America, the area now known as Peru served as the heartland for imperial states including the Wari and Inca, as well as numerous large chiefdoms such as the Moche, Nazca, and Chimú (Moseley 1993). The material remnants of these civilizations are still highly visible in the

Peruvian geographic and cultural landscape. There are hundreds of registered archaeological sites under the stewardship of the Peruvian Instituto Nacional de Cultura (INC), and Inca stonework still provides the foundations for colonial cathedrals and other buildings in Lima, Cuzco, and elsewhere (Cuadra et al. 2005; Hyslop 1990).

Peru, and the larger region of the Central Andes, is host to an extremely rich and complex cultural heritage, and its arid and high-altitude regions yield some of the best preservation of archaeological materials in the world. The region therefore hosts a multitude of local and international archaeologists and bioarchaeologists. Further, the role of archaeology in Peruvian national discourse is substantially greater, and qualitatively different, than that in the United States. In addition, the local actors involved differ substantially from the North American case as well. Unlike the United States, the Peruvian population today broadly consists of "Amerindian 45%, mestizo (mixed Amerindian and white) 37%, white 15%, black, Japanese, Chinese, and other 3%" (World Fact Book 2009). *Mestizos/as* are generally urban-dwelling, middle- and upper-class populations, while indigenous populations include Quechua and Aymara speakers. While there is a historical and continuing marginalization of indigenous communities and devaluation of *campesinos* ("Indian" peasants) by *mestizos/as* (Franco 2006; García 2005; van den Berghe and Primov 1977), there is nonetheless a much stronger indigenous presence in mainstream Peruvian society than in mainstream U.S. culture (de la Cadena 2000).

In addition, the visibility of Peruvian prehistory in modern Peru is often intentional, as international tourism constitutes a substantial portion of the national economy. In particular, *turismo místico* (mystical tourism) features prominently in economic development discourse as a major engine of prosperity and modernization (Hill 2007). Andean archaeological sites, artifacts, and remains are at the core of this burgeoning industry. Tourism is conceptualized at different levels of Peruvian society through the varied enactment of *incanismo*, defined by Hill (2007: 457) as "a cultural ethos or projective space of identification through which diverse social actors construct utopic visions of an ideal society and culture grounded in a romanticized historical memory of the Inca empire." In essence, numerous groups in Peru claim shared cultural and ethnic heritage with the Inca, portrayed in popular media as living in a Camelot-like golden age. This is especially widespread in the highland city of Cuzco and surrounding highland regions (van den Bergh and Flores Ochoa 2000). Studies of *incanismo* describe the numerous layers and forms that it takes, including ethnic nationalism, cultural patrimony, resistance to neoliberal development, and marketing strategy, with roots in early 20th-century indigenous movements (de la Cadena 2000; Hill 2007; Poole 1997; van den Bergh and Flores Ochoa 2000).

Importantly, *incanismo* has been a unifying ethos in populist resistance to privatization and Western influence, seen as neo-colonialism. In 2002, widespread national strikes and protests against the privatization of two electric utilities and even the famous Machu Picchu site culminated in hundreds of injuries and arrests, and millions of dollars in damages (Latinamerica Press 2002). In Cuzco, phrases on placards during these strikes and protests included "Cusco belongs to Peruvians, not to spoiled Yankees" and "[Peruvian president] Toledo, Bush's puppet" (Hill 2007: 440).

The implication of this for bioarchaeologists working in Peru, particularly those from the United States, is that the same *incanismo* that can create enthusiasm and support for studying and understanding ancient Peruvian life ways, can also generate substantial distrust and frustration with perceived injustices visited upon the Peruvian people by international powers. These same issues abound in the United States, where there is often substantial resistance on the part of indigenous groups toward archaeological research on ancestral remains (Bray 2001; Fine-Dare 2002). However, whereas North American ethical considerations have historically focused on a dual scheme of stakeholders involving physical anthropologists and Native Americans, in Peru we see a tripartite scheme with physical anthropologists, Peruvians within the dominant *mestizo/a* culture, and indigenous groups of the Quechua- and Aymara-speaking minority culture.

This tripartite scheme introduces additional complexity into a situation that is already difficult to navigate due to the absence of a NAGPRA-type law. Policies for archaeological study of any kind are determined by the regional offices of the INC, which are guided by international laws such as the Hague Convention of 1954 and the UNESCO Convention of 1970 (Merryman 1986); more recent is the Memorandum of Understanding for Cultural Patrimony between Peru and the United States, established in June 1997 and renewed every five years since. Even with these laws, looting and illegal collecting remain problematic (Alva 2001). While the Memorandum specifically addresses human remains that have been modified or in other ways show a cultural affiliation with ancient Andean groups, at this point there is no cohesive legal framework that determines how human remains should be curated and/or reburied, and how Peruvian groups can express their concerns regarding the treatment of ancient skeletal remains. As a result, policies often vary among the regional INC offices concerning the analysis, curation, and sample exportation of human remains.

In addition, the concerns, wishes, and considerations of the indigenous minority culture within this tripartite scheme have been largely absent in discourse surrounding human remains, possibly owing to a larger context of marginalization by *mestizo/a* populations. While Native American groups in the United States have made major strides in rebalancing the power differential from a colonial historical legacy, in Peru the indigenous minority groups are largely disenfranchised and constrained by communication limitations within the dominant discourse. Consequently, they have little authority to express their views in legal terms (see Watkins 2005:437), and discourse has *de facto* occurred between the dominant *mestizo/a* population and foreign researchers.

In essence, the preservation of cultural patrimony and veneration of indigenous heritage in Peru, often embodied as *incanismo*, has occurred largely without any meaningful contribution from its indigenous groups. This is particularly ironic given that *campesinos* are the closest living descendants of pre-Hispanic populations, and often continue ancient, pre-Hispanic agricultural practices (Altieri 1996) and religious traditions (Orlove 1979). Moreover, many of them work with excavation crews for archaeological digs, often for many successive field seasons. Their close and often long-term participation in archaeological fieldwork provides many indigenous Peruvians with an impressive corpus of archaeological knowledge. However, this training occurs outside university settings and does not result in degrees or other formal trappings of education. The result is that the views of *campesinos* are typically absent in discussions surrounding cultural patrimony and the value of ongoing research.

It is clear, then, that in Peru there are complex variables involved in negotiating ethical praxis, without the help of overarching legal guidelines. This presents a scenario that is qualitatively different from the post-NAGPRA paradigm that constitutes a large part of bioarchaeological training. Pragmatically speaking, the onus is therefore placed on the bioarchaeologist, who must act ethically based on his or her own internal guidelines and self-regulation, guided by previous education in ethical behavior and seeking to avoid mistakes made in the past.

In some cases, this approach has worked well and has resulted in some creative solutions to ethical dilemmas, described further below. In other situations, the absence of a legal framework has led to ethical quandaries, exemplified by the recent and ongoing lawsuit in preparation by the Peruvian government against Yale University. The lawsuit centers on the repatriation of the Machu Picchu artifact and human skeletal assemblages, which were collected in 1912 and 1915 by Yale archaeologist Hiram Bingham and brought back to New Haven for curation and study (Bingham 1979 [1930]; Eaton 1916). The Peruvian government has asserted that the Machu Picchu materials were sent to Yale as a temporary loan only, and has demanded their return; moreover, Peruvian officials argue that the inventories provided to them by the university do not encompass the full collection in the museum's holdings (Harman 2005). This fundamental disagreement has evoked strong reactions in Peru, as many have decried the university's reluctance to return what they see unequivocally as Peruvian cultural patrimony (Karp-Toledo 2008). Accordingly, analyses of the Machu Picchu collection at Yale prior to the announcement of the lawsuit (Turner 2008; Turner et al. 2009) are now being published and disseminated with a heightened awareness of the surrounding and ongoing controversy.

Key Areas in Peruvian Bioarchaeology

Historical trends

Some of the ethical problems involved in bioarchaeological research in Peru can be traced to the history of the discipline, which has paralleled research in the United States in its historical trajectory (Heizer 1974; Verano 1997a). Much like in North America, early 20th-century bioarchaeological explorations in Peru began with cases of medical oddities and a focus on skull collecting (Hrdlicka 1914; Tello 1913; Uhle 1909). Only later did the focus shift to a systematic, population approach (Verano 1997a:239).

The development of systematic bioarchaeological research in Peru has largely occurred since the 1980s. During this time, much has been learned about paleopathological conditions in the past (Allison 1984; Ortner et al. 1999; Verano 1992, 1997a), medical practices such as trepanation (Andrushko and Verano 2008; Nystrom 2007), ancient warfare practices (Kellner 2002; Murphy 2004; Tung

2007, 2008; Verano et al. 1999), prehistoric dietary patterns (Burger et al. 2003; Finucane et al. 2006; Hastorf and Johannessen 1993), cultural modification and mortuary practices (Dillehay 1995; Lozada 1998; Torres-Rouff 2003), and migration (Andrushko et al. 2009; Blom et al. 1998; Knudson 2004; Knudson and Buikstra 2007; Sutter 2000; Turner et al. 2009). As noted above, this proliferation of research has occurred without an overarching law governing the study of human remains. Therefore, the bioarchaeologists involved, most of whom are North American, have approached the ethical aspects of their research in different ways.

Partnerships and team research

It is becoming increasingly common that the most innovative research finds in Peru are the result of collaborative work and team projects involving Peruvian and foreign researchers. For example, a longstanding interdisciplinary project, centered on the bioarchaeology of the Osmore Valley of southern Peru, has brought together researchers from the United States and Peru to investigate mortuary practice (Bürgi et al. 1989), infectious disease prevalence (Reinhard and Buikstra 2003; Salo et al. 1994), diet (Tomczak 2003), coca leaf chewing (Cartmell et al. 1994; Indriati and Buikstra 2001; Springfield et al. 1993), residential mobility (Knudson et al. 2004), skeletal pathology (Blom et al. 2005), and cranial modification (Lozada et al. 1997), creating a body of knowledge in the region that is significant in both scope and depth. In addition, a number of highly productive partnerships between foreign and Peruvian researchers has resulted in studies of skeletal populations from the north coast (Baraybar and Shimada 1993; Klaus and Tam 2009; Verano 1998, 2001; Verano et al. 2000; Verano et al. 1999), central and southern highlands (Andrushko et al. 2006a,b; Bray et al. 2005; de la Vega et al. 2005; Finucane et al. 2006; Ochatoma et al. 2008; Shinoda et al. 2006; Tung et al. 2007; Wilson et al. 2007), northern highlands (Conlogue et al. 2004), and the central coast (Gaither et al. 2007; Murphy et al. 2007; Murphy et al. 2009).

These projects demonstrate the substantial progress that has been made in forging partnerships and collaborative teams in Peruvian research. Indeed, given the increasing specialization of bioarchaeological techniques worldwide, these partnerships are not only ethically desirable, but also analytically essential. Drawing on the varying areas of expertise among foreign and Peruvian bioarchaeologists and archaeologists, integrated research such as this permits stronger understandings of larger interpretive significance. However, a recent controversy over the site of Caral, located on the central coast of Peru and considered the oldest known urban complex in the Western Hemisphere (Shady Solis et al. 2001), highlights some problems involved in Peruvian/U.S. collaborations.

The Caral controversy centers on the alleged misappropriation of research results originally made by Ruth Shady Solis, a Peruvian archaeologist with the INC (Atwood 2005). Shady Solis, as director of the Caral Archaeological Project, invited Jonathan Haas and Winifred Creamer of the Field Museum and Northern Illinois University, respectively, to participate in a radiocarbon-dating project to affirm the antiquity of the site (Miller 2005). Haas and Creamer acquired funding and logistical support for the project, with the results published as a jointly-authored article in *Science* (Shady Solis et al. 2001). However, Haas and Creamer's respective websites appeared to attribute not only the radiocarbon results, but also the larger findings of Caral's significance, to the North American researchers alone. Information from these websites was then cited in several news articles and widely reprinted in newspapers. In an open letter, Shady Solis subsequently claimed that Haas and Creamer were receiving international attention for research completed by her team of Peruvian archaeologists (Shady Solis 2005). Haas replied with an e-mail that offered a public apology for the media's mistakes and "for the role we may have played in this misapprehension" (Miller 2005).

The lessons to be gleaned from the Caral controversy are numerous, and are directly relevant to our search for a more ethical research practice in Peruvian bioarchaeology. First, it is a cautionary tale of the ways in which media coverage can skew the proper attribution of research results, or the results themselves. Particularly given the proliferation of online news sites, a single source of misinformation may be transmitted globally without its veracity confirmed. Second, and more importantly, the controversy offers a view into the problematic issue of laboratory access and funding for Peruvian researchers – Shady Solis herself stated that the goal in collaborating with Haas and Creamer was to improve the chances of obtaining funds in the United States (Miller 2005). Moreover, the radiocarbon dating so necessary for the Caral project could not be completed in Peru, where laboratories offering such analyses are non-existent. Finally, the controversy illuminates the problem of disseminating Peruvian research to the international community. While Shady Solis had continually published on Caral since 1997, all publications prior to the *Science* article had been Spanish-language articles published in regional Peruvian and Spanish journals (e.g., Shady Solis 1997a; Shady Solis 1999, 2001; Shady Solis and Lopez 1999), with the addition of one French news article (Shady Solis 1997b). Unfortunately, the limited scope of dissemination for these journals was likely a precipitating factor in the controversy; had Shady Solis's work been widely read and disseminated, the international media's misattribution of her work to Haas and Creamer might have been avoided (Miller 2005).

Dissemination of findings

These problems apparent in Peruvian research dissemination cannot be overstated. The peer-review process for North American, European and international journals, often baffling to scholars within the United States, Canada, Europe, and elsewhere, often proves impossible to navigate for those who have not received training in that system. Language barriers and convoluted online submission processes offer additional obstacles. There may also be the differences in the ways that Peruvian and foreign scholars approach submitting and publishing their findings in peer-reviewed journals (Schreiber, personal communication).

Although these problems may seem daunting, strides are fortunately being made to increase the dissemination of Peruvian research. *Ñawpa Pacha*, a journal devoted to Andean archaeological research, encourages submissions in Spanish and provides reviews in the author's native language. *Chungará: Revista de Antropología Chilena*, published by Chile's University of Tarapaca, offers a forum for archaeological and bioarchaeological dissemination of research results in both Spanish and English (Watkins 2005:437). Both journals are regularly cited in international works. Other avenues for dissemination include Mexico's *Estudios de Antropología Biologica* (e.g., Verano 1997b), Chile's *Estudios Atacamenos* and France's *Bulletin de l'Institut français d'études andines*. The number of publishing venues has increased in the last ten years, a trend that will provide additional opportunities for dissemination by Andean scholars in the future. The proliferation of electronic articles by these and other journals also promises to increase their visibility in the international academic sphere.

Active publication by North American bioarchaeologists and their Peruvian (or Chilean) colleagues in journals such as *Ñawpa Pacha* and *Chungará* is one way to firmly cement collaborative partnerships between North American and Andean scholars, and to more fully incorporate these partnerships into scholarly dissemination. Admittedly, these journals have lower impact ratings (scored indicators of the size and breadth of readership) than *American Antiquity*, *American Journal of Physical Anthropology*, *Journal of Archaeological Science*, *Current Anthropology*, and others in which North American bioarchaeologists commonly publish (Thompson Reuters 2007). Consequently, these lower ratings could likely factor in deciding where to submit one's research manuscripts, given the emphasis in the United States and elsewhere to submit to journals with the widest dissemination. However, the hybrid English/Spanish journals provide a unique opportunity for bioarchaeologists, schooled in the "publish or perish" ethos of North American academia, to serve as bridges for Andean scholars to publish in both Andean journals and those North American and international journals with wider academic audiences. Indeed, authorship in both *Chungará* and *Ñawpa Pacha* includes a substantial representation of North American authors, in both English and Spanish articles. This points to a promising direction in collaborative research, as findings should not only be disseminated widely in multi-authored publication, but also consciously disseminated to Andean audiences.

Training of Peruvian bioarchaeologists

Another promising recent trend is the training of bioarchaeologists within Peru, although the impetus for such training comes from a tragic source – the need for physical anthropologists who can excavate and analyze bodies from mass graves in Peru. These mass graves are the result of political violence between two insurgent groups (the Shining Path and the Tupac Amaru Revolutionary Army) and the Peruvian Army carried out between 1982 and 2000. Approximately 69,000 people were killed and another 8,500–16,000 individuals disappeared during that time and remain missing, according to Peru's Truth and Reconciliation Commission (Comisión de la Verdad y Reconciliación 2003).

In response to this crisis, the Peruvian Forensic Anthropology Team (Equipo Peruano de Antropología Forense, EPAF) was formally created in 2001. Many of

EPAF members, including EPAF director Jose Pablo Baraybar, formerly worked as forensic anthropologists in other global instances of genocide such as Rwanda and Bosnia. Now, these anthropologists are applying their training to mass fatality sites throughout Peru, focusing on the epicenter of political violence in Ayacucho.

This new focus on forensic anthropology has created a demand for comprehensive training in physical anthropological data collection. Indeed, a critical aspect of EPAF's work involves training in the form of undergraduate courses, workshops, and internships. For graduate training in the discipline, the Instituto de Democracia y Derechos Humanos de la Pontificia Universidad Católica del Perú in Lima has recently begun offering a Master's degree program in Forensic Anthropology and Bioarchaeology, under the direction of Sonia Guillén. This comprehensive two-year program includes courses in Human Anatomy, Dental Anthropology, Biomolecular Anthropology, Criminalistics, and Paleopathology, taught by experts from Peru, Argentina, Spain, England, Finland, and the United States. The program is now producing a new generation of Peruvian anthropologists and bioarchaeologists trained in the latest methods by leading educators. With this new generation of Peruvian scholars, exciting possibilities for partnerships and collaborations abound. Even more notable, the new generation will no doubt inspire a renewed focus on ethical practices in Peruvian bioarchaeology and the need for standardization of these practices throughout the country.

Intrusive/Destructive analysis

Despite these significant developments in Peruvian bioarchaeological partnerships, training, and dissemination, there are areas of research that remain more difficult to ethically navigate than others. In particular, recent decades have seen a worldwide surge in bioarchaeological research that harnesses techniques from genetics, geochemistry, parasitology, and microbiology. These include trace element and isotopic analysis, ancient DNA extraction, and identification of parasitic microbes and food remnants (Arriaza 1995; Horrocks et al. 2004; Katzenberg and Harrison 1997). The advantage of these analyses is that they generate data directly from biological remains; independent of archaeological context, parameters of diet, disease, and migration are measured in human bone, teeth, mummified tissues, and gut contents. The disadvantage of these techniques in the eyes of museum curators, descent communities, and INC officials is that they are intrusive, involving the collection of samples from skeletonized or mummified individuals, and destructive, in that the collected samples are consumed by laboratory analyses. In addition, the laboratory facilities required for these analyses typically are unavailable in Peru, necessitating the exportation of samples to countries in which laboratory facilities are located. Moreover, the associated costs of analysis exceed most funding opportunities available in Peru, requiring funding from governmental agencies or foundations in wealthier countries. As such, these analyses tend to be carried out by foreign researchers using laboratories in, and funding from, their home countries. This can create understandable concern on the part of scholars and the public in Peru, as pieces of Peruvian skeletons and mummies are exported for use with little room for their own participation.

Of further concern is balancing the pros of essential data with the cons of irreparable alteration of the remains themselves. Fortunately, advances in these intrusive and destructive analytical techniques have meant a steady decline in the amount of tissue required for analyses of their substrates; in many cases, analyses can be successfully carried out on samples of tooth enamel the size of a grain of rice, bone the size of a postage stamp, and similarly small amounts of intestinal contents or soft tissues. By reducing the amount of required materials, researchers are able to better complete ethically responsible research, since the destructive nature of the analysis is greatly reduced. However, any intrusive sampling of human remains could be seen as an unacceptable violation of ancestral remains and cultural patrimony; indeed, these concerns are widespread in research on North American indigenous remains (Katzenberg 2001). Therefore, bioarchaeologists who specialize in these intrusive and destructive analyses must (and overwhelmingly do) use them strategically to address salient questions surrounding ancient life ways that cannot be addressed through other means.

Indeed, findings from a burgeoning body of research have significantly advanced studies of diet, disease, and migration in antiquity based on intrusive and/or destructive analyses of human skeletons and mummies. Intrusive analyses of mummified gut contents from Peru (Guhl et al. 2000; Guhl et al. 1999) and elsewhere (Ferreira et al. 2000; Madden et al. 2001) have been used to diagnose specific types of microbial infections and reconstruct specific diets from partially digested foods (Holden 1991). Destructive analyses of mummified human remains from Peru have also been instrumental in developing methodologies for ancient DNA extraction and amplification (Pääbo 1989) and absolute dating (Molin et al. 1998).

Recently, carbon and/or nitrogen isotopic studies of Andean skeletal populations have provided critical insights into subsistence variation through time (Aufderheide et al. 1994; Aufderheide and Santoro 1999; Benfer 1984; Benfer 1990; Sandness 1992), gendered or status-based diets (Hastorf 1990; 1996; Ubelaker et al. 1995; Williams 2004), and regionally divergent subsistence economies (Tomczak 2003). An emerging trend in Andean bioarchaeological research centers on identifying immigrants within skeletal populations. Strontium, oxygen, and lead isotopic data have been used to infer nuanced patterns of residential mobility linked to social complexity (Andrushko et al. 2009; Knudson 2008; Knudson and Price 2007; Knudson et al. 2004; Slovak et al. 2008; Tung and Knudson 2008; Turner et al. 2008; Verano and DeNiro 1993) and suggest causal factors in the prevalence of pathological conditions (Turner et al. 2009).

These destructive analyses provide data that would otherwise be unavailable in Peru, and elucidate patterns of movement, diet, disease, and population history in antiquity that would otherwise be invisible. Further, these data are critical in elucidating patterns on local and regional levels, permitting more effective comparisons of different populations; arguably, the insights gleaned from intrusive and destructive analyses can far outweigh the loss of osseous, dental, or soft tissue material involved. However, we suggest that the continuation of intrusive and destructive analyses on Peruvian human remains depends heavily on equitable partnership, inclusive dissemination of findings, and opportunities for Peruvian students and scholars to train in associated laboratories. In essence, intrusive and destructive research must be at the forefront of ethical practice precisely because the stakes surrounding it are higher.

Non-destructive analysis

Non-destructive analyses provide additional, complementary avenues toward a more ethical practice in Peruvian bioarchaeology. While non-destructive analyses – in the form of anthroposcopy and osteometry – currently make up the majority of data collection techniques for Peruvian bioarchaeology, these traditional methods are now being augmented by a newer technique borrowed from medicine. In this approach, known as paleoimaging, radiography, magnetic resonance, and endoscopy are used to examine skeletal and mummified remains in a wholly non-destructive manner (Beckett and Guillen 2000; Conlogue et al. 2008; Conlogue et al. 2004; Verano and Lombardi 1999). Radiography and magnetic resonance allow for the detection of pathological lesions that may not be visible to the naked eye, while endoscopic probing adds another dimension by providing up-close images of structures only hinted at on x-rays. Importantly, these techniques can be used to investigate the internal structures of mummies without destroying the integrity of the mummy bundle itself. This approach has led to the confirmation of tuberculosis in Peruvian mummies (Conlogue 2002) as well as the detection of spinal trauma (Bravo et al. 2001). Another important aspect of this approach is its portability: mobile x-ray equipment and scopes can be transported easily and field imaging facilities may be constructed using a variety of creative means (Conlogue and Nelson 1999:254–255; Conlogue et al. 2004). This portability makes it ideal for archaeological sites in Peru located far from cities where radiographic resources are commonly found.

We highlight the example of paleoimaging here not only because it enhances our research capabilities, but also because it furthers the goal of practicing Peruvian bioarchaeology in an ethically responsible manner. Skeletal remains and mummies are thoroughly researched, yet left unaltered and preserved for the future. We feel that this type of innovative thinking, coupled with a deep commitment to partnerships and promotion of Andean scholarship, will only improve the academic environment of Peruvian bioarchaeology in the future.

Conclusions

The importance of ethical practice is central to bioarchaeology in the United States, and therefore has become a central component of training in the field. However, the issues surrounding excavation, treatment, and stewardship of indigenous remains (indeed, human remains in general) vary between different countries, creating both opportunities and potential pitfalls for the bioarchaeologists involved. In this review, we highlighted what we feel to be important ethical considerations in the bioarchaeology of Peru, using historical trends and divergent cultural context as backdrops against which we discussed central themes and presented illustrative examples of ethically grounded research. In particular, the growing number of partnerships between foreign and Peruvian scholars, and publications in Andean journals, has fostered an ethos of inclusion and equity. An excellent future direction in this area would be to include Peruvian colleagues as co-principal investigators on North American grant proposals as well as publications and presentations. The development and refinement of newer analytical techniques, when incorporated into the training of Peruvian students and collaborative field research, appears similarly promising.

However, based on our review, we suggest that there are still gray areas to navigate and additional opportunities for North American and other foreign bioarchaeologists working in Peru. The most obvious gray area is that which surrounds intrusive and destructive analyses; as discussed above, we argue that these areas of research are both highly sensitive, given the nature of the sampling involved, and highly promising, as the data generated opens entirely new windows into ancient Andean life ways. One possible approach to resolving this intrinsically ambivalent situation is to provide ample opportunities for Peruvian students and practicing bioarchaeologists alike to receive laboratory training in these techniques from their foreign collaborators. Increased participation in the laboratory analyses themselves, data analysis, and publication on the part of Peruvian researchers could help to foster the same sense of inclusion and equity that exists in other collaborative research areas, and could be prudent additions to North American grant budgets.

A final suggestion for future directions in Peruvian bioarchaeology concerns the relationship between bioarchaeologists and the Peruvian public, and in particular, Quechua- and Aymara-speaking indigenous populations. In general, the controversies surrounding Caral and the Machu Picchu material culture at Yale University have been met with strong reactions on the parts of Peruvian *mestizos/as* and reflect a larger resistance to foreign influence. This dialogue, however, often excludes the voices of indigenous stakeholders in Peru. Therefore, we suggest that bioarchaeologists in Peru seek opportunities to disseminate the rationale and results of their research not only to Peruvian bioarchaeologists, but to the Peruvian public as well. Through newspapers, magazines, websites, video or other means, greater efforts on the part of foreign researchers and their Peruvian colleagues to reach lay audiences could provide valuable bridges that may prevent similar controversies in the future.

As it applies to Quechua- and Aymara-speaking communities, studies of *incanismo* and tourism in Peru have pointed to the glorification of ancient indigenous empires while ignoring the marginalization and denigration of rural indigenous *campesinos*. We assert that bioarchaeologists studying ancient Peruvian and greater Andean cultures should work to foster a sense of inclusion for modern indigenous groups in arenas surrounding cultural patrimony, heritage, and scholarship that extends beyond the tourism industry. Some small steps toward that goal include disseminating research findings in Quechua or Aymara, and using a variety of media such as radio or television to reach more remote audiences. In addition, we suggest that the contributions of indigenous *campesinos* as excavators and field assistants ought to receive more attention in the presentation and dissemination of findings; the fact that their training is informal makes it no less deserving of recognition.

The examples of ethical practice in Peru described here are by no means universally applicable to working in foreign contexts in general. However, we argue that neither are specific examples from North America. The rich scholarship that has stemmed from NAGPRA and other similar legislation has resulted in a more nuanced, socially conscious bioarchaeology. We argue that an important next step is to apply this paradigm to an increasingly diverse global stage. The principles of North American bioarchaeological ethics – respect, transparency, inclusion, collaboration – thereby serve as an essential foundation from which to shape a locally situated ethical praxis. In that sense, the themes that we have outlined here, using Peruvian cases, are applicable to any foreign context, so long as the particulars of a given foreign context is well understood.

In his discussion of *incanismo* and tourism in Cuzco, Hill (2007:453) recounts a conversation with a local woman about the concept of *ayni*, or reciprocity, and argues that *ayni* needs to be reinserted into mystical tourism. We would suggest that, in the absence of clear federal guidelines pertaining to human remains, the same ethos is a useful framework for the practice of bioarchaeology in Peru and elsewhere. Bioarchaeologists working in foreign countries with indigenous remains so often do, and must continue to, give back as much – if not more – as we receive from our research. This reciprocity must include both academic and public arenas, in order to move the field forward in an ethically sound manner.

ACKNOWLEDGMENTS

We are grateful to the editors for the opportunity to contribute to this volume, and for the insightful comments of the volume and series editors. We would also like to thank Elva Torres Pino, Katharina Schreiber, and John Verano for their helpful comments regarding this chapter, and our colleagues for generously sharing their field experiences with us.

REFERENCES

Allison, M. J. 1984 Paleopathology in Peruvian and Chilean Populations. *In* Paleopathology at the Origins of Agriculture. M. N. Cohen, and G. J. Armelagos, eds. Pp. 515–530. Orlando: Academic Press.

Altieri, M. 1996 Indigenous Knowledge Re-Valued in Andean Agriculture. ILEIA Newsletter 12(1):7–12.

Alva, W. 2001 The Destruction, Looting and Traffic of the Archaeological Heritage of Peru. *In* Trade in Illicit Antiquities: The Destruction of the World's Archaeological Heritage. N. J. Brodie, J. Doole, and C. Renfrew, eds. Pp. 89–96. Cambridge: McDonald Institute of Archaeological Research.

Andrushko, V. A., M. R. Buzon, A. Simonetti, and R. A. Creaser 2009 Strontium Isotope Evidence for Prehistoric Migration at Chokepukio, Valley of Cuzco, Peru. Latin American Antiquity 20:57–75.

Andrushko, V. A., E. C. Torres Pino, and V. Bellifemine 2006a The Tombs of Sacsahuaman: A Bioarchaeological Analysis of Elite Burials from the Capital of the Inca Empire. Annual Meetings of the Society for American Archaeology. San Juan.

Andrushko, V. A., E. C. Torres Pino, and V. Bellifemine 2006b The Burials At Sacsahuaman and Chokepukio: A Bioarchaeological Case Study of Imperialism from the Capital of the Inca Empire. Ñawpa Pacha 28:63–92.

Andrushko, V. A., and J. W. Verano 2008 Trepanation in the Cuzco Region of Peru: A View into an Ancient Andean Practice. American Journal of Physical Anthropology 137:4–13.

Anyon, R., T. J. Ferguson, L. Jackson, L. Lane, and P. Vicenti 1997 Native American Oral Tradition and Archaeology: Issues of Structure, Relevance, and Respect. *In* Native Americans and Archaeologists: Stepping Stones to Common Ground. N. Swidler, K. E. Dongoske, R. Anyon, and A. S. Downer, eds. Pp. 77–87. Walnut Creek: AltaMira Press.

Arriaza, B. T. 1995 Beyond Death: The Chinchorro Mummies of Ancient Chile. Washington, DC: Smithsonian Institution Press.

Atwood, R. 2005 A Monumental Feud. Archaeology 58(4):18–25.

Aufderheide A. C., M. A. Kelley, M. Rivera, L. Gray, L. L. Tieszen, E. Iversen, H. R. Krouse, and A. Carevic 1994 Contributions of Chemical Dietary Reconstruction to the Assessment of Adaptation by Ancient Highland Immigrants (Alto-Ramirez) to Coastal Conditions at Pisagua, North Chile. Journal of Archaeological Science 21(4): 515–524.

Aufderheide A. C., and C. M. Santoro 1999 Chemical Paleodietary Reconstruction: Human Populations at Late Prehistoric Sites in the Lluta Valley of Northern Chile. Revista Chilena De Historia Natural 72(2):237–250.

Baker, B. J., T. L. Varney, R. G. Wilkinson, M. Anderson, and M. A. Liston 2001 Repatriation and the Study of Human Remains. *In* The Future of the Past: Archaeologists, Native Americans, and Repatriation. T. L. Bray, ed. Pp. 69–89. New York: Garland Publishers.

Baraybar, J. P., and I. Shimada 1993 A Possible Case of Metastatic Carcinoma in a Middle Sican Burial from Batán Grande, Perú. International Journal of Osteoarchaeology 3:129–135.

Beckett, R. G., and S. Guillen 2000 Field Videoendoscopy: A Pilot Project at Centro Mallqui, El Algorrobal, Peru. Supplement to Paleopathology Newsletter June(110).

Benfer R. A. 1984 The Challenges and Rewards of Sedentism: The Preceramic Village of Paloma, Peru. *In* Paleopathology at the Origins of Agriculture. M. N. Cohen, and G. J. Armelagos, eds. Pp. 531–558. Orlando: Academic Press.

Benfer, R. A. J. 1990 The Preceramic Period Site of Paloma, Peru: Bioindications of Improving Adaptation to Sedentism. Latin American Antiquity 1(4):284–318.

Bingham, H. 1979 [1930] Machu Picchu: Citadel of the Incas. New York: Hacker Art Books.

Blom, D. E., J. E. Buikstra, L. Keng, P. D. Tomczak, E. Shoreman, and D. Stevens-Tuttle 2005 Anemia and Childhood Mortality: Latitudinal Patterning Along the Coast of Pre-Columbian Peru. American Journal of Physical Anthropology 127(2):152–169.

Blom, D. E., B. Hallgrimsson, L. Keng, M. C. Lozada, and J. E. Buikstra 1998 Tiwanaku "Colonization": Bioarchaeological Implications for Migration in the Moquegua Valley, Peru. World Archaeology 30(2):238–261.

Bowman, M. B. 1989 The Reburial of Native American Skeletal Remains: Approaches to the Resolution of a Conflict. Harvard Environmental Law Review 13:147–208.

Bravo, A. J., G. Conlogue, and S. Guillen 2001 Dead Men Walking: A Radiographic Study of Spinal Trauma in Peruvian Chachapoya Mummies. Radiology 221(Suppl): 475–476.

Bray, T. L. ed. 2001 The Future of the Past: Archaeologists, Native Americans and Repatriation. New York: Garland Publishing.

Bray, T. L., L. D. Minc, M. C. Ceruti, J. A. Chávez, R. Perea, and J. Reinhard 2005 A Compositional Analysis of Pottery Vessels Associated With the Inca Ritual of Capacocha. Journal of Anthropological Archaeology 24:82–100.

Buikstra, J. E. 1995 Tombs for the Living … or … for the Dead: The Osmore Ancestors. *In* Tombs for the Living: Andean Mortuary Practices. T. D. Dillehay, ed. Pp. 229–280. Washington, DC: Dumbarton Oaks Research Library and Collection.

Buikstra, J. E., and D. H. Ubelaker 1994 Standards for Data Collection from Human Skeletal Remains: Proceedings of a Seminar at the Field Museum of Natural History. Fayetteville: Arkansas Archeological Survey Press.

Burger, R. L., J. A. Lee-Thorp, and N. J. van der Merwe 2003 Rite and Crop in the Inca State Revisited: An Isotopic Perspective from Machu Picchu and Beyond. *In* The 1912 Yale Peruvian Scientific Expedition Collections from Machu Picchu. R. L. Burger, and L. C. Salazar, eds. Pp. 119–137. New Haven: Yale University Publications in Anthropology.

Bürgi, P., S. R. William, J. E. Buikstra, N. R. Clark, M. C. Lozada Cerna, and E. C. Torres Pino 1989 Aspects of Mortuary Differentiation at the Site of Estuquiña, Southern Peru. *In* Ecology, Settlement and History in the Osmore Drainage. D. S. Rice, and C. Stanish, eds. Pp. 347–369. Oxford: BAR.

Cantwell, A.-M. 2000 "Who Knows the Power of His Bones"; Reburial Redux. Annals of the New York Academy of Sciences 925:79–119.

Cantwell, M. F., M. McKenna, E. McCray, and I. M. Onorato 1998 Tuberculosis and Race/Ethnicity in the United States – Impact of Socioeconomic Status. American Journal of Respiratory and Critical Care Medicine 157(4):1016–1020.

Cartmell, L. W., A. C. Aufderheide, A. Springfield, J. E. Buikstra, B. Arriaza, and C. Weems 1994 Radioimmunoassay for Cocaine in Mummy Hair to Determine Antiquity and Demography of Coca Leaf Chewing Practices in Southern Peru and Northern Chile. Eres (Arqueología) 5(1):83–95.

Chester, B., P. Mahalish, and J. Davis 1999 Mental Health Needs Assessment of Off-Reservation American Indian People in Northern Arizona. American Indian and Alaska Native Mental Health Research 8(3):25–40.

Comisión de la Verdad y Reconciliación 2003 Informe final. In: Reconciliación Cdlvy, editor. Comisión de la verdad y Reconciliación. Lima.

Conlogue, G. 2002 More TB in Peruvian Mummies. Archaeology 55(2):14.

Conlogue, G., R. Beckett, Y. Bailey, J. Posh, D. Henderson, G. Double, and T. King 2008 Paleoimaging: The Use of Radiography, Magnetic Resonance, and Endoscopy to Examine Mummified Remains. Journal of Radiology Nursing 27:5–13.

Conlogue, G., and A. J. Nelson 1999. Polaroid Imaging at an Archaeological Site in Peru. Radiologic Technology 703:244–250.

Conlogue, G., A. J. Nelson, and S. Guillen 2004 The Application of Radiography to Field Studies in Physical Anthropology. Canadian Association of Radiologists Journal 55:254–257.

Cuadra, C., Y. Sato, J. Tokeshi, H. Kanno, J. Ogawa, M. B. Karkee, and J. Rojas 2005 Preliminary Evaluation of the Seismic Vulnerability of the Inca's Coricancha Temple Complex in Cusco. Transactions on the Built Environment 83:245–253.

Cybulski J. S. 2001 Current Challenges to Traditional Anthropological Applications of Human Osteology in Canada. *In* Out of the Past: The History of Human Osteology at the University of Toronto. L. Sawchuk, and S. Pfeiffer, eds. Scarborough: CITD Press, University of Toronto at Scarborough.

De Coteau, T. J., D. A. Hope, and J. Anderson 2003 Anxiety, Stress, and Health in Northern Plains Native Americans. Behavior Therapy 34(3):365–380.

de la Cadena, M. 2000 Indigenous Mestizos: The Politics of Race and Culture in Cuzco, Peru, 1919–1991. Durham: Duke University Press.

de la Vega, E., K. L. Frye, and T. A. Tung 2005 The Cave Burial from Molino-Chilacachi. *In* Advances in the Archaeology of the Titicaca Basin. C. Stanish, A. B. Cohen, and M. S. Aldenderfer, eds. Los Angeles: Cotsen Institute of Archaeology, University of California Press.

Dillehay, T. D. ed. 1995 Tombs for the Living: Andean Mortuary Practices. Washington DC: Dumbarton Oaks Research Library and Collection.

Dillinger, T. L., S. C. Jett, M. J. Macri, and L. E. Grivetti 1999 Feast or Famine? Supplemental Food Programs and Their Impacts on Two American Indian Communities in California. International Journal of Food Sciences and Nutrition 50(3):173–187.

Dongoske, K. E. 1996 The Native American Graves Protections and Repatriation Act: A New Beginning, Not the End, for Osteological Analysis – A Hopi Perspective. American Indian Quarterly 20(2):287–297.

Eaton, G. F. 1916 The Collection of Osteological Material from Machu Picchu. Memoirs of the Connecticut Academy of Arts and Sciences.

Ferguson, T. J. 1996 Native Americans and the Practice of Archaeology. Annual Review of Anthropology 25(63–79).

Ferreira, L. F., C. Britto, M. A. Cardoso, O. Fernandes, K. Reinhard, and A. Araújo 2000 Paleoparasitology of Chagas Disease Revealed by Infected Tissues from Chilean Mummies. Acta Tropica 75:79–84.

Fine-Dare, K. S. 2002 Grave Injustice: The American Indian Repatriation Movement and NAGPRA. Lincoln: University of Nebraska Press.

Finucane, B., P. M. Agurto, and W. H. Isbell 2006 Human and Animal Diet at Conchopata, Peru: Stable Isotope Evidence for Maize Agriculture and Animal Management Practices During the Middle Horizon. Journal of Archaeological Science 33:1766–1776.

Flores, G., H. Bauchner, A. R. Feinstein, and U. Nguyen 1999 The Impact of Ethnicity, Family Income, and Parental Education on Children's Health and Use of Health Services. American Journal of Public Health 89(7):1066–1071.

Franco, J. 2006 Alien to Modernity: The Rationalization of Discrimination. Journal of Latin American Cultural Studies 3:1–16.

Gaither, C., M. Murphy, G. Cock, and E. Goyochea 2007 Consequences of Conquest? Interpretation of Subadult Trauma at Puruchuco-Huaquerones, Peru. Poster Presented at the Annual Meeting of the Paleopathology Association. Philadelphia, PA.

García, M. E. 2005 Making Indigenous Citizens: Identities, Education, and Multicultural Development in Peru. Palo Alto: Stanford University Press.

Grossman, D. C., B. C. Milligan, and R. A. Deyo 1991 Risk-Factors for Suicide Attempts Among Navajo Adolescents. American Journal of Public Health 81(7):870–874.

Guhl, F., C. Jaramillo, G. A. Vallejo, F. Cárdenas-Arroyo, and A. Aufderheide 2000 Chagas Disease and Human Migration. Memorias del Instituto de Oswaldo Cruz 95(4):553–555.

Guhl, F., C. Jaramillo, G. A. Vallejo, R. Yockteng, F. Cárdenas-Arroyo, G. Fornaciari, B. Arriaza, and A. C. Aufderheide 1999 Isolation of *Trypanosoma cruzi* DNA in 4,000-year-old Mummified Human Tissue from Northern Chile. American Journal of Physical Anthropology 108:401–407.

Harman, D. 2005 90 Years Later, Peru Battles Yale Over Incan Artifacts. Christian Science Monitor December 29: www.csmonitor.com/2005/1229/p2001s2003-woam.html.

Harris, S., and B. L. Harper 2001 Lifestyles, Diets, and Native American Exposure Factors Related to Possible Lead Exposures and Toxicity. Environmental Research 86(2):140–148.

Hastorf, C. A. 1990 The Effect of the Inka State on Sausa Agricultural Production and Crop Consumption. American Antiquity 55(2):262–290.

Hastorf, C. A. 1996 Gender, Space and Food in Prehistory. *In* Contemporary Archaeology in Theory. R. W. Preucel, and I. Hodder, eds. Oxford: Blackwell.

Hastorf, C. A., and S. Johannessen 1993 Pre-Hispanic Political-Change and the Role of Maize in the Central Andes of Peru. American Anthropologist 95(1):115–138.

Heizer, R. F. 1974 A Question of Ethics in Archaeology – One Archaeologist's View. Journal of California Anthropology 1:145–151.

Hill, M. D. 2007 Contesting Patrimony: Cusco's Mystical Tourist Industry and the Politics of Incanismo. Ethnos 72(4):433–460.

Holden, C. 2004 Kennewick Man: Court Battle Ends, Bones Still Off-Limits. Science 305(5684):591.

Holden, T. G. 1991 Evidence of Prehistoric Diet from Northern Chile: Coprolites, Gut Contents, and Flotation Samples from the Tulan Quebrada. World Archaeology 22:320–331.

Horrocks, M., G. Irwin, M. Jones, and D. Sutton 2004 Starch Grains and Xylem Cells of Sweet Potato (*Ipomoea Batatas*) and Bracken (*Pteridium Esculentum*) in Archaeological Deposits from Northern New Zealand. Journal of Archaeological Science 31(3):251–258.

Hrdlicka, A. 1914 Anthropological Work in Peru in 1913, with Notes on the Pathology of the Ancient Peruvians. Smithsonian Miscellaneous Collections.

Hyslop, J. 1990 Inca Settlement Planning. Austin: University of Texas Press.

Indriati, E., and J. E. Buikstra 2001 Coca Chewing in Prehistoric Peru: Dental Evidence. American Journal of Physical Anthropology 114(3):242–257.

Johnson, G. 2002 Tradition, Authority and the Native American Graves Protection and Repatriation Act. Religion 32:355–381.

Jones, D. G., and R. J. Harris 1998 Archaeological Human Remains: Scientific, Cultural, and Ethical Considerations. Current Anthropology 39(2):253–264.

Karp-Toledo, E. 2008 The Lost Treasure of Machu Picchu. The New York Times Op/Ed, February 23, 2008, retrieved on August 1, 2009 from www.nytimes.com/2008/02/23/opinion/23karp-toledo.html.

Katzenberg, M. A. 2001 Destructive Analyses of Human Remains in the Age of NAGPRA and Related Legislation. *In* Out of the Past: The History of Human Osteology at the University of Toronto. L. Sawchuk, and S. Pfeiffer, eds. Scarborough: CITD Press, University of Toronto at Scarborough.

Katzenberg, M. A., and R. G. Harrison 1997 What's in a Bone? Recent Advances in Archaeological Bone Chemistry. Journal of Archaeological Research 5:265–293.

Kellner, C. M. 2002 Coping with Environmental and Social Challenges in Prehistoric Peru: Bioarchaeological Analyses of Nasca Populations. Ph.D. Dissertation. Santa Barbara: University of California.

Klaus, H. D., and M. E. Tam 2009 Contact in the Andes: Bioarchaeology of Systemic Stress in Colonial Móroppe, Peru. American Journal of Physical Anthropology 138(3):356–368.

Knudson, K. J. 2004 Tiwanaku Residential Mobility and Archaeological Chemistry: Strontium and Lead Isotope Analyses in the South Central Andes. Ph.D. Dissertation. Madison: University of Wisconsin.

Knudson, K. J. 2008 Oxygen Isotope Analysis in a Land of Environmental Extremes: The Advantages and Disadvantages of Isotopic Work in the Andes. Invited Paper Presented at the 73rd Annual Society for American Archaeology Meeting, Vancouver, British Columbia, March 26–30.

Knudson, K. J., and J. E. Buikstra 2007 Residential Mobility and Resource Use in the Chiribaya Polity of Southern Peru: Strontium Isotope Analysis of Archaeological Tooth Enamel and Bone. International Journal of Osteoarchaeology 17:563–580.

Knudson, K. J., and T. D. Price 2007 Utility of Multiple Chemical Techniques in Archaeological Residential Mobility Studies: Case Studies from Tiwanaku- and Chiribaya-Affiliated Sites in the Andes. American Journal of Physical Anthropology 132(1):25–39.

Knudson, K. J., T. D. Price, J. E. Buikstra, and D. E. Blom 2004 The Use of Strontium Isotope Analysis to Investigate Tiwanaku Migration and Mortuary Ritual in Bolivia and Peru. Archaeometry 46(1):5–18.

Lannan, R. W. 1998 Anthropology and Restless Spirits: The Native American Graves Protection and Repatriation Act, and the Unresolved Issues of Prehistoric Human Remains. Harvard Environmental Law Review 22:369–384.

Larsen, C. S., and P. Walker 2005 Ethics of Bioarchaeology. *In* Biological Anthropology and Ethics: From Repatriation to Genetic Identity. T. Turner, ed. Pp. 111–119. Albany: Statue University of New York Press.

Latinamerica Press. 2002. Peru and Privatization: "Just Say No". World Press Review 49(9): www.worldpress.org/Americas/657.cfm.

Lester, D. 1995 Suicide and Homicide Among Native-Americans – A Comment. Psychological Reports 77(1):10–10.

Lippert, D. 1997 In Front of the Mirror: Native Americans and Academic Archaeology. *In* Native Americans and Archaeologists: Stepping Stones to Common Ground. N. Swidler, K. E. Dongoske, R. Anyon, and A. S. Downer, eds. Pp. 120–127. Walnut Creek: AltaMira Press.

Lozada, M. C. 1998 The Senorio of Chiribaya: A Bioarchaeological Study in the Osmore Drainage of Southern Peru. Chicago: University of Chicago Press.

Lozada, M. C., D. E. Blom, and J. E. Buikstra 1997 The Practice of Artificial Deformation in Precolumbian Peru. El Chaski, Organo Periodistico de Pervian Arts Society: 7.

Madden M., W. L. Salo, J. Streitz, A. Aufderheide, G. Fornaciari, C. Jaramillo et al. 2001 Hybridization Screening of Very Short PCR Products for Paleoepidemiological Studies of Chagas' Disease. Biotechniques 30:102–104.

Martin, D. L. 1997 Violence Against Women in the La Plata River Valley (AD 1000–1300). *In* Troubled Times: Violence and Warfare in the Past. D. L. Martin, and D. W. Frayer, eds. Pp. 45–76. Amsterdam: Gordon and Breach.

McGinnis, S., and R. K. Davis 2001 Domestic Well Water Quality Within Tribal Lands of Eastern Nebraska. Environmental Geology 41(3–4):321–329.

Merryman, J. H. 1986 Two Ways of Thinking about Cultural Property. The American Journal of International Law 80:831–853.

Miller, K. 2005 Showdown at the O.K. Caral. Discover Magazine Online, retrieved on August 1, 2009 from http://discovermagazine.com/2005/sep/showdown-at-caral/article_view?b_start:int=1&-C=.

Molin G., A. G. Drusini, D. Pasqual, F. Martingnago, and G. Scarazzati 1998 Microchemical and Crystallographic Analysis of Human Bones from Nasca, Peru. A Possible Method of Direct Dating of Archaeological Skeletal Material. International Journal of Osteoarchaeology 8:38–44.

Moseley, M. E. 1993 The Incas and Their Ancestors. London: Thames and Hudson.

Murphy, M., G. Cock, and E. Goyochea 2007 Inca Resistance to Spanish Colonization: Violent Uprising and Nonviolent Resistance at Puruchuco-Huaquerones. Invited Paper Presented at the 72nd Annual Meeting of the Society for American Archaeology. Austin, TX. April 25–29.

Murphy, M., C. Gaither, E. Goyochea, G. Cock, and M. F. Boza 2009 Violent Confrontation and Spanish Conquest at Puruchuco-Huaquerones, Peru. Paper Presented at the 49th Meeting of the Institute of Andean Studies. Berkeley, CA.

Murphy, M. S. 2004 From Bare Bones to Mummified: Understanding Health and Disease in an Inca Community. Dissertation. Philadelphia: University of Pennsylvania.

Nystrom, K. C. 2007 Trepanation in the Chachapoya Region of Northern Perú. International Journal of Osteoarchaeology 17:39–51.

Ochatoma, J., T. A. Tung, and M. Cabrera 2008 Emergence of a Wari Military Class. Paper Presented at the 73rd Annual Meeting of the Society for American Archaeology. Vancouver, BC.

Ortner, D. J., E. H. Kimmerle, and M. Diez 1999 Probable Evidence of Scurvy in Subadults from Archaeological Sites in Peru. American Journal of Physical Anthropology 108:321–331.

Orlove, B. S. 1979 Two Rituals and Three Hypotheses: An Examination of Solstice Divination in Southern Highland Peru. Anthropological Quarterly 52(2):86–98.

Pääbo, S. 1989. Ancient DNA: Extraction, Characterization, Molecular Cloning, and Enzymatic Amplification. Proceedings of the National Academy of Science 86:1939–1943.

Peterson II J. E. 1990–1 Dance of the Dead: A Legal Tango for Control of Native American Skeletal Remains. American Indian Law Review 15:115–150.

Poole, D. 1997 Vision, Race, and Modernity: A Visual Economy of the Andean Image World. Princeton: Princeton University Press.

Powell, J. F., and J. C. Rose 1999 Chapter 2: Report on the Osteological Assessment of the "Kennewick Man" Skeleton (CENWW.97.Kennewick). National Park Service Archeology and Ethnography Program, U.S. Department of the Interior. Summary of Scientific Inquiries into Kennewick Man.

Ravesloot, J. C. 1997 Changing Native American Perceptions of Archaeology and Archaeologists. *In* Native Americans and Archaeologists: Stepping Stones to Common Ground. N. Swidler, K. E. Dongoske, R. Anyon, and A. S. Downer, eds. Pp. 172–178. Walnut Creek: AltaMira Press.

Reinhard, K. J., and J. E. Buikstra 2003 Louse Infestation of the Chiribaya Culture, Southern Peru: Variation in Prevalence by Age and Sex. Memorias del Instituto de Oswaldo Cruz 98(Suppl. 1):173–179.

Rice, D. G. 1997 The Seeds of Common Ground: Experimentations in Indian Consultation. *In* Native Americans and Archaeologists: Stepping Stones to Common Ground. N. Swidler, K. E. Dongoske, R. Anyon, and A. S. Downer, eds. Pp. 217–226. Walnut Creek: AltaMira Press.

Rose, J. C., T. J. Green, and V. D. Green 1996 NAGPRA is Forever: Osteology and the Repatriation of Skeletons. Annual Review of Anthropology 25:81–103.

Salo, W. L., A. C. Aufderheide, J. E. Buikstra, and T. A. Holcomb 1994 Identification of *Mycobacterium Tuberculosis* DNA in a Pre-Columbian Peruvian Mummy. Proceedings of the National Academy of Sciences 91:2091–2094.

Sandness, K. L. 1992 Temporal and Spatial Dietary Variability in the Prehistoric Lower and Middle Osmore Drainage: The Carbon and Nitrogen Isotope Evidence. MA Thesis. Lincoln: University of Nebraska at Lincoln.

Schreiber, K. J. 2009 Personal communication, February.

Shady Solis, R. 1997a La Ciudad Sagrada De Caral – Supe En Los Albores De La Civilización En El Perú. Fondo Editorial, UNMSM, Lima.

Shady Solis, R. 1997b Caral. La Cité Ensevelie. Revista Archéologie 340:58–65.

Shady Solis, R. 1999 Los Orígenes De La Civilización Y La Formación Del Estado En El Perú: Las Evidencias Arqueológicas De Caral-Supe (Primera Parte). Boletín del Museo de Arqueología y Antropología de la UNMSM 2(12):2–4.

Shady Solis, R. 2001 Caral, La Ciudadela Más Antigua De América. Revista Rumbos 5(29):72–76.

Shady Solis, R. 2005 Open Letter from Ruth Shady. Caral Civilization Peru: The Origins of Civilization in Peru, retrieved on August 1, 2009 from http://caralperu.typepad.com/caral_civilization_peru/2005/01/open_letter_fro.html.

Shady Solis, R., J. Haas, and W. Creamer 2001 Dating Caral, a Preceramic Site in the Supe Valley on the Central Coast of Peru. Science 292:723–726.

Shady Solis, R., and S. Lopez 1999 Ritual De Enterramiento De Un Recinto En El Sector Residencial A. En Caral-Supe. Boletín de Arqueología PUCP 3:187–212.

Shinoda, K.-i., N. Adachi, S. Guillen, and I. Shimada 2006 Mitochondrial DNA Analysis of Ancient Peruvian Highlanders. American Journal of Physical Anthropology 131: 98–107.

Slovak, N. M., Paytan A., and Wiegand B. A. 2008 Reconstructing Middle Horizon Mobility Patterns on the Coast of Peru Through Strontium Isotope Analysis. Journal of Archaeological Science 36(1):157–165.

Smith, L. 2004. Archaeological Theory and the Politics of Cultural Heritage. 1st Edition. London: Routledge.

Springfield, A. L. W., A. C. Aufderheide, J. E. Buikstra, and J. Ho 1993 Cocaine and Metabolites in the Hair of Ancient Peruvian Coca Leaf Chewers. Forensic Science International 63:269–275.

Sutter, R. C. 2000 Prehistoric Genetic and Culture Change: A Bioarchaeological Search for Pre-Inka Altiplano Colonies in the Coastal Valleys of Moquegua, Peru, and Azapa, Chile. Latin American Antiquity 11(1):43–70.

Swidler, N. ed. 1997 Native Americans and Archaeologists: Stepping Stones to Common Ground. Walnut Creek, CA: AltaMira Press.

Tello, J. C. 1913. Prehistoric Trephining Among the Jauyos of Peru. International Congress of Americanists, Proceedings of the XVIII Session, 1912. London: Harrison and Sons.

Thomas, D. H. 2000 Skull Wars: Kennewick Man, Archaeology and the Battle for Native Identity. New York: Basic Books.

Thompson Reuters 2007 Journal Citation Reports®. http://thomsonreuterscom/products_services/scientific/Journal_Citation_Reports.

Tomczak, P. D. 2003 Prehistoric Diet and Socioeconomic Relationships Within the Osmore Valley of Southern Peru. Journal of Anthropological Archaeology 22:262–278.

Torres-Rouff, C. 2003 Shaping Identity: Cranial Vault Modification in the Pre-Columbian Andes. Dissertation. Santa Barbara: University of California.

Tung, T. 2008 Dismembering Bodies for Display: A Bioarchaeological Study of Trophy Heads from the Wari Site of Conchopata, Peru. American Journal of Physical Anthropology 136:294–308.

Tung, T. A. 2007 Trauma and Violence in the Wari Empire of the Peruvian Andes: Warfare, Raids, and Ritual Fights. American Journal of Physical Anthropology 133(3): 941–956.

Tung, T. A., M. Cabrera, and J. Ochatoma 2007 Cabezas Trofeo Wari: Rituales del Cuerpo en el Recinto Ceremonial en "D" de Conchopata. Revista de Investigacion. Universidad Nacional de San Cristobal de Huamanga, Peru 15(2):216–227.

Tung, T. A., and M. Del Castillo 2005 Una Visión de la Salud Comunitaria en el Valle de Majes durante la Época Wari. Muerte y Evidencias Funerarias en los Andes Centrales: Avances y Perspectivas. Pp. 149–172. Lima: Universidad Nacional de Federico Villareal.

Tung, T. A., and K. J. Knudson 2008 Social Identities and Geographical Origins of Wari Trophy Heads from Conchopata, Peru. Current Anthropology 49(5):915–925.

Turner, B. L. 2008 The Servants of Machu Picchu: Life Histories and Population Dynamics in Late Horizon Peru. Unpublished Ph.D. Dissertation. Emory University.

Turner, B. L., G. D. Kamenov, J. D. Kingston, and G. J. Armelagos 2008 Insights into Immigration and Social Class at Machu Picchu, Peru Based on Oxygen, Strontium, and Lead Isotopic Analysis. Journal of Archaeological Science 36(2):317–332.

Turner, B. L., J. D. Kingston, and G. J. Armelagos 2009 Diet Versus Locale: Isotopic Support for Causal Roles in Pathological Conditions at Machu Picchu, Peru. American Journal of Physical Anthropology 132(S48):401–402.

Turner, B. L., J. D. Kingston, and J. T. Milanich 2005 Isotopic Evidence of Immigration Linked to Status During the Weeden Island and Suwanee Valley Periods in North Florida. Southeastern Archaeology 24(2):121–136.

Ubelaker D. H., and L. G. Grant 1989 Human Skeletal Remains: Preservation or Reburial? American Journal of Physical Anthropology 32(S10):249–287.

Ubelaker, D. H., M. A. Katzenberg, and L. G. Doyon 1995 Status and Diet in Precontact Highland Ecuador. American Journal of Physical Anthropology 97(4):403–411.

Uhle, M. 1909 Uhle Fieldnotes. Berkeley: University of California Hearst Museum.

van den Bergh, P., and J. Flores Ochoa 2000 Tourism and Nativistic Ideology in Cuzco, Peru. Annals of Tourism Research 27(1):7–26.

van den Berghe, P., and G. Primov 1977 Inequality in the Peruvian Andes: Class and Ethnicity in Cuzco. Columbia: University of Missouri Press.

Verano, J. W. 1992 Prehistoric Disease and Demography in the Andes. *In* Disease and Demography in the Americas. J. W. Verano, and D. H. Ubelaker, eds. Pp. 15–24. Washington, DC: Smithsonian Institution Press.

Verano, J. W. 1997 Advances in the Paleopathology of Andean South-America. Journal of World Prehistory 11(2):237–268.

Verano, J. W. 1998. Sacrificios Humanos, Desmembramientos y Modificaciones Culturales en Restos Osteológicos: Evidencias de las Temporadas Investigación 1995–6 en la Huaca de la Luna. *In* Investiaciones en la Huaca de la Luna 1996. S. Uceda, E. Mujica, and R. Morales, eds. Pp. 159–171. Trujillo: Universidad Nacional de Trujillo, Peru.

Verano, J. W. 2001 Análisis de Restos Oseos Humanos. *In* Desentierro y Reenterramiento de una Tumba de Elite Mochica en el Complejo el Brujo. R. Franco, C. Gálvez, and S. Vásquez, eds. Lima, Peru: Programma Arqueológico Complejo El Brujo Boletin No. 2.

Verano, J. W., L. S. Anderson, and R. Franco 2000 Foot Amputation by the Moche of Ancient Peru: Osteological Evidence and Archaeological Context. International Journal of Osteoarchaeology 10(3):177–188.

Verano, J. W., and M. J. DeNiro 1993 Locals or Foreigners? Morphological, Biometric and Isotopic Approaches to the Question of Group Affinity in Human Skeletal Remains Recovered from Unusual Archaeological Contexts. *In* Investigations of Ancient Human Tissue: Chemical Analyses in Anthropology. M. K. Sandford, ed. Pp. 361–386. Langhorne: Gordon and Breach.

Verano, J. W., and G. P. Lombardi 1999 Paleopatología En Sudamérica Andina. Bulletin de l'Institut français d'études andines 28(2):91–121.

Verano, J. W., S. Uceda, C. Chapdelaine, R. Tello, M. I. Paredes, and V. Pimentel 1999 Modified Human Skulls from the Urban Sector of the Pyramids of Moche, Northern Peru. Latin American Antiquity 10(1):59–70.

Walker, P. 2004 Caring for the Dead: Finding a Common Ground in Disputes Over Museum Collections of Human Remains. *In* Documenta Archaeobiologiae: Yearbook of the State Collection of Anthropology and Paleoanatomy. G. Grupe, and J. Peters, eds. Rabden: Verlag M. Leidorf.

Walker, P. L. 2000 Bioarchaeological Ethics: A Historical Perspective on the Value of Human Remains. *In* Biological Anthropology of the Human Skeleton. M. A. Katzenberg, and S. R. Saunders, eds. Pp. 3–39. New York: Wiley.

Watkins, J. 2005 Through Wary Eyes: Indigenous Perspectives on Archaeology. Annual Review of Anthropology 34:429–449.

Williams, J. S. 2004 Inca-Period Diet for the Central Coast of Peru: A Preliminary Report on the Isotopic Anlaysis of Human Bone Collagen from Puruchuco-Huaquerones. American Journal of Physical Anthropology 209–209.

Wilson, A. S., T. Taylor, M. C. Ceruti, J. A. Chavez, J. Reinhard, V. Grimes et al. 2007 Stable Isotope and DNA Evidence for Ritual Sequences in Inca Child Sacrifice. Proceedings of the National Academy of Sciences 104(42):16456–16461.

World Fact Book – Peru. Retrieved on August 1, 2009 from www.cia.gov/library/publications/the-world-factbook/geos/pe.html.

Zimmerman, L. J. 1989 Made Radical By My Own: An Archaeologists Learns to Accept Reburial. *In* Conflict in the Archaeology of Living Traditions. R. Layton, ed. Pp. 60–67. London: Unwin Hyman.

Zimmerman, L. J. 1997 Remythologizing the Relationship Between Indians and Archaeologists. *In* Native Americans and Archaeologists: Stepping Stones to Common Ground. N. Swidler, K. E. Dongoske, R. Anyon, and A. S. Downer, eds. Pp. 44–56. Walnut Creek, CA: AltaMira Press.

4

The Formation of Mortuary Deposits

Implications for Understanding Mortuary Behavior of Past Populations

Estella Weiss-Krejci

In the last 10,000 years, 100 billion people have died (Davies 1994:24). The majority of them have passed into oblivion and only a small part is actually present in the archaeological record. Many dead bodies still await recovery by archaeologists, but others will have forever disappeared due to natural destruction or alterations inflicted by humans. While at some archaeological sites, human remains constitute more or less representative samples of the population that lived and died there, at other sites they may only form a biased fraction (e.g., Carr and Knüsel 1997:168; Chapman 2005:37; Perlès 2001:273; Ucko 1969). Historic information can help in understanding the representativeness of the mortuary data, as in the case of ancient Egypt, classical Greece and Rome, historic Europe and North America (Antonaccio 1995; Daniell 1997; Morris 1992; Saunders et al. 1995; Taylor 2001). In prehistoric contexts, however, we need to be extremely cautious about assumptions regarding what the recovered human remains represent, what they mean, and by which processes they entered the archaeological record.

The last two decades have been characterized by considerable advances in the theory of mortuary archaeology (Beck 1995; Carr 1995; Rakita et al. 2005). Archaeologists have particularly reshaped their notions of what constitutes a "burial" in considering areas which lack inhumation burials (e.g., Carr and Knüsel 1997; Stodder 2005). However, at sites where the dead are deposited in familiar ways,

Social Bioarchaeology. Edited by Sabrina C. Agarwal and Bonnie A. Glencross
© 2011 Blackwell Publishing Ltd

recording and interpreting burials is still very much guided by our own cultural perceptions. Though special data sheets for human remains guarantee that all important aspects during the excavation are recorded (Roskams 2001:199–208), the problem often lies in the selection of the data which are included in burial analysis. As noted by Brown (1995:5), the focus on individual treatments is supported by the ease of their analyses and the domination of individual treatments in contemporary life. Assuming that one specific type of deposition, e.g., inhumation, constitutes the predominant proper burial practice by an ancient society, archaeologists possibly overlook other less obvious types of mortuary contexts. To study one aspect of mortuary ritual under the assumption that it represents the whole and to assume that the mortuary behavior of one specific group of people stands for the overall society can seriously flaw our understanding of the past (Bradley 1995; Brown 1995; Chapman 2005:37).

In the first part of this chapter I will elaborate on this problem. To establish a meaningful frame of reference in mortuary analysis requires a constant awareness for the formation processes of mortuary deposits. The term "formation processes" was coined by Michael Schiffer (1987) and describes the natural and cultural events which create an archaeological context. The discussion is guided by a graphic model (Figure 4.1), which I developed using historic and ethnographic examples. The formation of mortuary deposits is often the result of complex and unrelated events, which can take place at any moment between the death of an individual and the recovery of the human remains by the archaeologist.

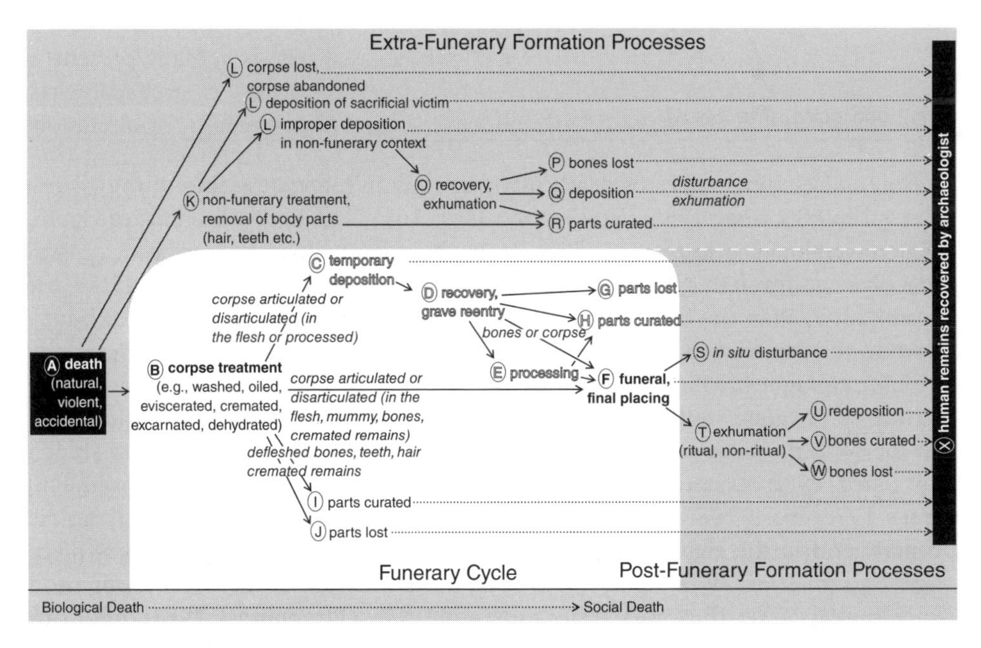

Figure 4.1 Schematic model showing the possible formation processes of archaeological deposits from the death of an individual until its discovery by an archaeologist. The funerary portion forms only a fraction.
Source: Author's drawing

In the second part of this chapter I will demonstrate the complex nature of mortuary deposits through a retrospective analysis of the data from the Maya site of Tikal. Tikal was investigated by the Tikal Project of the University of Pennsylvania Museum between 1956 and 1969 (Coe and Haviland 1982) and a variety of Guatemalan research projects which followed (Gómez and Vidal Lorenzo 1997; Laporte 2003; Laporte and Fialko 1995). Herein I will focus on the human remains data collected by the University Museum, relying both on published information (Becker 1999; Coe 1990; Haviland 1985; Moholy-Nagy 2003; 2008) as well as unpublished materials. I will argue that deposition of an unaltered corpse in a grave beneath a structure was probably not the dominant burial mode at Tikal but only that of a small part of the population. Traces of the remains of the majority of Tikal's inhabitants may be found in some of the deposits registered as miscellaneous bones and problematical deposits. However, these types of deposits do not only comprise the remains of people who received funerary treatment; post-funerary and extra-funerary processes may also account for similar assemblages.

Formation Processes of Mortuary Deposits

The formation of the archaeological mortuary record is a chronological process which always starts with the death of an individual and always ends at the point of the discovery and excavation by the archaeologist (Figure 4.1: Process A–Process X). Anything that happens in between these two events is shaped by cultural and natural processes. The funerary portion forms only a fraction of a larger cycle of formation.

Dying and death (Process A)

Reason and circumstances of death often directly influence the fate of the corpse (Hertz 1960:85; Shay 1985; Van Gennep 1960:161). The body of a murder victim, a fallen warrior or a human sacrifice, from their moment of biological death will most likely go through different processes (e.g., Coppet 1981; Knüsel 2005:58–61; Verano 2005) than the body of somebody who has died a peaceful death at home. "Bad death" can lead to "deviant" mortuary behavior (Aspöck 2008; 2009; Goody 1962:52; Sprague 2005:174; Weiss-Krejci 2008). Which exact types of death are considered as "good" and which ones "bad" may vary from one society to the other. But the concept of "good" and "bad" death is a cross-cultural phenomenon (Bloch and Parry 1982; Cátedra 1992; Humphreys 1981:261–263; Kluckhohn 1948; Straus 1978; Ward 1980). In order to ward off evil and ensure that the dying die properly and in the right place, death rituals often start before death takes place (e.g., Naquin 1988:39, Rospatt 2005). The determination of the exact moment in which biological death occurs varies cross-culturally (Humphreys 1981:265; Lock 1997). In pre-industrial societies it is frequently associated with the exhalation of the last breath, which can serve as a metaphor for the soul (Hirst 1986:107; Houston et al. 2006:146, Metcalf and Huntington 1991:87; Naquin 1988:39). With the exception of individuals who suffer "social death" during their life time

(Elliott 2008; Glaser and Strauss 1966; Mulkay 1993; Sweeting and Gilhooly 1997), the process of social dying starts as soon as biological death has been confirmed by the survivors (Hertz 1960:27; Morris 1992:9–10).

The funerary cycle

The funerary cycle is a transitional period, characterized by social dying on the part of the deceased, accompanied by bodily disintegration, and mourning and social reorganization (e.g., redistribution of property) on the part of the survivors (Hertz 1907; Humphreys 1981:263; Van Gennep 1909). The survivors enter this period through rites of separation, carry out various rites of transition and emerge through rites of re-integration (Van Gennep 1960:147). During the funerary cycle, social life is often suspended for those with the right or obligation to mourn. A number of phenomena such as adhering to a specific dress code, self mutilation, etc., are part of this process which can last months or years (Durkheim 1912; 2001:297; Eisenbruch 1984; Frazer 1886; Gluckmann 1937:122; Leach 1958; Malinowski 1926:48; Needham 1967; Radcliffe-Brown 1922; Turner 1967). At the end, usually a final rite is held which lifts regulation and prohibitions, signaling the restoration of normality. The deceased is finally removed from the sphere of the living, i.e., he or she has become socially dead. In contrast to Humphreys (1981:263), who considers the process of social dying as a complete cessation of all social actions directed towards the remains of the deceased, I associate social death with the end of the funerary cycle. This does not mean that all interactions of the living with the dead end at that point or that the deceased have moved beyond memory. They are often transformed into something else (e.g., an ancestor, a saint; Fortes 1965; Gluckmann 1937) with some of the bodily remains playing very active post-funerary roles (e.g., Johnson 1996; Verdery 1999).

In Figure 4.1 the funerary cycle starts with the treatment of the corpse (Process B) whereas final deposition (Process F) ends the cycle. I would like to emphasize that a one-to-one relationship between Van Gennep's (1909) rites of separation, transition and re-integration with specific funerary activities such as treatment and deposition of the corpse does not exist. The deposition of a dead person does not have to take place at the end of the funerary cycle. Although there is a chronological order which leads from treatment to deposition, the time which passes between these two events can be highly variable (Innes 1999:68; Hertz 1907; Van Gennep 1909; Watson 1988:208). In Islamic societies, ideally the body is buried within 24 hours (Innes 1999:68), which implies that the period of mourning outlasts those rites directly associated with the corpse. Aspöck (2009:184–185) has made the important observation that the deposition of a body in the flesh is actually part of the rites of separation because it usually takes place soon after death. In this instance the rites associated with the end of the funerary cycle and the end of mourning are characterized by other activities such as a funeral meal or the return to the usual dress code. On the other hand, in the case of cremation, the treatment (= the burning) of the corpse serves as the rite of separation whereas the collection of burned remains from a pyre and deposition in a grave play a role in rites of transition or re-integration (Aspöck 2009:184–185; Williams 2005; 2008).

Treatment and deposition can also constitute one process, for example when a cremated person is not removed from the pyre (McKinley 2005:10). In summary, the same activity, such as the deposition of a body, can represent a rite of separation in one society whereas it constitutes a rite of re-integration in another. Some activities and rites within the funerary cycle are directly associated with the corpse and the grave and may be archaeologically more visible than other rites, which have no direct relation to the human remains.

In a variety of societies (ancient Egypt, Ostyak of Siberia, some aristocratic houses of Europe, Toda of India, etc.) surrogate bodies represent the deceased's corpse until the funerary cycle is completed. These transitional bodies look very different from each other and comprise figurines, dolls, effigies, masks, hair locks, bone splinters, and empty coffins (Assmann 2005; Brückner 1966; Giesey 1960; Knipe 2005; Schukraft 1989:94–98; Van Gennep 1960:150; Weiss-Krejci 2005:167). It is often these surrogates which are deposited during the rites of re-integration at the end of the funerary cycle.

Corpse treatment (Process B)

If survivors take care of the body they usually treat it. Body preparation[1] may be minimal. Washing, oiling, perfuming, shrouding, or dressing don't alter the corpse and can be detected only under good conditions of preservation (e.g., Mallory and Mair 2000). However, survivors may employ procedures that leave detectable traces in order to avoid or accelerate processes of putrefaction; in other words methods which are either directed at preservation or at reduction (Sprague 2005:70). The most common body treatments are dehydration, chemical and thermal modification (Aufderheide 2003:43), cremation (Parker Pearson 2000:6–7; Schmidt and Symes 2008), and active excarnation (Weiss-Krejci 2005:160–162). Some methods of dehydration, chemical and thermal modification as well as active excarnation (skinning and reassembling the bones) can produce articulated corpses (Aufderheide 2003; Yarrow 1880b:figure 1) whereas others such as cremation and certain types of active excarnation (boiling and mutilating, consumption by vultures, etc.) result in disarticulation and fragmentation (Weiss-Krejci 2008:figure 10.2; Yarrow 1880b:figure 3). It is also important to be aware of the fact that bones which have burned in a green state are not always evidence for cremation. When corpses are smoked or roasted in front or above a fire the bones can also sometimes burn (Sempowski 1994:142–144). Dehydration, cremation, active excarnation, etc. can also form parallel treatments for different parts of one corpse. A body may be eviscerated and embalmed, but the entrails are burned. A corpse may be mutilated and a part defleshed, while the rest remains in the flesh. The separation of the corpse can lead to separate deposition of body parts and involves boney parts or soft tissue. Body parts such as skulls, teeth, hair, or artifacts made from human bone may not be buried at all but kept around among the living (Figure 4.1, Process I; Hallam and Hockey 2001:136–141; Stodder 2006). After processing of the corpse and before deposition whatever is left may immediately go through a second cycle of processing as in Tibet and Nepal where the cremated remains of the deceased are sometimes taken to the lamas who pulverize the fragments in a mortar, mix them with clay and mold them into clay tablets (Malville 2005:197). Also in the Himalayas, in the past human skulls were frequently taken from the defleshing grounds and manufactured into hand drums (Beer 2003:108; Malville 2005:198).

Which type of treatment is chosen for a corpse or its parts is often determined by environmental, social, economic, and ideological factors. Circumstances and place of death, the distance from the burial place, the climate and the prospective time to bury a body as well as vertical and horizontal status and age of the deceased all come into play (Binford 1971; Saxe 1970; Van Gennep 1960:146; Weiss-Krejci 2005). From a cross-cultural perspective no method carries exclusively positive or negative connotations. What is considered normal, acceptable or negative varies from one society to the other (Aspöck 2009:84–99; Kroeber 1927:313; Ucko 1969:270–271). Bodily mutilation, burning, and nontreatment of the corpse can be considered an honor or a punishment (Duncan 2005; Finucane 1981:57–58; Iserson 1994:238; O'Shea 1984:74; Radcliffe-Brown 1922:166; Weiss-Krejci 2008). In an entirely prehistoric context the correct interpretation of the symbolic meaning of any particular body treatment has to be undertaken with great caution.

Temporary deposition (Process C)

Temporary deposition of the corpse may form a stage in a program of mortuary treatment that includes exhumation (Brown 1995:16). As discussed by Hertz (1907) for the Ngaju Dayak of Southeast Asia (Figure 4.2), and also known from other world regions such as Greece (Danforth 1982), southern China (Freedman 1966; Tsu 2000), Australia (Tonkinson 2008), and North America (Feest 1997), the body is not taken to its final burial place immediately after death. Instead it is kept in temporary storage either in the ground, in a tree, on a scaffold, or in the house (Hutchinson and Aragon 2002; Kuhnt-Saptodewo and Kampffmeyer 1996; Kuhnt-Saptodewo and Rietz 2004; Metcalf and Huntington 1991; Yarrow 1880a,b). The intermediary period between death and final deposition can vary between a few months and many years (Feest 1997:427; Tsu 2000; Watson 1988:208). In this case, temporary storage is a process of passive excarnation by which the corpse is naturally defleshed through putrefaction. Once the corpse has decomposed the bones are taken and reburied in a place distinct from the storage place (Figure 4.1, Processes D–F). If not totally bare, before reburial the remaining flesh is detached through cleaning (Kuhnt-Saptodewo and Kampffmeyer 1996; Kuhnt-Saptodewo and Rietz 2004; Meyers 1971:96). In some instances burning of the dry bones can accompany reburial rituals (Process E). Burning of dry bones after temporary storage is the final mortuary procedure for the Thai kings and collective cremation is also an element of Balinese mortuary practices (Hutchinson and Aragon 2002:30; Metcalf and Huntington 1991:101, 137). Fractions of a cremated corpse can also go through continuous cycles of reburning (cremation of dry bones) as part of secondary rituals as well as post-funerary commemorative services (Beck 2005; Hutchinson and Aragon 2002:44). Even mummies can be subject to secondary treatment. The mummies of the Khasi chiefs of India were pickled in alcohol and stored in the house for years until their final cremation (Gurdon 1907).

Figure 4.2 Ngaju Dayak of Borneo preparing the remains of a temporarily stored relative for final funerary deposition. This ceremony of secondary burial is called *tiwah* and forms a collective rite which is performed by the entire village. After their exhumation from the ground the bones are washed beside a small stream and the remaining hair and nails are removed. The cleaning of the bones is later followed by the purification of the soul. This part of the ritual is performed by a priest who circles a burning bamboo stick around the bones and chants in a sacred language. At the end of the *tiwah* the bones are buried into funeral structures above the ground (*sandung*), which already hold the remains of predeceased family members (Kuhnt-Saptodewo and Rietz 2004). The *tiwah* during which the photograph was taken lasted 38 days and involved 23 families reburying 35 deceased family members. It is subject of the documentary *Bury me Twice* by Kuhnt-Saptodewo and Kampffmeyer (1996).
Source: Kuhnt-Saptodewo, S., and H. Rietz 2004 Helbigs Beobachtungen zu den Ritualen der Dayak und der heutige Stand der Forschung. *In* Karl Helbig. Wissenschafter und Schiffsheizer: Sein Lebenswerk aus heutiger Sicht. Rückblick zum 100. Geburtstag. W. Rutz and A. Sibeth, eds. Pp. 178–197. Hildesheim: Georg Olms Verlag. Photograph copyright and courtesy of Hanno Kampffmeyer and Jani Kuhnt-Saptodewo

The ethnographic literature suggests that funerals for the temporarily buried are not always completed (e.g., Tsu 2000). Secondary burial as among the Ngaju Dayak is a ceremony of consumption which entails an outlay of wealth (Kuhnt-Saptodewo and Kampffmeyer 1996; Miles 1965). The long time period that usually elapses between first disposal and the final funeral allows the survivors to accumulate the necessary resources. Although secondary burial "is the right thing to do" (Miles 1965:174) it is sometimes not performed owing to the lack of capital necessary for its sponsorship. As noted by Hutchinson and Aragon (2002:47), in a prehistoric context such temporary deposition spaces and incomplete burials carry an enormous potential for misinterpretation of funerary customs.

Temporary storage and exhumation can take place for a variety of other reasons, which are not the result of ritual considerations. The seasonality of tribal movement and environmental conditions are probably responsible for dismemberment, storage of bodies, and collective burial at Aymyrlyg in Tuva (Altai Mountains, Iron Age Uyuk culture, third to second century B.C.) (Murphy and Mallory 2000:393–394). The populations practiced a semi-nomadic form of economy including cyclic migration with fixed routes and set winter camps. Burial was only possible at specific times of the year, when the ground was not frozen and the tribe set up camp in the valley close to the burial site. When buildings, crypts, or tombs are not yet ready to house mortal remains, corpses may also be stored and later reburied (Weiss-Krejci 2004; 2005:165–166; Tsu 2000:20). If somebody dies in a distant area and needs to be transported back home, considerable delays between death and final deposition are possible (Weiss-Krejci 2001). Whether articulated corpses and bones are reburied often depends on the time that has elapsed since death.

Final deposition (Process F)

Final funerary deposition of the corpse or its parts can take place in a variety of locations such as in subterranean pits or subterranean burial chambers, in funerary monuments above the ground, in rivers or the ocean, in trees, in or under houses or in caves and, in modern times, even in space. Archaeologically some of these places are more visible than others. A corpse that has been laid to rest in a structure like a megalithic tomb has a good chance of being recovered by the archaeologist, whereas deposition in water will not create an easily identifiable burial deposit and will not likely be recovered (Bradley 1995; Schiffer 1987). Final funerary deposition can be a first deposition of an untreated and articulated corpse – archaeologists like to call this a primary burial or inhumation (Sprague 2005:66–69) – or a first deposition of articulated or disarticulated remains, which have been altered through dehydration, cremation, or active excarnation. It can also be a secondary or tertiary deposition of human remains which have gone through episodes of temporary storage, exhumation, and reburial (Figure 4.2). In the case of prolonged body storage, the borderline between the funeral and post-funeral sphere is blurred. In some instances, the final deposition of a dead person can take place long after the end of the mourning period (Watson 1988:208). The southern Chinese exhume the remains of the deceased after seven to ten years and store the bones in a large pottery urn. Many are lost after a few generations (Watson 1988:205). Only the bones of a very few are permanently buried into a brick tomb years later. While removal from the first into the secondary context is a common practice and part of the funerary rites, tertiary burial is reserved for important ancestors (Freedman 1966).

There are also instances where there is no final deposition of the corpse. When all of the funerary rituals had been completed, the mummified remains of the dead Inca rulers were put in town houses or royal sanctuaries and continued to actively participate in Cuzco's ceremonial political life, attending festivities and making visits (D'Altroy 2002:96–97). On the tomb an idol was placed, containing the deceased's fingernails and hair, which had been saved during lifetime. This idol was worshipped and could serve as a surrogate for the deceased lord in public affairs.

There also may be more than just one disposal place for a corpse. As an example, the hearts and intestines of European dynasts were often buried in a different location than the main body (Weiss-Krejci 2001; Westerhof 2008). Among the Dowayo of North Cameroon, the corpse is first buried in a rock tomb, but the head is removed some three weeks later, cleaned and placed in a pot in a tree (Barley 1981:151). Separation of the skull from the corpse is also of considerable antiquity in the Pacific (Stodder 2006). Cremation, in particular, provides many options for disposal of burned remains. They can be left in situ at the place where burning takes place, they can be distributed among various places, some parts buried and others kept for future rituals (Beck 2005; Parker Pearson 2000:6–7; Williams 2008). What is important for the archaeologist to recognize is that a dead person's remains, when being finally deposited during the funerary cycle can constitute anything from a complete and articulated body to disarticulated minute fragments.

Beyond the funerary cycle

Extra-funerary processes

If a person comes to rest without receiving funerary treatment this is an extra-funerary event. Persons who die alone far away from other humans will enter the archaeological record at that point and will likely be found in non-funerary contexts if found at all. Under certain environmental circumstances some of these bodies can preserve through spontaneous natural mummification (Aufderheide 2003:41) such as the Iceman (Figure 4.1, Process L) but, more likely – especially when left above the ground – they will disintegrate without a trace or be consumed by animals. Persons whose death is not confirmed because their bodies have not been recovered by the survivors, may take a very prolonged social death or never really "die" in a social sense (Zur 1998). That "biological death of an individual sets off a more prolonged social process of dying" (Morris 1992:9–10) only holds true when survivors have been able to acknowledge it.

Certain categories of persons do not receive a proper funeral because of circumstances of their life or death (criminals, witches, suicides, those struck by lightning, sacrificial victims, murder victims, etc.) (Aspöck 2009:86; Hertz 1960:85; Ucko 1969:271; Van Gennep 1960:161). The corpse of a murder victim – the same applies to the body of a sacrificial victim – may be treated in some form by the person who is responsible for the killing but its deposition is not necessarily a funeral (Processes K–M,N). One famous example of a person who has gone through an extra-funerary process is Che Guevara. He was killed in Bolivia (Process A) and his hair and hands cut and kept separately from the body (Process K). The corpse was hidden in the ground for 30 years together with the bodies of his companions (Process N). After discovery, the bodies were exhumed (Process O), investigated, and reburied in Cuba (Process Q) (Dosal 2004; Weiss-Krejci In press). Victims of political murder, massacres, and genocides are often given their last rites at a later point in time (Verdery 1999; Watson 1994:66; Weiss-Krejci 2008; Zur 1998). If too long a period of time elapses between the death and reburial and there is nobody around to grieve the dead, these reburial ceremonies are better understood as rites of commemoration. Reburial of people who are not properly buried can be highly political and public in nature. Famous individuals as well as anonymous dead often play an important role in the creation of collective, historic, and national identities (Ballinger 2002; Robben 2000; Verdery 1999; Weiss-Krejci In press). Examples include executed royals like Louis XVI, Marie Antoinette, and the Romanovs (Brown 1985:255–256; Weiss-Krejci In press) as well as political victims of Former Yugoslavia, Argentina, and Guatemala (Hayden 1994; Robben 2000; Zur 1998).

Post-funerary processes

Once the remains of a dead person are finally deposited as part of the mourning period and there exists no immediate necessity to remove them, the funerary deposition is completed. If, for some reason, the living decide to exhume these mortal remains at a later point in time, this constitutes a post-funerary process. The reasons for post-funerary disturbance (Figure 4.1, Process S), exhumation, and redeposition (Processes T–U) can include a wide range of ritual and nonritual behaviors. Among the most common causes are ancestral rites and ancestral appropriation, veneration and commemoration of important individuals, political disputes, rescue of bodies, relic cult, practicality, but also looting and desecration (Brown 1981; Brown 1985:250–253; Freedman 1966; McAnany 1995; Verdery 1999; Waterson 1995:208–210; Weiss-Krejci 2004; 2005:169; In press). As with temporary storage, bones that are exhumed after the funeral and at a much later point in time may be cleaned and burned. To interpret alterations on bones in this case can be extremely complex because the bones can often go through more than one process. For example, when tombs are looted, bones may get thrown out, destroyed, or burned. These bones may then be reburied ceremonially (Figure 4.3), thus completely masking the reasons for exhumation (Weiss-Krejci 2001:778). In Europe it is also a frequent custom to exhume bones from overcrowded churchyards and re-deposit them in charnel houses (Binski 1996:55; Daniell 1997:123). Some of these bones are used as *memento mori* (Figure 4.4) or as ornaments, as at the famous Sedlec ossuary at Kutna Hora, Moravia.

While ancestor veneration and other types of post-funeral relocations have received a lot of attention in archaeology (Antonaccio 1995; McAnany 1995; Verdery 1999; Weiss-Krejci 2001, 2004; Whitley 2002), the phenomenon of relics and sacred bone objects in prehistory is largely neglected. Relics represent people, but not necessarily ancestors. Bone relics can take any form between a complete body and bone splinters or dust. The list of relics made from human bodies is long and comprises objects such as Tibetan musical instruments (Malville 2005), the *kwaimatnie* of the Baruya of New Guinea (oblong parcels wrapped in a strip of bark cloak displayed solely on the occasion of initiation ceremonies and containing powerful objects such as eagles bones and a miraculous human forearm (Godelier 1999)), embalmed corpses of political persons such as Eva Perón and Lenin (Aufderheide 2003:159–160, 210–211; Robben 2000; Tumarkin 1983) and complete bodies and bone particles of Christian saints (Angenendt 1994, 2002;

Figure 4.3 Crystal cask in a wooden box at the Carthusian monastery of Mauerbach, Austria (post-funerary deposition); the container was used from 1683 to 1739 and holds the bones of King Frederick the Fair (d. 1330) from the House of Habsburg and his daughter Elizabeth (d. 1336). The bodies of King Frederick and his daughter were first exhumed by order of Emperor Maximilian I in 1514, rinsed with wine and temporarily stored in the vestry (Meyer 2000:70). When the Turks plundered the abbey in1529 the mortuary remains were thrown out but then collected by inhabitants of the area and transported to Vienna. One year later they were returned to Mauerbach and buried in a stone coffin between 1557 and 1561. In 1683 they were threatened by the Turks once again and when the attack was over, inserted into the container shown in this figure. After the dissolution of the monastery by Emperor Joseph II in 1782 the bones were brought to Vienna in 1789 and reburied into the Habsburg house vault at St. Stephen's, Vienna where they rest as of today in a small 18th-century coffin (Meyer 2000:70–72; Weiss-Krejci 2004:390).
Source: Figure from Gerbert, M., M. Herrgott, and R. Heer 1772 Taphographia Principum Austriae, Monumenta Augustae Domus Austriae Vol. 4/2. St. Blasien: St. Blasien Abbey, plate XII, engraving by Georg Nicolai

Aufderheide 2003:201–202; Johnson 1996). Relics are attributed a variety of ceremonial functions (initiation, legitimization of domination) and are often equipped with healing and apotropaic powers (Angenendt 2002; Brown 1981; Godelier 2004; Malville 2005; Schaeffer 2007:222). They can be unspectacular objects upon which enormous importance is placed by the people who use them. Divorced from their contexts they are without significance (Geary 1990:5).

Bones of the dead, which have been subject to post-funerary and extra-funerary processes, can be hundreds of years older than their recovery context. Especially mummified corpses are often curated for long periods without even losing their bodily articulation; they can be exhumed and reburied various times and moved back and forth between different locations (e.g., the Inca kings, D'Altroy 2002:96–97). Skeletons can also be reconstructed and appear articulated (e.g., the Catacomb saints, Johnson 1996). If the remains are not investigated by a physical anthropologist and not dated, re-deposition may be mistaken for deposition in the flesh. Articulated Bronze Age skeletons from Cladh Hallan, Scotland were buried several hundred years after death beneath roundhouses. The corpses provide evidence for mummification and manipulation of body parts (Parker Pearson et al. 2005, 2007). One male skeleton is composed of bones from three different individuals (Parker Pearson et al. 2005:534). The knee of a second individual (Figure 4.5) was broken

Figure 4.4 Niche with human skulls in the facade of the Kaiser-Heinrich Kapelle at Kirchdorf, Tyrol, Austria (post-funerary deposition). The medieval churchyard chapel was rebuilt after the fire of 1809 (Widmoser 1970:397). The skulls, which were exhumed from the local churchyard, serve as *memento mori*. The skull on the outer left in the back row is painted, the second skull from the left in the front row shows evidence for being sawed open during an autopsy.
Source: Author's own photograph

off prior to deposition but long after death when the bones had lost most of their collagen (Parker Pearson et al. 2007). Extraction of DNA shows that the body – originally identified as a female – is a composite body with a male head on a female torso (Mike Parker Pearson, Christine Cox, Terry Brown, and Keri Brown personal communication).

Contexts with human remains of multiple individuals

Quite frequently, multiple bodies are present in one mortuary context (Bloch 1971; Chambon 2003; Chapman 1981; Duday 1997; Parker Pearson 2000:figure 1.5; Weiss-Krejci 2004). These "collective" or "multiple" burials, can be the result of different processes and carry diverse meanings. People who die in quick succession and in close proximity to each other may end up in the same grave as a result of Processes A–B–F, A–M, or A–N (Daniell 1997:193; Fiorato et al. 2000; Guilaine and Zammit 2005:33–34; Zur 1998). People who die at different moments in time

Figure 4.5 Articulated Bronze Age skeleton from Clad Hallan, Sout Uist, Outer Hebrides, Scotland. The skeleton was found under the primary floor of a roundhouse (No. 2613). Analyses show that it is fabricated from the mummified remains of a female torso and a male head. The remains predate their deposition context by hundreds of years (non-funerary deposition).
Source: Parker Pearson et al. 2005, 2007; personal communication Mike Parker Pearson, Christine Cox, Terry Brown and Keri Brown. Photograph copyright and courtesy of Mike Parker Pearson

may share their final burial place with others who die before or after them. Older bodies are swept away or sorted to make room for new depositions (Process A–B–F–S) (Chase and Chase 1996; Weiss-Krejci 2004). Additionally children and spouses are sometimes stored to await burial with a parent or a husband or wife (Weiss-Krejci 2005:166; 2008:179) and those who die in between communal funerary celebrations sometimes are the subject of joint reburial (Beier 1997:447; Hutchinson and Aragon 2002) (Processes A–B–C–D–F and A–B–F). Finally, post-funeral re-depositions (Process F–T–U) can bring together individuals from different places, which likely died in different centuries (Weiss-Krejci 2004:389–391).

Rethinking Formation Processes of Mortuary Deposits at Tikal

In the second part of this chapter I am going to look at the mortuary sample from the ancient Maya site of Tikal from the perspective of formation processes. I have chosen Tikal for several reasons. The Tikal human remains sample is one of the largest in the Maya Lowlands making it suitable for statistical analysis. The bones have been studied by physical anthropologists (e.g., Haviland 1967; Wright 2002) and although the full results are not yet published, the preliminary data (Moholy-Nagy 2003:appendix F; Wright 2002) allow some general observations. Additionally, the provenience of the bones is well documented. The example of Tikal also shows how *ad hoc* categorizations of human remains may distort our understanding of interactions of past people with the dead. The categorizations, employed throughout the Maya area (e.g., Hurtado et al. 2007; Serafin and Peraza 2007; Tiesler 2007), are problematic and have met criticism in the past (Becker 1992; Kunen et al. 2002).

The site of Tikal

Tikal is located in the central Maya Lowlands, in the Department of the Peten, Guatemala. The site core is surrounded by perennial wetlands (*bajos*) and is built on a broad hill, which at its highest point reaches approximately 265 meters above sea level (Carr and Hazard 1961; Harrison 1999; Puleston 1983). Tikal was occupied from the Middle Preclassic to the Terminal Classic period (approx. 900 B.C. to A.D. 900). In the eighth century A.D., the site reached its height in population density with an estimated population between 11,000 for the central nine square kilometers (Figure 4.6) and approximately 62,000 when including the larger periphery (Culbert et al. 1990; Haviland 1970:193, 2003:129).

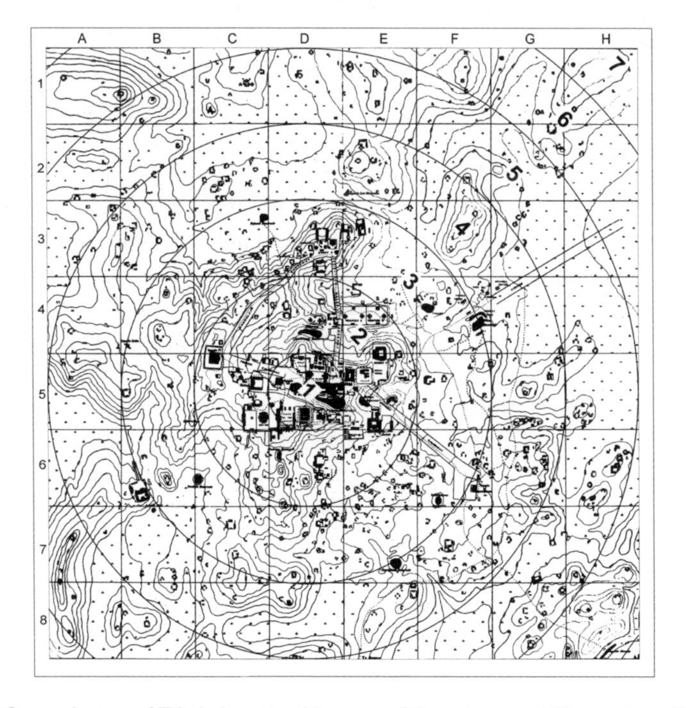

Figure 4.6 General map of Tikal showing 16 square kilometers and Zones 1 to 7. One square measures 500 × 500 meters. The Tikal map was made by Carr and Hazard (1961:figure 1). The division into various zones is based on Moholy-Nagy (2008:figure 58).
Source: Courtesy of the University of Pennsylvania Museum

Like elsewhere in the Maya Lowlands, Tikal has no communal cemeteries. Complete and incomplete human bodies are found in graves within structures or enclosed in the structure fill with the structures being residential and ceremonial, small and monumental in nature (Becker 1999; Coe 1990; Haviland 1985; Iglesias 2003a; Krejci and Culbert 1995; Laporte et al. 1992; Welsh 1988). Complete and fragmentary human skeletons also occur under monuments (Jones and Satterthwaite 1982) and in underground artificial chambers. Called *chultuns*, the function of the chambers is debated and they are variously interpreted as latrines, limestone quarries, fermentation chambers, sweat baths, weaving chambers, food storage chambers, and water cisterns (Dahlin and Litzinger 1986; Haviland 1963:505; Hunter-Tate 1994; Puleston 1971; Ricketson 1925:390; Scarborough et al. 1995:109). Additionally, some bodies are found in different types of ceramic containers such as bowls and jars (Iglesias 2003a:220), as well many fragmentary bones are distributed throughout the site.

Mortuary categorizations at Tikal

The Tikal Project of the University of Pennsylvania excavated the site core and its surrounding periphery in flexible units called lots that are grouped into spatially defined operations and suboperations. In total there are about 10,690 lots (Moholy-Nagy 2003:4). These lots are subdivided into two broad kinds of recovery contexts: "special deposits" and "general excavations" (Table 4.1). Special deposits are "features and materials that are thought to be the result of a single activity or of closely spaced sequential activities" (Moholy-Nagy 2003:4); general excavations is a category covering everything else. Special deposits are further divided into "burials," "caches," and "problematical deposits." Moholy-Nagy (2008:38) defines burial as "a set of specially-deposited human remains recovered from archaeological context, distinguishable from a cache or problematical deposit by the focus of past behavior upon a human subject." To Haviland a burial is "a culturally patterned way of carrying out postmortem requirements of the human soul or spirit which is expressed archaeologically as an association of grave, body and materials" (Haviland 1963:313).

Caches are regarded as concealed offerings (Becker 1992:191; Coe 1959:77), as "the material remains of consecratory and votive rituals performed by and for Tikal's rulers and elites" (Moholy-Nagy 2008:18). The category of problematical deposits was introduced by Coe (Moholy-Nagy 2008:67) for material remains whose past cultural function was uncertain. Some problematical deposits appear to be offerings, some more closely resemble burials, some are entirely enigmatic and a few are classified as special purpose dumps. Problematical deposits often include jumbled and incomplete human remains, sometimes with signs of burning and breakage (Iglesias 2003b; Moholy-Nagy 2003:4–6). They are thought "to represent more than one event" (Moholy-Nagy 2008:68), which means that they contain materials from different time periods and of diverse provenience, which are deposited together.

Table 4.1 Distribution of human remains at Tikal

	LOTS			*SPECIAL DEPOSITS*		
	Number of lots	*Lots with human remains*	*NIE**	*Number of deposits*	*Deposits with human remains*	*NIE**
SPECIAL DEPOSITS						
Burial	285	207 (27%)	250 (23%)	205	200	250
Burial (no lot number)				2		
Cache	240	36 (5%)	46 (4%)	205	35	46
Cache (no lot number)				4		
Problematical deposit	379	109 (15%)	216 (19%)	223	59	216
GENERAL EXCAVATION	9,786**	399 (53%)	594 (54%)			
TOTAL	10,690**	751 (100%)	1,106 (100%)	639	294	512

*NIE = Number of individual entries in Appendix F of Tikal Report 27B (Moholy-Nagy 2003), roughly correlating with the minimum number of human individuals. Only bones under code 30 "humans, unworked" and 40 "artifacts of human remains" in Variable 7 ("Material Category") have been considered (for codes see Moholy-Nagy 2008: Appendix 2).
**Approximate number.

While almost all the deposits classified as burials hold human remains, human bones in caches and problematical deposits are much rarer (Table 4.1). Apart from bones in special deposits many miscellaneous bones are scattered at random throughout the site. Because of their fragmentary and ambiguous nature none of these bones have been given any special consideration with respect to Tikal's mortuary behavior.

The distribution of bones at Tikal

Burials

The Tikal Project identified 207 burials, classified as chamber burials (19), crypt burials (11), and minor burials (177). Two burials do not have lot information and I removed them from the analysis; 205 burials correspond to 285 lots (Table 4.1). In five instances (four minor and one chamber burial) the bones were not collected and therefore not analyzed in the laboratory. Of the remaining 200 burials, 172 are single, 20 double, and eight held more than two individuals. In total, 200 burials hold 250 individuals. Although the majority of the single burials hold complete corpses, in a few cases body parts such as heads, hands, and legs are missing (Coe 1990:218; Haviland 1985:140).

In Table 4.1 I refer to the number of human individuals as NIE, which stands for the number of individual entries in Appendix F of Tikal Report 27B (Moholy-Nagy 2003), a spreadsheet with 1,106 entries for human remains (1,052 for unworked and 54 for worked human remains). The members of the Tikal Project of the University of Pennsylvania did not record the bones of almost complete and entirely complete skeletons by bone but instead only counted each individual once. On the other hand, miscellaneous bones are counted individually. Since there exists no estimate of a minimum number of individuals (MNI), the number of individual spreadsheet entries (NIE) in Appendix F (Moholy-Nagy 2003) forms the best available basis for discussion.[2]

Caches

Caches occur in structures and below monuments. Coe, who had a special interest in these types of deposits, identified 209 caches including some from older excavations (Moholy-Nagy 2008:appendix 10). Four caches do not have lot information and I removed them from the analysis. Thirty-five caches hold human remains corresponding to approximately 46 individuals (Table 4.1). Twenty-seven caches contain the remains of one human individual; six caches hold two individuals; and two caches hold more than two individuals each. Some caches hold complete bodies, some skulls (Figure 4.7), some only fragmentary human remains.

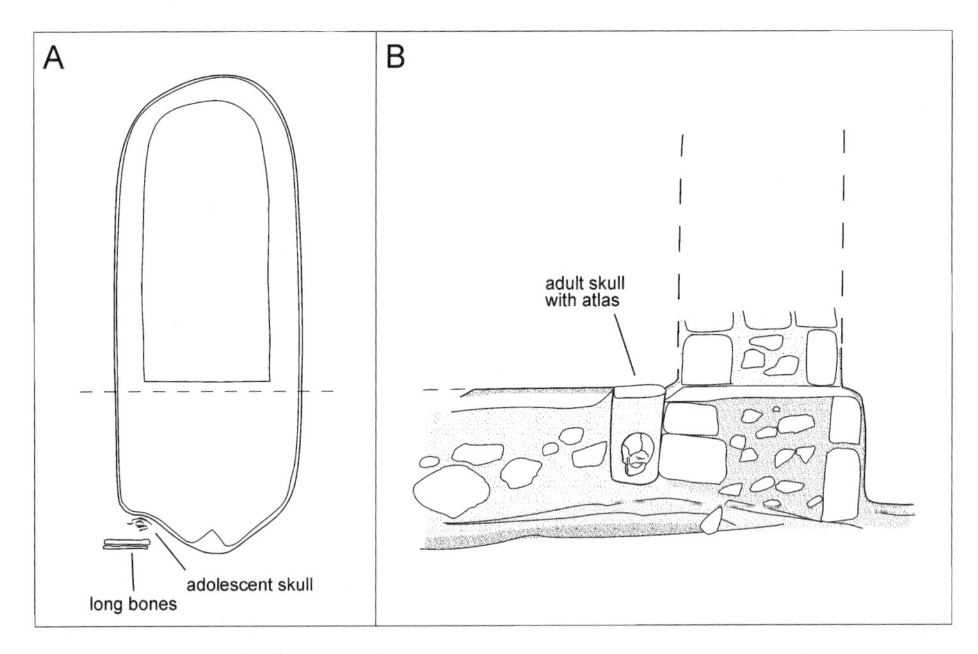

Figure 4.7 A Cache 32/137 from below Stela 16, Structure 5C-17; Group 5C-1 (after unpublished drawing by Sally Bates 1963); **B** Cache 165 from tunnel in Structure 6E-146, Group 6E-2 (excavated by Dennis Puleston 1964; detail of earlier construction phase, after unpublished drawing).
Source: Courtesy of the University of Pennsylvania Museum

Problematical deposits

In total 223 problematical deposits are identified. Of these, 59 hold human remains representing at least 216 fragmentary individuals (NIE) in Appendix F (Moholy-Nagy (2003) (Table 4.1). Twenty-nine problematical deposits contain a single individual, 11 deposits contain at least two individuals and 19 deposits hold more than two (Figure 4.8). However, the 216 NIE for problematical deposits are definitely an underestimation because several deposits which only receive one entry in Appendix F (Moholy-Nagy 2003) contain multiple individuals, such as Problematical Deposit 48 which holds the fragmentary remains of 24 people (Haviland, unpublished note on Human Bone Card 12C, Lot 26, Burial 7 alias PD 48). Six problematical deposits (PD 8, 72, 186, 224, 274 and 275) hold worked human bones (one tibia, two cut flatbone pieces, three rasping bones, one tie rod, and three so-called "pulidors").

Miscellaneous human bones

Human bones from general excavations are classified as miscellaneous bones. Of the 10,690 lots excavated by the University of Pennsylvania Tikal Project, at least 399 lots hold miscellaneous human bones corresponding to 594 individual spreadsheet entries (NIE) (Moholy-Nagy 2003:appendix F) (Table 4.1). Thirty of the

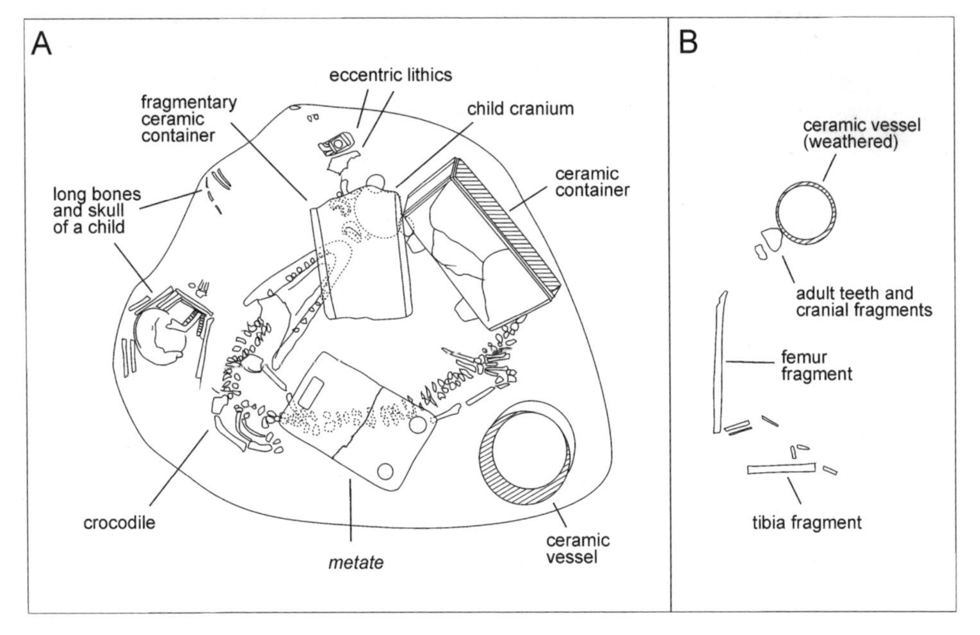

Figure 4.8 **A** Problematical Deposit 76, Structure 5C-17; Group 5C-1 (after unpublished drawing by Christopher Jones 1963); **B** Problematical Deposit 64 alias old Burial 51, Structure 2G-59-1st, Group 2G-1 (after Haviland 1963:figure 95).
Source: Courtesy of the University of Pennsylvania Museum

lots contain 50 worked human bone pieces (43 NIE). The number of miscellaneous human bones is probably even higher because there exists a large number of unidentified bones which are considered as pertaining to animals.

Problems with mortuary categorizations

For the members of the Tikal Project the classification of human remains was not easy because of the peculiar nature of some of these deposits. The records in the Tikal Archives in Philadelphia show that several caches and burials were renumbered as problematical deposits and problematical deposits turned into caches. As a result, the numbers used for special deposits by the Tikal Project today in no way reflect the actual number of deposits. Of 218 numbered burials only 207 remain; 241 cache numbers correspond to 209 cache deposits and of 275 numbered problematical deposits only 223 exist. Despite efforts to make amends, the classifications still remain highly inconsistent. As noted by Becker (1993:47), typologies remain on the level of field classifications.

There are several problems with these mortuary categorizations. First, burials, caches, and problematical deposits cannot always be cleanly separated. Instead of forming different categories, some seem to exist along the same continuum (Becker 1992; Kunen et al. 2002). Becker (1992) considers caches just another form of burial with both categories having similar cognitive meaning to the ancient Maya. He believes that burials and caches were considered as "earth offerings" to impregnate the earth and bring forth new life and continue the cycle of being. Becker's idea that there is no symbolic difference between caches and burials is supported by the fact that subadults in the University of Pennsylvania Tikal sample make up only one-quarter in deposits defined as burials, whereas in deposits classified as caches, children between 2 and 11 years of age make up one-half (Table 4.2). While

Table 4.2 Sex and age distribution at Tikal

	*Burials (NIE** = 250)*		*Caches (NIE = 46)*		*PDs* (NIE = 216)*		*Miscellaneous (NIE = 594)*		*Total (NIE = 1106)*	
Infants	11	5%	1	4%	4	7%	5	16%	21	6%
Children (2–11)	29	13%	12	52%	12	20%	6	19%	59	17%
Juveniles	15	6%	2	8%	5	8%	0	0%	22	6%
Adults	173	76%	8	34%	39	65%	20	65%	240	70%
Total age known	228	100%	23	100%	60	100%	31	100%	342	100%
Female	42	30%	0	0%	6	24%	3	100%	51	30%
Male	97	70%	3	100%	19	76%	0	0%	119	70%
Total gender known	139	100%	3	100%	25	100%	31	100%	170	100%

*PD = Problematical deposit.

**NIE = Number of individual entries in Appendix F of Tikal Report 27B (Moholy-Nagy 2003), roughly correlating with the minimum number of human individuals.

the high number of children in caches may suggest a special ritual role, the predominance of children in caches – usually smaller than graves – could be simply a function of size (Becker 1992:190). Children are often treated differentially from adults in a large number of societies (Aspöck 2009:84; Donnelly and Murphy 2008; Goody 1959:136; Weiss-Krejci 2008:183–185). The smaller size of caches not only favors the depositions of children, but also that of reburied people (after temporary storage or as part of post-funerary exhumation and reburial). Because their bodies are without flesh and their bones disarticulated there is no need for full-sized graves. Elsewhere, e.g., in Europe, reburied people are usually stored in much smaller graves or coffins (see Figure 4.3).

Secondly, a closer look also reveals that burials, caches, and problematical deposits can be homomorphous, i.e., there exist similarities in form within the same category, which do not result from the same processes. Each category has been used to describe deposits which can be funerary, extra-, and post-funerary (Figure 4.1). For the following discussion I have selected some exemplary topics to give an idea of the possible variability in formation processes of human bones at Tikal and make the problem of categorization apparent. I will discuss contexts with skulls, I address the problem of miscellaneous human remains and interpret deposits which hold human bone fragments in combination with monument fragments.

Skull deposits in Plaza Plan 1 and Plaza Plan 4 groups: a comparison

As I will show through a brief comparison between skulls from Plaza Plan 1 and Plaza Plan 4 groups the formation processes and meaning of caches with skulls at Tikal are variable.[3] Plaza Plans are architectural patterns identified by Becker (2003) and associated with particular functions. Plaza Plan 1 groups, also called twin pyramid complexes, are ceremonial groups composed of twin eastern and western stepped pyramids with the eastern pyramid having a variable number of pairs of plain stelae and altars, a southern range structure with nine doors, and a northern enclosure with one stela and altar pair (Jones 1969). They played a role in celebrating the ending of the *k'atun*, a period of roughly 20 years (Houston et al. 2006:81–84). Plaza Plan 4 groups are residential in function, with a series of rectangular structures arranged in an orderly fashion around a courtyard and a low square platform (central courtyard shrine) in the middle of the plaza (Becker 2008–2009). The layout and architecture of the central building suggests connections with Teotihuacan in Central Mexico (Coe 1965:40).

Of 13 skulls associated with four Plaza Plan 1 groups (5B-1, 3D-1, 5C-1, 4D-1, dating between A.D. 672 and 731), five are buried beneath stelae.[4] Four skulls belong to children between six and eight years of age. They exhibit postmortem loss or removal of some deciduous teeth, which should be present in individuals of that age. All four deposits include fragmentary long bones and two include teeth from a second adult individual. None of the skulls show evidence for removal from the body in the flesh. The first vertebrae are absent and the bones are weathered, probably due to extended postmortem exposure (William Haviland, unpublished information on Human Bone Cards, Tikal Archives). The fifth skull belongs to a person aged between 12 and 20 years (Cache 137/32 beneath Stela 16). This skull was also found together with long bones and was missing some teeth (Figure 4.7A). The cache deposit looks exactly like the image of a skull with long bones on Altar 5, which is paired with Stela 16. The text on Altar 5 talks about the death, burial, and exhumation of a noble lady (Grube and Schele 1994). The strong resemblance between the image on Altar 5 and the skull deposit beneath the associated stela strongly supports the idea that the skull and long bones pertain to a person, who was exhumed from elsewhere and re-deposited beneath the monument (Weiss-Krejci In press).

Skulls from Plaza Plan 4 groups are those of adolescents and adults. With the exception of the first vertebrae, postcranial bones are missing. In an Early Classic construction phase structure (probably dating between A.D. 400 and 600) of Building 6E-146 of Group 6E-II, three skulls in Caches 165, 166, and 170 (Figure 4.7B) still had the first vertebrae attached (Tikal Archives, unpublished information on Human Bones Cards). This suggests that the skulls do not pertain to exhumed disarticulated individuals but are the result of pre- or postmortem decapitation (Becker 2008–2009).

Although skulls are symbols of rebirth and fertility in Mesoamerica (Becker 2008–2009; Carlson 1981:193; McAnany 1995:47), the different contexts in which these skulls are encountered, the different age groups, and the obvious different conditions of the corpses from which they come, implies that these skull deposits are imbued with diverging meanings. As outlined in the first part of this chapter, the deposition of skulls can result from funerary deposition following decapitation or temporary storage, as well as extra-funerary or post-funerary processes. No doubt that the skulls in Plaza Plan 1 groups have gone through different formation processes than those of Plaza Plan 4, but an interpretation of the meaning of Tikal's skull deposits would require a lot of additional information. A series of functions are suggested; for the skulls in Plaza Plan 4 groups deposition after very brief temporary storage as part of the funerary cycle, trophy taking, and human sacrifice form possibilities; deposition after longer periods of storage as well as post-funerary processes and the use of skulls as apotropaic and mnemonic objects or even relics is more likely for Plaza Plan 1 stelae deposits (Angenendt 2002:27; Berrymen 2007; Duncan 2005; Hill 2006:91–93; Malville 2005:203; Massey 1989).

The problem of miscellaneous bones

Although miscellaneous human bones do not constitute intentional deposits, they can nevertheless provide important information about ancient Maya mortuary behavior. The sheer quantity of miscellaneous human bones from Tikal deserves attention. Of the 10,690 lots excavated by the University of Pennsylvania Tikal Project, 751 lots (7 percent) contain human bones. Of these, only 352 lots derive from special deposits whereas 399 lots with human remains are found in general contexts (Table 4.1). This means that lots with miscellaneous bones make up the majority (53.1 percent) of lots with human bones. If we consider the approximate number of human individuals (NIE), miscellaneous human remains from general excavations also appear more frequent than those from special deposits. Tikal's 751 lots with miscellaneous human remains correspond to 1,106 estimated individuals;

512 NIE (46.3 percent) are associated with special deposits, 594 (53.7 percent) derive from general contexts (Table 4.1).

If we separate caches and problematical deposits from burials, the percentage of human remains found in formal graves is even lower. As a matter of fact, individuals from burials constitute less than 30 percent; 27 percent from the perspective of lots and only 23 percent from the perspective of the individual entries in Moholy-Nagy's Appendix F (Table 4.1). Seen from this angle, Moholy-Nagy's (2008:38) statement that "the traditional burial pattern of Tikal was the primary burial of a single individual in an excavated repository," can hardly be sustained. More than two-thirds of the human remains at Tikal do not derive from contexts which are defined as burials.

Additionally, the overall number of burials at Tikal – 207 burials discovered by the University of Pennsylvania Project and a few hundred more found during later excavations (e.g., Laporte and Fialko 1995; Laporte et al. 1992) – is low, especially if one considers that population estimates for Late Classic Tikal hover around 200 persons per square kilometer for rural areas and 600 to 700 persons per square kilometer for central Tikal during the Late Classic Period (A.D. 600–900) (Culbert et al. 1990). Although sampling strategies, soil conditions, and disturbance by animals can contribute to the lack of bodies in graves (Duncan 2009; Schiffer 1987), it seems that the Tikal "burial" population does not represent a regular population (see also Chase and Chase 2004:204).

The idea that Tikal's mortuary population is a biased sample is further supported by sex and gender distributions. At Tikal the sex of 342 human individuals could be identified (Table 4.2). Males outnumber females two to one in so-called "burials". This imbalance also exists at other Maya Lowland sites such as Uaxactun and Cuello (Krejci and Culbert 1995; Ricketson and Ricketson 1937; Saul and Saul 1991; Smith 1950; Welsh 1988). There are no identified females in caches and no males in miscellaneous lots. There are three times as many males than females in problematical deposits. The overall ratio of females to males – for all time periods – is about two to five. The unbalanced ratio between males and females could suggest that females were treated differently from males.

Adults are the largest group represented; infants are probably underrepresented, which again is in line with other sites (Krejci and Culbert 1995; Ricketson and Ricketson 1937; Smith 1950; Welsh 1988). Adults make up three-quarters in deposits defined as burials. In problematical deposits and miscellaneous lots, adults are less frequent but still form the majority. The overall ratio between subadults and adults is about one to two. In deposits classified as caches on the other hand, and as discussed above, the number of adults is only one-third whereas children between 2 and 11 years of age make up one-half (Table 4.2).

One has to wonder if treatment and disposal of an unaltered corpse in the ground or under a structure forms the mortuary behavior of only a minority with a large quantity of the population being disposed elsewhere. As discussed in the first part of this chapter, anthropologists have provided abundant descriptions from all over the world of people being temporarily stored or finally deposited in locations such as trees, platforms, and storage pots, which leave rather fragmentary traces in the archaeological record (Lagercrantz 1991; Ucko 1969; Yarrow 1880a,b). The same applies to corpses which are processed and reduced to fragments during one-time or multistage rituals (Malville 2005). Various descriptions of mortuary treatment of corpses in colonial Maya sources (Tozzer 1941:131) could support the argument that final deposition of an unaltered, complete skeleton in the ground may have only been one of many ways among the ancient Maya to treat the dead. Some archaeological indicators for exposed bodies (Carr and Knüsel 1997:170), such as gnawing, weathered and scattered bones, etc. are present. For example, six bones (Moholy-Nagy 2003:appendix F) show evidence for gnawing. They all derive from Zones 1 and 2. However, the presence of miscellaneous bones is probably also the result of ritual and nonritual grave disturbance and post-funerary exhumation (Coe 1990:867–872). Again a lot of additional information about the bones is required to solve these problems.

Post-funerary deposits (Problematical Deposits 76, 64, and 22)

Problematical deposits are the most variable mortuary category at Tikal. As already mentioned, these types of special deposits are enigmatic and can result from entirely unrelated processes (funerary placement, offerings, ceremonial trash, etc.). The border between cache and problematical deposits is often murky. For illustration I will give three examples. Problematical Deposit 76 was found north of Stela 16 in Twin-Pyramid Group 5C-1 (Weiss-Krejci In press). The bottom of the pit was occupied by a large river crocodile with its head pointing north (Figure 4.8A). The crocodile was complete and articulated and apparently had been sacrificed and deposited on its back. In the same layer and to the north of the reptile's nose were several flint and obsidian eccentrics and a badly preserved youth's skull (Tikal Archives, Human Bones Card 43C-21). In the layer above the crocodile there were ceramic vessels (Culbert 1993:figure 133c), a metate, and a disarticulated incomplete child skull and long bones. The child is about six years of age. The bones were mixed with a few unidentified animal remains (Tikal Archives, Human Bones Card 43C-20). The whole deposit was covered by a broken but complete thin disk of chalky limestone (Jones 1963:34; Jones and Satterthwaite 1982:90). Problematical Deposit 76 is probably associated with the *k'atun* rites, which were held in twin-pyramid groups and the crocodile is a representative of the Starry Deer Crocodile (Stuart 2005). This mythical creature is of calendrical and cosmological significance embodying the concept of completion and subsequent renewal (Houston et al. 2006:92). It is often depicted vomiting torrential liquids decorated with various artifacts and bones (Velásquez García 2006). Problematical Deposit 76 could represent the physical remains of the reenactment of a mythological scene (Weiss-Krejci In press).

A very different deposit is Problematical Deposit 64, in the fill of the axis of Structure 2G-59 (Figure 4.8B). It contained the fragmentary remains of an adult and one ceramic vessel. Haviland (1963:74) suspects that these are the remains of a disturbed burial, which were reburied in the course of the construction of 2G-59-2nd. Such disturbance and reburial is common in residential contexts (e.g., Hammond et al. 1992:957).

Problematical Deposit 22 is yet of an entirely different nature. It was discovered in Tikal's North Acropolis, the burial place of the Tikal kings. It was made in the seventh century A.D. and contained carved Early Classic (fourth to fifth century

A.D.) stelae fragments (Jones and Satterthwaite 1982:figure 55) and unworked monument fragments. The deposit also includes human cranial, mandibular, and dental parts, along with major long bones and numerous ones of hands and feet, many burnt.[5] Additionally Early Classic and Intermediate Classic (sixth to seventh century A.D.) vessel sherds, eccentric flints, and faunal remains were found. According to Coe (1990:324–327) Problematical Deposit 22 holds the cleared-out remains of a burial. The deposit is located in the vicinity of Burial 48, which is ascribed to the 16th ruler Sihyaj Chan K'awiil II, who ruled in the early fifth century (Martin 2003). The tomb principal lacked skull, long bones, and the bones of hands and feet (Coe 1990:120). One of the sherds in Problematical Deposit 22 (Culbert 1993:figure 125a) mentions another king and affinal relative of that same dynasty; Kaloomte' Bahlam, the 19th king of Tikal who ruled with his queen Lady of Tikal in the first part of the sixth century (Martin 2003:20).

This and similar deposits with Early Classic royal monument fragments and fragmentary, exhumed human bones were made by the seventh-century Tikal kings in order to commemorate specific Early Classic kings and legitimize power. The seventh century was a time of crisis with major political transformations taking place in the Maya Lowlands (Martin 2003; Martin and Grube 2008). Such active interest in the dead in critical times is not uncommon throughout the world (Verdery 1999) and especially typical for dynasties who have a problem with genealogical legitimization. To a group of people like the ancient Maya aristocracy, which emphasized blood lines and invested a lot of energy into placing royal persons within long genealogies (Houston and McAnany 2003:37), the securing of the support of ancient authorities in their claim for rulership was certainly important. In the process of genealogical appropriation of fictive or even mythical ancestors, tomb visits, exhumation and reburial as well as tomb reuse often play an important role (Legner 2003:108–110; Weiss-Krejci 2004).

Discussion and Conclusions

To fully solve the question of what processes are involved in the formation of mortuary deposits at Tikal would require additional investigations of the human remains. Based on the overall distribution of bones at Tikal, as well as distribution of sex and age, Tikal's "burials" unlikely represent a regular population with modes of treatment and disposal other than the deposition of an unaltered corpse in the ground. Parts of the funeral population of Tikal have not survived in the archaeological record. Tikal's mortuary sample is shaped by the results of various funerary, post-funerary, and extra-funerary behaviors. Alterations on some bones suggest that corpses are processed as part of funerary rituals, but they could also point to extra-funerary processes such as sacrifice and the taking of trophies during warfare (Duncan 2005; Massey 1989:16; Tozzer 1941:131; Weiss-Krejci 2006). Post-funerary disturbance of graves, as well as exhumation and re-deposition of bones, are certainly also responsible for some of the problematical deposits, especially those with Early Classic grave goods (A.D. 250–600) found in Late Classic contexts (A.D. 600–900). Post-funerary behaviors such as ancestor enshrinement and rituals of veneration of the dead, multiple use of tombs, accidental disturbance of old graves, and looting and desecration all hold responsible (Chase and Chase 1996,

1998; Fitzsimmons 2006; Hammond et al. 1975, 1992:957; Harrison-Buck et al. 2007; Houston et al. 1998:19; McAnany 1995; 1998; Pendergast 1979:183–184; Stuart 1998; Weiss-Krejci 2004).

In this chapter I have tried to show that the formation of mortuary contexts is a complex process that makes our current mortuary categorizations obsolete. As the example of Tikal proves, *ad hoc* categorizations are not helpful to reveal human interactions with the dead. There exists continuity between different categories but at the same time each category subsumes deposits which can result from not only funerary but extra- or post-funerary processes too (Figure 4.1). In my opinion the use of the term "burial" as well as other mortuary categories such as cache or problematical deposit should be avoided because they classify field data prior to their analysis. To a certain degree categorizations are necessary in order to analyze data. However, we need to consider whether certain types of categorizations are really necessary and which ones are useful. We have to be constantly aware that what we collect and how we categorize our finds already constitutes an interpretation. What constitutes a deposit that is made as part of a funeral cannot be explained by our own cultural standards (Duncan 2005). The differentiation between funerary and non-funerary, funerary versus ancestral, ancestral versus sacrificial, and respectful versus ignoble treatment (e.g., Tiesler 2007; Welsh 1988) should not be based on the state of the bones alone. In the first part of the chapter, I demonstrated that a direct correlation between funerary and so-called non-funerary deposits and the completeness of the corpse does not exist. A dead person's remains, when being finally deposited during the funerary cycle, can constitute anything from a complete and articulated body to disarticulated fragments. On the other hand, a complete corpse in the grave may be the result of a funeral that is not complete (Miles 1965; Hutchinson and Aragon 2002). To be able to infer that a specific dead person in an archaeological deposit did or did not receive mortuary treatment in accordance with his/her normal social persona (Binford 1971; Saxe 1970) one must understand the nature of the society that produced the deposit and its ideas relating to the body, to death, and to proper burial (Gillespie 2002; Houston et al. 2006; Meskell and Joyce 2003). In order to correctly identify the remains of funerals it is necessary to identify the formation processes leading to the creation of a mortuary deposit.

Ad hoc classification of mortuary deposits can be harmful to archaeological science in many other ways. In 1999 a law was passed in Portugal (Dec° Lei 270/99 from July 15, Section 8) that bones from so-called mortuary contexts, (i.e., burials, tombs) have to be excavated in the presence of a physical anthropologist. This implies that isolated bones from settlements do not get the same attention (Cidália Duarte, personal communication). As in the past, fragmentary human bones from "non-mortuary contexts" will continue to end up in bags with animal bones only to be discovered by animal osteologists years after the excavation. I question whether a distinction in the field between human and animal bones is even necessary. First, it only makes sense when a specialist is excavating the deposit who knows the difference, which from my experience, in many countries, is not the case. Secondly, animals coexist with people and therefore form an essential part of the "cognitive mindscape" of a human population (Ingold 1988; O'Connor 1996:12). People in pre-industrial societies don't necessarily share our view of human superiority over animals (Ingold 1988:11). Animals are often used as metaphors, as idealized models of order and morality or of disorder (Tapper 1988:51), a symbolic role which may be found in the relationships between animals represented in physical remains and representations in artwork (Holt 1996:92).

At Tikal, there is no evidence that human bone was regarded differently than animal bone. Like humans, animals are concentrated towards the center of the city and both animal and human bones are used for artifact production (Moholy-Nagy 2003:63–66, appendix F; 2008:appendix 5). Animals also frequently occur in special deposits. Some of the animals in special deposits are complete; others only consist of isolated bones. As a matter of fact, some of the animal caches look exactly like burials. All this could indicate that animals were primarily of ritual importance and not just food (see also Emery 2004). Animals at Tikal and other Maya sites probably were much more than some useful resource to be exploited.

Instead of investing too much energy in classification and finding the "right" terminology we should focus on documenting as detailed as possible what we see in the field and think about the processes, which have created a specific archaeological deposit. This is an inductive process which starts with the formulation of several working hypotheses based on the field data and the verification or falsification during post-excavation investigations by specialists including the careful examination of bones (e.g., Duncan 2009). Especially dating the bones may reveal significant results (Parker Pearson et al. 2005). Thus it should be possible to develop more accurate scenarios to explain the formation processes of deposits with human remains and compare those deposits with each other, which really constitute the remains of funerals.

ACKNOWLEDGMENTS

I would like to express my thanks to the Austrian Science Fund FWF (grant P18949-G02) for the ongoing financial support to conduct this research including the funds for a research month in Philadelphia in 2008. My special thanks go to Christopher Jones and Alex Pezzati from the University of Pennsylvania Museum of Archaeology and Anthropology for giving me access to unpublished materials from the Tikal Archives. I also would like to thank Jani Kuhnt-Saptodewo and Mike Parker Pearson for providing photographs to include in this publication. Edeltraud Aspöck, Marshall Becker, Bill Duncan, Christopher Knüsel, Karl Herbert Mayer, Reinhold Mittersakschmöller, Mike Parker Pearson, and Vera Tiesler have provided relevant literature, information, and various kinds of assistance. Finally my thanks also go to the editors of this book for their very extensive comments and suggestions.

NOTES

1 According to Binford (1971:21) and Carr (1995:130), whose terminology I follow, body preparation consists in the washing of the corpse whereas

mummification, cremation, and mutilation are considered body treatments. Sprague's (2005:70) definition of body preparation, on the other hand, includes techniques of preservation such as mummification.

2 The quantification of the Tikal human remains and comparison of bones from special deposits with bones from general excavations is difficult because of the way the remains are recorded. Current records show just over 5,300 human bone specimens but this actually includes 283 complete and almost complete individuals as well as 141 skulls which were only counted as one bone specimen each (Moholy-Nagy 2003:66). In reality, the total number of individual specimens of human bones recovered by the University of Pennsylvania Tikal Project probably exceeds 10,000. Unfortunately the bones were not weighed either. As Moholy-Nagy (2003:67) remarks, "MNI and bone weight would have been a more helpful way to assess the relative importance of remains in different kinds of recovery contexts." In Tables 4.1 and 4.2 I also have taken into account the updates concerning special deposits in Appendices 3 and 4 of Tikal Report 27A (Moholy-Nagy 2008), originally published as Appendices C and D in Tikal Report 27B (Moholy-Nagy 2003).

3 Appendix F (Moholy-Nagy 2003) contains 141 entries for skulls. Two accompany complete corpses in Burials 15 and 217, eight are classified as caches and appear beneath stelae (Caches 53 and 124) and in structures (Caches 165, 166, 170, 173A,B and 206), 45 skulls are found in 14 problematical deposits (75, 76, 87, 89, 95, 110, 111, 114, 170, 224, 271, 273, 274, and 275), and the remaining 86 are classified as miscellaneous bones. Although the registry applies to deposits with skulls only, it is not consistent because it also includes some skulls associated with long bones (e.g., Cache 124). Other skulls with long bones (e.g., Cache 116) are registered as partially complete bodies in Appendix F (Moholy-Nagy 2003).

4 Skulls of children together with long-bone fragments were found in Cache 116 beneath Stela P41, Miscellaneous Bones Lot 9B/5 beneath Stela 30, Cache 124 beneath Stela P44, and Cache 152 beneath Stela P56. The skull below Stela P41 is missing three decidious canines and one molar, the skull below Stela P44 misses one canine, and the skull below Stela P56 one molar and one canine. The child skull below Stela P41 was accompanied by four adult molars; adult teeth are also associated with the child skull below Stela 30. The adolescent skull from Cache 137 alias Cache 32 (Stela 16) is missing one and a half molars, four premolars, one incisor, and one canine (unpublished information, William Haviland, Human Bone Cards, Tikal Archives).

5 Moholy-Nagy (2003:appendix F) lists 37 cases where the bones show evidence of burning. These bones are from nine problematical deposits (PD 22, 31, 50, 72, 74, 77, 96, 111, 275) and 18 general lots. There is one additional case (Cache 178 from Twin-Pyramid Group 4D-1) not mentioned in Appendix F that also contains burnt bones (finger bones and vertebrae) (Tikal Archives, Object Catalogue Card, Human Bone Op. 56G, Lot 5). The large majority of burned bones derive from Zones 1 and 2. Four are from Zone 3 and one from the Tikal periphery (Zone 9). The burnt body parts primarily consist of skulls, mandibles, teeth, long bones, phalanges, and pelvic elements. Unfortunately it is not known whether the burning of the bones occurred in the flesh or in a dry state and

whether the body was articulated at the time of the burning or not. Bones that burn in the flesh have a different appearance from bones that are dry (Rue et al. 1989:399; Sempowski 1994:141; Ubelaker 1989). This information would constitute an importance piece of evidence in the interpretation (Weiss-Krejci 2006:76–78). The association of the bones with older artifacts in problematical deposits (e.g., Problematical Deposit 22, Coe 1990: 324–327) suggests that the bones are reburied from elsewhere and that the burning happened after the exhumation and fulfilled a ritualistic purpose (Houston et al. 1998:19; Stuart 1998).

REFERENCES

Angenendt, A. 1994 Heilige und Reliquien: Die Geschichte ihres Kultes vom frühen Christentum bis zur Gegenwart. Munich: C. H. Beck.

Angenendt, A. 2002 Relics and Their Veneration in the Middle Ages. *In* The Invention of Saintliness. A. B. Mulder-Bakker, ed. Pp. 27–37. London: Routledge.

Antonaccio, C. M. 1995 An Archaeology of Ancestors: Tomb Cult and Hero Cult in Early Greece. London: Rowman and Littlefield.

Aspöck, E. 2008 What Actually is a "Deviant Burial"? Comparing German-Language and Anglophone Research on "Deviant Burials". *In* Deviant Burial in the Archaeological Record. E. M. Murphy, ed. Pp. 17–34. Oxford: Oxbow.

Aspöck, E. 2009 The Relativity of Normality: An Archaeological and Anthropological Study of Deviant Burials and Different Treatment at Death. Ph.D. Dissertation, University of Reading.

Assmann, J. 2005 Die Lebenden und die Toten. *In* Der Abschied von den Toten: Trauerrituale im Kulturvergleich. J. Assmann, F. Maciejewski, and A. Michaels, eds. Pp. 16–36. Göttingen: Wallstein Verlag.

Aufderheide, A. C. 2003 The Scientific Study of Mummies. Cambridge: Cambridge University Press.

Ballinger, P. 2002 History in Exile: Memory and Identity at the Border of the Balkans. Princeton: Princeton University Press.

Barley, N. 1981 The Dowayo Dance of Death. *In* Mortality and Immortality: The Anthropology and Archaeology of Death, S. C. Humphreys, and H. King, eds. Pp. 149–159. London: Academic Press.

Beck, L. A. ed. 1995 Regional Approaches to Mortuary Analysis. New York: Plenum Press.

Beck, L. A. 2005 Secondary Burial Practices in Hohokam Cremations. *In* Interacting with the Dead: Perspectives on Mortuary Archaeology for the New Millennium. G. F. M. Rakita, J. E. Buikstra, L. A. Beck and S. R. Williams, eds. Pp. 150–154. Gainesville: University Press of Florida.

Becker, M. J. 1992 Burials as Caches, Caches as Burials: A new Interpretation of the Meaning of Ritual Deposits Among the Classic Period Lowland Maya. *In* New Theories on the Ancient Maya. E. C. Danien, and R. J. Sharer, eds. Pp. 185–196, University Museum Monograph, 77. Philadelphia: The University Museum, University of Pennsylvania.

Becker, M. J. 1993 Earth Offerings Among the Classic Period Lowland Maya: Burials and Caches as Ritual Deposits. *In* Perspectivas Antropológicas en el Mundo Maya. M. J. Ponce de León and F. Perramon, eds. Pp. 45–74. Barcelona: Sociedad Española de Estudios Mayas.

Becker, M. J. 1999 Excavations in Residential Areas of Tikal: Groups with Shrines. University Museum Monographs, Tikal Report, 21. Philadelphia: The University Museum, University of Pennsylvania.

Becker, M. J. 2003 Plaza Plans at Tikal: A Research Strategy for Inferring Social Organization and Processes of Culture Change at Lowland Maya Sites. *In* Tikal: Dynasties, Foreigners and Affairs of State. J. A. Sabloff, ed. Pp. 253–280. Santa Fe: School of American Research Press.

Becker, M. J. 2008–2009 Skull Rituals and Plaza Plan 4 at Tikal: Lowland Maya Mortuary Patterns. The Codex 17/1–2:12–41.

Beer, R. 2003 The Handbook of Tibetan Buddhist Symbols. Chicago: Serindia Publications.

Beier, H.-J., 1997 Das tugu der Toba-Batak und die Bestattungsbefunde aus mitteleuropäischen Kollektivgräbern. Ethnographisch-Archäologische Zeitschrift 38:443–449.

Binford, L. R. 1971 Mortuary Practices: Their Study and Their Potential. *In* Approaches to the Social Dimensions of Mortuary Practices. J. A. Brown, ed. Pp. 6–29., Memoir, 25. New York: Society for American Archaeology.

Binski, P. 1996 Medieval Death: Ritual and Representation. London: British Museum Press.

Bloch, M. 1971 Placing the Dead: Tombs, Ancestral Villages, and Kinship Organization in Madagascar. London: Seminar Press.

Bloch, M., and J. Parry 1982 Introduction: Death and the Regeneration of Life. *In* Death and the Regeneration of Life. M. Bloch, and J. Parry, eds. Pp. 1–44. Cambridge: Cambridge University Press.

Bradley, R. 1995 Foreword: Trial and Error in the Study of Mortuary Practices – Exploring the Regional Dimension. *In* Regional Approaches to Mortuary Analysis. Lane A. Beck, ed. Pp. v–ix. New York: Plenum Press.

Brown, E. 1985 Burying and Unburying the Kings of France. *In* Persons in Groups: Social Behavior as Identity Formation in Medieval and Renaissance Europe. Medieval and Renaissance Texts and Studies, 36. R. C. Trexler, ed. Pp. 241–266. Binghamton: Medieval and Renaissance Texts and Studies.

Brown, J. A. 1995 On Mortuary Analysis – With Special Reference to the Saxe-Binford Research Program. *In* Regional Approaches to Mortuary Analysis. Lane A. Beck, ed. Pp. 3–26. New York: Plenum Press.

Brown, P. 1981 The Cult of the Saints: Its Rise and Function in Latin Christianity. Chicago: University of Chicago Press.

Brückner, W. 1966 Bildnis und Brauch: Studien zur Bildfunktion der Effigies. Berlin: Erich Schmidt.

Carlson, J. B. 1981 A Geomantic Model for the Interpretation of Mesoamerican Sites: An Essay in Cross-Cultural Comparison. *In* Mesoamerican Sites and World-Views, E. P. Benson, ed. pp. 143–215. Washington, DC: Dumbarton Oaks.

Carr, C. 1995 Mortuary Practices: Their Social, Philosophical-Religious, Circumstantial, and Physical Determinants. Journal of Anthropological Method and Theory 2:105–200.

Carr, G., and C. Knüsel 1997 The Ritual Framework of Excarnation by Exposure as the Mortuary Practice of the Early and Middle Iron Ages of Central Southern Britain. *In* Reconstructing Iron Age Societies: New Approaches to the British Iron Age. A. Gwilt, and C. Haselgrove, eds. Pp. 167–173. Oxbow Monograph, 71. Oxford: Oxbow Books.

Carr, R. F., and J. E. Hazard 1961 Map of the Ruins of Tikal, El Peten, Guatemala. University Museum Monographs, Tikal Report, 11. Philadelphia: The University Museum, University of Pennsylvania.

Cátedra, M. 1992 This World, Other Worlds: Sickness, Suicide, Death, and the Afterlife Among the Vaqueiros de Alzada of Spain. Chicago: University of Chicago Press.

Chambon, P. 2003 Les Morts dans les Sépultures Collectives Néolithiques en France. Du Cadavre aux Restes Ultimes. Paris: CNRS Editions.

Chapman, R. 1981 Archaeological Theory and Communal Burial in Prehistoric Europe. *In* Pattern of the Past: Studies in Honour of David Clarke. Ian Hodder, Glynn Isaak and Norman Hammond, eds. Pp. 387–411. Cambridge: Cambridge University Press.

Chapman, R. 2005 Mortuary Analysis: A Matter of Time? *In* Perspectives on Mortuary Archaeology for the New Millennium. G. F. M. Rakita, J. E. Buikstra, L. A. Beck, and S. R. Williams, eds. Pp. 25–40. Gainesville: University Press of Florida.

Chase, D. Z., and A. F. Chase 1996 Maya Multiples: Individuals, Entries, and Tombs in Structure A 34 of Caracol, Belize. Latin American Antiquity 7:61–79.

Chase, D. Z., and A. F. Chase 1998 The Architectural Context of Caches, Burials, and Other Ritual Activities for the Classic Period Maya (as Reflected at Caracol, Belize). *In* Function and Meaning in Classic Maya Architecture. S. D. Houston, ed. Pp. 299–332. Washington, DC: Dumbarton Oaks Research Library and Collection.

Chase, D. Z., and A. F. Chase 2004 Patrones de Enterramiento y Ciclos Residenciales en Caracol, Belize. *In* Culto Funerario en la Sociedad Maya. Memoria de Cuarta Mesa Redonda de Palenque. R. Cobos, Coord. Pp. 203–230. Mexico City: Instituo Nacional de Antropología e Historia.

Coe, W. R. 1959 Piedras Negras Archaeology: Artifacts, Caches, and Burials. University Museum Monographs, 4. Philadelphia: The University Museum, University of Pennsylvania.

Coe, W. R. 1965 Tikal: Ten Years of Study of a Maya Ruin in the Lowlands of Guatemala, Expedition 8: 5–56.

Coe, W. R. 1990 Excavations in the Great Plaza, North Terrace and North Acropolis of Tikal. University Museum Monographs, Tikal Report, 14. Philadelphia: The University Museum, University of Pennsylvania.

Coe, W.R., and W. A. Haviland 1982 Introduction to the Archaeology of Tikal, Guatemala. University Museum Monographs, Tikal Report, 12. Philadelphia: The University Museum, University of Pennsylvania.

Coppet, D. de, 1981 The Life-Giving Death. *In* Mortality and Immortality: The Anthropology and Archaeology of Death. S. C. Humphreys, and H. King, eds. Pp. 175–204. London: Academic Press.

Culbert, T. P. 1993 The Ceramics of Tikal: Vessels from the Burials, Caches, and Problematical Deposits. University Museum Monographs, Tikal Report, 25A. Philadelphia: The University Museum, University of Pennsylvania.

Culbert, T. P., L. J. Kosakowsky, R. E. Fry, and W. A. Haviland 1990 The Population of Tikal, Guatemala. *In* Precolumbian Population History in the Maya Lowlands. T. P. Culbert, and D. Rice, eds. Pp. 103–121. Albuquerque: University of New Mexico Press.

D'Altroy, T. N. 2002 The Incas. Oxford: Blackwell.

Dahlin, B. H., and W. J. Litzinger 1986 Old Bottle, New Wine: The Function of Chultuns in the Maya Lowlands. American Antiquity 51:721–736.

Danforth, L. M. 1982 The Death Rituals of Rural Greece. Princeton: Princeton University Press.

Daniell, C. 1997 Death and Burial in Medieval England: 1066–1550. London: Routledge.

Davies, J. 1994 One Hundred Billion Dead: A General Theology of Death. Ritual and Remembrance. *In* Responses to Death in Human Societies. J. Davis, ed. Pp. 24–39. Sheffield: Sheffield Academic Press.

Donnelly, C. J., and E. M. Murphy 2008. The Origins of *Cillíní* in Ireland. *In* Deviant Burial in the Archaeological Record. E. M. Murphy, ed. Pp. 191–223. Oxford: Oxbow.

Dosal, P. J. 2004 San Ernesto de la Higuera: The Resurrection of Che Guevara. *In* Death, Dismemberment, and Memory: Body Politics in Latin America. L. L. Johnson, ed. Pp. 317–341. Albuquerque: University of New Mexico Press.

Duday, H. 1997 Antropología Biológica de Campo, Tafonomía y Arqueología de la Muerte. *In* El Cuerpo Humano y Su Tratamiento Mortuorio. E. Malvido, G. Pereira, and V. Tiesler, eds. Pp. 91–126. Mexico City: Colleción Científica. Instituto Nacional de Antropología y Historia.

Duncan, W. N. 2005 Understanding Veneration and Violation in the Archaeological Record. *In* Interacting with the Dead: Perspectives on Mortuary Archaeology for the New Millennium. G. F. M. Rakita, J. E. Buikstra, L. A. Beck, and S. R. Williams, eds. Pp. 207–227. Gainesville: University Press of Florida.

Duncan, W. N. 2009 The Bioarchaeology of Ritual Violence at Zacpetén *In* The Kowoj: Identity, Migration, and Geopolitics in Late Postclassic Petén, Guatemala. P. M. Rice, and D. S. Rice, eds. Pp. 340–367. Boulder: University Press of Colorado.

Durkheim, É. 1912 Les Formes Elementaire de la Vie Religieuse. Paris: Félix Alcan.

Durkheim, É. 2001 The Elementary Forms of Religious Life: A New Translation by Carol Cosman. Oxford: Oxford University Press.

Eisenbruch, M., 1984 Cross-Cultural Aspects of Bereavement. I: A Conceptual Framework for Comparative Analysis. Culture, Medicine and Psychiatry 8:283–309.

Elliott, C. 2008 Social Death and Disenfranchised Grief: An Alyawarr Case Study. *In* Mortality, Mourning and Mortuary Practices in Indigenous Australia. K. Glaskin, M. Tonkinson, Y. Musharbash, and V. Burbank, eds. Pp. 103–119. Farnham: Ashgate Publishing.

Emery, K. F. 2004 Animals from the Maya Underworld: Reconstructing Elite Maya Ritual at the Cueva de los Quetzales, Guatemala. *In* Behavior Behind Bones: The Zooarchaeology of Ritual, Religion, Status and Identity. S. Jones O'Day, W. Van Neer, and A. Ervynck, eds. Pp. 101–113. Proceedings of the 9th Conference of the International Council of Archaeozoology, Durham August 2002. Oxford: Oxbow Books.

Feest, C. F. 1997 Zwischen den Welten – Zwischen den Disziplinen: Archäologische und Historisch-Ethnographische Perspektiven zu Bestattungsformen im Küstenland von Virginia und North Carolina. Ethnographisch-Archäologische Zeitschrift 38:419–432.

Finucane, R. C. 1981 Sacred Corpse, Profane Carrion: Social Ideals and Death Ritual in the Later Middle Ages. In Mirrors of Mortality. J. Whaley, ed. Pp. 40–60. London: Europa Publications Ltd.

Fiorato, V., A. Boylston, and C. Knüsel, eds. 2000 Blood Red Roses: The Archaeology of a Mass Grave from the Battle of Towton AD 1461. Oxford: Oxbow Books.

Fitzsimmons, J. L. 2006 Classic Maya Tomb Re-Entry. *In* Jaws of the Underworld: Life, Death, and Rebirth Among the Ancient Maya. P. R. Colas, G. LeFort, and B. Liljefors Persson, eds. Pp. 33–40, Acta Mesoamericana, 16. Markt Schwaben: Verlag Anton Saurwein.

Fortes, M. 1965 Some Reflections on Ancestor Worship in Africa. *In* African Systems of Thought. M. Fortes, and G. Dieterlen, eds. Pp. 122–142. London: Oxford University Press.

Frazer, J. G. 1886 On Certain Burial Customs as Illustrative of the Primitive Theory of the Soul. The Journal of the Anthropological Institute of Great Britain and Ireland 15:63–104.

Freedman, M. 1966 Chinese Lineage and Society: Fukien and Kwangtung. London School of Economics Monographs on Social Anthropology, 33. New York: The Athlone Press.

Geary, P. J. 1990 Furta Sacra: Thefts of Relics in the Central Middle Ages. Princeton: Princeton University Press.

Gerbert, M., M. Herrgott, and R. Heer 1772 Taphographia Principum Austriae, Monumenta Augustae Domus Austriae Vol. 4/2. St. Blasien: St. Blasien Abbey.

Giesey, R. E. 1960 The Royal Funeral Ceremony in Renaissance France. Geneva: Libraire E. Droz.

Gillespie, S. D. 2002 Body and Soul Among the Maya: Keeping the Spirits in Place. *In* The Space and Place of Death. H. Silverman, and D. B. Small, eds. Pp. 67–78, Archaeological Papers of the American Anthropological Association, 11. Arlington: American Anthropological Association.

Glaser, B. G., and Strauss, A. L. 1966 Awareness of Dying. London: Weidenfeld and Nicholson.

Gluckmann, M. 1937 Mortuary Customs and the Belief in Survival after Death Among the South-Eastern Bantu. Bantu Studies 11:117–136.

Godelier, M. 1999 The Enigma of the Gift. Chicago: University of Chicago Press.

Godelier, M. 2004 What Mauss Did Not Say: Things You Give, Things You Sell, and Things That Must Be Kept. *In* Values and Valuables: From the Sacred to the Symbolic. Society for Economic Anthropology Monograph Series, 21. C. Werner, and D. Bell, eds. pp. 3–20. Waltnut Creek: Altamira Press.

Gómez, O., and C. Vidal Lorenzo, 1997 El Templo V de Tikal: Su Excavación. *In* X Simposio de Investigaciones Arqueológicas en Guatemala, 2001. J. P. Laporte and H. Escobedo, eds. Pp. 309–323. Guatemala: Museo Nacional de Arqueología y Etnología.

Goody, J. 1959 Death and Social Control Among the LoDagaa. Man 59:134–138.

Goody, J. 1962 Death, Property and the Ancestors: A Study of the Mortuary Customs of the Lodagaa of West Africa. Stanford: Stanford University Press.

Grube, N., and L. Schele, 1994 Tikal Altar 5. Texas Notes on Precolumbian Art, Writing and Culture 66:1–6.

Guilaine, J., and J. Zammit 2005 The Origins of War: Violence in Prehistory. Oxford: Blackwell Publishing.

Gurdon, P. R. T. 1907 The Khasis. London: David Nutt.

Hallam, E., and J. Hockey 2001 Death, Memory and Material Culture. Oxford: Berg Publishers.

Hammond, N., K. Pretty, and F. P. Saul 1975 A Classic Maya Family Tomb. World Archaeology 7:57–78.

Hammond, N., A. Clarke, and F. Estrada Belli 1992 Middle Preclassic Maya Buildings and Burials at Cuello, Belize. Antiquity 66:955–964.

Harrison, P. D. 1999 The Lords of Tikal: Rulers of an Ancient Maya City. London: Thames and Hudson.

Harrison-Buck, E., P. A. McAnany, and R. Storey 2007 Empowered and Disempowered During the Late to Terminal Classic Transition: Maya Burial and Termination Rituals in the Sibun Valley, Belize. *In* New Perspectives on Human Sacrifice and Ritual Body Treatments in Ancient Maya Society. V. Tiesler, and A. Cucina, eds. Pp.74–101. New York: Springer.

Haviland, W. A. 1963 Excavation of Small Structures in the Northeast Quadrant of Tikal, Guatemala. Ph.D. Dissertation, University of Pennsylvania.

Haviland, W. A. 1967 Stature at Tikal, Guatemala: Implications for Ancient Maya Demography and Social Organization. American Antiquity 32:316–325.

Haviland, W. A. 1970 Tikal, Guatemala and Mesoamerican Urbanism. World Archaeology 2:186–198.

Haviland, W. A. 1985 Excavations in Small Residential Groups of Tikal: Groups 4F-1 and 4F-2. University Museum Monographs, Tikal Report, 19. Philadelphia: The University Museum, University of Pennsylvania.

Haviland, W. A. 2003 Settlement, Society, and Demography at Tikal. *In* Tikal: Dynasties, Foreigners and Affairs of State. J. A. Sabloff, ed. Pp. 11–142. Santa Fe: School of American Research Press.

Hayden, R. M. 1994 Recounting the Dead: The Rediscovery and Redefinition of Wartime Massacres in Late- and Post-Communist Yugoslavia. *In* Memory, History and Opposition under State Socialism. R. S. Watson, ed. Pp. 167–184. Santa Fe: School of American Research Press.

Hertz, R. 1907 Contribution à une Étude sur la Représentation Collective de la Mort. Année Sociologique 10:48–137.

Hertz, R. 1960 A Contribution to the Study of the Collective Representation of Death. *In* Death and the Right Hand. R. Needham, and C. Needham, eds. Pp. 27–86. London: Cohen and West.

Hill, E. 2006 Moche Skulls in Cross-Cultural Perspective. *In* Skull Collection, Modification and Decoration. M. Bonogofsky, ed. pp. 91–100. BAR International Series, 1539. Oxford: Archaeopress.

Hirst, M. M. 1986 Some Ideas About Dying and Death Among the Western Xhosa. Curare 4/85:103–116.

Holt, J. Z. 1996 Beyond Optimization: Alternative Ways of Examining Animal Exploitation. World Archaeology 28:89–109.

Houston, S. D., and P. A. McAnany, 2003 Bodies and Blood: Critiquing Social Construction in Maya Archaeology. Journal of Anthropological Archaeology 22:26–41.

Houston, S. D., D. Stuart, and K. Taube 2006 The Memory of Bones: Body, Being, and Experience Among the Classic Maya. Austin: University of Texas Press.

Houston, S. D., H. L. Escobedo, D. Forsyth, P. Hardin, D. Webster, and L. Wright 1998 On the River of Ruins: Explorations at Piedras Negras, Guatemala, 1997. Mexicon 20:16–22.

Humphreys, S. C. 1981 Death and Time. *In* Mortality and Immortality: The Anthropology and Archaeology of Death. S. C. Humphreys, and H. King, eds. Pp. 261–283. London: Academic Press.

Hunter-Tate, C. C. 1994 The Chultuns of Caracol. *In* Studies in the Archaeology of Caracol, Belize. D. Z. Chase, and A. F. Chase, eds. Pp. 64–75. Monograph, 7. San Francisco: Pre-Columbian Art Research Institute.

Hurtado C., A. C. B. Araceli, V. Tiesler, and W. J. Folan 2007 Sacred Spaces and Human Funerary and Nonfunerary Placements in Champotón, Campeche, During the Postclassic Period. *In* New Perspectives on Human Sacrifice and Ritual Body Treatments in Ancient Maya Society. V. Tiesler, and A. Cucina, eds. Pp. 209–231. New York: Springer.

Hutchinson, D. L., and L. V. Aragon 2002 Collective Burials and Community Memories: Interpreting the Placement of the Dead in the Southeastern and Mid-Atlantic United States with Reference to Ethnographic Cases from Indonesia. *In* The Space and Place of Death. H. Silverman, and D. B. Small, eds. Pp. 27–54, Archaeological Papers of the American Anthropological Association, 11. Arlington: American Anthropological Association.

Iglesias, P. de L. and M. Josefa 2003a Contenedores de Cuerpos, Cenizas y Almas: El Uso de las Urnas Funerarias en la Cultura Maya. *In* Antropología de la Eternidad: La Muerte en la Cultura Maya. A. Ciudad et al. eds. Pp. 209–254, Madrid: Sociedad Española de Estudios Mayas, Centro de Estudios Mayas.

Iglesias, P. de L. 2003b Problematical Deposits and the Problem of Interaction: The Material Culture of Tikal During the Early Classic Period. *In* The Maya and Teotihuacan: Reinterpreting Early Classic Interaction. G. E. Braswell, ed. Pp. 167–198. Austin: University of Texas Press.

Ingold, T. 1988 Introduction. *In* What is an Animal? Tim Ingold, ed. pp. 1–16. One World Archaeology, 1. London: Unwin Hyman.

Innes, B. 1999 Death and the Afterlife. London: Brown Partworks Ltd.

Iserson, K. V. 1994 Death to Dust: What Happens to Dead Bodies. Tucson: Galen Press.

Johnson, T. 1996 Holy Fabrications: The Catacomb Saints and the Counter Reformation in Bavaria. The Journal of Ecclesiastical History 47:274–297.

Jones, C. 1969 The Twin-Pyramid Group Pattern: A Classic Maya Architectural Assemblage at Tikal, Guatemala. Unpublished Ph.D. Dissertation. University of Pennsylvania.

Jones, C., and L. Satterthwaite 1982 The Monuments and Inscriptions of Tikal: The Carved Monuments. University Museum Monographs, Tikal Report, 33A. Philadelphia: The University Museum, University of Pennsylvania.

Kluckhohn, C. 1948 Conceptions of Death Among the Southwestern Indians: Ingersoll Lecture on the Immortality of Man, for the Academic Year 1947–1948, Harvard University. Divinity School Bulletin 66:5–19.

Knipe, D. M. 2005 Zur Rolle des "Provisorischen Körpers' für den Verstorbenen in Hindu-Istischen Bestattungen. *In* Der Abschied von den Toten: Trauerrituale im Kulturvergleich. J. Assmann, F. Maciejewski, and A. Michaels, eds. Pp. 62–81. Göttingen: Wallstein Verlag.

Knüsel, C. J. 2005 The Physical Evidence of Warfare – Subtle Stigmata? *In* Warfare, Violence and Slavery in Prehistory. Proceedings of a Prehistoric Society Conference at Sheffield University. BAR International Series, 1374. M. Parker Pearson and I. J. N. Thorpe, eds. Pp. 49–65. Oxford: Archaeopress.

Krejci, E., and T. P. Culbert 1995 Preclassic and Classic Burials and Caches in the Maya Lowlands. *In* The Emergence of Lowland Maya Civilization. N. Grube, ed. Pp. 103–116, Acta Mesoamericana, 8. Möckmühl: Verlag Anton Saurwein.

Kroeber, A. L. 1927 Disposal of the Dead. American Anthropologist 29:308–315.

Kuhnt-Saptodewo, S., and H. Kampffmeier 1996 Bury Me Twice: Death Ritual Among the Ngaju Dayak. Ethnographic Film D 1917. Göttingen: IWF.

Kuhnt-Saptodewo, S., and H. Rietz 2004 Helbigs Beobachtungen zu den Ritualen der Dayak und der Heutige Stand der Forschung. *In* Karl Helbig. Wissenschafter und Schiffsheizer: Sein Lebenswerk aus Heutiger Sicht. Rückblick zum 100. Geburtstag. W. Rutz and A. Sibeth, eds. Pp. 178–197. Hildesheim: Georg Olms Verlag.

Kunen, J. L., M. J. Galindo, and E. Chase 2002 Pits and Bones: Identifying Maya Ritual Behavior in the Archaeological Record. Ancient Mesoamerica 13:197–211.

Lagercrantz, S. 1991 The Dead Man in the Tree. Uppsala: Förutvarande Institutionen för Allmän och Jämförande Etnografi vid Uppsala Universitet.

Laporte, J. P. 2003 Thirty Years Later: Some Results of Recent Investigations in Tikal. *In* Tikal: Dynasties, Foreigners & Affairs of State. Jeremy A. Sabloff, ed. Pp. 281–318. Santa Fe: School of American Research Press.

Laporte, J. P., B. Hermes, L. de Zea, and M. J. Iglesias 1992 Nuevos Entierros y Escondites de Tikal, Eubfases Manik 3a y 3b. Cerámica de Cultura Maya 16:30–68.

Laporte, J. P., and V. Fialko 1995 Un reencuentro con Mundo Perdido, Tikal, Guatemala. Ancient Mesoamerica 6:41–94.

Leach, E. 1958 Magical Hair. Journal of the Royal Anthropological Institute 88:147–154.

Legner, A. 2003 Kölner Heilige und Heiligtümer: Ein Jahrtausend Europäischer Reliquienkultur. Cologne: Greven Verlag.

Lock, M. 1997 Displacing Suffering: The Reconstruction of Death in North America and Japan. *In* Social Suffering. A. Kleinmann, V. Das, and M. Lock, eds. Pp. 207–244. Berkeley: University of California Press.

Malinowski, B. 1926 Magic, Science and Religion. *In* Science, Religion and Reality. J. Needham, ed. Pp. 19–84. London: The Sheldon Press.

Mallory, J. P., and V. H. Mair 2000 The Tarim Mummies: Ancient China and the Mystery of the Earliest Peoples from the West. London: Thames and Hudson.

Malville, N. J. 2005 Mortuary Practices and Ritual Use of Human Bone in Tibet. *In* Interacting with the Dead: Perspectives on Mortuary Archaeology for the New Millennium. G. F. M. Rakita, J. E. Buikstra, L. A. Beck, and S. R. Williams, eds. Pp. 190–204. Gainesville: University Press of Florida.

Martin, S. 2003 In the Line of the Founder: A View of Dynastic Politics at Tikal. *In* Tikal: Dynasties, Foreigners and Affairs of State. J. A. Sabloff, ed. Pp. 3–45. Santa Fe: School of American Research Press.

Martin, S., and N. Grube 2008 Chronicle of the Maya Kings and Queens: Deciphering the Dynasties of the Ancient Maya. Revised Edition. London: Thames and Hudson.

Massey, V. K. 1989 The Human Skeletal Remains from a Terminal Classic Skull Pit at Colhá, Belize. Papers of the Colhá Project, 3, College Station: Texas Archeological Research Laboratory, University of Texas Press, Austin and Texas A&M University.

McAnany, P. 1995 Living with the Ancestors: Kinship and Kingship in Ancient Maya Society. Austin: University of Texas Press.

McAnany, P. 1998 Ancestors and the Classic Maya Built Environment. *In* Function and Meaning in Classic Maya Architecture. S. D. Houston, ed. pp. 271–298. Washington, DC: Dumbarton Oaks Research Library and Collection.

McKinley, J. I. 2005 Archaeology of Britain. *In* Encyclopedia of Cremation. D. J. Davies, ed. Pp. 9–15. Hants: Ashgate.

Meskell, L. M., and R. A. Joyce 2003 Embodied Lives: Figuring Ancient Maya and Egyptian Experience. London: Routledge.

Metcalf, P., and R. Huntington 1991 Celebrations of Death: The Anthropology of Mortuary Ritual. Cambridge: Cambridge University Press.

Meyer, R. J. 2000 Königs- und Kaiserbegräbnisse im Spätmittelalter. Beihefte zu J. F. Böhmer, Regesta Imperii, 19. Cologne: Böhlau.

Meyers, E. M. 1971 Jewish Ossuaries: Reburial and Rebirth. Biblica et Orientalia, 24. Rome: Biblical Institute Press.

Miles, D. 1965 Socio-Economic Aspects of Secondary Burial. Oceania 35:161–174.

Moholy-Nagy, H. 2003 Artifacts of Tikal: Utilitarian Artifacts and Unworked Material. Museum Monographs, Tikal Report, 27B. Philadelphia: The University Museum, University of Pennsylvania.

Moholy-Nagy, H. 2008 The Artifacts of Tikal: Ornamental and Ceremonial Artifacts and Unworked Material. Tikal Museum Monographs, Tikal Report, 27A. Philadelphia: The University Museum, University of Pennsylvania.

Morris, I. 1992 Death-Ritual and Social Structure in Classical Antiquity. Cambridge: Cambridge University Press.

Mulkay, M. 1993 Social Death in Britain. *In* The Sociology of Death: Theory, Culture, Practice. D. Clark, ed. Pp. 31–49. Oxford: Blackwell.

Murphy, E. M., and J. P. Mallory 2000 Herodotus and the Cannibals. Antiquity 74:388–394.

Naquin, S. 1988 Funerals in North China: Uniformity and Variation. *In* Death Ritual in Late Imperial and Modern China. J. L. Watson, and E. S. Rawski, eds. Pp. 37–70. Berkeley: University of California Press.

Needham, R. 1967 Percussion and Transition. Man, New Series 2:606–614.

O'Connor, T. P. 1996 A Critical Overview of Archaeological Animal Bone Studies. World Archaeology 28:5–19.

O'Shea, J. 1984 Mortuary Variability: An Archaeological Investigation. New York: Academic Press.

Parker Pearson, M. 2000 The Archaeology of Death and Burial. College Station: Texas A&M University Press.

Parker Pearson, M., A. Chamberlain, O. Craig, P. Marshall, J. Mulville, H. Smith et al. 2005 Evidence for Mummification in Bronze Age Britain. Antiquity 79:529–546.

Parker Pearson, M., A. Chamberlain, M. Collins, C. Cox, G. Craig, O. Craig et al. 2007 Further Evidence for Mummification in Bronze Age Britain. Antiquity 81/313. Project Gallery, www.antiquity.ac.uk.

Pendergast, D. M. 1979 Excavations at Altun Ha, Belize, 1964–1970. Publications in Archaeology, 1. Toronto: Royal Ontario Museum.
Perlès, C. 2001 The Early Neolithic in Greece: The First Farming Communities in Europe. Cambridge: Cambridge University Press.
Puleston, D. E. 1971 An Experimental Approach to the Function of Classic Maya Chultuns. American Antiquity 36:322–335.
Puleston, D. E. 1983 The Settlement Survey of Tikal. University Museum Monographs, Tikal Report, 13. Philadelphia: The University Museum, University of Pennsylvania.
Radcliffe-Brown, A. R. 1922 The Andaman Islanders. Cambridge: Cambridge University Press.
Rakita, G. F.M., J. E. Buikstra, L. A. Beck, and S. R. Williams, eds. 2005 Interacting With the Dead: Perspectives on Mortuary Archaeology for the New Millennium. Gainesville: University Press of Florida.
Ricketson, O. G. 1925 Burials in the Maya Area. American Anthropologist 27:381–401.
Ricketson, Jr., O. G., and E. B. Ricketson 1937 Uaxactun, Guatemala, Group E 1926–1931. Carnegie Institution of Washington, 477, Washington, DC: Carnegie Institution of Washington.
Robben, A. C. G. M. 2000 State Terror in the Netherworld: Disappearance and Reburial in Argentina. *In* Death Squad: The Anthropology of State Terror. J. A. Sluka, ed. Pp. 91–113. Philadelphia: University of Pennsylvania Press.
Roskams, S. 2001 Excavation. Cambridge Manuals in Archaeology. Cambridge: Cambridge University Press.
Rospatt, A. V. 2005 Der Nahende Tod: Altersrituale bei den Newars. *In* Der Abschied von den Toten: Trauerrituale im Kulturvergleich. J. Assmann, F. Maciejewski, and A. Michaels, eds. Pp. 199–222. Göttingen: Wallstein Verlag.
Rue, D. J., A.-C. Freter, and D. A. Ballinger 1989 The Caverns of Copan Revisited: Preclassic Sites in the Sesesmil River Valley, Copan, Honduras. Journal of Field Archaeology 16:395–404.
Saul, F. P., and J. M. Saul 1991 The Preclassic Population of Cuello. *In* Cuello: An Early Maya Community in Belize. N. Hammond, ed. Pp. 134–158. Cambridge: Cambridge University Press.
Saunders, S. R., A. Herring, L. Sawchuk, and G. Boyce 1995 The Nineteenth-Century Cemetery at St. Thomas' Anglican Church, Belleville: Skeletal Remains, Parish Records, and Censuses. *In* Grave Reflections: Portraying the Past Through Cemetery Studies. S. R. Saunders, and A. Herring, eds. Pp. 93–117. Toronto: Canadian Scholars' Press.
Saxe, A. A. 1970 Social Dimensions of Mortuary Practices. Ph.D. Dissertation, University of Michigan.
Scarborough, V. L., M. E. Becher, J. L. Baker, G. Harris, and F. Valdez, Jr. 1995 Water and Land at the Ancient Maya Community of La Milpa. Latin American Antiquity 6:98–119.
Schaeffer, K. A. 2007 Dying Like Milarépa: Death Accounts in a Tibetan Hagiographic Tradition. *In* The Buddhist Dead: Practices, Discourses, Representations. Kuroda Institute Studies in East Asian Buddhism, 20. B. J. Cuevas, and J. I. Stone, eds. Pp. 208–233. Honolulu: University Hawaii Press.
Schiffer, M. B. 1987 Formation Processes of the Archaeological Record. Albuquerque: University of New Mexico Press.
Schmidt, C. W., and S. A. Symes, eds. 2008 The Analysis of Burned Human Remains. London: Academic Press.
Schukraft, H. 1989 Die Grablegen des Hauses Württemberg. Stuttgart: Theiss.
Sempowski, M. L. 1994 Mortuary Practices at Teotihuacan. *In* Mortuary Practices and Skeletal Remains at Teotihuacan. M. L. Sempowski, and M. W. Spence, eds. Pp. 2–314. Urbanization at Teotihuacan, 3. Salt Lake City: University of Utah Press.

Serafin, S., and C. Peraza Lope, 2007 Human Sacrificial Rites Among the Maya of Mayapán: A Bioarchaeological Perspective. *In* New Perspectives on Human Sacrifice and Ritual Body Treatments in Ancient Maya Society. V. Tiesler, and A. Cucina, eds. Pp. 232–250. New York: Springer.

Shay, T. 1985 Differentiated Treatment of Deviancy at Death as Revealed in Anthropological and Archaeological Material. Journal of Anthropological Archaeology 4:221–241.

Smith, A. L. 1950 Uaxactun, Guatemala: Excavations of 1931–1937, Carnegie Institution of Washington, 588. Washington, DC: Carnegie Institution of Washington.

Sprague, R. 2005 Burial Terminology: A Guide for Researchers. Lanham: Altamira Press.

Stodder, A. L. W. 2005 The Bioarchaeology and Taphonomy of Mortuary Ritual on the Sepik Coast, Papua New Guinea. *In* Interacting with the Dead: Perspectives on Mortuary Archaeology for the New Millennium. G. F. M. Rakita, J. E. Buikstra, L. A. Beck, and S. R. Williams, eds. Pp. 228–250. Gainesville: University Press of Florida.

Stodder, A. L. W. 2006 The Taphonomy of Cranial Modification in Papua New Guinea: Implications for the Archaeology of Mortuary Ritual. *In* Skull Collection, Modification and Decoration. M. Bonogofsky, ed. Pp. 77–89. BAR International Series, 1539. Oxford: Archaeopress.

Straus, A. S. 1978 The Meaning of Death in Northern Cheyenne Culture. Plains Anthropologist 23:1–6.

Stuart, D. 1998 "The Fire Enters His House": Architecture and Ritual in Classic Maya Texts. *In* Function and Meaning in Classic Maya Architecture. S. D. Houston, ed. Pp. 373–425. Washington, DC: Dumbarton Oaks Research Library and Collection.

Stuart, D. 2005 The Inscriptions from Temple XIX at Palenque. San Francisco: Pre-Columbian Art Research Institute.

Sweeting, H., and M. Gilhooly 1997 Dementia and the Phenomenon of Social Death. Sociology of Health & Illness 19:93–117.

Tapper, R. 1989 Animality, Humanity, Morality, Society. *In* What is an Animal? Tim Ingold, ed. Pp. 47–62. London: Unwin Hyman.

Taylor, J. H. 2001 Death and the Afterlife in Ancient Egypt. London: The Trustees of the British Museum.

Tiesler, V. 2007 Funerary or Nonfunerary? New References in Identifying Ancient Maya Sacrificial and Postsacrificial Behaviors from Human Assemblages. *In* New Perspectives on Human Sacrifice and Ritual Body Treatments in Ancient Maya Society. V. Tiesler, and A. Cucina, eds. pp. 14–44. New York: Springer.

Tonkinson, M. 2008 Solidarity in Shared Loss: Death-related Observances Among the Martu of the Western Desert. *In* Mortality, Mourning and Mortuary Practices in Indigenous Australia. K. Glaskin, M. Tonkinson, Y. Musharbash, and V. Burbank, eds. Pp. 36–53. Farnham: Ashgate Publishing.

Tozzer, A. M. 1941 Landa's Relación de las Cosas de Yucatan. Papers of the Peabody Museum of American Archaeology and Ethnology, Cambridge: Peabody Museum of American Archaeology and Ethnology.

Tsu, T. Y. 2000 Toothless Ancestors, Felicitous Descendants: The Rite of Secondary Burial in South Taiwan. Asian Folklore Studies 59:1–22.

Tumarkin, N. 1993 Lenin Lives! The Lenin Cult in Soviet Russia. Cambridge: Harvard University Press.

Turner, V. 1967 The Forest of Symbols: Aspects of Ndembu Ritual. Ithaca: Cornell University Press.

Ubelaker, D. H. 1989 Human Skeletal Remains: Excavation, Analysis and Interpretation. Chicago: Taraxacum.

Ucko, P. J. 1969 Ethnography and Archaeological Interpretation of Funerary Remains. World Archaeology 1:262–280.

Van Gennep, A. 1909 Les Rites de Passage. Paris: Nourry.

Van Gennep, A. 1960 The Rites of Passage. Chicago: University of Chicago Press.

Velásquez García, E. 2006 The Maya Flood Myth and the Decapitation of the Cosmic Caiman. The Pari Journal 7/1:1–10.

Verano, J. W. 2005 Human Sacrifice and Postmortem Modification at the Pyramid of the Moon, Moche Valley Peru *In* Interacting with the Dead: Perspectives on Mortuary Archaeology for the New Millennium. G. F. M. Rakita, J. E. Buikstra, L. A. Beck, and S. R. Williams, eds. Pp. 277–289. Gainesville: University Press of Florida.

Verdery, K. 1999 The Political Lives of Dead Bodies: Reburial and Post-Socialist Change. New York: Columbia University Press.

Ward, A. E. 1980 Navajo Graves, Vol. 2 Albuquerque: Center for Anthropological Publication.

Waterson, R. 1995 Houses, Graves and the Limits of Kinship Groupings Among the Sa'dan Toraja. Bijdragen tot de Taal-, Land- en Volkenkunde 151:194–217.

Watson, R. S. 1988 Graves and Politics in Southeastern China. *In* Death Ritual in Late Imperial and Modern China. Studies on China, 8. J. L. Watson, and E. S. Rawski, eds. Pp. 203–227. Berkeley: University of California Press.

Watson, R. S. 1994 Making Secret Histories: Memory and Mourning in Post-Mao China. *In* Memory, History and Opposition Under State Socialism. R. S. Watson, ed. Pp. 65–85. Santa Fe: School of American Research Press.

Weiss-Krejci, E. 2001 Restless Corpses. "Secondary Burial" in the Babenberg and Habsburg Dynasties. Antiquity 75:769–780.

Weiss-Krejci, E. 2004 Mortuary Representations of the Noble House: A Cross-Cultural Comparison Between Collective Tombs of the Ancient Maya and Dynastic Europe. Journal of Social Archaeology 4:368–404.

Weiss-Krejci, E. 2005 Excarnation, Evisceration, and Exhumation in Medieval and Post-Medieval Europe. *In* Interacting with the Dead: Perspectives on Mortuary Archaeology for the New Millennium. G. F. M. Rakita, J. E. Buikstra, L. A. Beck, and S. R. Williams, eds. Pp. 155–172. Gainesville: University Press of Florida.

Weiss-Krejci, E. 2006 The Maya Corpse: Body Processing from Preclassic to Postclassic Times in the Maya Highlands and Lowlands. *In* Jaws of the Underworld: Life, Death, and Rebirth Among the Ancient Maya. P. R. Colas, G. LeFort, and B. L. Persson, eds. Pp. 71–86, Acta Mesoamericana, 16. Markt Schwaben: Verlag Anton Saurwein.

Weiss-Krejci, E. 2008 Unusual Life, Unusual Death and the Fate of the Corpse: A Case Study from Dynastic Europe. *In* Deviant Burial in the Archaeological Record. E. M. Murphy, ed. Pp. 169–190. Oxford: Oxbow.

Weiss-Krejci, E. In press The Role of Dead Bodies in Ancient Maya Politics: Cross-Cultural Reflections on the Meaning of Tikal Altar 5. *In* The Conception and Treatment of the Dead in Latin America: Between the Dead and the Living in Mesoamerica. J. Fitzsimmons, And I. Shimada, eds. Tucson: University of Arizona Press.

Welsh, W. B. M. 1988 An Analysis of Classic Lowland Maya Burials. BAR International Series, 409. Oxford: British Archaeological Reports.

Westerhof, D. 2008 Death and the Noble Body in Medieval England. Woodbridge: Boydell & Brewer.

Whitley, J. 2002 Too Many Ancestors. Antiquity 76:119–126.

Widmoser, E. 1970 Tirol von A bis Z. Innsbruck: Südtirol-Verlag.

Williams, H. 2005 Keeping the Dead at Arm's Length: Memory, Weaponry and Early Medieval Mortuary Technologies. Journal of Social Archaeology 5:253–275.

Williams, H. 2008 Towards an Archaeology of Cremation. *In* The Analysis of Burned Human Remains. C. W. Schmidt, and S. A. Symes, eds. Pp. 239–369. London: Academic Press.

Wright, L. E. 2002 The Inhabitants of Tikal: A Bioarchaeological Pilot Project. Famsi Grantee Report, www.famsi.org/reports/95050.

Yarrow, H. C. 1880a Introduction to the Study of Mortuary Customs Among the North American Indians. Smithsonian Institution, Bureau of Ethnology. Washington, DC: Government Printing Office.

Yarrow, H. C. 1880b A Further Contribution to the Study of the Mortuary Customs of the North American Indians. Smithsonian Institution, Bureau of Ethnology. Washington, DC: Government Printing Office.

Zur, J. N. 1998 Violent Memories: Mayan War Widows in Guatemala. Boulder: Westview Press.

5

Representativeness and Bias in Archaeological Skeletal Samples

Mary Jackes

Introduction

In discussing the term "bioarchaeology," Wright and Yoder (2003:44) write that "Our ability to make statements about [the past] is ... dependent on the representativeness of archaeological sampling of ancient human remains". We cannot, of course, sample ancient human remains in any meaningful way: our "population" and our "sample" are more or less the same thing. We can only work with what we are given by the chance of preservation and discovery and the multiple factors that determine whether excavations are complete and careful, whether human skeletons can be retained and curated satisfactorily and made available for study.

This raises a series of questions regarding whether or not our samples are valid representations of the populations from which they came. How can we recognize bias in the materials bioarchaeologists study, and how might we deal with any shortcomings? This chapter discusses the impact of social/cultural and demographic factors on bioarchaeological samples along with the effect of taphonomy and research techniques. Above all, it emphasizes the need to understand the context and demographic characteristics of our samples so that we can arrive at interpretations that have some relation to reality. The importance of comparison in evaluating the representativeness of samples will be evident from many of the historical and archaeological examples given.

Social Bioarchaeology. Edited by Sabrina C. Agarwal and Bonnie A. Glencross
© 2011 Blackwell Publishing Ltd

Bias: Can We Recognize It, Can We Compensate for It?

The human age-at-death distribution is biased, except in some exceptional circumstances in which all age groups have an equal probability of dying. Normally, the probability of dying is not uniform: deaths are not random with regard to age or sex. There are endogenous causes of perinatal death, especially in males (Reid 2000; Wells 2000), and as the nursing period ends, as breast milk ceases to shield the baby and sources of infection increase, there will be more deaths (Brown 2003; Hanson 2001; Rowland 1986; Wells, 2000). With the exception of massacres and catastrophes, deaths will often be mediated by age-specific onset diseases, by accident and conflict associated with young males, by reproductive hazards in young females, by the increasing rates of degenerative conditions among the elderly. As such, the dead are not an unbiased sample of the living population: the force of mortality, normally greatest on the young and the old, creates a U-shaped mortality curve. We will consider an unbiased cemetery sample to be one that closely approximates this "normal" U-shaped mortality by age curve, while also recognizing that the exact shape of the curve will differ according to fertility and mortality steered by varied biological and social/cultural determinants. For a preliminary idea of some of the many factors influencing our archaeological data, lacking historical control, we can examine a case where the nonbiological factors can be determined.

An historical example from Sydney, Australia

The cemetery records for St. John's Anglican Church in Ashfield, now an inner suburb of Sydney, Australia, are very informative. Connah (1993:151, figure 10.2) provides the data on the cemetery (age-at-death distribution by decade), commenting that the records demonstrate an increase in life expectancy at birth. Indeed they do – apparently. The first three decades during which the cemetery was used have a life expectancy at birth (e_0) value of about 31 years, the second three decades around 43.5 years, and the next 30 years have an e_0 value of just over 66 years. But is this a valid interpretation of the data?

The first 30 years of burials represent the beginning of growth in this area just outside the early colonial settlement of Sydney. People are moving in: the church is consecrated in the mid-1840s, and by the mid-1850s a railway station is established a few hundred meters from the church. Now begins a rush, leading into the second 30-year period, with the establishment of grand houses around the church, the richer of the new inhabitants in this English colonial society being Anglicans seeking to escape the increasingly congested center of Sydney. The third period (1900–1929) marks the start of a decline in use of the cemetery, years during which the wealthy Anglican community members, many of their sons lost in World War I, move away to the north of Sydney Harbour. Industries are beginning to come into Ashfield and the large single family houses are being replaced by denser settlement. Early in the third period a Roman Catholic church is built in Ashfield for incoming Irish working families.

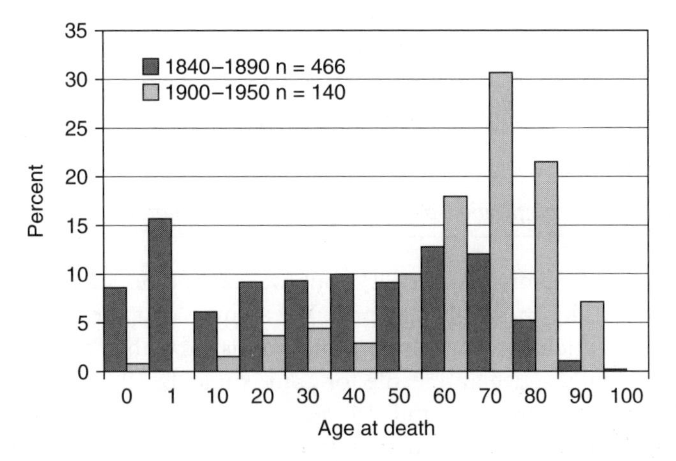

Figure 5.1 Comparison of age-at-death distributions of early and late period burials at St. John's Church, Ashfield, Sydney.
Source: Author's drawing, based on Connah, G. 1993 "Of The Hut I Builded": The Archaeology of Australia's History, figure 10.2. Cambridge: Cambridge University Press

In the first three decades young mobile people move in, establishing their families. The next 30 years show a decline in fertility as the immigrants live beyond the reproductive years. In the last three decades many members of the Anglican community leave for the newly opened up suburbs to the north: few young reproductive age people are left and a larger proportion of the population is over 50 years of age. In the first three decades, only 25 percent of the dead are over 49. In the decades 1900 to 1929 the percentage of the living over 49 has more than doubled and these older people (50 and over) make up 85 percent of the dead. St. John's cemetery represents a period of change during which people might choose not to be buried in a (now) inner city church graveyard, but rather in a new suburban cemetery (a public cemetery had been laid out further along the railway line at the end of the 1860s). Furthermore, the earlier growth of the area as a suburb of large houses is succeeded by social and economic change, with the influx of industry, multiple housing units, and people less likely to bury their dead in an Anglican cemetery.

From this one isolated example, we can see the operation of a number of factors: immigration of reproductive age individuals accompanied by few elderly people; catastrophic war-time deaths of young males, buried elsewhere; emigration of reproductive age individuals and their children, leaving the old behind; selection by religious affiliation; selection by occupation and social status; choice of a particular burial place. In this instance, changes over time in fertility and longevity cannot be assessed because the sample is biased. Figure 5.1, covering the entire period of the burial records, illustrates that examining age-based characteristics of this cemetery sample will give us information biased toward 19th-century juveniles and younger adults, while the examination of the old will be biased toward 20th-century characteristics. Cemeteries have beginnings, middle periods, and endings: in and out migration both are important, but other factors will also be in operation during the time in which the cemetery sample accumulates. The end result is a muddied picture of fertility and longevity. In the absence of chronological controls of archaeological cemetery samples, we must be very aware that our samples could be biased in the various ways illustrated by St. John's, Ashfield.

Recognition of sample composition bias through fertility estimates

There are biological limits on fertility, but we cannot be sure about the social constraints on fertility in the distant past. We assume that the populations of the past were not practicing contraception and we assume that the majority of the females were sexually active and fertile for approximately 30 years, from sometime after 15 years to sometime around 45 years. I have reviewed a number of the factors which militate against women having live born children at minimal birth intervals of, say, 15 months (9 months gestation plus 6 months lactation) and discussed the range of fertility levels that we might regard as biologically possible (Jackes 1994; 2009a; Jackes and Meiklejohn 2004; 2008; Jackes et al. 2008).

I estimate fertility levels as the total fertility rate (TFR) from cemetery populations (e.g., Jackes and Meiklejohn 2008; Jackes 2009a) based on two estimators. One is J:A, the ratio of juveniles aged 5–14 years to adults aged 20 and over (the juvenility index of Bocquet-Appel and Masset 1977). The second is MCM, or mean childhood mortality, the mean probability of death across three subadult age categories, 5–9, 10–14, and 15–19 years (Jackes 1986). I have used fertility data from external sources, that is, fertility values from the literature on historical fertility, to verify my estimates. I do not estimate fertility from historical life table values. Bocquet-Appel and Naji (2006:356, figure 6) have suggested that fertility rates derived from historical life tables are quite low. Based on records in Geneva, Switzerland, some of the most reliable historical demographic data that we have (Perrenoud 1984:55), this is true of fertility estimates from life table calculations. It is theoretically possible to estimate the TFR from ${}_{30}C_{15}$ (the proportion of the living population in the appropriate age classes) in a life table, but normally this will lead to underestimations of the true fertility rate, as seen in Figure 5.2. The life table underestimation varies with the level of population increase, approaching zero as the birth and death rates become more equal.

My estimates do not derive directly from model tables but indirectly by quadratic regression from 51 West model tables (Coale and Demeny 1983) and three United Nations model tables (United Nations 1982). In Figure 5.2 the stars illustrate that the estimated TF ranges are reasonable, especially in the earlier periods when time ranges for Perrenoud's fertility and mortality data are more equivalent (Perrenoud 1990:249, table 15.3 total progeny, adjusted age-specific fertility rates). The fertility estimates are most discrepant around 1770 following an influx of refugees and the beginning of contraception, especially among older women.

If it is granted (i) that we can estimate fertility rates with a certain accuracy from the estimators, and (ii) that we have a fairly clear idea of the biological constraints on the total fertility rate of human populations, we can at least weed out those samples in which the J:A ratio among the dead is beyond the bounds of biological possibility. I have built up a database of archaeological and historical age-at-death

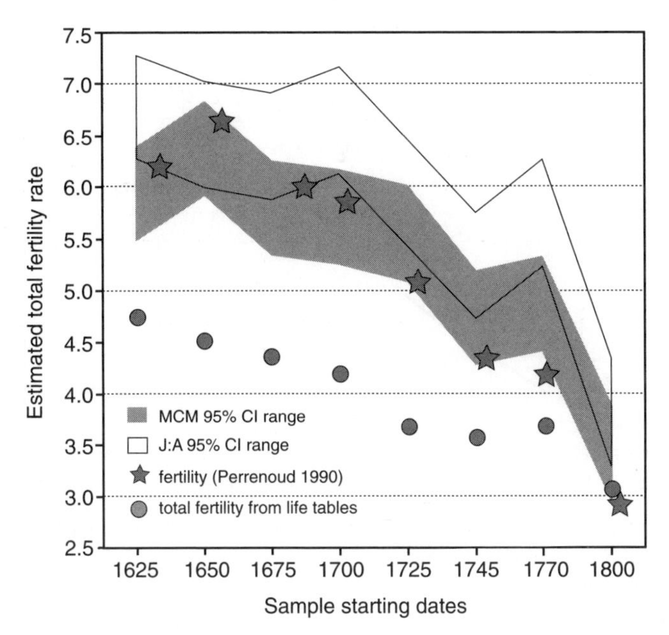

Figure 5.2 Estimates of historical fertility rates for Geneva shown as 95 percent CI ranges derived from J:A and MCM (data from Perrenoud 1984), compared with total fertility derived from life tables (circles) and fertility data (stars) given in Perrenoud (1990).
Source: Author's drawing, based on Perrenoud, A. 1990 Aspects of Fertility Decline in an Urban Setting: Rouen and Geneva. *In* A. Urbanization in History. A Process of Dynamic Interactions. J. van deWoude, J. de Vries, and A. Hayami, eds. Pp. 243–263. Oxford: Oxford University Press

distributions with sample sizes above 100 in order to test the reasonable limits of subadult age-at-death distributions. Proportions of juveniles and adults in archaeological samples are problematic because it is so often impossible to find out what percentage of the dead in a site might be represented by the excavated skeletons. If a site is not completely excavated, the dead may or may not represent a biased sample. Nonetheless, while we may never know the true nature of a sample, it may be possible to discern the presence of bias.

Over- and underrepresentation of age classes: examples from the bioarchaeological literature

We should suspect bias, and closely examine the context, in any archaeological cemetery sample in which the estimators indicate a total fertility rate of 12.5 or more.[1] In round terms, any J:A value over ~.380 and any MCM value over ~.135 needs to be examined for partial site excavation, for special-purpose burials, for incomplete life table totals, and for sample size (which should be 100 or more: Hoppa and Saunders 1998; Paine and Harpending 1996). Even if the estimates of fertility are lower than 12.5, it is necessary to consider whether figures approaching the maximum recorded or theoretical total fertility rates are reliable. The TFR is

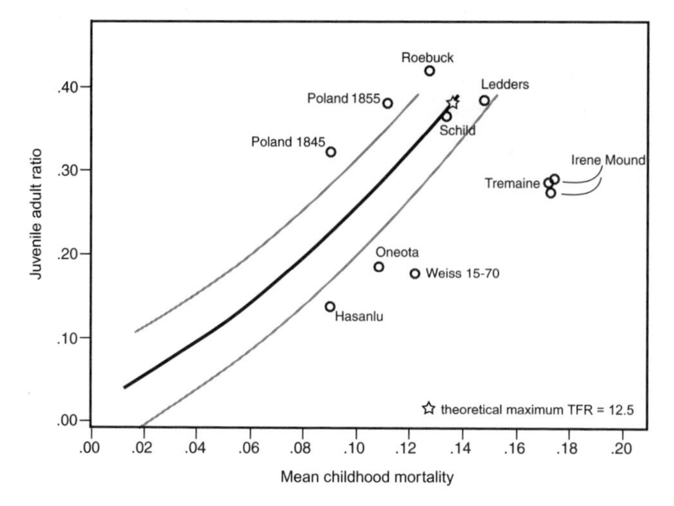

Figure 5.3 MCM and J:A, 95 percent CI for archaeological and historical data sets, showing positions of some of the samples discussed.
Source: Author's drawing

the average figure used for comparative purposes in demographic studies: it does not reflect the experience of individual women. Individual women or their husbands may be sterile and will begin and end their reproductive lives at different ages. During their reproductive lives death, disease, famine, miscarriages, stillbirths, deaths of children while still nursing, and disruption causing extended absence of their sexual partners may occur. Their fertility levels will vary over their reproductive life spans. The exact pattern of marriage, polygamy, monogamy, marriages of young women to older men, remarriage permitted or forbidden – all such factors will have an effect on fertility rates. The TFR is a "synthetic" rate, rather than a reflection of the experience of individual women. Discrepant values between the two estimators may indicate bias or error. Such discrepancy can be caused by a variety of factors and should not be ignored.

Where high MCM values cause sites to fall outside the 95 percent CI of the quadratic regression of J:A on MCM, one possibility is that the age estimates of late adolescents are wrong. An individual who is actually 20 may be estimated to be, say, 19. When this is a systematic error, causing too high a percentage of individuals in the 15–19 age category, an age distribution will fall too far to the right of the regression line (the *x*-axis is MCM, see Figure 5.3). For this reason, it is important to pay very close attention to age estimates around 19–21.

Age estimation errors during initial analysis should be considered, but when databases are derived from the literature, age redistribution can introduce error evidenced by a site falling beyond the 95 percent CI on the MCM axis. For example, the Oneota data in the Health and Nutrition in the Western Hemisphere (HNWH) database (Steckel et al. 2002, as used by McCaa 2002), is derived from Milner et al. 1989 and specifies that there were no deaths in the 10–14 age group. The Oneota data with $n = 264$ (Norris Farms 36 in Milner and Smith 1990:122–

123; see also Vradenburg 1999) recorded four deaths in that age group. The two sets of data are very different: one will fall beyond the 95 percent CI (see Figure 5.3) while the other will give a very reasonable TFR estimate range. This is a data-handling error of the type introduced into comparative studies by recasting of data from broad age categories. The Dickson Mounds site problems (Cohen et al. 1994:636; Jackes 1993) are compounded in the HNWH database use of information from Lallo et al. 1980, again with age distribution anomalies resulting from redistributing ages from broad age categories.

The HNWH database also includes Irene Mound where the TFR estimates range from 8.8 to 19.6 and individuals aged 15 to 19 constitute 27 percent of the entire sample. The discrepant nature of this is clearly illustrated by Larsen et al. (1991:194, figure 9, precontact agricultural)[2] and is more extreme than the model table Weiss 15–70 (Weiss 1973) in which the methodology led to an excessively high death rate in the 15–19 age category (Jackes and Meiklejohn 2008:231, figure 4). With sites like Tremaine (Vradenburg 1999) (Figure 5.3), we cannot automatically assume age estimation error, but might question whether the dead represent a demographically valid sample. An obvious example of a nonvalid sample is Hasanlu IV (Rathbun 1982) representing a period during which the settlement was sacked and burned (Dyson 1965:202). The skeletal sample appears to have contained "the bodies of defenders and looters alike" (Hawkes 1974:190) with 36 percent of Rathbun's total sample of 112 found in one of the several collapsed and burned buildings (BB II) (see Figure 5.3). Here we would expect an excess of young males.

While it is very rare for samples to lie to the left of the 95 percent CI range (with a *low* value for MCM relative to J:A), it can occur when late adolescents are underrepresented. This is the case for ten-year samplings from a Polish rural parish (Figure 5.3; Piontek and Henneberg 1981: corrected data; Piontek, personal communication, November 15, 1990), no doubt resulting from internal migration and political disruption.

There may be unexpectedly low numbers of children or of adults in a sample, and this type of bias could be discerned from the distribution of the fertility estimators.

Apparent low fertility when one might expect a non-contracepting population could indicate that children (not just very young children and infants) were being buried elsewhere and that there is, in fact, a relative overrepresentation of adults. It is too widely accepted to need discussion that infants will be underrepresented in skeletal samples: estimators exclude children under five years, but the possibility of differential burial of older children must be taken into account.

The reverse case, in which there seems to be an excessively high representation of children must also be considered. A very obvious case of too many juveniles is that of Nea Nikomedia in Greek Macedonia. The 95 percent CI for the two discordant fertility estimates (.168 and .683) would range from 17 to 27, far beyond acceptable demographic values. The skeletal sample is inappropriate for demographic analysis, despite having been used for comparative purposes (Angel 1971:70). A mortuary practice placing emphasis on the burial of juveniles within the walls of the village, with the archaeological excavation focusing on the settlement area, would lead to this type of result.

Table 5.1 Comparison of fertility estimates for Late Woodland mid-western sites

Site (Late Woodland)	*Source*	*TFR 95% CI*
Kuhlman Mounds	Atwell 1991	5.2 to 6.5
Joe Gay	Garner 1991	5.2 to 8.1
Koster	Droessler 1981	6.5 to 7.6
Ledders	Garner 1991	9.6 to 11.0
Schild	Droessler 1981	11.2 to 13
Ledders	Droessler 1981	13.7 to 15.9

Apparent high fertility may result from underrepresentation of adults because of exclusion of indeterminate age adults from life tables: the use of estimators allows us to avoid adult age assessment difficulties. In fact, adults may be missing for a number of reasons and much of the discussion to follow is of sites in which adults seem underrepresented as a result of the bioarchaeological analyses or because of factors in operation at the time of burial.

When there are groups of sites which do not provide clear answers on why their demographic parameters may be unrepresentative, detailed analysis by a regional specialist could clarify the situation. A series of sites in Illinois (Koster and especially Ledders and Schild; Droessler 1981, see Figure 5.3 and Table 5.1) led me to suggest several times (e.g., Jackes 1986) that the Late Woodland period was a time of exceptionally high fertility, but this is now questionable because, having a method of comparison, I can clearly identify discrepancies among sites. The Schild Late Woodland site was incompletely excavated and the Koster, Ledders, and Schild Late Woodland life tables were based on just under 75 percent of the excavated skeletons. If many of the excluded individuals were adults, then the fertility estimates would be too high. Gordon and Buikstra (1981), in studying the effect of soil pH on bone, examined 95 uncremated skeletons from the Ledders and Helton sites and noted that 16 percent would definitely or probably be excluded from age or sex assignment and a further 22 percent were cracked and fragmented to the point where detailed study was difficult or impossible. Unfortunately, we do not know the details of differential preservation of children and adults, but it is clear that full life tables would exclude many individuals. Added to this we have cremations, which were rare at Koster (Braun 1981; Tainter 1977) but more common for high-status individuals elsewhere in the area. Given the difference between the high fertility estimates for Schild and Ledders, and the Koster mid-level fertility estimate, in association with variability in diet (Buikstra and Milner 1991:324), it seems important that no definitive statements be made except in the context of focused regional studies (Jackes 2006) which include other sites, like Kuhlman Mounds.

Kuhlman Mounds Late Woodland (Table 5.1) has been studied in detail (Atwell 1991), both Mound 1 which has no cremations, and the total site, including cremations. Despite Atwell's (1991:167) opinion that Mound 1 "biases the overall shape of the site's age ... distribution", the estimates from Mound 1 alone (TFR 95 percent CI 5.2–6.5) are little different from the overall site (TFR 95 percent CI

5.4–6.9). Without a comparative method, it is not always obvious whether age-at-death distributions differ in significant ways. Because we can compare the very reasonable fertility estimates for Kuhlman Mounds with those for other Late Woodland sites in the area (Table 5.1), it is possible to suggest that Late Woodland fertility might well have been around 6. The suggested total fertility rate, though high, is much more plausible than a range of 11 to 16. Such an analysis would lead to the conclusion that some Late Woodland skeletal samples, specifically Schild and Ledders, are biased by underrepresentation of adults and should not be used as components of a study of Late Woodland demography.[3] Specialist appraisal of this suggestion is necessary.

A clear-cut case would be one where some adults were excluded from normal burial – a situation documented for Iroquoian Ontario (certain categories of deaths, by freezing, drowning and violence, excluded males from normal burial, Jackes 2009a:363). The Roebuck Site (see Figure 5.3), dated 1450–1530 (Pfeiffer and Fairgrieve 1994), is a large, palisaded village (Jamieson 1990) situated in an area which was inhabited at the time of Cartier's voyages down the St. Lawrence River in the mid-1530s and early 1540s, but was completely depopulated by 1603 when the area was again visited by French explorers. The reasons for this abandonment and the fate of the St. Lawrence Iroquoians are still under discussion. The human remains were studied long ago (Knowles 1937) and the material cannot be used for comparative purposes (contra Bocquet-Appel and Naji 2006). It is an exceptional site with 11–14.5 as 95 percent CI for the two fertility estimators, only 84 individuals in the sample and a slight underrepresentation of late adolescents. There appeared to be only four males among the 43 adults excavated and it has been assumed that many of the adult inhabitants of the village were not buried within the palisades and earthworks. Besides the formally buried individuals, there were many cranial and postcranial bones recovered from "refuse deposits". The highest MNI for these fragmentary remains is 35, based on mandibular portions, all but one of them adults and generally male in appearance. The presence of burnt, cut, and gnawed human bone is not unexpected and the stray bones should not be interpreted as representing the dead adult males of the village. The exclusion of the scattered bone from consideration is reasonable: the ethnohistorically documented Iroquoian practice in war was that captives, especially males, were taken back to warriors' villages to be put to death (Jackes 2004). By the same token, we may assume that adult males were lost to Roebuck village as a consequence of Iroquois raiding.

The importance of context is very evident from our final example of the value of estimators. This is an exceptional case in which ethnohistoric resources are used to determine that a cemetery sample is heavily biased and cannot be used in comparative studies. Grimsby, a Neutral Nation Iroquoian cemetery in southern Ontario, was excavated in the 1970s as a salvage operation. The entire site was excavated, so that we know the sample represents all individuals buried in the area. The ratio of children to adults indicates an extremely high fertility rate, but such an interpretation is not possible, since there is good ethnohistoric evidence that Iroquoians of that period, and specifically the Neutral, had very small families. The reasons for low fertility relate in part to the specifics of Iroquoian life (Jackes 1994). While premarital sex was allowed, self-restraint in public was enjoined and there was no privacy in the long houses; sexual abstinence was required during ritual occasions and men were away for long periods hunting, fishing, trading, or ambushing enemies. Marriage occurred upon pregnancy but only with appropriate kin and with permission. Lactation was accompanied by abstinence and the Jesuits who lived with the Ontario Iroquoians considered that the children were nursed for a very long time.[4]

Further consideration must be given to the period covered by the cemetery, 1620 to 1650. This was a time of documented war, famine, disease, and the type of social and cultural disruption that can only lead to lowered fertility (Jackes 2009a:357). Indeed, the latest burials in the cemetery appear to date to the very period when the Neutral Nation was completely destroyed by its enemies, its people killed, dispersed, or taken captive. And yet the cemetery sample indicates high fertility.

It was possible, by examining the various burials according to the time periods suggested by the European trade beads used as grave goods, to show that the middle period of the use of the cemetery was represented by very high numbers of women and children, consistent with the location having been a place of refuge from famine, disease, and war (see Jackes 2009a for supporting arguments, both archaeological and ethnohistorical). The TFR for this period would be in the range of 25–26, biologically impossible and impossible in the historical context. The final period also had high fertility and high mortality levels, with a broad range of deaths within family groupings, including not only young and disabled people, but many elders – the leaders of this society. Again, a gathering of people into an area in which there was actually no settlement seems plausible: the refugees must have included people from other Ontario Iroquoian nations, not just the Neutral.

The value of estimators, then, is that they allow us to get an idea of fertility levels and to judge whether some age distributions at death are so extreme that they must not be used for comparative purposes. Distributions that are suspect can be identified and examined in close detail for special circumstances leading to bias.

Error and Bias: Methodological Sources Relating to Current Practices for Estimating Adult Age at Death

If we can get some idea of subadult age-at-death distributions, the proportion of subadults in the sample and whether they can be taken as accurate representations of mortality in the society to which they belonged, can we also examine adult age-at-death distributions in some way?

The most problematic area of the mortality curve for skeletal biologists is from about age 30 onwards and we must recognize the constraint that this places upon us. In writing about the "osteological paradox" Wood et al. (1992) stated that the problems of adult age assessment had been solved, a proposition that seemed very questionable at the time (Jackes 1993). A decade later, Wright and Yoder (2003:49) in summarizing recent studies on the osteological paradox, were again reassuring us that the problem of adult age has been overcome, this time "by iteration against a mortality model that is used as an informed prior distribution." They argue that ages can now be estimated from the scores for the "individual components of … traditional age indicators," referring to methods of adult age assessment such as changes in the pubic symphysis. Thus, we should be able to reach a reasonable approximation of the real adult age distribution and also be able to recognize non-representative adult age-at-death distributions, for example from preservation bias. Furthermore, our statistical manipulations should not add bias to our researches. Can we accept these propositions as true?

The Bayesian approach: taking into account the influence of the reference samples' age structure

Adult skeletons of unknown age have been given estimated ages based on comparisons with standard morphological features – for example, changes in the pubic symphysis over the life of an adult – formalized into indicator stages from the characteristics of a known-age "reference" population. For many years now it has been evident that simply applying the mean age from a reference population for a certain indicator stage will not provide useful age estimates (Jackes 1985), and that smoothing and adjustment were in the past undertaken without sufficient deliberation. An early suggestion of smoothing by use of probabilities over the 95 percent confidence interval (Jackes 1985) was designed to point out the inadequacies of the use of the mean as a point estimate, and to highlight the way in which the probability ranges for each indicator stage overlap. Accurate age estimates will not be produced by this method of smoothing (Jackes 2000: figures 15.7 and 15.8). Attention was drawn to some basic problems of age estimates and to the fact that the method controlled the age estimates: the unknown age distribution of the sample does not control the age estimates (Jackes 1985). Acceptance was growing of the forceful argument that the reference population age distribution, rather than details of the method of age estimation, is a determining factor in age estimation. Bocquet-Appel (1986) summarized this well, stating that each reference population has its own "*a priori* distribution" dependent upon the history of the collection of known-age skeletons, noting that Masset (1971) had suggested that an *a priori* uniform distribution was necessary.

While there is some truth to the idea that the underlying problem in adult age determination is statistical rather than biological as Bocquet-Appel states (1986:127), there is also a slowly strengthening undercurrent of opinion (e.g., Masset 1989; see also Hoppa 2000; Jackes 1992, 2000; various authors in Hoppa and Vaupel 2002 – for summary see Hens *et al.* 2008; Jackes 2002; Kimmerle et al. 2008 with reference to females only) that indicator stages may have different trajectories of change over different populations. New techniques are now tested over several populations and found to be appropriate for some and inappropriate for others (Schmitt et al. 2002). While it is now fully accepted that Howell (1976) was correct to emphasize the underlying pattern of human mortality as uniform over space and time, there was a tendency to expand Howell's idea to encompass not only uniformity of a basic human mortality pattern, but also a uniformity of change in indicators of skeletal age. As Müller et al. (2002:4) point out with regard to age indicators, "invariance is reminiscent of the assumption of uniformitarianism" proposed by Howell. Reminiscent, but not the same. Invariance is stated by Müller and co-authors to be the "minimum assumption" necessary for palaeodemographic research, but it is certainly an assumption which requires detailed and critical study. Frankenberg and Konigsberg (2006:253) say that it is an "untestable assumption" and, indeed, it is dismissed as something that has "needlessly" occupied our attention because "many of the perceived differences in aging between samples derive from the different age structures of the study populations" (Konigsberg et al. 2008:542). And here is the whole problem: there is bias in the original reference samples, and the age structures of our archaeological samples presumably differ, as do our study populations, thereby confounding our study of whether age indicators are invariant across populations, across differing times, places, lifestyles, nutritional regimes, activity patterns. We seem to have come full circle.

In the interim, however, we can examine the adjustment of adult age distributions, based on approaches which are not constrained by assumptions of normal distributions, in contrast to distribution over the 95 percent probability range. A number of paleodemographic researchers have proposed a Bayesian approach as a means of estimating an adult age-at-death distribution, and Chamberlain (2000) laid out the essentials for applying basic Bayesian methods to the problems of estimating the ages of adult skeletons recovered from archaeological contexts. Setting out the problem for a Bayesian approach, Chamberlain's contribution in attempting to simplify the matter has led to its being recently employed (Storey 2007).

The Bayesian approach applies to situations of uncertainty, and therefore it is intuitively an excellent approach for paleodemography, since paleodemography is an area in which there is only one certainty: the individuals with whom we are dealing *are* dead. But there are aspects of the skeleton which allow the researcher to say "it is likely" that this individual was an adult who was young, middle aged, or elderly, and these likelihoods ($p(y \mid \theta)$) derive from the skeletal biologist's knowledge or experience of certain age indicators. The data are the age indicators from the skeletons (y), while the unknown age values to be determined are denoted by theta (θ). The Bayesian approach lies in a formulation such as: what we want to know is the probability that an individual is of a certain age, given that his skeletal characteristics (age indicators) are at a certain stage, expressed as *p(age | indicator stage)*, where | means "given" – thus $p(\theta \mid y)$. There is an underlying difficulty, that the two attributes (here stage and age) are assumed to be independent, and naturally this assumption cannot be upheld. Indeed, if the stage were independent of age, then the stage would be of no value in indicating age. However, the correlation of age and "age indicators" is imperfect, so it has been considered worthwhile by a number of researchers to attempt a Bayesian approach to the problem of age estimation.

Besides the age indicators (y), the other element in the Bayesian formula is $p(\theta)$, in this context age-specific mortality or the probability of death at certain ages. Chamberlain points out, and this is emphasized by several contributors to Hoppa and Vaupel (2002), that the probability of death at each age is an important factor in the age | stage formulation. As noted earlier, age-specific mortality is not uniform across all adult ages. There is a lower probability of death at some adult ages, and a higher one at other adult ages. Thus the specific adult mortality pattern (the variation within the overall basic human pattern of mortality), one of the unknowns which a skeletal biologist seeks to determine for an archaeological population, is itself an important component of the probability of an unknown adult skeleton having a certain age. Chamberlain proposes the use of model data, and Boldsen et al. (2002) the use of appropriate historical data. The age-specific probability of death in the chosen data set constitutes $p(\theta)$, which is called the "prior" probability.

Thus, to our knowledge (derived from the reference population of known-age individuals) of the age distribution of age indicators ($p(y)$), we can add our knowledge of human age-specific mortality ($p(\theta)$), by using perhaps a model age-at-death distribution, and in this way we can gain some idea of the probability ($p(\theta|y)$) of a certain age (the unknown parameter θ) for a skeleton with a specific age indicator morphology. We do this by multiplying the likelihood $p(y|\theta)$ by the "prior" belief as to the age-at-death distribution ($p(\theta)$), to give a "posterior" probability $p(\theta|y)$ of the age given the indicator stage (Litton and Buck 1995; Lucy et al. 1996). We can then use this "next step" or "posterior" probability to suggest the age distribution in a sample of which nothing is known, beyond the morphological indications of age.

This is theoretically interesting, but does it produce satisfactory results? Since we are dealing with so many unknowns, paleodemographers must always be skeptical. A straightforward test should therefore be applied to allow us to see whether Chamberlain (2000:107) is correct in his suggestion that a Bayesian approach "is simple yet effective in removing the influence of the age structure of the skeletal reference population". Chamberlain works with the contingency table of the original data from the Suchey–Brooks pubic symphysis (Brooks and Suchey 1990) reference sample. The reference data are biased toward late adolescents and young adults, because the sample derives from forensic cases in a late 20th-century American city, therefore not representing normal attrition by death. The issue is then whether the Bayesian approach can remove the effect of this biased reference sample (given in Chamberlain 2000:108, table 1).

Here we will use known-age data from two sources and archaeological data from one source in order to test the approach. The known-age samples are Spitalfields, derived from a London church crypt (generally 18th century), and Coimbra, a Portuguese (generally 20th century) anatomical sample: the archaeological sample is from early mediaeval Germany (see Jackes 2000:425,430 for details). The two known-age samples are highly significantly different from each other, as are the German and Coimbra indicator stage distributions. Spitalfields and the German distributions are not statistically different. We should therefore find interesting similarities and differences in the estimated age distributions.

In Table 5.2 we have the data to be tested, and we can use not only the reference, uniform, and model prior probabilities of death, but we can also test the efficiency of the method in allowing us to estimate ages at death by using the actual known distribution of ages at death, as given in the last column in Table 5.3, expressed simply as a ratio of number in age category to sample total. In other words, the multiplication of the sample priors by the reference sample distribution across the age and stage matrix gives the values in Table 5.3.

Table 5.4 gives the posterior probabilities from which to derive the age-at-death estimates for Coimbra females. Generating an age category by stage matrix for the Coimbra females is a simple matter of multiplying the posterior probability by the appropriate Coimbra stage sample. For example (using figures from the first cells in Table 5.4 and Table 5.2) $.98 \times 15$ gives the number of individuals estimated to

Table 5.2 Suchey–Brooks pubic symphysis stage known age and archaeological samples to be tested

Stage 1	*Stage 2*	*Stage 3*	*Stage 4*	*Stage 5*	*Stage 6*	*Total*
Coimbra female pubic symphysis stages $p(x)$						
15	2	12	27	30	14	100
Spitalfields female pubic symphysis stages $p(x)$						
3	4	9	17	14	9	56
German archaeological female pubic symphysis stages $p(x)$						
9	1	11	14	9	9	53

Table 5.3 Coimbra female pubic symphysis stages and female age distribution. The last column shows the probability of death $p(\theta)$ or sample priors and derives from the actual known distribution of ages at death

10-year age categories	*Stage 1*	*Stage 2*	*Stage 3*	*Stage 4*	*Stage 5*	*Stage 6*	*Total: Coimbra mortality $p(\theta)$*
15	0.07	0.04	0.01	0.00	0.00	0.00	0.120
25	0.00	0.03	0.05	0.03	0.01	0.00	0.120
35	0.00	0.01	0.02	0.06	0.06	0.02	0.180
45	0.00	0.00	0.01	0.02	0.05	0.06	0.140
55	0.00	0.00	0.00	0.01	0.03	0.08	0.130
65	0.00	0.00	0.00	0.02	0.01	0.11	0.140
75	0.00	0.00	0.00	0.00	0.06	0.04	0.100
85	0.00	0.00	0.00	0.00	0.02	0.05	0.070
Total	0.07	0.08	0.10	0.14	0.25	0.36	1

Source: Santos personal communication, February 11, 1999.

Table 5.4 Posterior probabilities for the Coimbra female pubic symphysis sample

10-year age categories	*Stage 1*	*Stage 2*	*Stage 3*	*Stage 4*	*Stage 5*	*Stage 6*
15	0.98	0.45	0.15	0.00	0.00	0.00
25	0.02	0.38	0.47	0.21	0.05	0.00
35	0.00	0.17	0.24	0.45	0.24	0.05
45	0.00	0.00	0.10	0.13	0.21	0.17
55	0.00	0.00	0.04	0.06	0.14	0.23
65	0.00	0.00	0.00	0.15	0.04	0.30
75	0.00	0.00	0.00	0.00	0.24	0.11
85	0.00	0.00	0.00	0.00	0.07	0.15
Total	1	1	1	1	1	1

Table 5.5 Comparison of Coimbra female known age and age estimates (n = 100), derived by several different methods. The age-at-death distribution of the pubic symphysis reference sample is shown in the last column

10-year age categories	*Known age*	*Estimated from known age posteriors*	*Estimated from reference posteriors*	*Estimated from uniform posteriors*	*Estimated from model posteriors*	*Los Angeles reference age distribution (%)*
15	12	17	19	18	17	30
25	12	14	25	16	18	25
35	18	23	21	18	21	14
45	14	13	14	12	15	11
55	13	10	12	9	13	11
65	14	10	5	9	10	5
75	10	9	3	11	5	2
85	7	4	2	7	0	1
Mean age	51.6	46.6	40.7	47.9	43.4	37.1
DI*		12.00	22.36	11.00	15.80	31.60

*Dissimilarity Index expresses the sum of the absolute differences between two distributions, here the difference from the known age: the possible range as calculated here is 1–100.

fall both within the age category 15–24 *and* the Suchey–Brooks Stage 1 age indicator phase. And summing these multiplication products across will provide the distribution of ages at death (Table 5.5, the column headed "Estimated from known-age posteriors").

Table 5.5 demonstrates that no generated adult age distribution approximates the real adult age distribution. Our tests for Table 5.5 include using: (i) a distribution exactly like the real age; (ii) a distribution exactly like the Los Angeles forensic sample, biased toward younger adults; (iii) a test assuming a uniform adult age-at-death distribution; (iv) a test based on the assumption that the West 1 model adequately reflects the age-at-death distribution. Were we to know with absolute certainty the real age distribution, we could still not produce an acceptable estimate of the mean age at death. We can see that the mean age at death estimated from the reference data is much closer to the actual Los Angeles female sample mean age at death (given in the last column in Table 5.5) than it is to the known Coimbra mean age at death. The Coimbra age at death is fairly uniformly distributed across the age samples, so the uniform estimates perform best: the Coimbra sample was selected as a test of age-assessment methods and does not reflect a normal adult death assemblage. Of course, a uniform age-at-death distribution for adults is not a reasonable choice in data manipulation (di Bacco et al. 1999).

What happens with Spitalfields, a real distribution in which deaths are concentrated within the 45–74 age range? Table 5.6 records that the Dissimilarity Index is highest when the Los Angeles reference sample distribution is used: the Spitalfields female distribution has a mean age at death which is 20 years older than the Los Angeles reference sample. The assumption among osteologists in the past was that mean adult ages at death were young so that it does not matter if we use "seed

Table 5.6 Comparison of the known age Spitalfields female pubic symphysis (n = 56) with the estimates derived from several different methods

10-year age categories	*Known age*	*Estimated from known age posteriors*	*Estimated from reference posteriors*	*Estimated from uniform posteriors*	*Estimated from model posteriors*	*95% probability distribution*
15	4	6	7	6	6	9
25	3	7	17	11	13	12
35	4	7	12	11	13	12
45	13	13	8	7	8	10
55	11	8	7	5	8	7
65	12	9	3	6	6	3
75	5	4	1	5	3	1
85	4	2	1	4	0	0
Mean age	57.4	50.8	41.1	49.0	44.6	40.8
DI* from known age		16.1	44.6	40.8	36.5	41.5

*Dissimilarity Index expresses the sum of the absolute differences between two distributions, here the difference from the known age: the possible range as calculated here is 1–100.

data" for our reconstructions reflecting this young adult mean age at death (Weiss 1973 exemplified this and the assumption influenced tests of age-assessment methods, see Jackes 1992:196). The assumption that people rarely survived past 50 and that the adult mean age at death was very low becomes a self-fulfilling prophecy when it controls methods. Table 5.6 makes it clear that using a sample with a distribution similar to that of the Los Angeles reference sample which has a mean age at death of 37 (Table 5.5, last column), will give a mean age which is barely more accurate than that calculated from the 95 percent probability ranges of the reference sample indicator stage distribution. The 95 percent probability range will provide nothing more than a slight reduction of inaccuracy over using the mean age of each stage in the reference sample (Jackes 1992:198).

The bias of the reference sample age distribution and the consequent bias of the method derived from Los Angeles pubic morphology are reduced by this Bayesian method, but the correct ages cannot be arrived at, even when we have absolutely precise and accurate knowledge of the age distribution of the sample. Of course, we do NOT know the shape of the adult age distribution in archaeological samples. The only data we have, archaeologically, is the distribution of pubes across the Suchey–Brooks pubic symphyses indicator stages. The age estimation technique obviously exemplifies the frequently cited difficulties of under-aging of older individuals, resulting in a heaping of individuals in younger age classes. The pattern has recently been confirmed in a test on a sample of 390 known-age Sardinian pubes, resulting in "a shift in both sexes from slight overestimation of age to underestimating age after age 40. "Age predictions over age 60 are drastically underestimated by 25.2 years" (Hens et al. 2008:1041). The pattern of under-aging old adults and over-aging young adults was identified by Masset in his work on

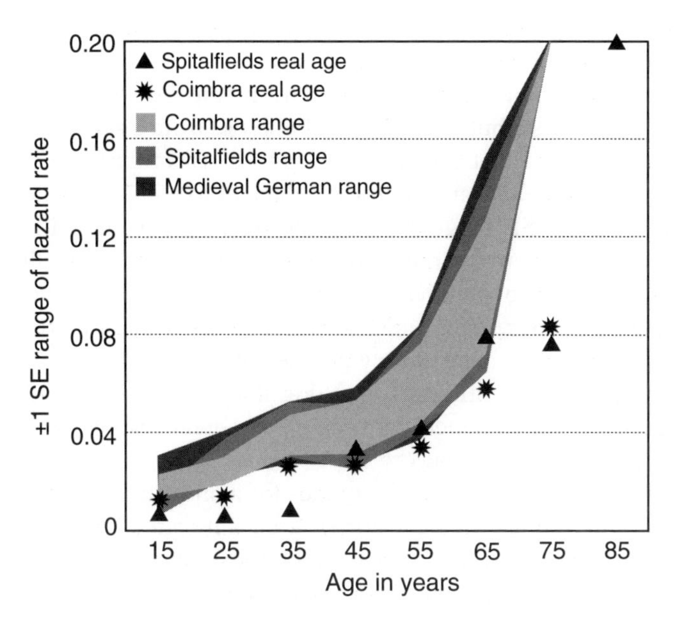

Figure 5.4 Comparison of the hazard function calculated directly from the known age distributions (Spitalfields triangles; Coimbra stars) with the Bayesian estimated standard error range of the hazard rate for Coimbra, Spitalfields, and medieval German archaeological samples. The estimated ranges are more or less identical and do not include the significantly different known-age distributions.
Source: Author's drawing

Portuguese skulls as "attraction of the middle" (see e.g., Masset 1989:82), and suggested that probability matrices would eliminate this error (Masset 1990:113), but he specified the need for standardization of age distributions.

It seems then that developing age-at-death distributions from age indicators and attempting to redistribute those distributions by choosing a model or historical example will lead to mortality patterns that reflect not the real age at death, but the methods of age assessment and, more importantly, the methods of data manipulation, specifically here the model data used. In Figure 5.4 we see that our known, very different, distributions and an unknown archaeological distribution will be rendered basically identical by the methods employed: this is illustrated by the standard error of the hazard rate. The hazard rate is derived from the probability of dying (q) and surviving within each age interval, taking the width of the age interval into account: $2q/n(1 + p)$, where n is the width of the age interval. In Figure 5.4 the actual known-age hazard functions for Spitalfields and Coimbra females are plotted and are clearly somewhat different from each other because the Coimbra sample has a fairly uniform probability distribution, while the Spitalfields distribution is weighted toward individuals aged 45 to 70. In both cases the *actual* known-age hazard rate falls outside the range estimated from model probabilities. But the further important point made by this diagram is not that the *actual* hazard functions do not lie within the SE ranges, but that the two SE ranges derived from West 1 r = 0 model adult age distribution are more or less identical, despite representing two quite different age-at-death distributions.

It is important to be clear about this. If we manipulate two completely different actual age-at-death distributions using probabilities from a specific mortality model, those two manipulated distributions will end up resembling the mortality model, not reality. If we now add archaeological data (Table 5.2) and derive the hazard rate expressed as the SE range by using the West 1 r = 0 model mortality profile, we will find the result to be the same (Figure 5.4). The mortality pattern from which we derive the probabilities will overwhelm the effect of the indicator stage data. In the past, we have been satisfied with age-assessment methods that did not indicate the presence of many elderly people within archaeological sites. Since the Bayesian priors are by definition something that we choose because they fit with preconceptions, we are in danger of forcing paleodemography into the mold of our ideas about the past, adding bias to bias.

To sum up, since the majority of methods for aging adults tend to heap individuals within certain younger age categories (well illustrated by Aykroyd et al. 1999: figure 1), it has been usual to envisage a high mortality among young adults. If the Bayesian approach of using what is considered most likely for the probabilities, and the researcher's best guess is similar to West 1 r = 0, then we would get a hazard rate consistent with that shown in Figure 5.4. Instead of age determinations controlled by the reference sample bias, we are adding methodological residue onto "the great unknown" of the actual age-at-death of an archaeological skeletal sample of adults.

As Boldsen et al. (2002:78) say, "any assumption about an informative prior can create a tautological circle". In that the estimated age distribution of the target sample will come to reflect or reproduce major features derived from exogenous data on mortality, Boldsen and his colleagues are correct. We will be imposing another layer of assumption on that already derived from the assumption of "invariance". We need a clear demonstration of the accuracy of bioarchaeological age indicator stage data manipulation before we can be certain of even an estimated adult mean age at death (Jackes 2000, 2003).

Taphonomy, macroscopic age indicators, and demography

Taphonomy refers to processes that act on organic matter after death. Taphonomical processes may lead to bias caused by differential preservation in a collection of bones. An important aspect of the biased representation of skeletal elements among sites is that age indicators may be unevenly represented, and this is particularly difficult to deal with when multifactorial age assessments are used. While not universally accepted (Saunders et al. 1992), most osteologists would probably agree that "... 'multifactorial' methods for age estimation are preferable to the use of a single ordinal categorical system" (Konigsberg et al. 2008:556). Unfortunately, what may be applicable in forensic situations, or in theoretical discussions on age assessment and paleodemography, can be very different in the situations bioarchaeologists confront.

Because preservation differs among sites, the methods used to age adults cannot be fully comparable across sites. For example, only about 50 percent of known-age individuals in the Spitalfields coffin burial collection could be examined for pubic symphysis age indicators (Jackes 2000). The lack of pubic symphyses can be even more extreme in collections made before modern excavation techniques came into common use. Excavation techniques and the value placed on different types of archaeological materials, leading to differential retention and curatorial care, no doubt explain why precisely one pubic symphysis is now to be found among the 19th-century collection from the Portuguese site of Cabeço da Arruda, despite the fact that over 70 individuals aged 15 and more were excavated (Jackes and Meiklejohn 2004). However, it may be no more dangerous to the accuracy of our interpretations to have no representation of a certain age indicator than to have partial representation.

A very clear example of the bias introduced by partial representation is provided by paleodemographic analyses of Ossossané, a fully excavated Huron ossuary from the first half of the 17th century which was studied for adult age distribution several times before the material was reburied. As originally analyzed (Katzenberg and White 1979) the demography was based on a sample size of 249. The right innominates were used, and although many more than 249 were counted, the life table was drawn up using only the innominates to which ages were given (see Jackes 1985 for summaries on methods). The $n = 249$ life table has been used in comparative studies with acknowledgment (Jackes 1986:35) or no acknowledgment (Bocquet-Appel and Naji 2006) that the study was preliminary. Jimenez and Melbye (1987) used the right mandibular P_3 socket to establish an MNI and arrived at a total of 419. Laroque (1991:241) gave the MNI based on mandibles as 447.

In order to show the effect of the differences in the three studies, we can specify the demographic estimator values (Table 5.7). As demonstrated above, this technique is of particular value in allowing comparison across data sets despite the use of different methods of analysis. Restricting the demographic analysis to those adults who could be aged by pubic indicators resulted in an extremely biased interpretation which we know from other sources was inaccurate: the Ossossané TFR estimate should probably be in the range of 4 to 5 (Jackes 1994, 2009a).

Recognition of this effect has been slow. As Wittwer-Backofen et al. (2008:385) say, "Differential preservation of skeletal elements is one of the most underrated confounders when it comes to age estimation within the bioarcheological context." Finally, we are beginning to get forceful statements such as "... when age markers are assessed independently, significant shifts in the resultant age structure can be observed. Both the number and selection of traits, and the state of preservation might impact the final paleodemographic reconstruction" (Wittwer-Backofen et al. 2008:394). This statement is, however, not forceful enough – "might" should be replaced with "will" in this quotation.

Table 5.7 Demographic estimators for Ossossané ossuary

Source for MNI and ages	*MCM*	*J:A*	*Estimated TFR*
Katzenberg and White 1979	.120	.283	9–10.7
Jimenez and Melbye 1987	.080	.118	~4.5–6.5
Larocque 1991	.053	.128	4.4–4.7

Diagenesis, microscopic aging techniques, and demography

We discussed above some of the problems relating to macroscopic techniques of adult age estimates. One aspect of taphonomical biasing is relevant to microscopic methods of age assessment. Diagenesis refers to specific changes undergone by bones and teeth after death. The changes can be chemical or they can be alteration of the microstructure. At one time there was intense interest in age assessment based on histological changes in bone cortex (Robling and Stout 2000 provide an excellent overview of publications from 1964 onwards). Since the standard deviation of the age prediction by regression is usually high (Maat et al. 2006), the method cannot give precise ages. With further testing, the evaluations of the technique have become much more critical (e.g., Lynnerup et al. 2006; Paine and Brenton 2006).

While forensic specialists see less hope of accuracy in ages at death, bioarchaeologists also see the limitations caused by bacterial destruction of cortical microstructure. Jans et al. (2004), in a large-scale study, found that 75 percent of human bone from a variety of contexts was biologically altered. Work in Portugal has shown that bone from sandy floored caves is much less likely to be altered than bone from limestone caves or from open-air sites (Jackes et al. 2001:418). Site-specific variability in the extent to which adults can be aged by examining cortical microstructure will again reduce the comparability of results across sites.

Since cortical bone diagenesis will not necessarily affect all individuals equally, there is the chance of compounding bias by differential destruction of older individuals' bones. Cortical width decreases and cortical porosity increases with age, especially in females (Cooper et al., 2007), making the elderly, and especially elderly females, more vulnerable to postmortem biological destruction of bone cortex. Add to this the under-aging of older adults by osteon counting (Walker et al., 1994) and it is unlikely that archaeological adult age-at-death distributions will be accurate or unbiased.

I have supported estimating age from microscopic examination of tooth roots using cemental annulation counts for the purpose of calibrating attrition levels in archaeological dentitions, specifically from Casa da Moura, a Neolithic Portuguese site (Jackes 1992, 2000). However, I do not believe that cross-sectioning tooth roots will provide accurate assessments of the age of all individuals in a sample. In my experience, even the roots of beautifully preserved and almost unworn large male upper canines may be unreadable because of destruction by bacteria. My experience is not unique: two dedicated laboratories were unable to estimate ages for over 50 percent of the individuals they had sampled (Geusa et al. 1998, data at section 5.2) from among people buried at the Isola Sacra necropolis, Ostia Antica (second and third centuries A.D.). Individuals assessed as under 20 years of age had the highest level of unread cemental annulations. Furthermore, the age estimates differed by up to 29 years between the two laboratories (Figure 5.5), one using phase contrast microscopy (Munich) and the other complex digital image processing (Rome). Wittwer-Backofen et al. (2008:388) have also noted "surprising" differences and "unexpectedly low [consistency]" in cemental annulation counts.

The lack of consistency in counting illustrated by Figure 5.5 is not surprising based on my experience using thin sections, 25–30 micrometers thick, generally

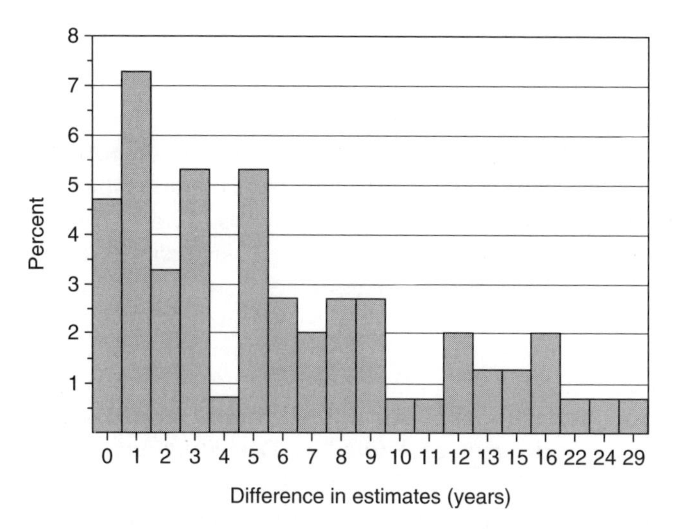

Figure 5.5 Data on cemental annulation analyses of 150 individuals from Isola Sacra necropolis, Ostia Antica, Italy. Of the 150 sectioned tooth roots, 53.3 percent were deemed unreadable by one or both laboratories. Age estimates based on range midpoints.
Source: Author's drawing, derived from Geusa, G., L. Bondioli, E. Capucci, A. Cipriano, G. Grupe, C. Savorè, and R. Macchiarelli 1998 Osteodental Biology of the People of Portus Romae (Necropolis of Isola Sacra, 2nd-3rd Cent. AD). II. Dental Cementum Annulations and Age at Death Estimates Digital Archives of Human Paleobiology. Rome: Pigorini Museum

from the juncture of the middle and lower thirds of each canine root. The problem lies not simply in visualizing the annulations. Renz and Radlanski (2006) have noted that the number of lines differs according to the portion of the root examined, along the whole of the middle third of the root, as well as according to the root face. We had multiple sections from the same region of each root, oriented so that we could specify which face of the root was being examined. There is absolutely no doubt that root faces for the one tooth can differ in the number of countable annulations. Since cementum disruption occurs with heavy attrition and cemental annulation analysis is likely to underestimate the age of old people (Meinl et al. 2008), annulation counting appears to have many problems, but a major drawback for bioarchaeologists is that many sections will be unreadable.

To summarize the argument to this point, it is important that we gain as accurate an idea as possible of the numbers of individuals in archaeological burial sites. We must base our paleodemographic studies on counts of *all* individuals because our results are not comparable across different archaeological samples if we use only those individuals who can be aged by particular techniques or combinations of techniques. Taphonomical factors on the macroscopic level are of the greatest importance in our analyses and we must acknowledge this in order to avoid biasing our findings as a result of unrecognized taphonomical factors. Furthermore, the diagenesis affecting cortical bone and dental cementum means that we cannot rely on microscopic methods in our efforts to gain a full picture of the age-at-death distribution of a cemetery sample any more than we can rely on gross morphological age assessment. In both cases we need to be aware of the ways in which postmortem changes will bias our samples.

Taphonomical bias has implications beyond questions of minimum numbers of individuals and age-at-death distributions. We need to examine how to study age-related characteristics. Dental pathology is an especially important aspect of age-related change and I will focus on the analysis of dentitions, illustrating methods and problems. Jackes and Lubell (1996) and Jackes (2009b) discuss some specific factors, including postmortem alterations, which bias caries assessments and those issues will not be touched on in this chapter.

Teeth, Attrition, and Seriation: An Experimental Approach to Inclusive and Comparable Estimates of Adult Age Distributions

How then can we avoid biasing our data by relying on particular methods of adult age assessment and by excluding those to whom we cannot give ages by these methods? My own method, which I have applied whenever possible since 1984, depends on seriation of mandibles based on examination of attrition, with attention primarily directed to attrition of the posterior teeth. Two European Mesolithic sites, Moita do Sebastião and Cabeço da Arruda in Portugal, exemplify the value of using mandibles to provide a proxy for adult age at death distributions comparable across skeletal collections. The Mesolithic material was first excavated in the 1880s and cannot be dealt with entirely as single burials for a variety of reasons: burial practices led to mixed burials; apparently not all bones were removed at excavation; retained material was curated in two (eventually three) locations, in some cases with post-cranials and skulls being located apart; some material was lost early on; material at two institutions went through fires; skeletons were curated in adjacent drawers without individual skeletal elements being marked for identification so that bones were transferred among individuals (Jackes and Meiklejohn 2004, 2008; Jackes and Alvim 2006).[5]

The minimum number of individuals at Moita and Arruda was determined on the seriation of the mandibles. This approach permits the reconstruction of fragments that are broken and separated, it makes it possible to recognize which individuals are represented by isolated fragments, and thus gives a higher MNI than if individual skeletal elements are simply counted (for example, right distal humeri or left petrous portion of temporal or even one particular tooth socket). A number of elements were recorded, but none gave as high an MNI as mandibular seriation. Seriation can also be used to develop proxy age-at-death distributions for adults who originate from different types of inhumations. The total information available is, of course, much more complete when derived from single articulated skeletons, but information from a variety of burial practices can be used when the methods are comparable. Thus, one can compare Mesolithic inhumation data with Neolithic disarticulated skeletons laid on cave floors such as occurred at Casa da Moura (a Neolithic ossuary cave in central Portugal).

If each study can be taken only in isolation and we cannot compare sets of data, then we are unable to test methods, to extrapolate from our results and to study the human past across changes of time, subsistence, geography, disease patterns, and cultures. "… heterogeneity is itself a major focus of bioarchaeological research" note Wright and Yoder (2003:46), but if we cannot ensure comparability of methods we cannot identify similarities and differences and our conclusions will have little meaning. This is especially important in regard to our study of age-related change and dental pathology is a central aspect of age-related change.

Dental pathology: a study in age-related characteristics

In asking whether our osteological samples are representative we have so far ignored a fundamental question raised at the beginning of this chapter: "are the dead an unbiased sample of the living?" In terms of age, the answer is obvious – no, the dead are dead, they did not survive beyond a certain age. Further, in cases like accident and violence the dead are dead because they were more at risk of death, or the dead may have suffered deprivation or illness, or perhaps they were frailer because they had suffered but survived such insults in the past. We will return to this question below, but in order to understand it fully we need to consider the types of evidence that might demonstrate that those who die are less healthy than those who survive. And we need to consider the contexts which would allow us to arrive at a clear interpretation of skeletal changes.

An obvious approach is to study skeletons in contexts in which we hold at least some of the variables constant – geological background or the gene pool, for example – while varying others, such as diet. In other cases, we could look at a similar diet, but in different locations, or observe change through time that is not mediated by an alteration of diet. The change in subsistence brought about by a growing dependence on horticulture is an ideal test case: Cohen and Armelagos (1984) asked contributors to their volume on the agricultural transition to examine specific features of skeletons. They concluded that the contributions allowed them to discern "clear trends … which have a significant bearing on discussions of comparative health … associated with the Neolithic Revolution" (1984:585), specifically (1984:594) that the "indicators fairly clearly suggest an overall decline in the quality … of human life associated with the adoption of agriculture".

A new volume (Cohen and Crane-Kramer 2007), 20 years on, is a reprise of this test case. The book starts with a retrospective review by Cook (2007) who, in making a strong case for the importance of local and regional studies rather than broad comparisons, questions the use of specific categories of data (e.g., stature, periosteal reactions) apart from dental caries, as contributing much to studies of subsistence change.

By focusing on caries we can also partly avoid the difficulties which stem from one undoubted fact: as Wood et al. (1992:344) said, "… the only 20-year-olds we observe in a skeletal sample are those who died at age 20." How are we to interpret what we see? Are we to interpret the presence of skeletal signs of nutritional or infectious insults as evidence of a population healthy enough to survive or of a population in poor health?

The discussion around the osteological paradox has centered on the transition to horticulture because Cohen and Armelagos (1984) argued that an increase in skeletal lesions indicates an increase in ill-health with agriculture, whereas Wood et al. (1992) would argue that people were able to survive ill-health for long enough to allow skeletal responses to infection to manifest themselves. As Cohen et al. (1994:630) ask are we seeing "relative stress … [or] … relative resilience"?

The question here is "are deaths random or selected with regard to dental pathology?" The cumulative nature of dental pathology must be emphasized. Conditions that accumulate into old age will occur at a higher rate among the dead than among the living population simply because older adults have a higher probability of dying than younger adults. So it is likely that, as samples, the dead have a higher rate of dental pathology than the living. However, it may be that dental pathology rates themselves are not entirely random. It is well known that caries is multifactorial and that heritability is quite high (Bretz et al. 2005; Slayton et al. 2005; Vieira et al. 2008). While environmental and genetic factors must be involved, an interesting outcome of the Dunedin Multidisplinary Health and Development Study is that carious lesions develop in teeth at a relatively constant rate over a person's lifetime (Broadbent et al. 2008:72). Dental caries is "chronic [and] cumulative" (Broadbent et al. 2008:69).

Figure 5.6A shows us the pathology rates for the two Portuguese Mesolithic sites of Moita do Sebastião and Cabeço da Arruda based on seriated mandibles of individuals aged 15 and older. The category (x) axis gives a distribution of lower molar sockets across ten groupings, each with a more or less equal representation of observable sockets, around 25 for Moita and 30 for Arruda. The rate of dental pathology is higher in Moita than in Arruda dentitions (also demonstrated by attrition grade analyses: Jackes and Lubell 1996; Meiklejohn and Zvelebil 1991). Moita has pathology throughout late adolescence and young adulthood, while Arruda has much less, and an early peak at category 5 is wholly contributed by two Arruda individuals, both apparently male. In both samples dental pathology increases over the life time of the individuals, but in Moita it is more common in very old age (when the dental crowns have been worn to the cemento-enamel junction). There are so few teeth in which to observe caries at this stage that it is not possible to prove a significant difference (two caries in four lower molars for Moita and three in 16 for Arruda), but for both abscessing ($P = .022$) and pre-mortem tooth loss ($P = .008$) dental pathology is significantly more severe at Moita in category 10.

In these two sites, we have a paradoxical situation. Dental pathology, in the form of caries, is believed to have an inverse relationship to dental attrition. In the case of Moita, however, Figure 5.6B demonstrates that caries is established very early in the late adolescent/young adult dentition of a group with a high rate of attrition. For this reason, not all of the abscessing and tooth loss can be attributed to heavy attrition – caries is obviously an important contributing factor. Meiklejohn et al. (1992) have independently discussed the fact that Moita attrition and caries do not have the expected inverse relationship. They did not, however, refer to a further paradoxical situation: Moita has a higher rate of occlusal caries than Arruda (Jackes 2009b; Jackes and Lubell 1996), whereas the expectation is that heavy attrition reduces the likelihood of occlusal caries developing. In this example of analysis, we can see that it is possible to compare across samples and to confirm differences, despite the uncertainties of age assessment in archaeological material. This study is valuable in showing that dental pathology may be a marker of dietary differences

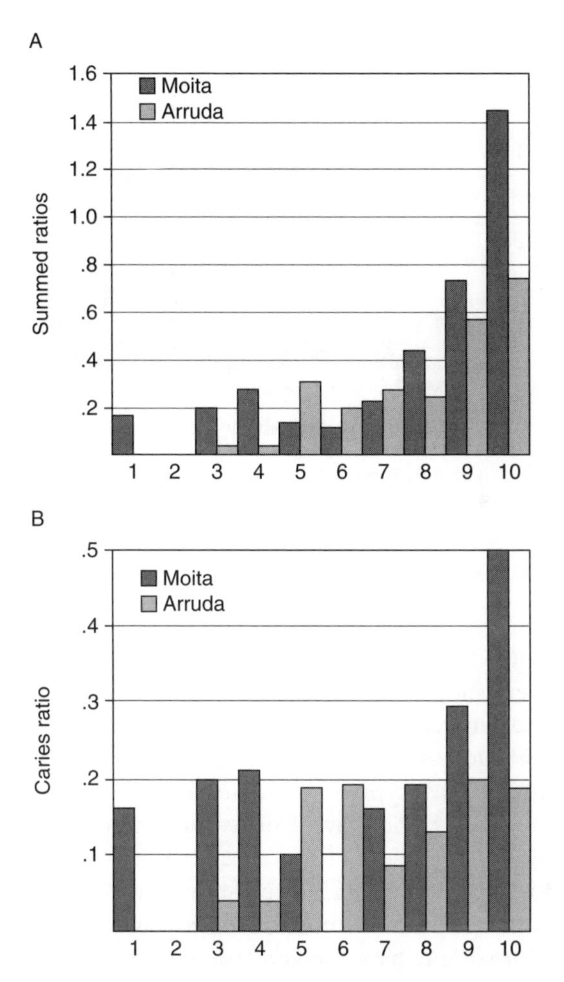

Figure 5.6 Comparison of dental pathology in two Mesolithic shell midden burial sites in central Portugal. The seriated mandibles are equally distributed across the ten categories based on representation of molar sockets. **A** The frequency of caries, of abscessing, and of pre-mortem tooth loss summed to represent pathology rate on the *y* axis. **B** The caries rate (*y* axis) – number of teeth with caries as a proportion of number of intact teeth in each category.
Source: Author's drawing

(Jackes and Meiklejohn 2008) when other variables are held relatively constant. The higher rate of pathology occurs at Moita not simply as a result of attrition and despite the possibility that a larger proportion of people survived into old age at Arruda (Jackes and Lubell 1999; Jackes 2009b).

Moita and Arruda have an overlapping time range and lie within a few kilometers of each other on either side of what was a narrow estuary in the late Mesolithic. There is no particular reason that they should differ from each other, but as in other features (Jackes and Lubell, 1999), there is evidence indicating that those living at Moita had higher rates in dental pathology than those living at Arruda and that the difference is real, not the product of bias in the cemetery samples.

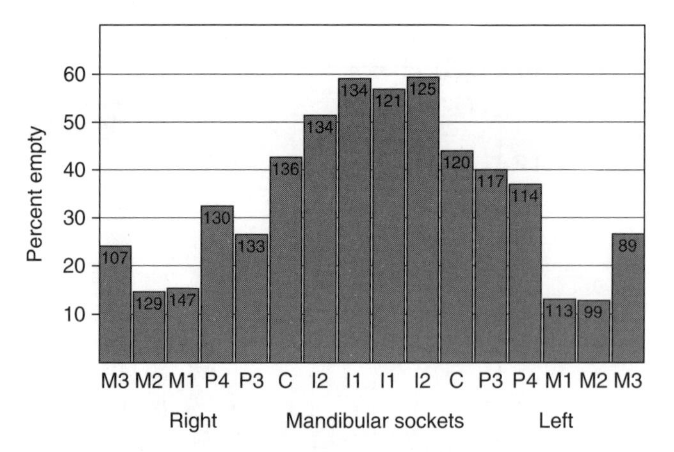

Figure 5.7 Percentage of Casa da Moura mandibular sockets empty due to postmortem tooth loss. Figures in bars refer to sample sizes for each tooth socket (highest n = 147).
Source: Author's drawing

The discussion above was based upon mandibles and specifically mandibular molars. There is a very simple reason for this – mandibles, mandibular sockets, and mandibular teeth are generally better represented than their maxillary counterparts (Jackes and Lubell 1996). And molars and molar sockets are the best preserved elements of mandibles. Of the preserved mandibular sockets at Casa da Moura, the Neolithic Portuguese burial cave from which disarticulated skeletons were excavated in the 1860s, 35 percent of sockets are empty, meaning that they contain no intact teeth and show no evidence of pre-mortem tooth loss. Figure 5.7 demonstrates that the incidence of postmortem tooth loss is very uneven across tooth types and that incisor sockets have the greatest chance of being empty. The simple explanation is that their small, straight, and short roots do not provide firm attachment, whereas the multiple and sometimes curved roots of lower first and second molars result in a higher retention rate after burial. The sample sizes are given for each of the sockets and it is immediately clear that they are higher on the right, the mandibular first molar socket being very much more likely to survive (the side difference is significantly different, $P = .001$). The explanation for this might be that skulls were more often laid on the cave floor on the right side, so that there would be less damage to the right mandibular rami from people moving or rearranging skeletons or from animals disturbing the bones.

The preservation of mandibular sockets and molars is general across most archaeological sites (Jackes and Lubell 1996) and a good example of this is provided by an Ontario ossuary, Kleinburg, carefully and fully excavated, curated, and analyzed. In this site, as recorded ethnohistorically for the Huron Nation (Jackes 2009a), primary burial was followed by secondary burial in such a way that one might expect full representation of every skeletal element. And yet, as is evident from Figure 5.8A, the mandibles will provide a more complete picture than the maxillae of all aspects of the dentition, of the MNI and the age distribution, of pathology and metrical and nonmetrical characteristics. Figure 5.8B, showing the

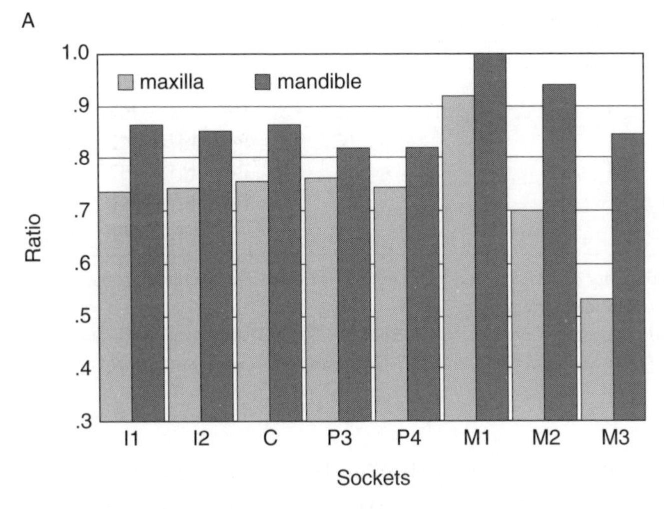

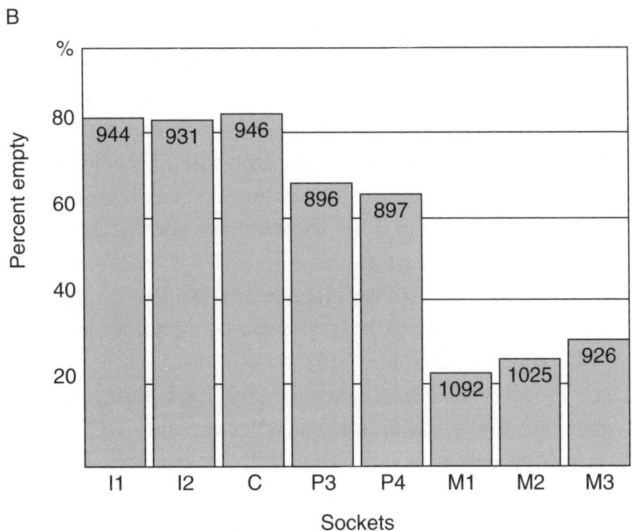

Figure 5.8 Kleinburg ossuary, Ontario. **A** Representation of permanent tooth sockets, including congenital absence, pre-mortem and postmortem tooth loss, as a ratio of the socket with highest presence (pooled sides M_1 socket, n = 1,092). **B** Percentage of permanent mandibular tooth sockets empty due to postmortem tooth loss (pooled sides mandibular sockets, n = 7,657).
Source: Author's drawing, data from Patterson, D. K. 1984 Diachronic Study of Dental Palaeopathology and Attritional Status of Prehistoric Ontario Pre-Iroquois and Iroquois Populations. Mercury Series No. 122, Archaeological Survey of Canada. Ottawa: National Museum of Man

percentage of empty mandibular sockets, demonstrates that the first and second lower molars are most likely to be retained in their sockets, as with the Portuguese sites (Jackes and Lubell 1996). Since they are also the most likely to be preserved among the loose teeth, analyses are best focused on mandibular molars for comparative studies.

Table 5.8 Summed probabilities of caries by age 16 in mandibular teeth

Mandibular teeth used in analysis														*P/n*	*n*	*P/s*	*ns*
M2	M1	P4	P3	C	I2	I1	I1	I2	C	P3	P4	M1	M2	.149	14	.033	64
M2	M1	P4	P3	C					C	P3	P4	M1	M2	.204	10	.043	48
M2	M1	P4	P3							P3	P4	M1	M2	.254	8	.051	40
M2	M1	P4									P4	M1	M2	.331	6	.066	30
M2	M1											M1	M2	.432	4	.086	20

P/n = summed probability of caries by teeth; *n* = number of teeth; *P/s* = summed probability of caries by surfaces; *ns* = number of susceptible surfaces.
Source: Calculated from Batchelor, P. A., and A. Sheiham 2004 Grouping of Tooth Surfaces by Susceptibility to Caries: A Study in 5–16-year-old Children. BMC Oral Health 4:2, www.biomedcentral.com/1472-6831/4/2.

It is important that dental pathology be studied in a careful and limited way because taphonomy biases dental samples in archaeological sites and the probabilities of caries differ across tooth classes (Batchelor and Sheiham 2004). The information given in Table 5.8 is for mandibular teeth, which have a higher probability of caries than maxillary teeth. The summed probability of caries across the five crown surfaces (four for canines and incisors which are not considered to have susceptible occlusal surfaces) is for U.S. children aged 5 to 16. Whether the probabilities for caries per tooth representation group are expressed per tooth or calculated on crown surfaces (five, and four for anterior teeth) the difference between the full set of mandibular teeth or the molars only is very clear. Caries rates are clearly shown to depend on the representation of tooth types in the sample. Therefore it is essential that caries be reported by tooth type or that tooth types included in analyses be specified.

To summarize, samples will be biased in different ways, as a result of burial practices and other factors such as excavation care and screening, conditions of curation and laboratory methods and methods of analysis, but taphonomy is of utmost importance in the reporting of dental pathology (Jackes and Lubell 1996; Jackes 2009b). Granted that mandibular sockets and teeth are most likely to be represented and that molar teeth are the most likely of those mandibular teeth to be well represented in a sample, it is best to present dental pathology for mandibular molars. Comparisons can only be made on a limited basis and in fact should be presented in association with seriation by wear. Since dental wear is specific to samples (even to individuals) this is not a perfect solution, especially since seriation becomes more difficult to apply as rates of dental pathology increase. Nevertheless, seriation of mandibles provides a means of distributing individuals by an age-dependent characteristic. It is of great importance that dental pathology be examined within the context of age and the composition of the sample in terms of age will be an important consideration for dental pathology.

A graphic illustration is provided by two Ontario sites already mentioned, the Neutral cemetery at Grimsby and the Huron ossuary at Kleinburg. It was working on these two sites – one just pre-contact and one post-contact, dated to within decades of each other and about 78 kilometers apart as the crow flies – that first introduced me to the possibility of methodological problems with adult age assess-

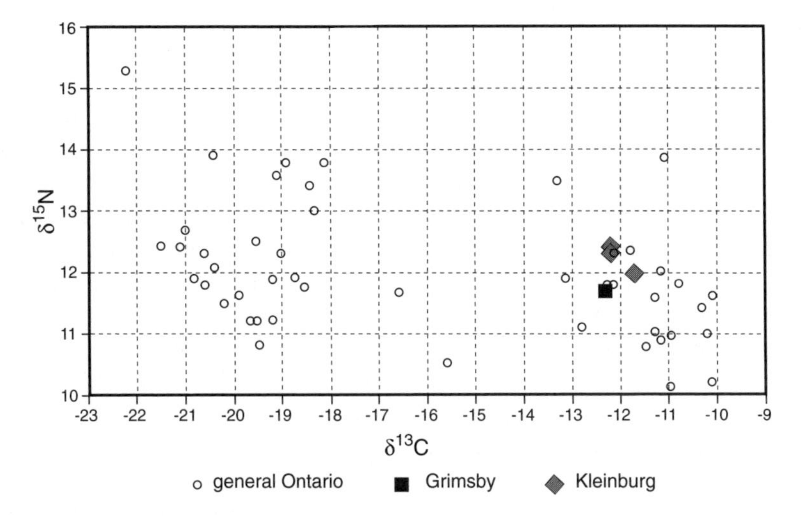

Figure 5.9 Stable isotope values for Ontario sites showing the placement of Grimsby within the data set of Harrison and Katzenberg (2003).
Source: Author's drawing, based on Harrison, R. G., and M. A. Katzenberg 2003 Paleodiet Studies using Stable Carbon Isotopes from Bone Apatite and Collagen: Examples from Southern Ontario and San Nicolas Island, California. Journal of Anthropological Archaeology 22:227–244

ment and, just as important, the possibility that samples may not be directly comparable. One problem was that there seemed to be a great deal more dental pathology at Kleinburg than at Grimsby. For example, examination of the permanent mandibular second molars for occlusal caries alone demonstrated that the difference was as extreme as 68 percent for Kleinburg and 34 percent for Grimsby. The Grimsby skeletons had been excavated as a salvage operation and had to be reburied within three months of the start of the analysis, so there seemed little chance of explaining the discrepant results by more detailed studies. One possibility was that the stable isotope values were very different: fortunately, it was possible to analyze one fragment of skull that had inadvertently escaped reburial. This analysis demonstrated that dietary differences were probably not a major factor (Figure 5.9) and it seemed likely that the explanation lay in the age-at-death distribution of the sample (Jackes 1988:125, table 116). The ratio of adults over 25 years of age to the total sample of those over 10 years of age differed remarkably between the two sites. The full explanation for this had to wait until comparative methods were developed and fully tested (Jackes 2009a). As discussed above, the Grimsby sample was highly biased in terms of age and the caries rate is not comparable with Kleinburg because of sample differences.

Do the Dead Constitute a Biased Sample? A Different Dental Perspective

We return now to the problem posed above: "Are the dead an unbiased sample of the living?" Here we are not comparing the age structure of the dead and living populations. Since "[t]he dying, on average are less healthy than the rest of the living" (DeWitte and Wood 2008:1436), we are again asking about features considered to be "skeletal indicators of health" (e.g., Steckel and Rose 2002). The "osteological paradox" introduces the concept that deaths are not random with regard to individual differences. Individual differences could relate to variations in allele frequencies relative to certain diseases, but the discussion in osteology is specifically related to the idea that some individuals are more "frail" and that these individuals can be identified in a cemetery sample by skeletal indicators of frailty. DeWitte and Wood (2008) have recently attempted to test skeletal indicators by comparing two samples of skeletons, one illustrating catastrophic mortality from the Black Death and the other, normal slow death rates from the same period. Their results indicate that linear enamel hypoplasia (LEH) of the mandibular canine most clearly indicates frailty (elevated risk of death) in normal attritional and in epidemic mortality situations (with an extremely high standard error, however, especially in regard to attritional mortality). They note that one underlying cause of the presence of the skeletal indicators and of the excess mortality associated with them is likely to be poor nutrition. We might question the assumption that a London plague pit is free of the bias introduced by the immigration of young rural people, (especially males?) and question the comparison of London with two Danish towns, still small in the 21st century. Nevertheless, we may take away the conclusion that the presence among the dead of skeletal signs of frailty indicates, at the very least, that some members of a group were subject to some type of insult, with malnutrition being one possibility.

While DeWitte and Wood (2008) obviously accept that most skeletal indicators are in fact indicators that individuals had the strength to survive past insult, they conclude that those who have survived are at increased risk of death, especially under conditions of normal attritional death. This would suggest that there is excess mortality (at younger ages) of those who show skeletal stress markers. Can we say something about the rate of stressors (in a homogeneous, unbiased sample of sufficient size), given the possibility of higher rates of mortality among survivors of stress?

We can assume from DeWitte and Wood's study that the skeletal indicator most clearly associated with increased likelihood of death is linear enamel hypoplasia (LEH). Is it possible to study LEH in older adults? In the sample of Portuguese Neolithic loose teeth from Casa da Moura, LEH might be best recorded only in younger adults because rapid wear removes crowns (Figure 5.10) and wear affects the labial and lingual surfaces as well as the occlusal surfaces (anterior teeth must have been used as tools). LEH could be worn away and unobservable unless it is very marked. However, in a first analysis of wear, 625 loose upper and lower canines were examined, and LEH that could be fitted into categories of "slight lines/pitting" or "deep lines/lines of pits" was recorded. The analysis demonstrated that although "slight" LEH increases in frequency across wear levels, "deep" LEH is also observed at higher wear levels. This was confirmed in a reanalysis of wear levels of 615 canines, in which both "slight" (31 percent) and "deep" (8 percent) LEH reached its highest frequency at wear level 4, but was very little lower at wear level 5. By wear level 6, "deep" LEH is reduced in frequency. This raises the question: does LEH in fact have any correlation with mortality at Casa da Moura? It seems more

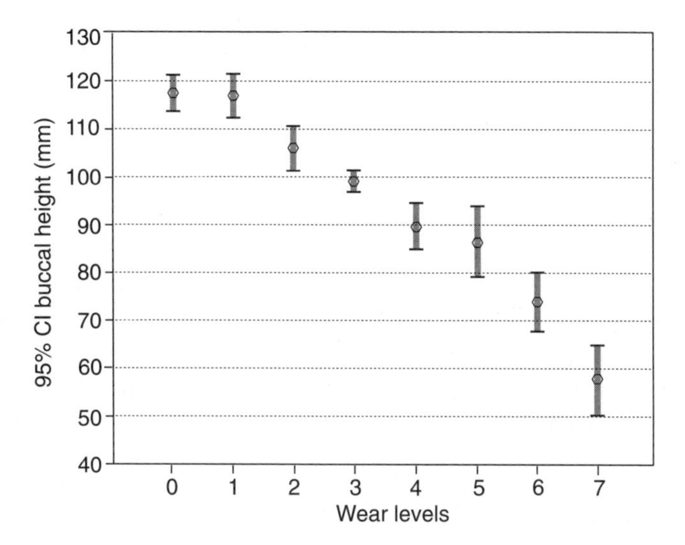

Figure 5.10 Casa da Moura mandibular canine buccal crown heights (n = 219, sides pooled).
Source: Author's drawing

likely that there is no selective mortality of younger individuals exhibiting LEH: mortality seems fairly neutral with regard to the presence of LEH. Across the two separate studies of Casa da Moura canine wear levels, about 30 percent display LEH; 25 percent and below for wear levels 0–2; 30 percent for 3 and 35 percent and above for 4–6; under 25 percent again for wear level 7; none in 8 (the crown being more or less completely worn away).

In summary, DeWitte and Wood's study suggests excess mortality among those who have suffered but survived past insult indicated by the presence of LEH. While we could postulate a bias in favor of LEH as observable in younger skeletons with higher rates than in older skeletons, this seems not to be true in one site with quite high wear levels. Since Casa da Moura is a loose tooth ossuary sample, we could not, of course, examine individuals, but we can study individual elements, right upper canines, for example (upper canines are used because the mandibular canine sample size is much lower). The mean percentage across sides of LEH is 30 percent in *unworn* upper canines, but lower for early wear levels. The mean percentage across sides reverts to 30 percent for upper canines in each of the second analysis wear categories 3 to 6. Thus, for Casa da Moura upper canines, we might expect about 30 percent of the population to have LEH, that slightly fewer late adolescents and young adults with LEH die, and that generally deaths are not selected for by whatever causes LEH. Whatever is the cause of lines or lines of pits appearing in developing upper canine labial enamel does not seem to predispose individuals to early death.

We have illustrated our discussion on whether the dead themselves represent a biased sample (beyond the age structure of the dead) on two different types of dental data: dental pathology, cumulative through life and likely to appear at higher rates among the oldest of the dead; LEH, established early in life and likely to have disappeared from the worn teeth of the oldest of the dead. In each case, the suggestion is that we have gained some insight into the incidences of these dental features, not only among the dead, but also among the living.

Summary: Not Just One Type of Bias …

Are bioarchaeological samples of human burials biased? Yes. They are biased by being representative of the dead, not the living. Are the dead biased? In one sense, no, because everyone dies. But when, why, how, and where each person dies is another matter. And, as we have seen, that presents us with difficult issues that really cannot be solved universally.

Bioarchaeologists face multiple problems. We must consider the burial practices, the determinants of differential preservation, the excavation history, the post-excavation curation for each of our samples. But above all, we must try to understand the context and the demographic characteristics of the population that has provided our sample. Our only hope of arriving at an understanding of all these factors is to use methods which will allow detailed and careful comparisons among samples that *are* comparable. Samples must be comparable, not only in terms of chronology and geography, but in terms of size and constituents, so that meaningful interpretations can be drawn from them.

We must also be careful not to use methods which could add bias in such a way that our results cannot have real meaning. In the 1980s it was a struggle to get anything published that pointed out that skeletal age assessment added methodological bias. Progress has been made, but we still see analysis of age-dependent factors by our methodologically biased age categories. We cannot ignore the circularity of categorizing skeletons on age-dependent factors and then analyzing those same age-dependent factors by the derived age categories. We need to choose our age-proxy characteristics carefully. In the 1980s it was almost unheard of for anyone to suggest that some skeletal samples might be quite unsuitable for demographic analysis. Even raising that question almost ensured that publication referees would be very negative in their assessments. Today published reports on age-dependent features still draw conclusions from apparent differences without any acknowledgment that sample characteristics may determine those differences. Dental pathology is still reported as crude rates without reference to the age structure of the sample or differential preservation of tooth types, let alone the myriad other factors that bear on teeth throughout the human life span. Papers are published which draw sweeping conclusions encompassing thousands of years and continent-wide geographic spreads based on flawed samples.

We have problems enough in trying to deal with the unknown age distribution of the dead, without adding methodological bias deriving from faulty techniques of age assessment or preconceptions of past populations. Interpreting differences among samples is difficult enough, without muddying our results by refusing to examine closely the nature of those samples. It is only by constant testing of methodologies and assumptions that we will improve our methods and our understanding of the past.

ACKNOWLEDGMENTS

I am grateful to the editors for inviting my contribution and for their helpful comments. I thank David Lubell for assistance with my research and Chris Meiklejohn, who as my collaborator in our work on Portuguese Mesolithic sites, is always willing to provide information. Dr. M. M. Ramalho, Serviços Geológicos, Lisbon, Portugal, authorized access to Portuguese Mesolithic and Neolithic material. Work on Grimsby derived from the excavation by the late Walter Kenyon of the Royal Ontario Museum. My knowledge and understanding of Huron ossuaries was filtered through Jerry Melbye's excavation at Kleinburg and years of discussions with Susan Pfeiffer while we worked on the Kleinburg bones.

NOTES

1 The Hutterites studied by Eaton and Mayer (1954) practiced universal marriage and no contraception of any type in the first half of the 20th century. The study led to the proposal that the maximum TFR for human beings would be between 12 and 14. This is not the biologically possible maximum fertility for individuals but it is well above recorded total fertility rates. Hern (1977:363) records the highest known TFR of 9.94 in the Shipibo village of Paococha in the Peruvian Amazon in the 1960s resulting from contact with Western influences and declining rates of polygyny. In Figure 5.3 the star marks the theoretically high TFR estimate of 12.5 derived from J:A ~.380 and MCM ~.135.
2 Bocquet-Appel and Naji (2006) appear to have used the same data as is in the HNWH database for Irene Mound (n = 170). The Irene Mound data as presented by Russell et al. (1990:43, table 3–4, n = 142) is also problematic, with a TFR range of 8.1 to 19.8 (see Figure 5.3 for both data points). A total of 280 individuals was excavated from the site (Powell 1990). Tremaine (Vradenburg 1999), another site used by Bocquet-Appel and Naji (2006), is very like Irene Mound with a TFR range of 8.6 to 18.4: 25 percent of individuals were given ages from 15 to 19 (see Figure 5.3).
3 Note that Bocquet-Appel and Naji (2006) have an error in the figures for Schild Late Woodland 5–19 age group: P should be .352, not .253 (Droessler 1981). Kuhlman Mounds P = .410 appears to be an error resulting from the 15–24 age group (Garner 1991) being treated as 15–19. P should be .303.
4 Where weaning is delayed, lactational amenorrhea will be a reasonably effective deterrent to conception well beyond six months postpartum (up to 97 percent effective at 12 months – Ramos et al. 1996), even without nursing abstinence.
5 Stodder (2008) includes what happens to bones after excavation within her broad-ranging study of taphonomy.

REFERENCES

Angel, J. L. 1971 The People of Lerna: Analysis of a Prehistoric Aegean Population. American School of Classical Studies at Athens. Washington, DC: Smithsonian.

Atwell, K. A. 1991 Paleodemographic Analysis. *In* Kuhlman Mound Group and Late Woodland Mortuary Behavior in the Mississippi River Valley of West-central Illinois. K. A. Atwell and M. D. Conner, eds. Pp. 164–179. Kampsville, IL: Center for American Archeology.

Aykroyd, R. G., D. Lucy, A. M. Pollard, and C.A. Roberts 1999 Nasty, Brutish, But Not Necessarily Short: A Reconsideration of the Statistical Methods Used to Calculate Age at Death from Adult Human Skeletal and Dental Age Indicators. American Antiquity 64:55–70.

Bacco, M. di, V. Ardito and E. Pacciani 1999 Age-at-Death Diagnosis and Age-at-Death Distribution Estimate: Two Different Problems with Many Aspects in Common. International Journal of Anthropology 14:161–169.

Batchelor, P. A., and A. Sheiham 2004 Grouping of Tooth Surfaces by Susceptibility to Caries: A Study in 5–16-year-old Children. BMC Oral Health 4:2, www.biomedcentral.com/1472-6831/4/2.

Bocquet-Appel, J.-P. 1986 Once Upon a Time: Palaeodemography. *In* Innovative Trends in der Prähistorischen Anthropologie. Bernd Herrmann, ed. Pp. 127–133, Mitteilungen der Berliner Gesellschaft für Anthropologie, Ethnologie und Urgeschichte No. 7.

Bocquet-Appel, J.-P., and C. Masset 1977 Estimateurs en Paléodémographie. L'Homme 17:65–90.

Bocquet-Appel, J.-P., and S. Naji 2006 Testing the Hypothesis of a Worldwide Neolithic Demographic Transition: Corroboration from American Cemeteries. Current Anthropology 47:341–365.

Boldsen, J. L., G. R. Milner, L. W. Konigsberg, and J. W. Wood 2002 Transition Analysis: A New Method for Estimating Age-Indicator Methods. *In* Paleodemography Age Distributions from Skeletal Samples. R. D. Hoppa, and J. W. Vaupel, eds. Pp. 73–106. Cambridge Studies in Biological and Evolutionary Anthropology No. 31. Cambridge University Press.

Braun, D. P. 1981 A Critique of Some Recent North American Mortuary Studies. American Antiquity 48:398–416.

Bretz, W. A., P. M. Corby, N .J. Schork, M. T. Robinson, M. Coelho, S. Costa et al. 2005 Longitudinal Analysis of Heritability for Dental Caries Traits. Journal of Dental Research 84:1047–1051.

Broadbent, J. M., W. M. Thomson, and R. J. Poulton 2008 Trajectory Patterns of Dental Caries Experience in the Permanent Dentition to the Fourth Decade of Life. Journal of Dental Research. 87:69–72.

Brooks, S. T., and J. M. Suchey 1990 Skeletal Age Determination Based on the Os Pubis: A Comparison of the Acsadi-Nemeskeri and Suchey-Brooks Methods. Human Evolution 5:227–238.

Brown, K. H. 2003 Symposium: Nutrition and Infection, Prologue and Progress Since 1968 Diarrhea and Malnutrition. Journal of Nutrition 133:328S–332S.

Buikstra, J. E., and G. R. Milner 1991 Isotopic and Archaeological Interpretations of Diet in the Central Mississippi Valley. Journal of Archaeological Science 18:319–330.

Chamberlain, A. T. 2000 Problems and Prospects in Palaeodemography. *In* Human Osteology in Archaeology and Forensic Medicine. M. Cox, and S. Mays, eds. pp. 101–115. London: Greenwich Medical Media.

Coale, A. J., and P. Demeney 1983 Regional Model Life tables and Stable Populations. 2nd Edition. New York: Academic Press.

Cohen, M. N., and G. Armelagos, eds. 1984 Paleopathology at the Origins of Agriculture. New York: Academic Press.

Cohen, M. N., and G. M. M. Crane-Kramer, eds. 2007 Ancient Health: Skeletal Indicators of Agricultural and Economic Intensification. Gainesville: University Press of Florida.

Cohen, M. N., J. W. Wood, and G. R. Milner 1994 The Osteological Paradox Reconsidered. Current Anthropology 35:629–637.

Cook, D. C. 2007 Maize and Mississippians in the American Midwest: Twenty Years Later. *In* Ancient Health: Skeletal Indicators of Agricultural and Economic Intensification. M. N. Cohen, and G. M. M. Crane-Kramer, eds. Pp. 10–19. Gainesville: University Press of Florida.

Connah, G. 1993 "Of The Hut I Builded": The Archaeology of Australia's History. Cambridge: Cambridge University Press.

Cooper, D. M. L., C. D. L. Thomas, J. G. Clement, A. L Turinsky, C. W. Sensen, and B. Hallgrimsson 2007 Age-Dependent Change in the 3D Structure of Cortical Porosity at the Human Femoral Midshaft. Bone 40:957–965.

DeWitte, S. N., and J. W. Wood 2008 Selectivity of Black Death Mortality with Respect to Preexisting Health. Proceedings of the National Academy of Sciences 105:1436–1441.

Droessler, J. 1981 Craniometry and Biological Distance: Biocultural Continuity and Change at the Late-Woodland- Mississippian Interface. Evanston, IL: Center for American Archeology Northwestern University.

Dyson, R. H. 1965 Problems of Protohistoric Iran as Seen from Hasanlu. Journal of Near Eastern Studies 24:193–217.

Eaton, J. W., and A. J. Mayer 1953 The Social Biology of Very High Fertility Among the Hutterites: The Demography of a Unique Population. Human Biology 25:206–264.

Frankenberg, S. R., and L. W. Konigsberg 2006 A Brief History of Paleodemography from Hoot to Hazards Analysis. *In* Bioarchaeology: The Contextual Analysis of Human Remains. J. E. Buikstra, and L. A. Beck, eds. Pp. 227–262. Emerald Group Publishing.

Garner, C. 1991 Comparative Demography and Disease. *In* The Kuhlman Mound Group and Late Woodland Mortuary Behavior in the Mississippi River Valley of West-Central Illinois. Karen Atwell and Michael D. Conner, eds. Pp. 180–207 Kampsville, IL: Center for American Archeology.

Geusa, G., L. Bondioli, E. Capucci, A. Cipriano, G. Grupe, C. Savorè, and R. Macchiarelli 1998 Osteodental Biology of the People of Portus Romae (Necropolis of Isola Sacra, 2nd-3rd Cent. AD). II. Dental Cementum Annulations and Age at Death Estimates Digital Archives of Human Paleobiology. Rome: Pigorini Museum.

Gordon, C. C., and J. E. Buikstra 1981 Soil pH, Bone Preservation, and Sampling Bias at Mortuary Sites. American Antiquity 46:566–571.

Harrison, R. G., and M. A. Katzenberg 2003 Paleodiet Studies Using Stable Carbon Isotopes from Bone Apatite and Collagen: Examples from Southern Ontario and San Nicolas Island, California. Journal of Anthropological Archaeology 22:227–244.

Hanson, L. Å. 2001 The Mother-Offspring Dyad and the Immune System. Acta Paediatrica 89:252–258.

Hawkes, J. 1974 Atlas of Ancient Archaeology. London: Heinemann.

Hens, S. M., E. Rastelli, and G. Belcastro 2008 Age Estimation from the Human Os Coxa: A Test on a Documented Italian Collection. Journal of Forensic Sciences 53:1040–1043.

Hern, W. M. 1977 High Fertility in a Peruvian Amazon Indian Village. Human Ecology 5:355–368.

Hoppa, R. D. 2000 Population Variation in Osteological Aging Criteria: An Example from the Pubic Symphysis. American Journal of Physical Anthropology 111:185–191.

Hoppa, R. D., and S. R. Saunders 1998 The MAD Legacy: How Meaningful is Mean Age-at-Death in Skeletal Samples. Human Evolution 13:1–14.

Hoppa, R. D. and J. W. Vaupel, eds. 2002 Paleodemography Age Distributions from Skeletal Samples. Cambridge Studies in Biological and Evolutionary Anthropology No. 31. New York: Cambridge University Press.

Howell, N. 1976 Toward a Uniformitarian Theory of Human Paleodemography. *In* The Demographic Evolution of Human Populations. R. H. Ward, and K. M. Weiss, eds. pp. 25–40. London: Academic Press.

Jackes, M. 1985 Pubic Symphysis Age Distributions. American Journal of Physical Anthropology 68:281–299.

Jackes, M. 1986 The Mortality of Ontario Archaeological Populations. Canadian Journal of Anthropology 5(2):33–48.

Jackes, M. 1988 The Osteology of the Grimsby Site. Department of Anthropology, University of Alberta, Edmonton.

Jackes, M. 1992 Paleodemography: Problems and Techniques. *In* Skeletal Biology of Past Peoples: Research Methods. S. R. Saunders, and M. A. Katzenberg, eds. Pp. 189–224. New York: Wiley-Liss.

Jackes, M. 1993 On Paradox and Osteology. Current Anthropology 34:434–439.

Jackes, M. 1994 Birth Rates and Bones. *In* Strength in Diversity: A Reader in Physical Anthropology. A. Herring, and L. Chan, eds. Pp. 155–185. Toronto: Canadian Scholar's Press.

Jackes, M. 2000 Building the Bases for Paleodemographic Analysis: Adult Age Determination. *In* Biological Anthropology of the Human Skeleton. M. A. Katzenberg, and S. R. Saunders, eds. Pp. 407–456. New York: John Wiley & Sons.

Jackes, M. 2002 *Review of* Palaeodemography: Age Distributions from Skeletal Samples. R. D. Hoppa, and J. W. Vaupel, eds. American Journal of Human Biology 14:792–795.

Jackes, M. 2003 Testing a Method: Paleodemography and Proportional fitting. American Journal of Physical Anthropology 121:385–386.

Jackes, M. 2004 Osteological Evidence for Mesolithic and Neolithic Violence: Problems of Interpretation. *In* Evidence and Meaning of Violent Interactions in Mesolithic Europe. M. Roksandic, ed. Pp. 23–39. BAR International Series 1237. Oxford: Archaeopress.

Jackes, M. 2006 Comment on Jean-Pierre Bocquet-Appel and Stephan Naji 2006. Current Anthropology 47:352–353.

Jackes, M. 2009a The Mid Seventeenth Century Collapse of Iroquoian Ontario: Examining the Last Great Burial Place of the Neutral Nation. *In* Vers une Anthropologie des Catastrophes (Actes des 9e Journées D'Anthropologie de Valbonne). Luc Buchet, Catherine Rigeade, Isabelle Séguy and Michel Signoli, eds. pp. 347–373. Antibes/Paris: Éditions APDA/INED.

Jackes, M. 2009b Teeth and the Past in Portugal: Pathology and the Mesolithic-Neolithic Transition. *In* Interdisciplinary Dental Morphology. T. Koppe, G. Meyer, and K. W. Alt, eds. Basel: Karger.

Jackes, M., and P. Alvim 2006 Reconstructing Moita do Sebastião, the First Step. *In* Do Epipalaeolítico ao Calcolítico na Península Iberica. N. F. Bicho, and H. Veríssimo, eds. pp. 13–25. Actas do IV Congresso de Arqueologia Peninsular. Promontoria Monográfica 04. Faro: Universidade do Algarve.

Jackes, M., and D. Lubell 1996 Dental Pathology and Diet: Second Thoughts. *In* Nature et Culture: Actes du Colloque International de Liège, 13–17 Decembre 1993. M. Otte, ed. Pp. 457–480. Etudes et Recherches Archéologiques de L'Université de Liège, no 68.

Jackes, M., and D. Lubell 1999 Human Biological Variability in the Portuguese Mesolithic. Arqueologia 24:25–42.

Jackes, M., and C. Meiklejohn 2004 Building a Method for the Study of the Mesolithic-Neolithic Transition in Portugal. Documenta Praehistorica XXXI Neolithic Studies 11:89–111.

Jackes, M., and C. Meiklejohn 2008 The Paleodemography of Central Portugal and the Mesolithic-Neolithic Transition. *In* Recent Advances in Paleodemography: Data, Techniques, Patterns. J.-P. Bocquet-Appel, ed. Pp. 209–258. Dordrecht: Springer.

Jackes, M., R. Sherburne, D. Lubell, C. Barker, and M. Wayman 2001 Destruction of Microstructure in Archaeological Bone: a Case Study from Portugal. International Journal of Osteoarchaeology 11:415–432.

Jackes, M., M. Roksandic, and C. Meiklejohn 2008 The Demography of the Djerdap Mesolithic-Neolithic Transition. *In* The Iron Gates in Prehistory: New Perspectives. C. Bonsall, V. Boroneant, and I. Radovanovic, eds. Pp. 77–88. BAR International Series 1893. Oxford: Archaeopress.

Jamieson, J. B. 1990 The Archaeology of the St. Lawrence Iroquoians. *In* The Archaeology of Southern Ontario to AD 1650. C. J. Ellis, and N. Ferris, eds. Pp. 385–404. London Chapter, Ontario Archaeological Society, Occasional Publication No. 5.

Jans, M. M. E., C. M. Nielsen-Marsh, C. I. Smith, M. J. Collins, and H. Kars 2004 Characterisation of Microbial Attack on Archaeological Bone. Journal of Archaeological Science 31:87–95.

Jimenez, S., and F. J. Melbye 1987 Ossossané Revisited. Paper Presented at the 15th Annual Meeting of the Canadian Association for Physical Anthropology. Barrie, Ontario, Canada.

Katzenberg, M. A., and R. White 1979 A Paleodemographic Analysis of the Os Coxae from Ossossané Ossuary. Canadian Review of Physical Anthropology 1:10–28.

Kimmerle, E. H., L. W. Konigsberg, R. L. Jantz, and J. P. Baraybar 2008 Analysis of Age-at-Death Estimation Through the Use of Pubic Symphyseal Data. Journal of Forensic Sciences 53:558–568.

Knowles, F. H. S. 1937 Physical Anthropology of the Roebuck Iroquois with Comparative Data from Other Indian Tribes. National Museum of Canada Bulletin No. 87, Anthropological Series No. 22.

Konigsberg, L. W., N. P. Herrmann, D. J. Wescott, and E. H. Kimmerle 2008 Estimation and Evidence in Forensic Anthropology: Age-at-Death. Journal of Forensic Sciences 53:541–557.

Lallo, J., J. C. Rose, and G. J. Armelagos 1980. An Ecological Interpretation of Variation in Mortality Within Three Prehistoric American Indian Populations from Dickson. *In* Early Native Americans. D. L. Browman, ed. Pp. 203–238. The Hague: Mouton Publishers.

Larocque, R. 1990 Une Étude Ethnohistorique et Paléoanthropologique des Epidémies en Huronie. Ph.D. Dissertation, University of Montreal.

Larsen, C. S. ed. 1991 The Archaeology of Mission Santa Catalina de Guale: 2. Biocultural Interpretations of a Population in Transition. Anthropological Papers of the American Museum of Natural History No. 68. New York: American Museum of Natural History.

Litton C. D., and C. E. Buck 1995 The Bayesian Approach to the Interpretation of Archaeological Data. Archaeometry 37:1–24.

Lucy D., R. G. Aykroyd, A. M. Pollard, and T. Solheim 1996 A Bayesian Approach to Adult Human Age Estimation from Dental Observations by Johanson's Age Changes. Journal of Forensic Sciences 41:189–194.

Lynnerup, N., B. Frohlich, and J. Thomsen 2006 Assessment of Age at Death by Microscopy: Unbiased Quantification of Secondary Osteons in Femoral Cross Sections. Forensic Science International 159:S100–S103.

Maat, G. J. R., A. Maes, M. J. Aarents, and N. J. D. Nagelkerke 2006 Histological Age Prediction from the Femur in a Contemporary Dutch Sample. Journal of Forensic Sciences 51:230–237.

Masset, C. 1971. Erreurs Systematiques dans la Determination de L'Age par les Sutures Craniennes Bulletins et Mémoires de la Société d'Anthropologie de Paris 7:85–105.

Masset, C. 1989 Age Estimation on the Basis of Cranial Sutures. *In* Age Markers in the Human Skeleton. M. Y. işcan, ed. Pp. 71–103. Springfield, IL: Charles C. Thomas.

Masset, C. 1990 Ou en est la Paléodémographie? Bulletins et Mémoires de la Société d'Anthropologie de Paris. N.s. 2:109–122.

McCaa, R. 2002 Paleodemography of the Americas: From Ancient Times to Colonialism and Beyond. *In* The Backbone of History: Health and Nutrition in the Western Hemisphere. R. H. Steckel, and J. C. Rose, eds. Pp. 94–124. New York: Cambridge University Press. See also www.hist.umn.edu/~rmccaa/paleo98/paletbl.htm.

Meiklejohn, C., and M. Zvelebil 1991 Health Status of European Populations at the Agricultural Transition and Implications for the Causes and Mechanisms of the Adoption of Farming; *In* Health in Past Societies. H. Bush and M. Zvelebil, eds. Pp. 129–146. Oxford: British Archaeological Reports, International Series 567.

Meiklejohn, C., J. M. Wyman, and C. T. Schentag 1992 Caries and Attrition: Dependent or Independent Variables? International Journal of Anthropology 7:17–22.

Meinl, A., C. D. Huber, S. Tangl, G. M. Gruber, M. Teschler-Nicola, and G. Watzek 2008 Comparison of the Validity of Three Dental Methods for the Estimation of Age at Death. Forensic Science International 178:96–105.

Milner, G. R., D. A. Humpf, and H. C. Harpending 1989 Pattern Matching of Age-at-Death Distributions in Paleodemographic Analysis. American Journal of Physical Anthropology 30:49–58.

Milner, G. R., and V. G. Smith 1990 Oneota Human Skeletal Remains. *In* Archaeological Investigations at the Morton. Village and Norris Farms 36 Cemetery. S. K. Santure, A. D. Harn, and D. Esarey, eds. Pp. 111–148. Reports of Investigations No. 45. Springfield: Illinois State Museum.

Müller, H. G., B. Love, and R. D. Hoppa 2002 Semiparametric Method for Estimating Paleodemographic Profiles from Age Indicator Data. American Journal of Physical Anthropology 117:1–14.

Paine, R. R., and B. P. Brenton 2006 Dietary Health Does Affect Histological Age Assessment: An Evaluation of the Stout and Paine (1992) Age Estimation Equation Using Secondary Osteons from the Rib. Journal of Forensic Sciences 51:489–492.

Paine, R. R., and H. C. Harpending 1996 Assessing the Reliability of Paleodemographic Fertility Estimators using Simulated Skeletal Distributions. American Journal of Physical Anthropology 101:151–159.

Patterson, D. K. 1984 Diachronic Study of Dental Palaeopathology and Attritional Status of Prehistoric Ontario Pre-Iroquois and Iroquois Populations. Mercury Series No. 122, Archaeological Survey of Canada. Ottawa: National Museum of Man.

Perrenoud, A. 1984 Mortality Decline in a Long-Term Perspective. *In* Pre-Industrial Population Change: The Mortality Decline and Short-Term Population Movements. T. Bengtsson, G. Fridlizius, and R. Ohlsson, eds. Pp. 41–69. Stockholm: Almqvist and Wiksell International.

Perrenoud, A. 1990 Aspects of Fertility Decline in an Urban Setting: Rouen and Geneva. *In* A. Urbanization in History. A Process of Dynamic Interactions. J. van deWoude, J. de Vries, and A. Hayami, eds. Pp. 243–263. Oxford: Oxford University Press.

Pfeiffer, S., and S. I. Fairgrieve 1994 Evidence from Ossuaries: The Effect of Contact on the Health of Iroquoians? *In* In the Wake of Contact: Biological Response to Conquest. Clark Spenser Larsen and George R. Milner, eds. pp. 47–61. New York: Wiley-Liss.

Piontek, J., and M. Henneberg 1981 Mortality Changes in a Polish Rural Community (1350–1972) and Estimation of the Evolutionary Significance. American Journal of Physical Anthropology 54:129–138.

Powell, M. L. 1990 On the Eve of the Conquest: Life and Death at Irene Mound, Georgia. *In* The Archaeology of Mission Santa Catalina de Guale: 2. Biocultural Interpretations of a Population in Transition. C. S. Larsen ed. pp. 26–35. Anthropological Papers of the American Museum of Natural No. 68. New York: American Museum of Natural History.

Ramos, R., K. I. Kennedy, and C. M. Visness 1996 Effectiveness of Lactational Amenorrhoea in Prevention of Pregnancy in Manila, the Philippines: Non-Comparative Prospective Trial. British Medical Journal 313:909–912.

Rathbun, T. A. 1982 Morphological Affinities and Demography of Metal Age Southwest Asian Populations. American Journal of Physical Anthropology 59:42–60.

Reid, A. 2000 Neonatal Mortality and Stillbirths in Early Twentieth Century Derbyshire, England. Population Studies 55:213–232.

Renz, H., and R. J. Radlanski 2006 Incremental Lines in Root Cementum of Human Teeth – A Reliable Age Marker? Homo 57:29–50.

Robling A. G., and S. D. Stout 2000 Histomorphometry of Human Cortical Bone: Applications to Age Estimation. *In* Biological Anthropology of the Human Skeleton. M. A. Katzenberg, and S. R. Saunders, eds. Pp. 187–213. New York: John Wiley and Sons.

Rowland, M. G. M. 1986 The Weanling's Dilemma: Are We Making Progress? Acta Pædiatrica 75(s323):33–42.

Russell, K. F., I. Choi, and C. S. Larsen 1990 The Paleodemography of Santa Catalina de Guale. *In* The Archaeology of Mission Santa Catalina de Guale: 2. Biocultural Interpretations of a Population in Transition. Clark Spenser Larsen, ed. pp. 36–49. Anthropological Papers of the American Museum of Natural History No. 68. New York: American Museum of Natural History.

Saunders, S. R., C. Fitzgerald, T. Rogers, C. Dudar, and H. McKillop 1992 Test of Several Methods of Skeletal Age Estimation Using a Documented Archaeological Sample. Canadian Society of Forensic Science Journal 25:97–118.

Schmitt, A., P. Murail, E. Cunha, and D. Rougé 2002 Variability of the Pattern of Aging on the Human Skeleton: Evidence from Bone Indicators and Implications on Age at Death Estimation. Journal of Forensic Sciences 47:1203–1209.

Slayton, R. L., M. E. Cooper, and M. L. Marazita 2005 Tuftelin, Mutans Streptococci, and Dental Caries Susceptibility. Journal of Dental Research 84:711–714.

Steckel, R. H., P. W. Sciulli, and J. C. Rose 2002 A Health Index from Skeletal Remains. *In* The Backbone of History: Health and Nutrition in the Western Hemisphere. R. H. Steckel, and J. C. Rose, eds. Pp. 61–93. New York: Cambridge University Press.

Steckel, R. H., and J. C. Rose, eds. 2002 The Backbone of History: Health and Nutrition in the Western Hemisphere. New York: Cambridge University Press.

Stodder, A. L. W. 2008 Taphonomy and the Nature of Archaeological Assemblages. *In* Biological Anthropology of the Human Skeleton. M. A. Katzenberg, and S. R. Saunders, eds. Pp. 71–114. New York: Wiley-Liss.

Storey, R. 2007 An Elusive Paleodemography? A Comparison of Two Methods for Estimating the Adult Age Distribution of Deaths at Late Classic Copan, Honduras. American Journal of Physical Anthropology 132:40–47.

Tainter, J. A. 1977 Woodland Social Change in West-Central Illinois. Mid-Continental Journal of Archaeology 2:67–98.

United Nations 1982 Model Life Tables for Developing Countries. Department of International Economic and Social Affairs Population Studies No. 77. New York: United Nations.

Vieira, A. R., M. L. Marazita, and T. Goldstein-McHenry 2008 Genome-Wide Scan Finds Suggestive Caries Loci. Journal of Dental Research 87:435–439.

Vradenburg, J. A. 1999 Skeletal Biology of Late-Prehistoric La Crosse-Region Oneota Populations. The Missouri Archaeologist 60:107–164.

Walker, R. A., C. O. Lovejoy, and R. S. Meindl 1994 Histomorphological and Geometric Properties of Human Femoral Cortex in Individuals over 50: Implications for Histomorphological Determination of Age-at-Death. American Journal of Human Biology 6:659–667.

Weiss, K. M. 1973 Demographic Models for Anthropology. Memoirs of the Society for American Archaeology No. 27. American Antiquity 38.

Wells, J. C. K. 2000 Natural Selection and Sex Differences in Morbidity and Mortality in Early Life. Journal of Theoretical Biology 202:65–76.

Wittwer-Backofen, U., J. Buckberry, A. Czarnetzki, S. Doppler, G. Grupe, G. Hotz et al. 2008 Basics in Paleodemography: A Comparison of Age Indicators Applied to the Early Medieval Skeletal Sample of Lauchheim. American Journal of Physical Anthropology 137:384–396.

Wood, J. W., G. R. Milner, H. C. Harpending, and K. M. Weiss 1992 The Osteological Paradox. Current Anthropology 33:343–370.

Wright, L. E., and C. J. Yoder 2003 Recent Progress in Bioarchaeology: Approaches to the Osteological Paradox. Journal of Archaeological Research 11:43–70.

Part II

Social Identity: Bioarchaeology of Sex, Gender, Ethnicity, and Disability

6

Sex and Gender in Bioarchaeological Research

Theory, Method, and Interpretation

Sandra E. Hollimon

Theoretical Issues: Relationships Between/Among Sex and Gender

Gender is only one of many axes of identity that may be involved in the construction of personhood. Sexed and/or gendered bodies are always simultaneously combined constructions of age, class, ethnicity, race, and social status (Joyce 2005). Therefore, bioarchaeologists conduct studies about death, as well as the complexities of social life and individual lives. Of particular interest are the ways that other cultures may have constructed gendered identities at different stages in the life course (Geller 2008; see also Agarwal and Beauchesne, Halcrow and Tayles, and Sofaer this volume). For a summary of recent developments in the bioarchaeology of social identities, see Knudson and Stojanowski (2008).

Recent research has begun to explicitly address the relationship between biological sex and cultural constructions of gender (Arnold 2002; Díaz-Andreu 2005; Gere 1999; Geller 2005, 2008; Sofaer 2006a,b; Stone and Walrath 2006; Walker and Collins Cook 1998; Worthman 1995). Many of these works critically examine the widespread and implicit assumption that biological sex equates to cultural gender (e.g., Conkey and Gero 1997; Gibbs 1987; Gilchrist 1999; Nordblah and Yates 1990; Sørenson 2000). An important contribution comes from theories of embodiment, which analyze the production and experience of lived bodies in the past by examining the juxtaposition of traces of body practices, idealized representations, and evidence of the effects of habitual gestures, postures, and consumption practices on the physical body (Joyce 2005, 2008; see also Perry and Joyce 2001).

Social Bioarchaeology. Edited by Sabrina C. Agarwal and Bonnie A. Glencross
© 2011 Blackwell Publishing Ltd

Leading this research are the studies by Joyce in Mesoamerica (e.g., Joyce 2002, 2005, 2006) and Meskell in Egypt (e.g., Meskell 1996, 1998; 2001). Meskell and Joyce (2003) collaborated on a project that compared ancient cultures of these two regions and the ways in which gender and other social identities were embodied by individuals in the past. These studies not only situate the body in social and cultural contexts, they are also among the relatively few archaeological studies that critically examine and deconstruct the category "masculinity" (e.g., Joyce 2004; Knapp and Meskell 1997; Yates 1993; see also Alberti 2006).

In addition, some researchers question the influence that culturally constructed methods of sex determination may play in the interpretations of bioarchaeological data. They echo the suggestions of post-processual archaeologists (e.g., Conkey 2007; Engelstad 2007; Marshall 2008) who note that the idea of a purely objective approach denies the political and social contexts of this research (see Claassen 1992, 2001; Geller 2005; Gilchrist 1997; Lucy 1997; Pyburn 2004; Walker 1995). For example, Geller (2008) notes that some bioarchaeologists are unaware or unwilling to acknowledge that their interpretations about the past are presentist, androcentric, and/or heteronormative (see also Schmidt and Voss 2000). In a defense of osteological methods of sex determination, Sofaer (2006a: 96) has noted that such criticism correctly points out the potential role of social construction, but it is not necessary to suggest that "observable differences between men and women are some sort of irrelevant mirage, or that osteological determinations do not form a useful axis of analysis" (see also Geller 2008).

In the attempt to bridge the seeming theoretical and methodological divides between science-based osteology and socially informed archaeological interpretation, recent developments have been proposed for the study of human skeletal remains from the past. One such approach is to view the body as material culture (Sofaer 2006a). As such, it may be possible to interpret archaeological evidence through the ways that the social lives of people were expressed in the creation of their bodies (Sofaer 2006a).

The biocultural interactions of sex and gender equip bioarchaeology perhaps uniquely to identify nonbinary genders in the archaeological record (see Geller 2005). For example, Perry (2004) considers the bioarchaeological evidence of a gender-based division of labor in skeletal remains from the prehispanic Southwest. Her study examined musculoskeletal stress markers (MSMs) that indicate bone remodeling due to repetitive motions. Perry observed typical patterns of MSMs on female and male skeletons, as well as female patterns on biologically male skeletons. A biological male that performed similar work as normative women may have been a person who was socially recognized as a third gender. A similar approach was employed by Hollimon (1996) who identified two possible third-gender males in a skeletal sample from coastal California. These individuals were relatively young (under 20 years old) but displayed advanced spinal arthritis that was typical of older females in this population. Repeated stress from digging graves, an occupation staffed by third-gender males and postmenopausal women, could have produced this pattern of degenerative joint disease. Ethnographic and ethnohistoric evidence indicates that the undertaking profession was the domain of individuals who were nonprocreative in their sexual behavior (see Hollimon 1997). In addition, these two individuals had burial accompaniments of basketry impressions and digging stick weights, part of the undertaker's toolkit; they were the only male skeletons in the sample that contained both of these items.

It has been suggested that nonbinary genders have a great time depth in North America, and that the colonizing populations who migrated from northeastern Asia recognized more than two genders (Hollimon 2001b, 2006; Kirkpatrick 2000). Numerous other examples of nonbinary genders can be found worldwide (e.g., Mesoamerica, Stockett 2005; Klein 2001; see Weston 1993), and bioarchaeological analyses may provide valuable approaches to identifying such individuals in the archaeological record.

This survey reviews bioarchaeological research that explicitly considers sex and/or gender as analytical categories in an effort to provide a representative survey of this research. One measure of the accessibility of this literature is the inclusion of the keywords sex or gender in the titles of these works. The lack of these words can be a general reflection of the inattention to or relative unimportance of sex and/or gender as explicit research foci. Therefore, one criterion for inclusion is the presence of the word "gender" in the title or keywords of a publication.

There are observable differences among publications and gender bias in citation practices (see Hutson 2002). The *American Journal of Physical Anthropology* does not publish many articles concerning gender, but archaeological journals, especially those concerned with social analyses, often do. In addition, recent analyses of sex and gender in bioarchaeology have been compiled in edited volumes, often the result of conferences devoted to gendered approaches in archaeological research. While the internet has increased accessibility to bioarchaeological research published in peer-reviewed journals, the full content of chapters in edited volumes is still relatively inaccessible.

The subject of gender bias in the discipline of bioarchaeology as a whole is beyond the scope of this chapter. However, a useful summary of many issues surrounding this important subject can be found in Geller's (2008, 2009) discussions of feminist-inspired theory in bioarchaeology. Another description is provided in the work by several authors that draws attention to the "unsung heroines" of bioarchaeological research (Powell et al. 2006).

It has long been a standard practice in osteological analysis to identify the sex of skeletons, in the same way that ages are calculated to construct demographic profiles of the sample. However, it has only been a recent development in the discipline of bioarchaeology to consider sex and gender in the context of the culture(s) under analysis. Therefore, the majority of the literature reviewed here postdates the groundbreaking call for archaeological gender analysis by Conkey and Spector (1984). While this survey cannot possibly be exhaustive, it does provide a fairly comprehensive sample of these analyses. It is admittedly biased toward English-language works and analyses of North American skeletal populations. Also, I have somewhat arbitrarily categorized analyses as to the principal research focus. Therefore, I have not repeated the citation of works that may fit multiple categories. For example, many of these studies involved mortuary analysis, but if I considered the main interpretive point to be about health and disease, it was included in the latter category.

Major Research Themes

Mortuary analysis

Increasingly, bioarchaeologists are encouraging the critical consideration of the theoretical relationships between sexed skeletons and the socially constructed identities of past lives (e.g., Cannon 2005; Gillespie 2001; Goldstein 2006; see above); for a brief history of gender and mortuary archaeology, see Arnold (2006). Important early contributions to gendered archaeological mortuary analysis include studies that consider the role of gender in the formation of the state in Korea (Nelson 1993, 1997) and among the Maya (Haviland 1997). Mesoamerica is a region with a relatively long history of gendered mortuary analyses. The ancient Maya have been studied by examining elite burials (e.g., Bell 2002; González Cruz 2004; McCafferty and McCafferty 1994; Tiesler et al. 2004), expressions of female power in burial contexts (Ardren 2002), and gender and health (Danforth et al. 1997; Monaghan 2001). The numerous works of Joyce (e.g., 1999, 2001, 2002) include several mortuary analyses.

Other examples include studies of the Near East, such as prehistoric Nubia (Nordström 1996), pre-dynastic Egypt (Savage 2000), and Ancient Cyprus (Bolger 2003; Bolger and Serwint 2002; Keswani 2004). Several studies of European prehistory counter the idea that gender-based divisions of labor are clearly marked in mortuary contexts (Arnold 1991; Gräslund 2001; Johnsson et al. 2000; Schmidt 2000; Sofaer Derevenski 2002; Strassburg 2000; Strömberg 1993; Toms 1998; Tuohy 2000; Vida Navarro 1992; Weglian 2001).

Mortuary studies from East Asia are represented by those in prehistoric Thailand (Bacus 2007; Higham 2002; Higham and Bannanurag 1990; Higham and Thosarat 1994) and in Neolithic China (Jiao 2001). Ancient Eurasian nomads have been examined in mortuary contexts, especially with regard to gender and warrior roles (e.g., Berseneva 2008; Legrand 2008; Linduff 2008; Rubinson 2002, 2008; Shelach 2008).

In North America, mortuary data have been analyzed among the Inuit (Crass 2000, 2001), in prehistoric California (Gamble et al. 2002), Massachusetts and Kentucky (Doucette 2001), Florida (Hamlin 2001), and the Great Lakes region (O'Gorman 2001). The construction of gender hierarchies in the prehispanic Southwest was examined by Neitzel (2000), and the expression of gender ideology in mortuary ritual in protohistoric North Carolina was explored by Rodning (2001). Also, there have been re-examinations of assemblages that changed interpretations of mortuary behavior in prehistoric Ohio (Konigsberg 1985) and Kentucky (Milner and Jeffries 1987). In historical archaeology, Wilson and Cabak (2004) have examined African American mortuary data to discuss gender differences in the experience of slavery.

The authors of several publications have discussed nonbinary genders and the potential for mortuary analyses to identify such individuals in the archaeological record (e.g., Claassen 1992; Duke 1991; Hollimon 1991, 1992, 2006; Pate 2004; Voss 2005). Examples include Arnold's (1991) discussion of "the deposed Princess of Vix," a spectacular female Iron Age burial from Germany (see also

Knüsel 2002); Weglian's (2001) description of "atypical" burials from the Neolithic/Bronze Age transition in Germany; and Whelan's (1991) analysis of a 19th-century Dakota cemetery that may have included third-gender males. In each case, the identification of nonbinary genders relied on the statistical analysis of variations in mortuary treatment, not simply noting that a biologically sexed burial contained artifacts that "belong to the opposite sex." Voss (2005) notes that archaeologists must make a firm distinction between *social* deviance and *statistical* deviation.

Activity reconstruction, division of labor and occupational specialization

Bioarchaeological data have been used to reconstruct past activities, especially with regard to gender-based divisions of labor (e.g., Cohen and Bennett 1993). A number of methods have been employed, including the examination of activity-induced pathologies, such as degenerative joint disease (DJD), tooth wear, biomechanical studies of robusticity and flexion, relationships between inferred workload and increased mortality, MSMs and trauma.

Degenerative joint disease (DJD) is a pathological condition that was among the earliest lines of evidence used for reconstructing past activities. The two most central etiological influences are mechanical factors and age (Jurmain 1999:50). Pathological changes in the bone include osteophytes and eburnation, a polish resulting from bone-on-bone contact (Jurmain 1999:26–35). Interpretative assumptions and methodological problems may lead to inaccurate behavioral reconstructions, but careful analysis may provide insight about past activity patterns (see Jurmain 1999:107–109 for a discussion of these issues).

In a classic study, Merbs (1983) examines activity-induced pathological conditions in a Canadian Inuit population in order to make inferences about past behaviors. DJD patterns are often used to infer activities related to food processing, and the term "metate elbow" was coined to describe the wear-and-tear associated with the use of this grinding equipment among agricultural populations (e.g., Bridges 1985). Merbs and Euler (1985) inferred habitual corn grinding by an Anasazi-period female from Bright Angel Ruin on the basis of this pattern of degenerative change. Bartelink (2001) has also examined DJD of the elbow to make inferences about a gender-based division of labor and resource intensification in prehistoric populations of the San Francisco Bay area.

Similarly, Bridges (1994) examined degenerative conditions in cervical vertebrae in prehistoric Southwestern populations and interpreted this as evidence of the possible use of a tumpline to carry heavy burdens. Although ethnographic evidence suggests that females were the primary burden carriers (e.g., Swanton 1979), Bridges (1994) concluded that males must have used tumplines or engaged regularly in some activities that put severe stress on the neck because rates were equal among females and males.

In some instances, it appears that there was a lack of gender-based task assignment, at least for certain activities. Miller (1985) found evidence for lateral epicondylitis in the elbow of both males and females from the prehistoric (A.D. 1000–1400), Arizona site of Nuvakwewtaqa (Chavez Pass). He interpreted this as evidence of the use of two-handed metates by both sexes because it was found equally in the right and left elbows.

Numerous examples of DJD analyses include the examination of pre- and post-Fur Trade female health in Native American groups (Lovell and Dublenko 1999; Reinhard et al. 1994), health changes at Dickson Mounds, Illinois between A.D. 950–1300 (Goodman et al. 1984), and DJD patterns that changed between foraging and agricultural populations in the American Southeast (Bridges 1991). Analyses of DJD in prehistoric or pre-contact populations include samples from the Central Ohio River Valley (Cassidy 1984), Mississippi (Eisenberg 1986), Georgia (Larsen 1984), and California (Walker and Hollimon 1989).

Examinations of DJD in historic period samples has provided another line of evidence for reconstructing past activities, in conjunction with historical sources. For example, Rathbun (1987) examined an enslaved African American population from a South Carolina plantation. He found significant DJD in most joints in this sample, and interpreted this as evidence of extremely hard physical labor to which these people were subjected. More specifically, Rathbun (1987) inferred that the pattern of Schmorl's nodes, indicating intervertebral disk herniations, was more frequent in males due to constant heavy lifting. A possible gender-based division of labor was suggested by different patterns of DJD in females (knees and shoulders) and males (elbow and hip).

Studies of sex differences in tooth wear have examined the probable use of teeth as tools in some populations. For example, the Mesolithic cemetery at Skateholm, Sweden, yielded female skeletons with much greater anterior tooth wear than that seen on males, suggesting non-masticatory use of the teeth (Frayer 1988). Other examples include women's buffalo hide processing among the Omaha (Reinhard et al. 1994), San women's plant fiber rope manufacture (Morris 1992), and a similar manufacturing process among Great Basin men (Larsen 1985). Sex differences in the use of teeth-as-tools include Australian Aborigines (Molnar et al. 1989; Richards 1984), and prehistoric populations from Greenland (Pedersen and Jakobsen 1989).

Another method of reconstructing past activities has included studies that examine robusticity, flexion, and other aspects of mechanical loading (Ruff 2000; for a critique of these approaches, see Bice 2003). Hamilton (1982) investigated robusticity as one measure of sexual dimorphism in modern *Homo sapiens*. Ruff (1987) analyzed sexual dimorphism in the lower limb structure with regard to subsistence strategies and gender-based divisions of labor. He found that prehistoric agricultural populations like Pecos Pueblo were intermediate in lower limb bending strength compared to foragers and modern samples. Ruff interpreted this as evidence that Pecos society had a gender-based division of labor that maintained different lower limb bending loads compared to modern populations.

Similar analyses have been conducted with populations from the Georgia Bight (Fresia et al. 1990) and the Pickwick Basin, Alabama, (Bridges 1985, 1989). These analyses focused on bilateral asymmetry in humerus size and strength, and in both regions, there was a more significant decline in females than males when agriculture was adopted. Bridges (1985) interpreted this as evidence of increased participation by females in agricultural activities as compared to the biomechanical requirements of foraging. Bridges suggested that males had similar upper limb activity levels in both the foraging and agricultural periods. In the Georgia Bight, bilateral asymmetry decreased more significantly among males. Fresia and colleagues (1990) inferred that males participated more frequently in agricultural activities when they were forced to work in the Spanish missions. In a South African slave population, Ledger and colleagues (2000) found significant humeral asymmetry in females, which they attributed to the use of digging sticks to harvest tubers.

In North America, analyses have also examined the past gender-based activities of indigenous populations (Boyd 1996). These include studies from the Great Plains (Ruff 1994; Westcott 2001; Westcott and Cunningham 2006), the Midwest (Bridges et al. 2000), the Southwest (Akins 1986, 2000; Chapman 1997; Fink 1989; Lambert 1999; Martin et al. 2001; Miller 1985; Spielmann 1995; Stodder 1987), the Great Basin (Brooks et al. 1988; Larsen 1985, 1995a; Ruff 1999), California (Gjerdrum et al. 2003; Weiss 2009), British Columbia (Weiss 2009), as well as coastal Georgia (Ruff and Larsen 1990) and pre- (Hamlin 2001) and post-Spanish Florida (Ruff and Larsen 2001). Stock and Pfeiffer (2004) compared robusticity in late Paleolithic females and males from two different ecological settings in southern Africa.

In contrast to these studies which critically examine sex, gender and division of labor, humeral bilateral asymmetry in Bronze Age samples from Eastern Europe were interpreted by Sládek and colleagues (2007) as evidence of male extra-domestic agricultural labor, while the lack of asymmetry in females derived from domestic labor. Sparacello and Marchi (2008) examined sex differences in Italian Neolithic and medieval populations with regard to humeral and femoral robusticity. On the basis of the humeral data, they concluded that males were involved in more physically demanding tasks than females, but the femoral data are more enigmatic. However, Geller (2005:122–123) cautions that these interpretations often imply that women are physically disadvantaged, tied to home bases, and/or responsible for domestic chores and childcare responsibilities (e.g., Aranda et al. 2009; Kuhn and Stiner 2006). As such, interpretations that do not critically and creatively think about the relationships among sex, gender, and labor universalize and naturalize the idea that biology is destiny.

Examinations of shifts in division of labor during major cultural transitions have also been conducted in the Mesolithic Near East (Molleson 1989, 1994, 2007; see also Crabtree 2006), 16th–19th century Britain (Sofaer Derevenski 2000b), and the Bronze Age Levant (Peterson 2000b). Some researchers have focused specifically on the related effects of increased workload and higher mortality rates after major social changes. For example, Šlaus (2000) examined medieval Croatian populations with regard to sex differences in stress levels and their effects on mortality patterns. Native American women of the Great Plains who experienced increased work loads following the introduction of the fur trade also had higher mortality rates than their pre-fur trade ancestors (Owsley and Bruwelheide 1997; Scheiber 2008).

Musculoskeletal stress markers (MSMs), as an example of activity-induced pathology, may be examined to make inferences about past activities. Churchill and Morris (1998) examined forest and savanna populations of prehistoric Khoisan peoples. They noted differences in MSMs between forest and savanna males, but not among females; they attributed this to significantly different activities on the part of males in the two environments, but overall similarities for females in both environments. These markers have also been examined with regard to sex differences among enslaved ironworkers in Virginia (Kelley and Angel 1987) and Hudson Bay Eskimos (Inuit) during major changes in subsistence strategy (Hawkey and Merbs 1995).

The examination of gender and occupational specialization using bioarchaeological data has been increasing in frequency. Examples include the study of auditory exostoses to make inferences about exposure to cold water, whether during ritual bathing practices among prehistoric populations in the American Southeast (Lambert 2001, 2002), or during occupationally-related activities in prehistoric Chilean populations (Standen et al. 1997). Activity-induced skeletal changes have been observed among Maya traders (Maggiano et al. 2008), medieval Muslim populations in Spain and Anglo-Saxon Britain (Pomeroy and Skrakrwesi 2008), and Neolithic Sweden (Molnar 2008a,b). Skeletal samples from the United Kingdom have often been examined and interpreted with regard to activity-induced pathologies, such as fracture trauma among medieval British farmers (Judd and Roberts 1999), skeletal changes associated with medieval archery (Rhodes and Knüsel 2005), and injuries experienced by 18th–19th century British men in various occupations (Mays 2001).

Intentional body modification

Cultural practices that modify the body are being examined with regard to sex and gender by a growing number of bioarchaeologists. The theoretical positions that the body is a medium of cultural display or is itself material culture are increasingly employed in bioarchaeological research (see above). Most of these studies have examined dental or cranial modification.

Becker (2000) describes intentional tooth evulsion among elite Etruscan women. These women replaced missing teeth with gold dental appliances, and Becker suggests that this was a status display in Etruscan society. On the prehistoric Northwest coast, Cybulski (1992, 1994) examined labret scars among females over the last 500 years. While the scars were present, the labrets themselves were not, suggesting to Cybulski that an "heirloom" practice was observed in which this symbol of wealth and matrilineal family status was inherited by surviving female family members. Cybulski interpreted the labret scars of both males and females from earlier time periods as evidence of patrilineal and/or bilateral kinship systems.

Cranial modification is another form of intentional change to the body that has been interpreted with regard to expressions of identity. Lorentz (2008) describes head-shaping practices in the Ancient Near East as a means of developing or denoting gender differentiation and/or social status. Cranial modification could also be an important signal of ethnic, regional, or local identity. For example, Torres-Rouff (2002) suggests that the lack of gender differences in the Atacama population of Chile was due to the fact that cranial modification was a childrearing strategy that was not gender-based. It appears to have been a sign of ethnic identity within the Tiwanaku empire rather than an expression of gendered identity. Logan and colleagues (2003) examined cranial modification in 19th-century Osage populations of the Great Plains. The absence of modification was interpreted as evidence of changes in cradle-boarding practices as result of more frequent marriages between Osage women and white men during this period (Logan et al. 2003).

Health and disease

The examination of skeletal populations for patterns of disease has a long history in bioarchaeology (e.g., Cook 1984). However, the explicit consideration of biocultural interactions with regard to sex and gender is a relatively recent phenomenon (see Grauer and Stuart-Macadam 1998). Examples include the interpretation of sex differences in enamel hypoplasia in human and nonhuman primates (Guatelli-Steinberg and Lukacs 1999) and the examination of caries rates among females and males at the transition to agriculture (Lukacs 1996, 2008). Lukacs (2008) argues that the impact of dietary change on women's oral health was intensified by the increased demands on women's reproductive systems, including the increase in fertility, that accompanied the rise of agriculture. He suggests that these factors contribute to the observed gender differential in dental caries (Lukacs 2008). Martin (2000) has considered health and disease patterns, especially those associated with corn grinding and consumption, with regard to gender-based division of labor in prehispanic Southwest populations.

A major focus in the study of health and disease has been the examination of pre- and post-contact populations in areas colonized by Europeans. Examples include the American Southeast (Larsen 1998) and Peru (Klaus and Tam 2008; Klaus et al. 2009).

Other examples include the influence of gender on the health of Maya populations (Danforth et al. 1997), Native North American groups of the Great Plains and California (Hollimon 1992, 2000; Kerr 2004), the Neolithic Near East (Peterson 2000a), and the Indus Valley (Hemphill et al. 1991). As with studies of activity-induced pathologies, numerous prehistoric, medieval, and modern skeletal populations from the United Kingdom have been examined for evidence of other pathological conditions, such as infectious diseases and trauma (e.g., Grauer and Roberts 1996; Redfern 2005; Sullivan 2004, 2005). DeWitte (2009) examined the effects of sex on the risk of dying from the plague in medieval London and determined that sex played no role in the risk of mortality from the Black Death.

Stable isotope analyses

The development of techniques such as stable isotope analysis has allowed for a more precise measure of the nutritional status of archaeological skeletal populations. These types of analyses have included populations of the Cahokia polity (Ambrose et al. 2003), the Maya (Danforth 1999; Gerry and Chesson 2000; White 2005), the prehistoric Western Cape of South Africa (Lee Thorp et al. 1989; Sealy et al. 1992; Wadley 1998), Mesolithic Europe (Schulting and Richards 2001), medieval Britain (Müldner and Richards 2007; Privat et al. 2002), and Early Historic Orkney, Scotland (Barrett and Richards 2004). Other bone chemistry studies have demonstrated differences in the diets of women and men in the American Southwest (Ezzo 1993; Whittlesey 2002) and Great Plains (Habicht-Mauche et al. 1994).

Several studies have discussed that access to and consumption of cariogenic foods differed by sex. This was most likely due to a gender-based division of labor in subsistence activities. Examinations have included pre- and post-agricultural populations in the American Southeast (Driscoll 2001; Larsen 1983, 1995; Larsen et al. 1991; Hoshower and Milanich 1993; Kestle 1988; Reeves 2000; Williamson 1998) and Japan (Temple and Larsen 2007), as well as prehistoric groups from coastal California that experienced intensified use of marine resources over time (Walker and Erlandson 1986).

Researchers have also considered the social, political, and ritual "value" of foods beyond the purely nutritional qualities of female and male diets. White (2005) examined Maya female and male dietary differences through carbon and nitrogen isotope analysis. She concluded that greater protein consumption by males meant that males had greater access to socially valued and ideologically important foods (White 2005). Similarly, Hastorf (1991) interpreted isotopic data in a Peruvian skeletal sample as evidence of differential access to certain foods, including maize beer. Some males apparently consumed the beer as part of ritual, political, and social practices that resulted from expansion of the Inca empire (Hastorf 1991).

Examination of isotope signatures has also refined the study of mobility and postmarital residence (e.g., Stojanowski and Schillaci 2006). Earlier studies relied primarily on patterns of nonmetric traits in cemetery populations (e.g., Hawikku in New Mexico – Corrucini 1972, 1998; Howell and Kintigh 1996, 1998). In prehistoric populations from Illinois, Konigsberg (1988) and Konigsberg and Buikstra (1995) inferred virilocality due to increased Middle and Late Woodland female mobility compared to that of males. In contrast, they found that during a period of agricultural intensification, male migration increased, presumably due to uxorilocality. Tomczak and Powell (2003) inferred patrilocality at the Windover site in Florida on the basis of nonmetric variations in dentition. Recent analyses employing isotopic data include inferred uxorilocality in prehistoric coastal California (Walker and DeNiro 1986) and the Northern Great Plains (Schneider and Blakeslee 1990), a reassessment of matrilocality in Chacoan society (Schillaci and Stojanowski 2002, 2003), and a study of a prehistoric population in eastern Canada (Williamson and Pfeiffer 2003).

An interesting study highlights the use of stable isotopes to identify "foreigners" in a skeletal population. Blakeslee (1992) describes three adult females from a prehistoric site in the American Midwest who displayed little or no maize consumption, unlike other individuals at this site that showed heavy ingestion of maize. On the basis of mortuary and dietary evidence, Schurr suggested that females were captured, recruited, or exchanged from other groups for the purpose of acquiring additional labor.

Mobility after major economic shifts or social disruptions has also been examined. For example, the mobility of women after the plague in Britain was studied by Müldner and Richards (2007). Bentley and colleagues have examined mobility patterns after the adoption of agriculture in Europe (Bentley et al. 2003) and Southeast Asia (Bentley et al. 2005, 2007).

Violence and warfare

Bioarchaeological research about violence and warfare is not new, but the explicit focus on sex and gender is a relatively recent development. Typically, the demographic profile of traumatic injuries was interpreted along fairly androcentric lines. Evidence of trauma in male skeletons was thought to be due to combat, and female skeletons showing such wounds were considered victims of either raids or domestic violence (e.g., Jiménez-Brobeil et al. 2009). For example, a recent volume (Chacon and Dye 2007) that examines practices of trophy-taking of human body parts in the Americas contains no index entry for "men" or "man" (the unmarked category) while "women" (the marked category) comprises an index entry. This implies that the males of these societies were assumed to be the perpetrators and/or victims of such practices (the unmarked category) while involvement of women in any capacity would be unusual, noteworthy, and requiring explanation (the marked category).

More recent interpretations of this evidence examine these assumptions more critically. They do not automatically assume that male injuries represent warriors and that female injuries represent victims. For example, employing ethnographic and historical information, some researchers have considered the possibility that some traumatic injuries suffered by females may have been related to combat (e.g., Berseneva 2008; Davis-Kimball 1997; Hanks 2008; Hollimon 2001a; Legrand 2008).

This highlights the point that ideas about who may fight, who may kill, and who may be killed are culturally defined. Therefore, all forms of contextual information, including ethnographic and ethnohistoric sources, should be employed to interpret the demographic patterns of violence-related trauma. In the kinds of warfare observed in modern small-scale societies, in- or outgroup membership is likely of greater significance than age or gender. In other words, everyone who is not "us" (i.e., "them"), is within bounds for violent treatment, whether young or old, female or male (or neither); these become dehumanized "others" (Walker 2001). For example, evidence from the massacre site of Crow Creek in South Dakota included the remains of at least 486 victims of a mass killing dating to A.D. 1325 (Willey 1990). Remains of women and children, as well as men were present, and the overwhelming majority (~95 percent) of the intact skulls showed signs of scalping marks. There was also evidence of widespread perimortem mutilation in the form of decapitation and dismemberment. Similar evidence from the Great Plains includes the remains of five individuals from the Fay Tolton site, who showed evidence of repeated attacks, including a 5–7-year-old child who had survived being scalped (Hollimon and Owsley 1995). Perimortem trauma, including the removal of hands and decapitation, suggests trophy-taking by the assailants (Hollimon and Owsley 1995).

Walker (1997, 2001) considers the cultural patterning of violence and makes cross-cultural comparisons of some bioarchaeological evidence of violent interactions (see also Novak 2006). He noted that there also appear to be variations through time in the demographic profiles of injured populations (Walker 1997). For example, ancient massacre sites frequently contain the remains of many females and children, but a modern pattern consists of more male traumatic injuries (Walker 2001). A case of the former pattern comes from a Tiwanaku outpost in Chile, in which females were injured at rates comparable to males (Torres-Rouff and Junqueira 2006). In contrast, the age and sex distributions of people with fatal projectile point wounds from the prehistoric California coast is similar to that seen in modern homicide victims, with nearly 20 percent of the 15–26-year-old males having projectile point injuries (Lambert 1997). Shifts in demographic profiles of violence over time were also examined in the American Southeast by Smith (2003).

Another example is Robb's (1997) survey of several Italian skeletal collections, in which he found that after the Neolithic period, male cranial traumatic injuries dramatically increased in comparison to females, and by the Iron Age, trauma of all kinds was much more common among males than females. Robb concluded that this was due to the development of gender roles that prescribed violent behavior for males and reinforced an existing gender-based division of labor in which women were not expected to perform activities such as warfare.

Other bioarchaeological studies of gender and violence include examinations of sacrifice among the Maya (White et al. 2002, 2004) and in the Wari empire of the Peruvian Andes (Tung 2007, 2008; Tung and Knudson 2008). Interesting bioarchaeological studies that include considerations of gender ideology are the analyses of women warriors of the Eurasian steppe (Davis-Kimball 1997), and of human finger and hand bone necklaces in the cultural construction of warrior-manhood among Plains Indians of North America (Owsley et al. 2007).

Regional analyses of prehistoric violence have been conducted in the American Southeast (Bridges 1996; Lahren and Berryman 1984; Smith 1996, 1997, 2003) and Northeast (Milner and Smith 1990; Milner et al. 1991; Ryan and Milner 2006; Wilkinson 1997), and document a probable Spanish–Native American conflict in the Southeast (Blakely and Matthews 1990). The skeletal evidence of warfare from the western side of the North American continent has been examined in the Arctic and Subarctic (Maschner and Reedy-Maschner 2007), prehistoric California (Andrushko et al. 2005, 2010; Jurmain and Bellifemine 1997; Jurmain et al. 2009; Lambert 1997; Walker 1989), Oregon (Hemphill 1992), and the Southwest (Kuckelman et al. 2002; Martin 1997; Martin and Akins 2001), and the Mississippi Valley (Ross-Stallings 2007). In particular, the work of Martin and colleagues at the site of La Plata (Martin et al. 2001) has provided evidence of a probable underclass composed of outgroup females who were routinely physically abused. This may have been a more widespread practice in the prehispanic Southwest, where raiding for females has been inferred from the demographic profiles of skeletal samples (e.g., Kohler and Turner 2006). Similarly, a group of prehistoric males from western Kentucky who showed unhealed perimortem cuts suggesting intentional trauma were interpreted as an "at-risk" portion of the population (Mensforth and Baker 1995); Mensforth (2001) interpreted these patterns of sublethal trauma in male skeletons as examples of codified social contests between men. Smith (1997) found a comparable pattern among a group of prehistoric males from western Tennessee.

The relationship between group demographic profiles and marriage practices was explored by Stojanowski (2005). While his study did not address warfare *per se*, he describes changes in practices of endogamy among indigenous populations of

Florida during the 17th century. He examined dental microevolution and concluded that groups who had previously practiced linguistic-group endogamy and inter-linguistic-group warfare shifted their practices after significant population loss and movement. These were in large part the result of the introduction of Spanish missions in the area after A.D. 1650.

Future Research

Queer theory

Intriguing directions for future bioarchaeological research including the analyses of gender are possible with the use of queer theory (see Geller 2005). This standpoint is based on opposition to the normative, and is not restricted to gender or to sexual practices (see Dowson 2000, 2006; Voss 2000, 2008); queer is by definition what is at odds with the normal, the legitimate, the dominant (Halperin 1995:62). As such, it is not a search for homosexuals in the archaeological record it is rather a consideration of any persons or practices considered marginal in their own cultural contexts. Queer studies of past cultures can be focused on ways in which the normative and deviant have been defined, not specifically in sexual behavior but in all social structures (Ardren 2008). In this way, deviance is reflexively related to the normative by continually referencing each other (see Voss 2000). As Voss (2008) notes, one core project in queer theory has been a critique of the conventional divisions among physical sex, cultural gender, and sexuality, instead exploring the ways that sex, gender, and sexuality are mutually constituted. Bioarchaeology may then be unique in its ability to identify multiple social factors that are incorporated in identity, including, but not limited to, gender. As examples, the work of Perry (2004) and Hollimon (1996) identified possible third-gender individuals in skeletal samples from the prehispanic Southwest and California, respectively. These studies incorporated interpretations of bodily changes that were likely related to gender-based division of labor, age, and perhaps other culturally meaningful social identities, such as ritual specialists.

Future research along these lines might examine the ways in which the life course affected these identities. For example, were rites of passage experienced by those who were recognized as third-gender persons? As adolescents or young adults, did these persons engage in gender-based division of labor or occupational specialization that might result in activity-induced pathologies? Might these patterns be recognizable in a skeletal population? The relatively young (under 20 years old) biological males from the California sample displayed advanced spinal arthritis that was typical of older females in this population. Repeated stress from digging graves, an occupation staffed by third-gender males and postmenopausal women, could have produced this pattern of degenerative joint disease. This evidence suggests that the third-gender persons were intiated into this profession during adolescence. It is therefore possible to view these bodily changes as markers of identity that were a result of culturally endorsed events, such as initiation rites related to gender, social age, occupation, group membership, or social rank (see Geller 2008).

Ancient DNA

A number of methods in bioarchaeological analysis hold great promise for future study of gender. Principal among these is the use of ancient DNA (Brown 1998a,b, 2000; Stone 2000; Stone and Stoneking 1999; Stone et al. 1996). Not only does this material offer the possibility of identifying chromosomal sex from nuclear DNA, it may also provide the opportunity to identify nonbinary chromosome arrangements among phenotypically "normal" female and male skeletons (Brown 1998b). This type of analysis could begin to explore the relationships between biological sex and cultural gender by accurately identifying XX, XY, and other chromosome karyotypes; other aspects of social identity, as inferred from mortuary treatment, for example, could follow these identifications. Such methods could be used in conjunction with refined methods in subadult sex estimation (e.g., Cunhaa et al. 2000; Effros 2000; Faerman et al. 1995, 1998; Kondo and Townsend 2004; Lassen and Hummel 1996; Stone et al. 1996; Stone and Stoneking1999; see below).

In addition, mitochondrial DNA (inherited only from the mother) recovered from archaeological skeletal populations can be used to reconstruct matrilineages. Such analysis has been performed to assess prehistoric postmarital residence practices at Norris Farms, Illinois (Stone 2006).

Gender and the life course

A number of scholars have used mortuary data to examine age, sex, and gender identities within the context of the life course; specifically when, where, and how did cultures mark gender? (e.g., Gilchrist 2000, 2004; Joyce 2000; Rega 1997, 2000; Sofaer Derevenski 1997a,b, 2000a; Stoodley 1999, 2000). These studies ask the question, "when does a person become gendered?" Socialization processes contribute to the production of persons into viable and gendered community members (Claassen 2002; Geller 2005), and recent research has begun to consider these cultural practices (e.g., Crown 2002; Sobolik 2002). For example, analyzing bioarchaeological data from the American Southwest, Whittlesey (1996, 2002) inferred very poor maternal health in the high rate of fetal and infant mortality, and suggested that this may have been related to cultural differences in the diets of women and men. Storey (1998) documented patterns of stature differences between elite and non-elite males at the Late Classic Maya site of Copán. There were apparently no differences in the treatment of female and male children at this site, but status distinctions had a differential impact on the growth and development of males during adolescence (Storey 1998).

Identity

Bioarchaeological data can be used to reconstruct past social processes in a manner simply not possible using archaeological or historical data sets alone (Knudson and Stojanowski 2008). Information from the human skeleton can be combined with historical documents and material culture to understand social processes on multiple scales. Social pluralism based on ethnic, linguistic, or religious identities have been experienced many times in human (pre)history, and bioarchaeology can contribute greatly to our understanding of these cultural dynamics (e.g., Whitehouse 2002).

The overlapping axes of identity are highlighted in an example of this type of bioarchaeological research from historic California. In February, 1999, a visitor at Fort Ross State Historic Park in California noticed skeletal remains eroding from the bank of Fort Ross Creek (Hollimon and Murley 2000). After determining that the remains were of archaeological origin, mitochondrial DNA analysis was performed (Hollimon and Murley 2000). The results suggest that this man had Native Alaskan ancestry on his maternal side, and that he had likely lived in California by virtue of the presence of the Russian-American Company (1812–1841). "Fort Ross Man," as he came to be known, apparently ate the foods of the local native people, the Kashaya Pomo, as indicated by the wear on his teeth (Hollimon and Murley 2000). This suggests that he may have had a local native "wife" who prepared this food, but his burial comformed neither to Kashaya nor to Russian Orthodox practices (Goldstein 1995; Osborn 1997). These interpretive challenges led me to consider the various gender, ethnic, class, and religious identities that this man may have chosen for himself, or had imposed upon him by others. I also consider the culturally constructed ethnic identities that resulted from the interaction of Russians and the indigenous peoples of the North Pacific (see Vinkovetsky 2002).

In their review of recent developments in bioarchaeological analyses, Knudson and Stojanowski (2008) call on researchers to continue to apply their developing methodologies to larger questions. They note that bioarchaeologists can contribute important long-term perspectives to studies of social identities, especially those based on health status, age, sex, and gender. Because bioarchaeologists investigate past bodies in terms of historical and cultural contextualization and biocultural interactions (Geller 2005), we are well equipped to contribute to the understanding of sex and gender in the archaeological record.

ACKNOWLEDGMENTS

This work is dedicated to the memory of Phillip L. Walker, who was a bioarchaeologist extraordinaire, a wonderful mentor, and a dear friend. I thank Sabrina Agarwal for inviting my participation in this volume.

REFERENCES

Akins, N. J. 1986 A Biocultural Approach to Human Burials from Chaco Canyon, New Mexico. Santa Fe: National Park Service.

Akins, N. J. 2000 Human Skeletal Remains from the Pena Blanca Archaeological Project. Santa Fe: Museum of New Mexico, Office of Archaeological Studies.

Alberti, B. 2006 Archaeology, Men, and Masculinities. *In* Handbook of Gender Archaeology. S. M. Nelson, ed. Pp. 401–434. Walnut Creek: AltaMira Press.

Ambrose, S. H., J. Buikstra, and H. W. Krueger 2003 Status and Gender Differences in Diet at Mound 72, Cahokia, Revealed by Isotopic Analysis of Bone. Journal of Anthropological Archaeology 22:217–226.

Andrushko, V., K. A.S. Latham, D. L. Grady, A. G. Pastron, and P. L. Walker 2005 Bioarchaeological Evidence for Trophy-Taking in Prehistoric Central California. American Journal of Physical Anthropology 127:375–384.

Andrushko, V., A. W. Schwitalla, and P. L. Walker 2010 Trophy-Taking and Dismemberment as Warfare Strategies in Prehistoric Central California. American Journal of Physical Anthropology 141(1):83–96.

Aranda, G., S. Montón Subías, M. Sánchez-Romero, and E. Alarcón 2009 Death and Everyday Life: The Argaric Societies from Southeast Iberia. Journal of Social Archaeology 9:139–162.

Ardren, T. 2002 Death Became Her: Images of Female Power from Yaxuna Burials. *In* Ancient Maya Women. T. Ardren, ed. Pp. 68–88. Walnut Creek: AltaMira Press.

Ardren, T. 2008 Studies of Gender in the Prehispanic Americas. Journal of Archaeological Research 16:1–35.

Arnold, B. 1991 The Deposed Princess of Vix: The Need for an Engendered European Prehistory. *In* The Archaeology of Gender. Proceedings of the 22nd Annual Chacmool Conference. D. Walde, and N. D. Willows, eds. Pp. 366–374. Calgary: Department of Archaeology, University of Calgary.

Arnold, B. 2001 The Limits of Agency in the Analysis of Elite Celtic Iron Age Burials. Journal of Social Archaeology 1(2):211–223.

Arnold, B. 2002 "Sein und Werden": Gender as Process in Mortuary Ritual. *In* In Pursuit of Gender: Worldwide Archaeological Approaches. S. M. Nelson, and M. Rosen-Ayalon, eds. Pp. 239–256. Walnut Creek: AltaMira Press.

Arnold, B. 2006 Gender and Archaeological Mortuary Analysis. *In* Handbook of Gender in Archaeology. Sarah M. N., ed., Pp. 137–170. Walnut Creek: AltaMira Press.

Bacus, E. 2007 Expressing Gender in Bronze Age North East Thailand: The Case of Nok Nok Tha. *In* Archaeology and Women: Ancient and Modern Issues. S. Hamilton, R. D. Whitehouse, and K. I. Wright, eds. Pp. 312–334. Walnut Creek: Left Coast Press.

Barrett, J. H., and M. P. Richards 2004 Identity, Gender, Religion and Economy: New Isotope and Radiocarbon Evidence for Marine Resource Intensification in Early Historic Orkney, Scotland, UK. European Journal of Archaeology 7:249–271.

Bartelink, E. 2001 Elbow Arthritis in the Prehistoric San Francisco Bay: A Bioarchaeological Interpretation of Resource Intensification and the Sexual Division of Labor. M.A. Thesis, California State University, Chico.

Becker, M. J. 2000 Reconstructing the Lives of South Etruscan Women. *In* Reading the Body: Representations and Remains in the Archaeological Record. A. E. Rautman, ed. Pp. 55–67. Philadelphia: University of Pennsylvania Press.

Bell, E. E. 2002 Engendering a Dynasty: A Royal Woman in the Margarita Tomb, Copan. *In* Ancient Maya Women. T. Ardren, ed. Pp. 89–104. Walnut Creek: AltaMira Press.

Bentley, R. A., L. Chikhi, and T. D. Price 2003 The Neolithic Transition in Europe: Comparing Broad Scale Genetic and Local Scale Isotopic Evidence. Antiquity 77:63–66.

Bentley, R. A., M. Pietrusewsky, M. T. Douglas, and T. C. Atkinson 2005 Matrilocality During the Prehistoric Transition to Agriculture in Thailand? Antiquity 79:865–881.

Bentley, R. A., N. Tayles, C. Higham, C. Macpherson, and T. C. Atkinson 2007 Shifting Gender Relations at Khok Phanom Di, Thailand: Isotopic Evidence from the Skeletons. Current Anthropology 48:301–314.

Berseneva, N. 2008 Women and Children in the Sargat Culture. *In* Are All Warriors Male? Gender Roles on the Ancient Eurasian Steppe. K. M. Linduff, and K. S. Rubinson, eds. Pp. 131–151. Walnut Creek: AltaMira Press.

Bice, G. 2003 Reconstructing Behavior from Archaeological Skeletal Remains: A Critical Analysis of the Biomechanical Model. Ph.D. Dissertation, Michigan State University, East Lansing.

Blakely, R. L., and D. S. Matthews 1990 Bioarchaeological Evidence for a Spanish-Native American Conflict in the Sixteenth-Century Southeast. American Antiquity 55:718–744.

Bolger, D. L. 2003 Gender in Ancient Cyprus: Narratives of Social Change on a Mediterranean Island. Walnut Creek: AltaMira Press.

Bolger, D. L., and N. J. Serwint eds. 2002 Engendering Aphrodite: Women and Society in Ancient Cyprus. Boston: American Society for Oriental Research.

Boyd, D. C. 1996 Skeletal Correlates of Human Behavior in the Americas. Journal of Archaeological Method and Theory 3:189–251.

Bridges, P. S. 1985 Structural Changes of the Arms Associated with the Habitual Grinding of Corn. American Journal of Physical Anthropology 66:149–150.

Bridges, P. S. 1989 Changes in Activities with the Shift to Agriculture in the Southeastern United States. Current Anthropology 30:385–393.

Bridges, P. S. 1991 Degenerative Joint Disease in Hunter-Gatherers and Agriculturalists from the Southeastern United States. American Journal of Physical Anthropology 85:379–391.

Bridges, P. S. 1994 Vertebral Arthritis and Physical Activities in the Prehistoric Southwestern United States. American Journal of Physical Anthropology 93:83–93.

Bridges, P. S. 1996 Warfare and Mortality at Koger's Island, Alabama. International Journal of Osteoarchaeology 6:66–75.

Bridges, P. S., J. H. Blitz, and M. C. Solano 2000 Changes in Long Bone Diaphyseal Strength with Horticultural Intensification in West-Central Illinois. American Journal of Physical Anthropology 112:217–238.

Brooks, S. T., M. B. Haldeman, and R. A. Brooks 1988 Osteological Analysis of the Stillwater Skeletal Series, Stillwater Marsh, Churchill County, Nevada. Portland: U.S. Fish and Wildlife Service Cultural Resources Series 2.

Brown, K. A. 1998a Gender and Sex: Distinguishing the Difference with Ancient DNA. *In* Gender and Italian Archaeology: Challenging Stereotypes. R. D. Whitehouse, ed. Pp. 35–44. London: Accordia Research Institute/University College London.

Brown, K. A. 1998b Gender and Sex – What Can Ancient DNA Tell Us? Ancient Biomolecules 2:3–15.

Brown, K. A. 2000 Ancient DNA Applications in Human Osteoarchaeology: Achievements, Problems and Potential. *In* Human Osteology in Archaeology and Forensic Science. M. Cox, and S. Mays, eds. Pp. 455–473. London: Greenwich Medical Media.

Cannon, A. 2005 Gender, Agency, and Mortuary Fashion. *In* Interacting with the Dead: Perspectives on Mortuary Archaeology for the New Millennium. G. F.M. Rakita, J. E. Buikstra, L. A. Beck, and S. R. Williams, eds. Pp. 41–65. Gainesville: University of Florida Press.

Cassidy, C. M. 1984 Skeletal Evidence for Prehistoric Subsistence Change in the Central Ohio River Valley. *In* Paleopathology at the Origins of Agriculture. M. N. Cohen, and G. J. Armelagos, eds. Pp. 307–346. New York: Academic Press.

Chacon, R. J., and D. H. Dye, eds. 2007 The Taking and Displaying of Human Body Parts as Trophies by Amerindians. New York: Springer.

Chapman, N. E. M. 1997 Evidence for Spanish Influence on Activity Induced Musculoskeletal Stress Markers at Pecos Pueblo. International Journal of Osteoarchaeology 7:497–506.

Churchill, S. E., and A. G. Morris 1998 Muscle Marking Morphology and Labour Intensity in Prehistoric Khoisan Foragers. International Journal of Osteoarchaeology 8:390–411.

Claassen, C. 1992 Questioning Gender: An Introduction. *In* Exploring Gender Through Archaeology: Selected Papers from the (1991) Boone Conference. C. Claassen, ed. Pp. 1–9. Madison: Prehistory Press.

Claassen, C. 2001 Challenges for Regendering Southeastern Prehistory. *In* Archaeological Studies of Gender in the Southeastern United States. J. M. Eastman, and C. B. Rodning, eds. Pp. 10–26. Gainesville: University of Florida Press.

Claassen, C. 2002 Mother's Workloads and Children's Labor During the Woodland Period. *In* In Pursuit of Gender: Worldwide Archaeological Approaches. S. M. Nelson and M. Rosen-Ayalon, eds. Pp. 225–234. Walnut Creek: AltaMira Press.

Cohen, M. N., and S. Bennett 1993 Skeletal Evidence for Sex Roles and Gender Hierarchies in Prehistory. *In* Sex Roles and Gender Hierarchies. B. D. Miller, ed. Pp. 273–296. Cambridge: Cambridge University Press.

Conkey, M. W. 2007 Questioning Theory: Is There a Gender of Theory in Archaeology? Journal of Archaeological Method and Theory 14:217–234.

Conkey, M. W., and J. Gero 1997 Programme to Practice: Gender and Feminism in Archaeology. Annual Review of Anthropology 26:411–437.

Conkey, M. W., and J. Spector 1984 Archaeology and the Study of Gender. Advances in Archaeological Method and Theory 7:1–38.

Cook, D. C. 1984 Subsistence and Health in the Lower Illinois Valley: Osteological Evidence. *In* Paleopathology at the Origins of Agriculture. M. N. Cohen, and G. J. Armelagos, eds. Pp. 237–270. New York: Academic Press.

Corrucini, R. S. 1972 The Biological Relationships of Some Prehistoric and Historic Pueblo Populations. American Journal of Physical Anthropology 37:373–388.

Corrucini, R. S. 1998 On Hawikku Cemetery Kin Groupings. American Antiquity 63:161–163.

Crabtree, P. 2006 Women, Gender, and Pastoralism. *In* Handbook of Gender Archaeology. S. M. Nelson, ed., Pp. 571–592. Walnut Creek: AltaMira Press.

Crass, B. 2000 Gender in Inuit Burial Practices. *In* Reading the Body: Representations and Remains in the Archaeological Record. A. E. Rautman, ed. Pp. 68–76. Philadelphia: University of Pennsylvania Press.

Crass, B. 2001 Gender and Mortuary Analysis: What Can Grave Goods Really Tell Us? *In* Gender and the Archaeology of Death. B. Arnold, and N. Wicker, eds. Pp. 105–118. Walnut Creek: AltaMira Press.

Crown, P. L. 2002 Learning and Teaching in the Prehispanic American Southwest. *In* Children in the Prehistoric Puebloan Southwest. K. A. Kamp, ed. Pp. 108–124. Salt Lake City: University of Utah Press.

Cunhaa, E., M.-L. Filyc, I. Clissonc, A. L. Santosd, A. M. Silvad, C. Umbelinod, P. Césare, A. Corte-Reale, E. Crubézy, and B. Ludesh 2000 Children at the Convent: Comparing Historical Data, Morphology and DNA Extracted from Ancient Tissues for Sex Diagnosis at Santa Clara-a-Velha (Coimbra, Portugal). Journal of Archaeological Science 27(10):949–952.

Cybulski, J. S. 1992 A Greenville Burial Ground: Human Remains and Mortuary Elements in British Columbia Coast Prehistory. Canadian Museum of Civilization Mercury Series Archaeological Survey Paper 146. Ottowa: Canadian Museum of Civilization.

Cybulski, J. S. 1994 Culture Change, Demographic History, and Health and Disease on the Northwest Coast. *In* In the Wake of Contact: Biological Responses to Conquest. C. S. Larsen, and G. R. Milner, eds. Pp. 75–85. New York: Wiley-Liss.

Danforth, M. E. 1999 Nutrition and Politics in Prehistory. Annual Review of Anthropology 28:1–25.

Danforth, M. E., K. Jacobi, and M. N. Cohen 1997 Gender and Health Among the Colonial Maya of Tipu, Belize. Ancient Mesoamerica 8:13–22.

Davis-Kimball, J. 1997 Sauro-Sarmation Nomadic Women: New Gender Identities. Journal of Indo-European Studies 25:327–343.

Díaz-Andreu, M. 2005 Gender Identity. *In* The Archaeology of Identity: Approaches to Gender, Age, Status, Ethnicity, and Religion. M. Díaz-Andreu, S. Lucy, S. Babič, and D. N. Edwards, eds. Pp. 13–42. London: Routledge.

DeWitte, S. N. 2009 The Effect of Sex on Risk of Mortality During the Black Death in London, A.D. 1349–1350. American Journal of Physical Anthropology, early view http://0-www3.interscience.wiley.com.

Doucette, D. L. 2001 Decoding the Gender Bias: Inferences of Atlatls in Female Mortuary Contexts. *In* Gender and the Archaeology of Death. B. Arnold, and N. Wicker, eds. Pp. 159–177. Walnut Creek: AltaMira Press.

Dowson, T. ed. 2000 Why Queer Archaeology? An Introduction. World Archaeology 32: 161–165.

Dowson, T. 2006 Archaeologists, Feminists, and Queers: Sexual Politics in the Construction of the Past. *In* Feminist Anthropology: Past, Present, and Future. P. L. Geller, and M. K. Stockett eds. Pp. 89–102. Philadelphia: University of Pennsylvania Press.

Driscoll, E. M. 2001 Bioarchaeology, Mortuary Patterning, and Social Organization at Town Creek. Ph.D. Dissertation, University of North Carolina, Chapel Hill.

Duke, P. 1991 Recognizing Gender in Plains Hunting Groups: Is It Possible or Even Necessary? *In* The Archaeology of Gender. Proceedings of the 22nd Annual Chacmool Conference. D. Walde, and N. D. Willows, eds. Pp. 280–283. Calgary: Department of Archaeology, University of Calgary.

Effros, B. 2000 Skeletal Sex and Gender in Merovingian Mortuary Archaeology. Antiquity 74: 632–639.

Eisenberg, L. E. 1986 Adaptation in a "Marginal" Mississippian Population from Middle Tennessee: Biocultural Insights from Paleopathology. Ph.D. Dissertation, New York University.

Engelstad, E. 2007 Much More Than Gender. Journal of Archaeological Method and Theory 14:217–234.

Ezzo, J. A. 1993 Human Adaptation at Grasshopper Pueblo, Arizona: Social and Ecological Perspectives. Ann Arbor: Archaeological Series No. 4, International Monographs in Prehistory.

Faerman, M., D. Filon, G. Kahila, C. L. Greenblatt, P. Smith, and A. Oppenheim 1995 Sex Identification of Archaeological Human Remains Based on Amplification of the X and Y Amelogenin Alleles. Gene 167:327–332.

Faerman, M., G. Kahila Bar-Gal, D. Filon, C. L. Greenblatt, L. Stager, A. Oppenheim, and P. Smith 1998 Determining the Sex of Infanticide Victims from the Late Roman Era Through Ancient DNA Analysis. Journal of Archaeological Science 25: 861–865.

Fink, T. M. 1989 The Human Skeletal Remains from the Grand Canal Ruins, AZ T:12:14 (ASU) and AZ T:12:16 (ASU). *In* Archaeological Investigations at the Grand Canal Ruins: A Classic Period Site in Phoenix, Arizona. D. R. Mitchell, ed. Pp. 619–704. Soil Systems Publications in Archaeology 12. Phoenix: Soil Systems.

Frayer, D. W. 1988 Sex Differences in Tooth Wear at Skateholm. Mesolithic Miscellany 9:11–12.

Fresia, A. E., C. B. Ruff, and C. S. Larsen 1990 Temporal Decline in Bilateral Assymetry of the Upper Limb on the Georgia Coast. *In* The Archaeology of Santa Catalina de Guale: 2. Biocultural Interpretations of a Population in Transition. C. S. Larsen, ed. Pp. 121–131. New York: Anthropological Papers of the Museum of Natural History.

Gamble, L. H., P. L. Walker, and G. S. Russell 2001 An Integrative Approach to Mortuary Analysis: Social and Symbolic Dimensions of Chumash Burial Practices. American Antiquity 66:185–212.

Geller, P. L. 2005 Skeletal Analysis and Theoretical Complications. World Archaeology 37:597–609.

Geller, P. L. 2006 Altering Identities: Body Modifications and the Pre-Columbian Maya. *In* The Social Archaeology of Funerary Remains. R. Gowland, and C. Knüsel, eds. Pp. 168–178. Oxford: Oxbow Books.

Geller, P. L. 2008 Conceiving Sex: Fomenting a Feminist Bioarchaeology. Journal of Social Archaeology 8:113–138.

Geller, P. L. 2009 Identity and Difference: Complicating Gender in Archaeology. Annual Review of Anthropology 38:65–81.

Gere, C. 1999 Bones That Matter: Sex Determination in Paleodemography 1948–1995. Studies in History and Philosophy of Biological and Biomedical Sciences 30:455–471.

Gerry, J. P., and M. S. Chesson 2000 Classic Maya Diet and Gender Relationships. *In* Gender and Material Culture in Archaeological Perspective. M. Donald, and L. Hurcombe, eds. Pp. 250–265. New York: St. Martin's Press.

Gibbs, L. 1987 Identifying Gender Representation in the Archaeological Record: A Contextual Study. *In* The Archaeology of Contextual Meanings. I. Hodder, ed. Pp. 79–89. Cambridge: Cambridge University Press.

Gilchrist, R. 1997 Ambivalent Bodies: Some Theoretical and Methodological Concerns on a Burgeoning Archaeological Pursuit. *In* Invisible People and Processes: Writing Gender and Childhood into European Archaeology. J. Moore, and E. Scott, eds. Pp. 42–58. London: Leicester University Press.

Gilchrist, R. 1999 Gender and Archaeology: Contesting the Past. London: Routledge.

Gilchrist, R. ed. 2000 Human Lifecycles. World Archaeology 31:325–328.

Gilchrist, R. 2004 Archaeology and the Life Course: A Time and Age for Gender. *In* A Companion to Social Archaeology. L. Meskell, and R. Preucel, eds. Pp. 142–160. Malden, MA: Blackwell.

Gillespie, S. D. 2001 Personhood, Agency, and Mortuary Ritual: A Case Study from the Ancient Maya. Journal of Anthropological Archaeology 20:73–112.

Gjerdrum, T., P. L. Walker, and V. A. Andrushko 2003 Humeral Retroversion: An Activity Pattern Index in Prehistoric Southern California. American Journal of Physical Anthropology Supplement 36:100–101.

Goldstein, L. 1995 Politics, Law, Pragmatics, and Human Burial Excavations: An Example from Northern California. *In* Bodies of Evidence. A. L. Grauer, ed. Pp. 3–17. New York: John Wiley and Sons.

Goldstein, L. 2006 Mortuary Analysis and Bioarchaeology. *In* Bioarchaeology: The Contextual Analysis of Human Remains. J. E. Buikstra, and L. A. Beck, eds. Pp. 375–387. Burlington, MA: Academic Press.

González Cruz, A. 2004 The Red Queen's Masks: Symbols of Power. Arqueología Mexicana 16:22–25.

Goodman, A., D. Martin, G. J. Armelagos, and G. Clark 1984 Health Changes at Dickson Mounds, Illinois (AD 950-1300). *In* Paleopathology at the Origins of Agriculture. M. N. Cohen, and G. J. Armelagos, eds. Pp. 271–306. New York: Academic Press.

Gräslund, A.-S. 2001 The Position of Iron Age Women: Evidence from Graves and Rune Stones. *In* Gender and the Archaeology of Death. B. Arnold, and N. Wicker, eds. Pp. 81–102. Walnut Creek: AltaMira Press.

Grauer, A. L., and C. A. Roberts 1996 Paleoepidemiology, Healing, and Possible Treatment of Trauma in the Medieval Cemetery Population of St. Helen-on-the-Walls, York, England. American Journal of Physical Anthropology 100:531–544.

Grauer, A., and P. Stuart-Macadam, eds. 1998 Sex and Gender in Paleopathological Perspective. Cambridge: University of Cambridge Press.

Guatelli-Steinberg, D., and J. R. Lukacs 1999 Interpreting Sex Differences in Enamel Hypoplasia in Human and Non-Human Primates: Developmental, Environmental, and Cultural Considerations. Yearbook of Physical Anthropology 42:73–126.

Habicht-Mauche, J. A., A. A. Levendosky, and M. J. Schoeninger 1994 Antelope Creek Phase Subsistence: The Bone Chemistry Evidence. *In* Skeletal Biology of the Great Plains. D. W. Owsley, and R. L. Jantz, eds. Pp. 291–304. Washington, DC: Smithsonian Institution Press.

Halperin, D. M. 1995 Saint Foucault: Towards a Gay Hagiography. New York: Oxford University Press.

Hamilton, M. E. 1982 Sexual Dimorphism in Skeletal Samples. *In* Sexual Dimorphism in *Homo sapiens*: A Question of Size. R. Hall, ed. Pp. 107–163. New York: Praeger.

Hamlin, C. 2001 Sharing the Load: Gender and Task Division at the Windover Site. *In* Gender and the Archaeology of Death. B. Arnold, and N. Wicker, eds. Pp. 119–135. Walnut Creek: AltaMira Press.

Hanks, B. 2008 Reconsidering Warfare, Status, and Gender in the Eurasian Steppe Iron Age. *In* Are All Warriors Male? Gender Roles on the Ancient Eurasian Steppe. K. M. Linduff, and K. S. Rubinson, eds. Pp. 15–34. Walnut Creek: AltaMira Press.

Hastorf, C. A. 1991 Gender, Space, and Food in Prehistory. *In* Engendering Archaeology: Women and Prehistory. J. W. Gero, and M. W. Conkey, eds. Pp. 132–159. Oxford: Blackwell.

Haviland, W. A. 1997 The Rise and Fall of Sexual Inequality: Death and Gender at Tikal, Guatemala. Ancient Mesoamerica 8:1–12.

Hawkey, D. E., and C. F. Merbs 1995 Activity-Induced Musculoskeletal Stress Markers (MSM) in and Subsistence Strategy Changes Among Hudson Bay Eskimos. International Journal of Osteoarcheaology 5:324–338.

Hemphill, B. E. 1992 An Osteological Analysis of Human Remains from Malheur Lake, Oregon. 3 vols. U.S. Fish and Wildlife Service Cultural Resource Series 6.

Hemphill, B. E., J. R. Lukacs, and K. A. R. Kennedy 1991 Biological Adaptations and Affinities of Bronze Age Harappans. *In* Harappa Excavations 1986–1990: A Multidisciplinary Approach to the Third Millennium Urbanism. R. H. Meadow, ed., Pp. 137–182. Madison: Prehistory Press.

Higham, C. 2002 Women in the Prehistory of Mainland Southeast Asia. *In* In Pursuit of Gender: Worldwide Archaeological Approaches. S. M. Nelson, and M. Rosen-Ayalon, eds. Pp. 207–224. Walnut Creek: AltaMira Press.

Higham, C., and R. Bannanurag 1990 The Princess and the Pots. New Scientist 26:50–55.

Higham, C., and R. Thosarat 1994 Thailand's Good Mound. Natural History 12:60–66.

Hollimon, S. E. 1992 Health Consequences of Sexual Division of Labor Among Native Americans: The Chumash of California and the Arikara of the Northern Plains. *In* Exploring Gender Through Archaeology. C. Claassen, ed. Pp. 81–88. Madison: Prehistory Press.

Hollimon, S. E. 1996 Sex, Gender and Health Among the Chumash of the Santa Barbara Channel Area. Proceedings of the Society for California Archaeology 9:205–208.

Hollimon, S. E. 1997 The Third Gender in Native California: Two-Spirit Undertakers Among the Chumash and Their Neighbors. *In* Women in Prehistory: North America and Mesoamerica. C. Claassen, and R. A. Joyce, eds. Pp. 171–188. Philadelphia: University of Pennsylvania Press.

Hollimon, S. E. 2000 Sex, Health, and Gender Roles Among the Arikara of the Northern Plains. *In* Reading the Body: Representations and Remains in the Archaeological Record. A. E. Rautman, ed., Pp. 25–37. Philadelphia: University of Pennsylvania Press.

Hollimon, S. E. 2001a Warfare and Gender in the Northern Plains: Osteological Evidence of Trauma Reconsidered. *In* Gender and the Archaeology of Death. B. Arnold, and N. Wicker, eds. Pp. 179–193. Walnut Creek: AltaMira Press.

Hollimon, S. E. 2001b The Gendered Peopling of North America: Addressing the Antiquity of Systems of Multiple Genders. *In* The Archaeology of Shamanism. N. S. Price, ed. Pp. 123–134. New York: Routledge.

Hollimon, S. E. 2006 The Archaeology of Nonbinary Genders in Native North American Societies. *In* Handbook of Gender Archaeology. S. M. Nelson, ed. Pp. 435–450. Walnut Creek: AltaMira Press.

Hollimon, S. E., and D. F. Murley 2001 Stranger in a Strange Land: The Fort Ross Burial Isolate. Society for California Archaeology Newsletter 35:28–30.

Hollimon, S. E., and D. W. Owsley 1994 Osteological Evidence of Violence at the Fay Tolton Site, South Dakota. *In* Skeletal Biology of the Great Plains. D. W. Owsley, and R. L. Jantz, eds. Pp. 345–353. Washington, DC: Smithsonian Institution Press.

Hoshower, L. M., and J. T. Milanich 1993 Excavations in the Fig Springs Mission Burial Area. *In* The Spanish Mission of La Florida. B. G. McEwan ed. Pp. 217–243. Gainesville: University of Florida Press.

Howell, T. L., and K. W. Kintigh 1996 Archaeological Identification of Kin Groups Using Mortuary and Biological Data: An Example from the American Southwest. American Antiquity 61:537–554.

Howell, T. L., and K. W. Kintigh 1998 Determining Gender and Kinship at at Hawikku: A Reply to Corruccini. American Antiquity 63:164–167.

Hutson, S. R. 2002 Gendered Citation Practices in American Antiquity and Other Archaeological Journals. American Antiquity 67:331–342.

Jiao, T. 2001 Gender Studies in Chinese Neolithic Archaeology. *In* Gender and the Archaeology of Death. B. Arnold, and N. L. Wicker, eds. Pp. 51–62. Walnut Creek: AltaMira Press.

Jiménez-Brobeil, S. A., Ph. Du Souich and I. al Oumaoui 2009 Possible Relationship of Cranial Traumatic Injuries with Violence in the South-East Iberian Peninsula from the Neolithic to the Bronze Age. American Journal of Physical Anthropology 140: 465–475.

Johnsson, C., K. Ross, and S. Welinder 2000 Gender, Material Culture, Ritual, and Gender System: A Prehistoric Example Based on Sickles. *In* Gender and Material Culture in Archaeological Perspective. M. Donald, and L. Hurcombe, eds. Pp. 169–184. London: Macmillan.

Joyce, R. A. 1999 Social Dimensions of Preclassic Burials. *In* Ritual Behavior, Social Identity and Cosmology in Pre-Classic Mesoamerica. D. C. Grove, and R. A. Joyce, eds. Pp. 15–47. Washington, DC: Dumbarton Oaks.

Joyce, R. A. 2000 Girling the Girl and Boying the Boy: The Production of Adulthood in Ancient Mesoamerica. World Archaeology 31(3):473–483.

Joyce, R. A. 2001 Burying the Dead at Tlatilco: Social Memory and Social Identities. *In* New Perspectives on Mortuary Analysis. M. Chesson, ed., Pp. 12–26. Washington, DC: American Anthropological Association.

Joyce, R. A. 2002 Beauty, Sexuality, Body Ornamentation and Gender in Ancient Meso-America. *In* In Pursuit of Gender: Worldwide Archaeological Approaches. S. M. Nelson, and M. Rosen-Ayalon, eds. Pp. 81–91. Walnut Creek: AltaMira Press.

Joyce, R. A. 2004 Embodied Subjectivity: Gender, Femininity, Masculinity, Sexuality. *In* A Companion to Social Archaeology. L. Meskell, and R. W. Preucel, eds. Pp. 82–95. Malden, MA: Blackwell.

Joyce, R. A. 2005 Archaeology of the Body. Annual Review of Anthropology 34:139–158.

Joyce, R. A. 2006 Feminist Theories of Embodiment and Anthropological Imagination: Making Bodies Matter. *In* Feminist Anthropology: Past, Present, and Future. P. L., Geller and M. K. Stockett eds. Pp. 43–54. Philadelphia: University of Pennsylvania Press.

Joyce, R. A. 2008 Ancient Bodies, Ancient Lives: Sex, Gender, and Archaeology. London: Thames and Hudson.

Judd, M. A., and C. A. Roberts 1999 Fracture Trauma in a Medieval British Farming Village. American Journal of Physical Anthropology 109:229–243.

Jurmain, R. 1991 Degenerative Changes in Peripheral Joints as Indicators of Mechanical Stress: Opportunities and Limitations. International Journal of Osteoarchaeology 1:247–252.

Jurmain, R. 1999 Stories from the Skeleton: Behavioral Reconstruction in Human Osteology. Amsterdam: Gordon and Breach.

Jurmain, R., and V. I. Bellifemine 1997 Patterns of Cranial Trauma in a Prehistoric Population from Central California. International Journal of Osteoarchaeology 7: 43–50.

Jurmain, R., and L. Kilgore 1998 Sex-Related Patterns of Trauma in Humans and African Apes. Sex and Gender in Paleopathological Perspective 11–26.

Jurmain, R., E. J. Bartelink, A. Leventhal, V. Bellifemine, I. Nechayev, M. Atwood, and Di. DiGiuseppe 2009 Paleoepidemiological Patterns of Interpersonal Aggression in a Prehistoric Central California Population From CA-ALA-329. American Journal of Physical Anthropology 140:462–473.

Kelley, J. O., and J. L. Angel 1987 Life Stresses of Slavery. American Journal of Physical Anthropology 74:199–211.

Kerr, S. 2004 Insular Interactions in the California Bight: A Bioarchaeological View from the Southern Channel Islands. Ph.D. Dissertation, University of California, Santa Barbara.

Kestle, S. 1988 Subsistence and Sex Roles. *In* The King Site: Continuity and Contact in Sixteenth Century Georgia. R. L. Blakely, ed. Pp. 63–72. Athens: University of Georgia Press.

Keswani, P. 2004 Mortuary Ritual and Society in Bronze Age Cyprus. London: Equinox.

Kirkpatrick, R. C. 2000 The Evolution of Human Homosexual Behavior. Current Anthropology 41:385–398.

Klaus, H. D., and M. E. Tam 2008 Contact in the Andes: Bioarchaeology of Systemic Stress in Colonial Mórrope, Peru. American Journal of Physical Anthropology, early view http://0-www3.interscience.wiley.com

Klaus, H. D., C. S. Larsen, and M. E. Tam 2009 Economic Intensification and Degenerative Joint Disease: Life and Labor on the Postcontact North Coast of Peru. American Journal of Physical Anthropology, early view http://0-www3.interscience.wiley.com

Klein, C. F. ed. 2001 Gender in Pre-Hispanic America. Washington, DC: Dumbarton Oaks.

Knapp, B., and L. Meskell 1997 Bodies of Evidence on Prehistoric Cyprus. Cambridge Archaeological Journal 7:183–204.

Knudson, K. J., and C. M. Stojanowski 2008 New Directions in Bioarchaeology: Recent Contributions to the Study of Human Social Identities. Journal of Archaeological Research 16:397–432.

Knüsel, C. J. 2002 More Circe than Cassandra: The Princess of Vix in Ritualized Social Context. European Journal of Archaeology 5:276–309.

Kohler, T. A., and K. K. Turner 2006 Raiding for Women in the Pre-Hispanic Northern Pueblo Southwest? Current Anthropology 47:1035–1045.

Kondo, S., and G. C. Townsend 2004 Sexual Dimorphism in Crown Units of Mandibular Deciduous and Permanent Molars in Australian Aborigines. Homo 55:53–64.

Konigsberg, L. W. 1985 Demography and Mortuary Practice at Seip Mound One. Midcontinental Journal of Archaeology 10:123–148.

Konigsberg, L. W., and J. E. Buikstra 1995 Regional Approaches to the Investigation of Past Human Biocultural Structure. *In* Regional Approaches to Mortuary Practices. L. A. Beck, ed. Pp. 191–220. New York: Plenum.

Kuckelman, K., R. Lightfoot, and D. Martin 2002 The Bioarchaeology and Taphonomy of Violence at Castle Rock and Sand Canyon Pueblos, Southwestern Colorado. American Antiquity 67:486–513.

Kuhn, S. L., and M. C. Stiner 2006 What's a Mother to Do? The Division of Labor Among Neandertals and Modern Humans in Eurasia. Current Anthropology 47: 953–980.

Lahren, C. H., and H. E. Berryman 1984 Fracture Patterns and Status at Chucallisa (40SY1): A Biocultural Approach. Tennessee Anthropologist 9:15–21.

Lambert, P. M. 1997 Patterns of Violence in Prehistoric Hunter-Gatherer Societies of Coastal Southern California. *In* Troubled Times: Violence and Warfare in the Past. D. L. Martin, and D. W. Frayer, eds. Pp. 77–109. Amsterdam: Gordon and Breach.

Lambert, P. M. 1999 Human Skeletal Remains. *In* The Puebloan Occupation of the Ute Mountain Piedmont, Vol. 5: Environmental and Bioarchaeological Studies. B. R. Billman, ed. Pp. 111–161. Soil Systems Publications in Archaeology 22. Phoenix: Soil Systems.

Lambert, P. M. 2001 Auditory Exostoses: A Clue to Gender in Prehistoric and Historic Farming Communities of North Carolina and Virginia. *In* Archaeological Studies of Gender in the Southeastern United States. J. M. Eastman, and C. B. Rodning, eds. Pp. 152–172. Gainesville: University of Florida Press.

Lambert, P. M. 2002 Bioarchaeology at Coweeta Creek: Continuity and Change in Native Health and Lifeways in Protohistoric Western North Carolina. Southeastern Archaeology 21:36–48.

Larsen, C. S. 1983 Behavioural Implications in Temporal Change in Cariogenesis. Journal of Archaeological Science 10:1–8.

Larsen, C. S. 1984 Health and Disease in Prehistoric Georgia: The Transition to Agriculture. *In* Paleopathology at the Origins of Agriculture. M. N. Cohen, and G. J. Armelagos, eds. Pp. 367–392. New York: Academic Press.

Larsen, C. S. 1985 Dental Modifications and Tool Use in the Western Great Basin. American Journal of Physical Anthropology 67:393–402.

Larsen, C. S. 1995a Prehistoric Human Biology of the Carson Desert: A Bioarchaeological Investigation of a Hunter-Gatherer Lifeway. *In* Bioarchaeology of the Stillwater Marsh: Prehistoric Human Adaptation in the Western Great Basin. C. S. Larsen, and R. L. Kelly, eds. Pp. 33–40. New York: American Museum of Natural History Papers 77.

Larsen, C. S. 1995b Biological Changes in Human Populations with Agriculture. Annual Review of Anthropology 24:185–213.

Larsen, C. S. 1998 Gender, Health and Activity in Foragers and Farmers in the American Southeast: Implications for Social Organization in the Georgia Bight. *In* Sex and Gender in Paleopathological Perspective. A. Grauer, and P. Stuart-Macadam, eds. Pp. 165–187. Cambridge: University of Cambridge Press.

Larsen, C. S., R. Shavit, and M. C. Griffin 1991 Dental Caries Evidence for Dietary Change: An Archaeological Context. *In* Advances in Dental Anthroplogy. M. A. Kelly, and C. S. Larsen, eds. Pp. 179–202. New York: Wiley-Liss.

Lassen, C., and S. Hummel 1996 PCR Based Sex Identification of Ancient Human Bones by Amplification of X- and Y-Chromosomal Sequences: A Comparison. Ancient Biomolecules 1:25–33.

Ledger, M., L.-M. Holtzhausen, D. Constant, and A. G. Morris 2000 Biomechanical Beam Analysis of Long Bones from a Late 18th Century Slave Cemetery in Cape Town, South Africa. American Journal of Physical Anthropology 112:207–216.

Lee Thorp, J. A., J. C. Sealy, and N. J. Van Der Merwe 1989 Stable Carbon Isotope Ratio Differences Between Bone Collagen and Bone Appatite, and Their Relationship to Diet. Journal of Archaeological Science 16:585–599.

Legrand, S. 2008 Sorting Out Men and Women in the Karasuk Culture. *In* Are All Warriors Male? Gender Roles on the Ancient Eurasian Steppe. K. M. Linduff, and K. S. Rubinson, eds. Pp. 153–174. Walnut Creek: AltaMira Press.

Linduff, K. M. 2008 The Gender of Luxury and Power Among the Xiongnu in Eastern Eurasia. *In* Are All Warriors Male? Gender Roles on the Ancient Eurasian Steppe. K. M. Linduff, and K. S. Rubinson, eds. Pp. 175–194. Walnut Creek: AltaMira Press.

Logan, M. H., C. S. Sparks, and R. L. Jantz 2003 Cranial Modification Among 19th Century Osages: Admixture and Loss of an Ethnic Marker. Plains Anthropologist 48:209–224.

Lorentz, K. 2008 From Life Course to Longue Durée: Headshaping as Gendered Capital? *In* Gender Through Time in the Ancient Near East. D. Bolger, ed. Pp. 281–311. Walnut Creek: AltaMira Press.

Lovell, N. C., and A. A. Dublenko 1999 Further Aspects of Fur Trade Life Depicted in the Skeleton. International Journal of Osteoarchaeology 9:248–259.

Lucy, S. J. 1997 Housewives, Warriors, and Slaves? Sex and Gender in Anglo-Saxon Burials. *In* Invisible People and Processes: Writing Gender and Childhood into European Archaeology. J. Moore, and E. Scott, eds. Pp. 150–168. London: Leicester University Press.

Lukacs, J. R. 1996 Sex Differences in Dental Caries Rates with the Origin of Agriculture in South Asia. Current Anthropology 37:147–153.

Lukacs, J. R. 2008 Fertility and Agriculture Accentuate Sex Differences in Dental Caries Rates. Current Anthropology 49:901–914.

Maggiano I. S., M. Schultz, H. Kierdorf, T. Sierra Sosa, C. M. Maggiano, and V. T. Blos 2008 Cross-sectional Analysis of Long Bones, Occupational Activities and Long-Distance Trade of the Classic Maya from Xcambó-Archaeological and Osteological Evidence. American Journal of Physical Anthropology 136:470–477.

Marshall, Y. 2008 Archaeological Possibilities for Feminist Theories of Transition and Transformation. Feminist Theory 9:25–45.

Martin, D. L. 1997 Violence Against Women in the La Plata River Valley (A.D. 100–1300). *In* Troubled Times: Violence and Warfare in the Past. D. L. Martin, and D. W. Frayer, eds. Pp. 45–75. Amsterdam: Gordon and Breach.

Martin, D. L. 2000 Bodies and Lives: Biological Indicators of Health Differentials and Division of Labor by Sex. *In* Women and Men in the Prehispanic Southwest: Labor, Power, and Prestige. P. L. Crown, ed. Pp. 267–300. Santa Fe: School of American Research Press.

Martin, D. L., and N. J. Akins 2001 Unequal Treatment in Death: Trauma and Mortuary Behavior at La Plata (A.D. 100–1300). *In* Ancient Burial Practices in the American Southwest: Archaeology, Physical Anthropology, and Native American Perspectives. D. R. Mitchell, and J. L. Brunson-Hadley, eds. Pp. 223–248. Albuquerque: University of New Mexico Press.

Martin, D. L., N. J. Akins, A. H. Goodman, W. Toll, and A. C. Swedlund 2001 Harmony and Discord: Bioarchaeology of the La Plata Valley. Santa Fe: Museum of New Mexico Press.

Maschner, H. D. G., and K. Reedy-Maschner 2007 Heads, Women, and the Baubles of Prestige: Trophies of War in the Arctic and Subarctic. *In* The Taking and Displaying of Human Body Parts as Trophies by Amerindians. R. J. Chacon and D. H. Dye, eds. Pp. 32–44. New York: Springer.

Mays, S. 2001 Effects of Age and Occupation on Cortical Bone in a Group of 18th–19th Century British Men. American Journal of Physical Anthropology 116:34–44.

McCafferty, S., and G. McCafferty 1994 Engendering Tomb 7 at Monte Alban. Current Anthropology 35:143–166.

Mensforth, R. P. 2001 Warfare and Trophy Taking in the Archaic Period. *In* Archaic Transitions in Ohio and Kentucky Prehistory. O. H. Prufer, S. E. Pedde, and R. S. Meindl, eds. Pp. 110–138. Kent: Kent State University Press.

Mensforth, R. P., and G. Baker 1995 Evidence of Violent Injury, Dismemberment, and Trophy-taking Behaviors Among Inhabitants of the Late Archaic Ward Site (Mc-ll) from Kentucky. Paper Presented at the 2nd Annual Meeting of the Midwest Bioarchaeology and Forensic Anthropology Association, Northern Illinois University, DeKalb.

Merbs, C. F. 1983 Patterns of Activity-Induced Pathology in a Canadian Inuit Population. Ottawa: Archaeological Survey of Canada. National Museum of Man Mercury Series 119.

Merbs, C. F., and R. C. Euler 1985 Atlanto-Occipital Fusion and Spondylolisthesis in an Anasazi Skeleton from Bright Angel Ruin, Grand Canyon National Park, Arizona. American Journal of Physical Anthropology 67:381–391.

Meskell, L. 1996 The Somatisation of Archaeology: Institutions, Discourses, Corporeality. Norwegian Archaeological Review 29:1–16.

Meskell, L. 1998 The Irresistible Body and the Seduction of Archaeology. *In* Changing Bodies, Changing Meanings: Studies on the Human Body in Antiquity. D. Montserrat, ed. Pp. 139–161. London: Routledge.

Meskell, L. 2001 Archaeologies of Identity, *In* Archaeological Theory Today. I. Hodder, ed. Pp. 187–213. Cambridge: Polity Press.

Meskell, L., and R. A. Joyce 2003 Embodied Lives: Figuring Ancient Maya and Egyptian Experience. London: Routledge.

Miller, R. J. 1985 Lateral Epicondylitis in the Prehistoric Indian Population from Nuvakwewtaqa (Chavez Pass), Arizona. *In* Health and Disease in the Prehistoric Southwest. C. F. Merbs, and R. J. Miller, eds. Pp. 391–399. Anthropological Research Papers 34. Tempe: University of Arizona Press.

Milner, G. R., and R. W. Jeffries 1987 A Reevaluation of the WPA Excavation of the Robbins Mound in Boone County, Kentucky. *In* Current Archaeological Research in Kentucky: Vol. 1. D. Pollack, ed. Pp. 33–42. Frankfort: Kentucky Heritage Council.

Milner, G. R., and V. G. Smith 1990 Oneota Skeletal Remains. *In* Archaeological Investigations at the Morton Village and Norris Farms 36 Cemetery. S. K. Santure, A. D. Harn, and D. Esarey, eds. Pp. 111–148. Springfield: Illinois State Museum Reports of Investigations 45.

Milner, G. R., E. Anderson, and V. G. Smith 1991 Warfare in Late Prehistoric West-Central Illinois. American Antiquity 56:581–603.

Molleson, T. 1989 Seed Preparation in the Mesolithic: the Osteological Evidence. Antiquity 63:356–362.

Molleson, T. 1994 The Eloquent Bones of Abu Hureyra. Scientific American 271:70–75.

Molleson, T. 2007 Bones of Work at the Origins of Labour. *In* Archaeology and Women: Ancient and Modern Issues. S. Hamilton, R. D. Whitehouse, and K. I. Wright, eds. Pp. 185–198. Walnut Creek: Left Coast Press.

Molnar, P. 2008a Patterns of Physical Activity and Material Culture on Gotland, Sweden, During the Middle Neolithic. International Journal of Osteoarchaeology early view http://0-www3.interscience.wiley.com

Molnar, P. 2008b Dental Wear and Oral Pathology: Possible Evidence and Consequences of Habitual Use of Teeth in a Swedish Neolithic Sample. American Journal of Physical Anthropology 136:423–431.

Molnar, S., L. Richards, J. McKee, and I. Molnar 1989 Tooth Wear in Australian Aboriginal Populations from the River Murray Valley. American Journal of Physical Anthropology 79:185–196.

Monaghan, J. 2001 Physiology, Production, and Gendered Difference: The Evidence of Mixtec and Other Mesoamerican Societies. *In* Gender in Pre-Hispanic America. C. F. Klein, ed. Pp. 285–304. Washington, DC: Dumbarton Oaks.

Morris, A. G. 1992 The Skeletons of Contact: A Study of Protohistoric Burials from the Lower Orange River Valley, South Africa. Johannesburg: University of Witwatersrand Press.

Müldner, G., and M. P. Richards 2007 Diet and Diversity at Later Medieval Fishergate: The Isotopic Evidence. American Journal of Physical Anthropology 134:162–174.

Neitzel, J. E. 2000 Gender Hierarchies: A Comparative Analysis of Mortuary Data. *In* Women and Men in the Prehispanic Southwest: Labor, Power, and Prestige. P. L. Crown, ed. Pp. 137–168. Santa Fe: School of American Research Press.

Nelson, S. M. 1993 Gender Hierarchy and the Queens of Silla. *In* Sex and Gender Hierarchies. B. D. Miller, ed. Pp. 297–315. Cambridge: Cambridge University Press.

Nelson, S. M. 1997 Gender in Archaeology: Analyzing Power and Prestige. Walnut Creek: AltaMira Press.

Nordblah, J., and T. Yates 1990 This Perfect Body, this Virgin Text: Between Sex and Gender in Archaeology. *In* Archaeology After Structuralism. I. Bapty, and T. Yates, eds. Pp. 222–237. London: Routledge.

Nordström, H.-Å 1996 The Nubian A-Group: Ranking Funerary Remains. Norwegian Archaeological Review 29:17–39.

Novak, S. A. 2006 Beneath the Façade: A Skeletal Model of Domestic Violence. *In* The Social Archaeology of Human Remains. R. Gowland, and C. Knüsel, eds. Pp. 238–252. Oxford: Oxbow Books.

O'Gorman, J. A. 2001 Life, Death, and the Longhouse: A Gendered View of Oneota Social Organization. *In* Gender and the Archaeology of Death. B. Arnold, and N. Wicker, eds. Pp. 23–49. Walnut Creek: AltaMira Press.

Osborn, S. 1997 Death in the Daily Life of the Ross Colony: Mortuary Behavior in Frontier Russian America. Ph.D. Dissertation, University of Milwaukee-Wisconsin.

Owsley D. W., and K. S. Bruwelheide 1997 Bioarchaeological Research in Northeastern Colorado, Northern Kansas, Nebraska, and South Dakota. *In* Bioarchaeology of the North Central United States. D. W. Owsley, and J. C. Rose, eds. Pp. 7–56. Fayetteville: Arkansas Archeological Survey Research Series 49.

Owsley, D. W., K. S. Bruwelheide, L. E. Burgess, and W. T. Billeck 2007 Human Finger and Hand Bone Necklaces from the Plains and Great Basin. *In* The Taking and Displaying of Human Body Parts as Trophies by Amerindians. R. J. Chacon, and D. H. Dye, eds. Pp. 124–166. New York: Springer.

Pate, L. 2004 The Use and Abuse of Ethnographic Analogies in Interpretations of Gender Systems at Cahokia. *In* Ungendering Civilization. K. A. Pyburn, ed. Pp. 71–93. New York: Routledge.

Pedersen, P. O., and J. Jakobsen 1989 Teeth and Jaws of the Qilakitsoq Mummies. *In* The Mummies from Qilakitsoq – Eskimos in the 15th Century. J. P. Hart Hansen, and H. C. Gulløv, eds. Pp. 112–130. Meddelelser om Grønland, Man and Society 12.

Perry, E. M. 2004 Bioarchaeology of Labor and Gender in the Prehispanic American Southwest. Ph.D. Dissertation, University of Arizona.

Perry, E. M., and R. A. Joyce 2001 Providing a Past for "Bodies that Matter": Judith Butler's Impact on the Archaeology of Gender. International Journal of Sexuality and Gender Studies 6:63–76.

Peterson, J. 2000a Sexual Revolutions: Gender and Labor at the Dawn of Agriculture. Walnut Creek: AltaMira Press.

Peterson, J. 2000b Labor Patterns in the Southern Levant in the Early Bronze Age. *In* Reading the Body: Representations and Remains in the Archaeological Record. A. E. Rautman, ed. Pp. 38–54. Philadelphia: University of Pennsylvania Press.

Pomeroy, E., and S. R. Zakrzewski 2008 Sexual Dimorphism in Diaphyseal Cross-sectional Shape in the Medieval Muslim Population of Écija, Spain, and Anglo-Saxon Great Chesterford, UK. International Journal of Osteoarchaeology 19:50–65.

Powell, M. L., D. Collins Cook, G. Bogdan, J. E. Buikstra, M. M. Castro, P. D. Horne et al. 2006 Invisible Hands: Women in Bioarchaeology. *In* Bioarchaeology: The Contextual Analysis of Human Remains. J. E. Buikstra, and L. A. Beck, eds. Pp. 131–194. Burlington, MA: Academic Press.

Privat, K. L., T. C. O'Connell, and M. P. Richards 2002 Stable Isotope Analysis of Human and Faunal Remains from the Anglo-Saxon Cemetery at Berinsfield, Oxfordshire: Dietary and Social Implications. Journal of Archaeological Science 29:779–790.

Pyburn, K. A. 2004 Ungendering the Maya. *In* Ungendering Civilization. K. A. Pyburn, ed. Pp. 216–233. New York: Routledge.

Rathbun, T. A. 1987 Health and Disease at a South Carolina Plantation: 1840–1870. American Journal of Physical Anthropology 74:239–253.

Redfern, R. 2005 A Gendered Analysis of Health from the Iron Age to the End of the Romano-British Period in Dorset, England (Mid to Late 8th Century BC to the End of the 4th Century AD). Ph.D. Dissertation, University of Birmingham.

Reeves, M. 2000 Dental Health at Early Historic Fusihatchee Town: Biocultural Implications of Contact in Alabama. *In* Bioarchaeological Studies of Life in the Age of Agriculture: A View from the Southeast. P. M. Lambert, ed. Pp. 78–95. Tuscaloosa: University of Alabama Press.

Rega, E. 1996 Age, Gender and Biological Reality in the Early Bronze Age Cemetery at Mokrin. *In* Invisible People and Processes: Writing Gender and Childhood into European Archaeology. J. Moore, and E. Scott, eds. Pp. 229–247. London: Leicester University Press.

Rega, E. 2000 The Gendering of Children in the EBA Cemetery at Mokrin. *In* Gender and Material Culture in Archaeological Perspective. M. Donald, and L. Hurcombe, eds. Pp. 238–249. New York: St. Martin's Press.

Reinhard, K. J., L. L. Tieszen, K. L. Sandness, L. M. Beiningen, E. Miller, A. Mohammad Ghazi et al. 1994 Trade, Contact, and Female Health in Northeast Nebraska. *In* In the Wake of Contact: Biological Responses to Conquest. C. S. Larsen, and G. R. Milner, eds. Pp. 63–74. New York: Wiley-Liss.

Rhodes, J. A., and C. J. Knüsel 2005 Activity-Related Skeletal Change in Medieval Humeri: Cross-Sectional and Architectural Alterations. American Journal of Physical Anthropology 128:536–546.

Richards, L. C. 1984 Principal Axis Analysis of Dental Attrition Data from Two Australian Aboriginal Populations. American Journal of Physical Anthropology 65:5–13.

Robb, J. 1997 Violence and Gender in Early Italy. *In* Troubled Times: Violence and Warfare in the Past. D. L. Martin, and D. W. Frayer, eds. Pp. 111–144. Amsterdam: Gordon and Breach.

Rodning, C. B. 2001 Mortuary Ritual and Gender Ideology in Protohistoric Southwestern North Carolina. *In* Archaeological Studies of Gender in the Southeastern United States. J. M. Eastman, and C. B. Rodning, eds. Pp. 77–100. Gainesville: University of Florida Press.

Ross-Stallings, N. A. 2007 Trophy Taking in the Central and Lower Mississippi Valley. *In* The Taking and Displaying of Human Body Parts as Trophies by Amerindians. R. J. Chacon, and D. H. Dye, eds. Pp. 339–370. New York: Springer.

Rubinson, K. 2002 Through the Looking Glass: Reflections on Mirrors, Gender, and Use Among Nomads. *In* In Pursuit of Gender: Worldwide Archaeological Approaches. S. M. Nelson, and M. Rosen-Ayalon, eds. Pp. 67–72. Walnut Creek: AltaMira Press.

Rubinson, K. 2008 Tillya Tepe: Aspects of Gender and Cultural Identity. *In* Are All Warriors Male? Gender Roles on the Ancient Eurasian Steppe. K. M. Linduff, and K. S. Rubinson, eds. Pp.51–63. Walnut Creek: AltaMira Press.

Ruff, C. B. 1987 Sexual Dimorphism in the Human Lower Limb Structure: Relationship to Subsistence Strategy and Sexual Division of Labor. Journal of Human Evolution 16: 391–416.

Ruff, C. B. 1994 Biomechanical Analysis of Northern and Southern Plains Femora: Behavioral Implications. *In* Skeletal Biology of the Great Plains. D. W. Owsley, and R. L. Jantz, eds. Pp. 235–245. Washington, DC: Smithsonian Institution Press.

Ruff, C. B. 1999 Skeletal Structure and Behavioral Patterns of Prehistoric Great Basin Populations. *In* Prehistoric Lifeways in the Great Basin Wetlands: Bioarchaeological Reconstruction and Interpretation. B. E. Hemphill, and C. S. Larsen, eds. Pp. 290–320. Salt Lake City: University of Utah Press.

Ruff, C. B. 2000 Biomechanical Analysis of Archaeological Human Skeletons. *In* Biological Anthropology of the Human Skeleton. M. A. Katzenberg, and S. R. Saunders, eds. Pp. 71–102. New York: Wiley-Liss.

Ruff, C. B., and C. S. Larsen 1990 Postcranial Biomechanical Adaptations to Subsistence Strategy Changes on the Georgia Coast. *In* The Archaeology of Santa Catalina de Guale: 2. Biocultural Interpretations of a Population in Transition. C. S. Larsen, ed., Pp. 94–120. New York: Anthropological Papers of the Museum of Natural History.

Ruff, C. B., and C. S. Larsen 2001 Reconstructing Behavior in Spanish Florida: The Biomechanical Evidence. *In* Bioarchaeology of La Florida: Human Biology of Frontier Northern New Spain. C. S. Larsen, ed. Pp. 113–145. Gainesville: University of Florida Press.

Ryan, T. M., and G. R. Milner 2006 Osteological Applications of High-Resolution Computed Tomography: A Prehistoric Arrow Injury. Journal of Archaeological Science 33:871–879.

Savage, S. H. 2000 The Status of Women in Predynastic Egypt as Revealed Through Mortuary Analysis. *In* Reading the Body: Representations and Remains in the Archaeological Record. A. E. Rautman, ed. Pp. 77–92. Philadelphia: University of Pennsylvania Press.

Scheiber, L. L. 2008 Life and Death on the Northwestern Plains: Mortuary Practices and Cultural Transformations. *In* Skeletal Biology and Bioarchaeology of the Northwestern Plains. G. W. Gill, and R. L. Weathermon, eds. Pp. 22–41. Salt Lake City: University of Utah Press.

Schillaci, M., and C. Stojanowski 2002 A Reassessment of Matrilocality in Chacoan Culture. American Antiquity 67:343–356.

Schillaci, M., and C. Stojanowski 2003 Postmarital Residence and Biological Variation at Pueblo Bonito. American Journal of Physical Anthropology 120:1–15.

Schmidt, R. A. 2000 Shamans and Northern Cosmology: The Direct Historical Approach to Mesolithic Sexuality. *In* Archaeologies of Sexuality. R. A. Schmidt, and B. L. Voss, eds. Pp. 220–235. London: Routledge.

Schmidt, R. A., and B. L. Voss eds. 2000 Archaeologies of Sexuality. London: Routledge.

Schneider, K. N., and D. J. Blakeslee 1990 Evaluating Residence Patterns Among Prehistoric Populations: Clues from Dental Enamel Composition. Human Biology 62:71–83.

Schulting, R. J., and M. P. Richards 2001 Dating Women and Becoming Farmers: New Paleodietary and AMS Evidence from the Breton Mesolithic Cemeteries of Téviec and Hoëdic. Journal of Anthropological Archaeology 20:314–344.

Schurr, M. R. 1992 Isotopic and Mortuary Variability in a Middle Mississippian Population. American Antiquity 57:300–320.

Sealy, J. C., M. K. Patrick, A. G. Morris, and D. Alder 1992 Diet and Dental Caries Among Later Stone Age Inhabitants of the Cape Province, South Africa. American Journal of Physical Anthropology 88:123–134.

Shelach, G. 2008 He Who Eats the Horse, She Who Rides It? Symbols of Gender Identity on the Eastern Edges of the Eurasian Steppe. *In* Are All Warriors Male?: Gender Roles on the Ancient Eurasian Steppe. K. M. Linduff, and K. S. Rubinson, eds. Pp. 93–109. Walnut Creek: AltaMira Press.

Sládek, V., M. Berner, D. Sosna, and R. Sailer 2007 Human Manipulative Behavior in the Central European Late Eneolithic and Early Bronze Age: Humeral Bilateral Asymmetry. American Journal of Physical Anthropology 133:669–681.

Šlaus, M. 2000 Biocultural Analysis of Sex Differences in Mortality Profiles and Stress Levels in the Late Medieval Population from Nova Rača, Croatia. American Journal of Physical Anthropology 111:193–209.

Smith, M. O. 1996 "Parry" Fractures and Female-Directed Interpersonal Violence: Implications from the Late Archaic period of West Tennessee. International Journal of Osteoarchaeology 6:261.1–262.8.

Smith, M. O. 1997 Osteological Indications of Warfare in the Archaic Period of the Western Tennessee Valley. Violence and Warfare in the Past. D. L. Martin, and D. W. Frayer, eds. Pp. 241–265. Amsterdam: Gordon and Breach.

Smith, M. O. 2003 Beyond Palisades: The Nature and Frequency of Late Prehistoric Deliberate Violent Trauma in the Chickamauga Reservoir of East Tennessee. American Journal of Physical Anthropology 121:303–318.

Sobolik, K. D. 2002 Children's Health in the Prehistoric Southwest. *In* Children in the Prehistoric Puebloan Southwest. K. A. Kamp, ed. Pp. 125–151. Salt Lake City: University of Utah Press.

Sofaer Derevenski, J. 1997a Engendering Children, Engendering Archaeology. *In* Invisible People and Processes: Writing Gender and Childhood into European Archaeology. J. Moore, and E. Scott, eds. Pp. 192–202. London: Leicester University Press.

Sofaer Derevenski, J. 1997b Linking Gender and Age as Social Variables. Ethnographisch-Archäeologischen Zeitschrift 38:485–493.

Sofaer Derevenski, J. 1998 Gender Archaeology as Contextual Archaeology: A Critical Examination of the Tensions Between Method and Theory in the Archaeology of Gender. Ph.D. Dissertation, University of Cambridge.

Sofaer Derevenski, J. 2000a Rings of Life: the Role of Early Metalwork in Mediating the Gendered Life Course. World Archaeology 31:389–406.

Sofaer Derevenski, J. 2000b Sex Differences in Activity-related Osseous Change in the Spine and the Gendered Division of Labor at Ensay and Wharram Percy, UK. American Journal of Physical Anthropology 111:333–354.

Sofaer Derevenski, J. 2002 Engendering Context: Context as Gendered Practice in the Early Bronze Age of the Upper Thames Valley, UK. European Journal of Archaeology 5:191–211.

Sofaer, J. 2006a The Body as Material Culture: A Theoretical Osteoarchaeology. Cambridge: University of Cambridge Press.

Sofaer, J. 2006b Gender, Bioarchaeology and Human Ontogeny. *In* The Social Archaeology of Human Remains. R. Gowland, and C. Knüsel, eds., Pp. 155–167. Oxford: Oxbow Books.

Sørenson, M.-L. 2000 Gender Archaeology. Cambridge: Polity Press.

Sparacello, V., and D. Marchi 2008 Mobility and Subsistence Economy: A Diachronic Comparison Between Two Groups Settled in the Same Geographical Area (Liguria, Italy). American Journal of Physical Anthropology 136:485–496.

Spielmann, K. A. 1995 Glimpses of Gender in the Prehistoric Southwest. Journal of Anthropological Research 51:91–102.

Standen V. G., B. T. Arriaza, and C. M. Santoro 1997 External Auditory Exostosis in Prehistoric Chilean Populations: A Test of the Cold Water Hypothesis. American Journal of Physical Anthropology 103:119–129.

Stock, J. C., and S. K. Pfeiffer 2004 Long Bone Robusticity and Subsistence Behaviour Among Later Stone Age Foragers of the Forest and Fynbos Biomes of South Africa. Journal of Archaeological Science 31:999–1013.

Stockett, M. 2005 On the Importance of Difference: Re-envisioning Sex and Gender in Ancient Mesoamerica. World Archaeology 37:566–578.

Stodder, A. L. W. 1987 The Physical Anthropology and Mortuary Behavior of the Dolores Anasazi: An Early Pueblo Population in Local and Regional Context. *In* Dolores Archaeological Program Supporting Studies: Settlement and Environment. K. L. Peterson, and J. D. Orcutt, eds. Pp. 339–504. Denver: U.S. Bureau of Reclamation Engineering and Research Center.

Stojanowski, C. M. 2005 Spanish Colonial Effects on Native American Mating Structure and Genetic Variability in Northern and Central Florida: Evidence from Apalachee and Western Timucua. American Journal of Physical Anthropology 128:273–286.

Stojanowski, C. M., and Mi. A. Schillaci, 2006 Phenotypic Approaches for Understanding Patterns of Intracemetery Biological Variation. Yearbook of Physical Anthropology 131: 49–88.

Stone, A. C. 2000 Ancient DNA from Skeletal Remains. *In* Biological Anthropology of the Human Skeleton. M. A. Katzenberg, and S. R. Saunders, eds. Pp. 351–371. New York: Wiley-Liss.

Stone, A. C., and M. Stoneking 1999 Analysis of Ancient DNA from a Prehistoric Amerindian Cemetery. Philosophical Transactions: Biological Sciences 354:153–159.

Stone, A. C., G. R. Milner, S. Pääbo, and M. Stoneking 1996 Sex Determination of Ancient Human Skeletons Using DNA. American Journal of Physical Anthropology 99:231–238.

Stone, P. K., and D. Walrath 2006 The Gendered Skeleton: Anthropological Interpretations of the Bony Pelvis. *In* The Social Archaeology of Funerary Remains. R. Gowland, and C. Knüsel, eds. Pp. 168–178. Oxford: Oxbow Books.

Stoodley, N. 1999 The Spindle and the Spear: A Critical Enquiry into the Construction and Meaning of Gender in the Early Anglo-Saxon Burial Rite. British Archaeological Reports British Series 288. Oxford: Archaeopress.

Stoodley, N. 2000 From the Cradle to the Grave: Age Organization and the Early Anglo-Saxon Burial Rite. World Archaeology 31:456–472.

Storey, R. 1998 Mothers and Daughters of a Patrilineal Civilization: The Health of Females Among the Late Classic Maya of Copán, Honduras. *In* Sex and Gender in Paleopathological Perspective. A. Grauer, and P. Stuart-Macadam, eds. Pp. 133–148. Cambridge: University of Cambridge Press.

Strassburg J. 2000 Shamanic Shadows: One Hundred Generations of Undead Subversion in Southern Scandinavia, 7,000–4,000 BC. Stockholm: Stockholm University Studies in Archaeology 20.

Strömberg, A. 1993 Male or Female? A Methodological Study of Grave Gifts as Sex-Indicators in Iron Age Burials from Athens. Jonsered: Paul Åströms Förlag.

Sullivan, A. 2004 Reconstructing Relationships Among Mortality, Status, and Gender at the Medieval Gilbertine Priory of St. Andrew, Fishergate, York. American Journal of Physical Anthropology 124:330–345.

Sullivan, A. 2005 Prevalence and Etiology of Acquired Anemia in Medieval York, England. American Journal of Physical Anthropology 128:252–272.

Swanton, J. R. 1979 The Indians of the Southeastern United States. Washington, DC: Smithsonian Institution.

Temple, D. H., and C. S. Larsen 2007 Dental Caries Prevalence as Evidence for Agriculture and Subsistence Variation During the Yayoi Period in Prehistoric Japan: Biocultural Interpretations of an Economy in Transition. American Journal of Physical Anthropology 134:501–512.

Tiesler, V., A. Cucina, and A. R. Pacheco 2004 Who Was the Red Queen? Identity of the Female Maya Dignitary from the Sarcophagus Tomb of Temple XIII, Palenque, Mexico. Homo – Journal of Comparative Human Biology 55:65–76.

Tomczak, P. D., and J. F. Powell 2003 Postmarital Residence Practices in the Windover Population: Sex-Based Dental Variation as an Indicator of Patrilocality. American Antiquity 68:93–108.

Toms, J. 1998 The Construciton of Gender in Early Iron Age Etruria. *In* Gender and Italian Archaeology. R. Whitehouse, ed. Pp. 157–179. London: Accordia Research Institute/Institute of Archaeology.

Torres-Rouff, C. 2002 Cranial Vault Modification and Ethnicity in Middle Horizon San Pedro de Atacama, Chile. Current Anthropology 43:163–171.

Torres-Rouff, C. 2008 The Influence of Tiwanaku on Life in the Chilean Atacama: Mortuary and Bodily Perspectives. American Anthropologist 110:325–337.

Torres-Rouff, C., and M. A. Costa Junqueira 2006 Interpersonal Violence in Prehistoric San Pedro de Atacama, Chile: Behavioral Implications of Environmental Stress. American Journal of Physical Anthropology 130:60–70.

Tung, T. A. 2007 Trauma and Violence in the Wari Empire of the Peruvian Andes: Warfare, Raids, and Ritual Fights. American Journal of Physical Anthropology 133:941–956.

Tung, T. A. 2008 Dismembering Bodies for Display: A Bioarchaeological Study of Trophy Heads from the Wari Site of Conchopata, Peru. American Journal of Physical Anthropology 136:294–308.

Tung, T. A., and K. J. Knudson 2008 Social Identities and Geographical Origins of Wari Trophy Heads from Conchopata, Peru. Current Anthropology 49:915–925.

Tuohy, T. 2000 Long Handled Weaving Combs: Problems in Determining the Gender of Tool-Maker or Tool-User. *In* Gender and Material Culture in Archaeological Perspective. M. Donald, and L. Hurcombe, eds. Pp. 137–152. London: Macmillan.

Vida Navarro M. C. 1992 Warriors and Weavers: Sex and Gender in Early Iron Age Graves from Pontecagnano. Journal of the Accordia Research Center 3:67–100.

Vinkovetsky, I. 2002 Native Americans and the Russian Empire, 1904–1867. Ph.D. Dissertation, University of California, Berkeley.

Voss, B. L. 2000 Feminisms, Queer Theories, and the Archaeological Study of Past Sexualities. World Archaeology 32:180–192.

Voss, B. L. 2005 Sexual Subjects: Identity and Taxonomy in Archaeological Research. *In* The Archaeology of Plural and Changing Identities: Beyond Identification. E. C. Casella, and C. Fowler, eds. Pp. 55–78. New York: Springer.

Voss, B. L. 2008 Sexuality Studies in Archaeology. Annual Review of Anthropology 37: 317–336.

Wadley, L. 1998 The Invisible Meat Providers: Women in the Stone Age of South Africa. *In* Gender in African Prehistory. S. Kent, ed. Pp. 69–81. Walnut Creek: AltaMira Press.

Walker, P. L. 1989 Cranial Injuries as Evidence of Violence in Prehistoric Southern California. American Journal of Physical Anthropology 80:313–323.

Walker, P. L. 1995 Problems of Preservation and Sexism in Sexing: Some Lessons from Historical Collections for Palaeodemographers. *In* Grave Reflections: Portraying the Past Through Cemetery Studies. S. R. Saunders, and A. Herring, eds. Pp. 31–47. Toronto: Canadian Scholars' Press.

Walker, P. L. 1997 Wife Beating, Boxing, and Broken Noses: Skeletal Evidence for the Cultural Patterning of Violence. *In* Troubled Times: Violence and Warfare in the Past. D. L. Martin, and D. W. Frayer, eds. Pp. 145–179. Amsterdam: Gordon and Breach.

Walker, P. L. 2001 A Bioarchaeological Perspective on the History of Violence. Annual Review of Anthropology 30:573–96.

Walker, P. L., and D. Collins Cook 1998 Gender and Sex: Vive la Difference. American Journal of Physical Anthropology 106:255–259.

Walker, P. L., and M. J. DeNiro 1986 Stable Nitrogen and Carbon Isotope Ratios in Bone Collagen as Indices of Prehistoric Dietary Dependence on Marine and Terrestrial Resources in Southern California. American Journal of Physical Anthropology 71:51–61.

Walker, P. L., and J. M. Erlandson 1986 Dental Evidence for Prehistoric Dietary Change on the Northern Channel Islands, California. American Antiquity 51:375–383.

Walker, P. L., and S. E. Hollimon 1989 Changes in Osteoarthritis Associated with the Development of a Maritime Economy Among Southern California Indians. International Journal of Anthropology 4:171–183.

Weglian, E. 2001 Grave Goods Do Not a Gender Make: A Case Study from Singen am Hohentwiel, Germany. *In* Gender and the Archaeology of Death. B. Arnold, and N. L. Wicker, eds. Pp. 137–155. Walnut Creek: AltaMira Press.

Weiss, E. 2009 Sex Differences in Humeral Bilateral Assymetry in Two Hunter-Gatherer Populations: California Amerinds and British Columbia Amerinds. American Journal of Physical Anthropology 140:19–24.

Westcott, D. J. 2001 Structural Variation in the Humerus and Femur in the American Plains and Adjacent Regions: Differences in Subsistence Strategy and Physical Terrain. Ph.D. Dissertation, University of Tennessee, Knoxville.

Westcott, D. J., and D. L. Cunningham 2006 Temporal Changes in Arikara Humeral and Femoral Cross-sectional Geometry Associated with Horticultural Intensification. Journal of Archaeological Science 33:1022–1036.

Weston, K. 1993 Lesbian/Gay Studies in the House of Anthropology. Annual Review of Anthropology 22:33–67.

Whelan, M. K. 1991 Gender and Historical Archaeology: Eastern Dakota Patterns in the 19th Century. Historical Archaeology 25:17–32.

White, C. D. 2005 Gendered Food Behaviour Among the Maya: Time, Place, Status and Ritual. Journal of Social Archaeology 5:356–382.

White, C. D., M. W. Spence, F. J. Longstaffe, H. Stuart-Williams, and K. R. Law 2002 Geographic Identities of the Sacrificial Victims from the Feathered Serpent Pyramid, Teotihuacan: Implications for the Nature of State Power. Latin American Antiquity 13:217–236.

White, C. D., R. Storey, F. J. Longstaffe, and M. W. Spence 2004 Immigration, Assimilation, and Status in the Ancient City of Teotihuacan: Stable Isotopic Evidence from Tlajinga 22. Latin American Antiquity 15:176–198.

Whitehouse, R. D. 2002 Gender in the South Italian Neolithic: A Combinatory Approach. *In* In Pursuit of Gender: Worldwide Archaeological Approaches. S. M. Nelson and M. Rosen-Ayalon, eds. Pp. 15–42. Walnut Creek: AltaMira Press.

Whittlesey S. M. 1996 Engendering the Mogollon Past: Theory and Mortuary Data from Grasshopper Pueblo. *In* Sixty Years of Mogollon Archaeology. S. M. Whittlesey, ed. Pp. 39–48. Tucson: SRI Press.

Whittlesey S. M. 2002 The Cradle of Death: Mortuary Practices, Bioarchaeology, and the Children of Grasshopper Pueblo. *In* Children in the Prehistoric Puebloan Southwest. K. A. Kamp, ed. Pp. 152–168. Salt Lake City: University of Utah Press.

Wilkinson, R. G. 1997 Violence Against Women: Raiding and Abduction in Prehistoric Michigan. *In* Troubled Times: Violence and Warfare in the Past. D. L. Martin, and D. W. Frayer, eds. Pp. 21–43. Amsterdam: Gordon and Breach.

Willey, P. S. 1990 Prehistoric Warfare on the Great Plains: Skeletal Analysis of the Crow Creek Massacre Victims. New York: Garland.

Williamson, M. A. 1998 Regional Variation in Health and Lifeways Among Late Prehistoric Georgia Agriculturalists. Ph.D. Dissertation, Purdue University.

Williamson, R. F., and S. Pfeiffer, eds. 2003 Bones of the Ancestors: The Archaeology and Osteobiography of the Moatfield Ossuary. Mercury Series Archaeological Paper 163. Gatineau, Quebec: Canadian Museum of Civilization.

Wilson, K. J., and M. Cabak 2004 Feminine Voices from Beyond the Grave: What Burials Can Tell Us About Gender Differences Among Historic African Americans. *In* Engendering African American Archaeology: A Southern Perspective. J. Galle, and A. Young, eds. Pp. 263–286. Knoxville: University of Tennessee Press.

Worthman, C. M. 1995 Hormones, Sex and Gender. Annual Review of Anthropology 24:593–616.

Yates, T. 1993 Frameworks for an Archaeology of the Body. *In* Interpretive Archaeology. C. Tilley, ed. Pp. 31–72. Oxford: Berg.

7

Population Migration, Variation, and Identity

An Islamic Population in Iberia

Sonia Zakrzewski

Bioarchaeological Identity

How may populations or groups be recognized and identified? Archaeological studies of identity have frequently comprised analyses of ethnicity (e.g., Jones 1997; Lucy 2005a), gender (e.g., Díaz-Andreu 2005; Sofaer 2006; Walde and Willows 1991), or age (e.g., Lucy 2005b; Moore and Scott 1997; Sofaer 2006). Other topics incorporated in such studies of identity have included class, status and rank (e.g., Babić 2005; Wason 1994), and sexuality (e.g., Schmidt and Voss 2000). Many of these studies could be considered as single-issue studies of identity (Meskell 2001). The multiple strands of identity are rarely considered together, and even more rarely include the role of religion in identity construction (Insoll 2005). Excluding aspects of health (e.g., Steckel and Rose 2002), these other aspects of identity have seldom been considered as an integrated whole within bioarchaeology. Notable exceptions that do examine the multiplicity of identity include certain papers within Lucas Powell et al. (1991), Grauer and Stuart-Macadam (1998), and more recently Gowland and Knüsel (2006) and Knudson and Stojanowski (2009). In this chapter I address aspects of the bioarchaeological recognition of social identity, with particular focus here placed upon the roles of religion and ethnicity.

Social identity includes aspects of ethnicity. Recognizing ethnicity requires the identification of biological affinities and, although ethnicity clearly cannot be mapped directly onto population affinity, from these ideas, ethnic groupings may be hypothesized. It may seem relatively simple to undertake studies that allow population affinity deductions to be made, but, in reality, this process is complex as group identification itself is hard. Through the use of the body and the skeleton,

Social Bioarchaeology. Edited by Sabrina C. Agarwal and Bonnie A. Glencross
© 2011 Blackwell Publishing Ltd

bioarchaeology has the potential to identify such groups much more directly than other forms of anthropology and archaeology. Archaeologists and anthropologists have, in the past, relied on typologies to define artifact "cultures," such as stone tool cultures. They have then, using a culture-historical approach, associated these artifacts types with their manufacturers to form culturally defined groups and hence human populations (Johnson 1999; Trigger 1989). In contrast, groups can be identified from certain attributes or changes to their bodies using bioarchaeological techniques. Obviously grave goods such as jewelry may be used to recognize human-derived groups within funerary contexts. But humans are not necessarily buried with material culture that demarcates ethnic grouping. Humans are also able to manipulate or modify their bodies in order to imprint upon themselves (or their offspring) a marker of their group membership, for example through head or foot binding, or through tattooing. These are artificial changes made to the body that enable group membership to be identified by those who understand them or are able to "read" them. There are, however, other bodily traits (including skeletal and dental traits) that also provide an indication of group membership and hence biological affinity or ethnicity. These latter traits, such as measurements of specific portions of the body or the presence of specific minor skeletal or dental anomalies, cannot (generally) be manipulated by the individual or by their parent through bodily or cultural modification. It is upon these traits that this chapter will focus. The traits typically have both a genetic and an environmental component (for detailed discussion see Larsen 1997) and hence cannot provide a definitive answer as to biological group, but they can provide a guide to biological affinity and hence ethnicity and identity. Bioarchaeological study enables the identification of subgroups within an overall human population by the identification of such skeletal features within the broader range of human variation. This, in turn, then permits further and more detailed identification of biological affinity.

The Bioarchaeology of Ethnicity

An implicit assumption underlying the culture-historical approach to archaeology is that bounded, homogeneous archaeological "culture" entities correspond with particular peoples, ethnic groups, tribes and/or races (Jones 1997). Linked to this is the (frequently unasserted) hypothesis that the transmission of cultural traits or items is assumed to be a function of the degree of interaction between either individuals or groups, and thus that a high degree of homogeneity in material culture can be considered as the product of regular contact and interaction. With the development of processual archaeology, ethnicity came to be viewed as an aspect of social organization involving the active maintenance of cultural boundaries through social interaction (Hodder 1982; Trigger 1989). Ethnicity could thus be viewed as one part of social process, similar to subsistence, religion, economy, or politics. Because of these multiple social processes, the definition of group membership is problematic. The development of post-processual archaeology has led to an archaeological understanding that artifact typologies or languages or other categorical entities do not map perfectly onto social groups. Hodder (1982) demonstrated from an ethnoarchaeological study of the Baringo of Kenya that aspects of material culture could be related to tribal groupings, but other aspects cross-cut these tribes despite prolonged interaction. This cross-cut nature of the material culture was maintained over generations despite repeated interactions, implying that ethnicity might better be considered as a personal choice rather than as one of the features of identity (Jones 1997). The implication is that ethnicity then exists only as an active part of personal identity and hence leaves no fixed markers, and thus might not be recognizable by external observers. This understanding of ethnicity would not be bioarchaeologically visible as, by its very definition, it would be ephemeral and not imprinted upon the skeletal or dental body.

Within anthropology, the study of ethnicity had a different history. Darwinian ideas led to the development of social Darwinism, and hence to a move from a hierarchical classification of racial types to a framework based upon an evolutionary trajectory of cultural stages (Stocking 1968; Trigger 1989). For some in the late 19th and early 20th centuries, such as Ripley (e.g., 1899) and Deniker (e.g., 1913), this led to a hierarchy of "races" and attempts at identification and classification of relatively fixed hereditary, physical types ("races") whereas others, most notably Boas (e.g., 1912), rejected this idea (Stocking 1968). In a recent history of bioarchaeology in the Americas, Buikstra notes that "biological distance investigations, particularly those on a continental scale and focused primarily on the origin of American Indians, continued to dominate the field" (Buikstra 2006a: 19). Many of these studies considered the variability of modern humans as a matter of description rather than of interpretation (Collins Cook 2006), with Hrdlička noted particularly for his interest in determining the potential range of normal variation (Buikstra 2006a). The actual interpretation of the morphological variability developed slightly later, with Hooton (1918) layering adaptation onto morphological expression (Collins Cook 2006). These studies were still primarily interested in the origins and migrations of groups of people and hence of potentially larger populations or ethnic groupings. More recently, biological studies have used modern genetic data, such as allele frequencies (e.g., Cavalli-Sforza et al. 1994; Sokal et al. 1993), to map population distributions and hence past population movements. Rather than recognizing internal group genetic diversity, this type of study frequently assumes a reasonably simple and direct link between ethnic groups and aspects of genetic variation (Mirza and Dungworth 1995). This is not currently the broadly accepted position, with the Human Genome Project noting that,

> DNA studies do not indicate that separate classifiable subspecies (races) exist within modern humans. While different genes for physical traits such as skin and hair color can be identified between individuals, no consistent patterns of genes across the human genome exist to distinguish one race from another. There also is no genetic basis for divisions of human ethnicity. People who have lived in the same geographic region for many generations may have some alleles in common, but no allele will be found in all members of one population and in no members of any other. Indeed, it has been proven that there is more genetic variation within races than exists between them (www.ornl.gov/sci/techresources/Human_Genome/elsi/humanmigration.shtml).

Furthermore, humans show only relatively modest levels of genetic differentiation among populations when compared with large-bodied mammals, and this level of genetic differentiation is below the threshold usually employed to identify subspecies within nonhuman species (Templeton 1998). It is not the intention of the author to enter into a prolonged discussion over the existence or not of biological races (or possible subspecies) within the human species (for such a biological discussion, see Cartmill (1998), and readers are recommended to review the American Association of Physical Anthropologists' "Statement on Biological Aspects of Race" (AAPA 1996)) or the applicability of anthroposcopic "race recognition" methods employed within certain sectors of forensic anthropology.

For the purposes of the current study, ethnicity is considered without reference to "race" or coloration, and follows a concept of ethnicity whereby subtle social categorizations are involved, with ethnic group identification requiring constant reiteration through both everyday actions and discursive practice. Ethnicity is thus fluid and malleable in nature, and dependent upon social relationships and boundaries that need repeated redefinition (Lucy 2005a), yet is a categorical identity within the individual social persona (Beck 1995). The dynamic nature of ethnicity can be demonstrated, for example, by gender differences in expression. For example, within certain French Catalan villages, male ethnic identities remain constant through life whereas females move through life between French and Catalan identities and languages (O'Brien 1994). In addition, ethnicity requires external definition and categorization in association with other aspects of identity. For example, in both Mauritius and Trinidad, an individual is not classified by others as simply male or middle-class but as an *Indian male* or *colored middle-class* (Eriksen 1991). The current study considers ethnic identity in relation to religion within a past population. And, as a human group does not exist in isolation, it links this past skeletal diversity and variability with concepts of ethnicity and population affinity within a dynamic system associated with medieval migration and religious conversion.

Religion has an important role within the construction of identity and ethnicity. Religion can comprise both perceived ethnic groupings and aspects of activity. Perceived ethnic groupings may include the maternal inheritance of Jewish identity. The aspects of behavior that identify religious affiliation are those that are ritually performed (Insoll 2004), although "within many Muslim societies distinctions can be made between practices and beliefs that may be classified as religion ... and customary practices" (Edwards 2005: 123). Religion thus permits differentiation between *us* and *them* to be identified by the living contemporaneous individuals. This concept of separation of *us* and *them* and the integration of *me* within *us* have been noted in the recognition of perceived ethnic boundaries (for more detail, see Beck 1995) and hence religious activity can both create and maintain this separation (Edwards 2005). An integrated and reflexive approach to religion has developed within archaeology (for discussion, see Edwards 2005), but has generally been most conspicuous in bioarchaeology with regard to requests for the repatriation and/or reburial of human remains from indigenous groups in North America and Australasia (Buikstra 2006b; Fforde et al. 2002). Bioarchaeology has the potential to embrace religion to a greater extent by being viewed as part of a broader anthropological entity with religious concerns integrated into bioarchaeology research projects.

As religious entities are socially constructed through repeated practice and tradition (Edwards 2005; Insoll 2004), religion therefore becomes a highly important factor in the identification and definition of "boundaries" between different social and cultural groups. Although Gellner (1981) argues for the existence of an essential Islamic social structure, Asad (1986: 14) argues that "Islam is neither a distinctive social structure nor a heterogeneous collection of beliefs, artefacts, customs, and morals. It is a tradition." Within Islam, however, religion supersedes other aspects of identity, such as ethnicity and class, in the creation and maintenance of the *Ummah* (the "Community of the Believers") (Insoll 1999). Using bioarchaeological evidence, this chapter will examine the role of religion in the creation of identity and ethnicity in Islamic Iberia. Medieval Spain presents a clear example of a region where religion and population migration interact in the creation of ethnic or identity groups. This chapter therefore describes and evaluates a variety of bioarchaeological methods of assessment of ethnicity and identity within a geographically, temporally, and religiously delineated population sample.

Identifying Migration

Ethnicity and religion both exist within a framework that separates *us* from *them*. As noted earlier, ethnicity is linked to, but has no direct causal relationship with, population affinity. It therefore is impacted upon by mobility and migration, both in terms of individuals and of groups. Theoretical frameworks that focus upon migration, mobility and movement in association with the creation of social identity boundaries can inform investigations of group organization and social dynamics. In order to identify movement and migration in the bioarchaeological record, it is, by implication, necessary to consider the methods that enable the recognition or identification of biological separation both within and between population groupings. Some biological and physical differences exist between humans on a global geographic scale (Howells 1973, 1989, 1995; Lahr 1996), and between groups within nonhuman species (Ridley 1993). As noted earlier, although attention has been placed upon physical and biological definitions of populations within humans, much of this has, in the past, been laden with colonial and racist overtones. Little attention has been placed upon recognizing and thus defining specific groups within larger populations. Studies that have considered biological differences between groups have generally done so in terms of differences between vastly geographically disparate populations (e.g., Howells 1973; Lahr 1996) rather than attempting to ascertain identity or ethnicity, particularly in bioarchaeological samples.

By their very migrating nature, migrants are usually predicted to differ (to some biological degree) both from the indigenous individuals within their new locale and potentially from the majority of the people within their original grouping. Migrants are therefore individuals who cross populations and hence potentially may also cross ethnic or religious boundaries. Certain classic anthropological studies (e.g., Boas 1912) have indicated that stature, cranial and other anthropometric measurements are modified in the offspring of migrants (see discussions in Gravlee et al. 2003a, 2003b; Relethford 2004; Sparks and Jantz 2002). Ewing (1950) demonstrated that the migrants who abandoned traditional practices of swaddling babies had children who grew into adults with narrower heads than the parents themselves, thereby showing potential cultural effects upon biological plasticity. The plastic nature of bone is the cause of problems that have meant that bioarchaeological attention has been focused upon biodistance studies with reasonably large temporal or geographic distances between the groups being studied (e.g., Brothwell and Krzanowski 1974; González-José et al. 2001; Hanihara 1992, 1996; Howells 1973, 1989, 1995; Irish 1997, 2005; Ishida and Dodo 1993; Relethford 1994; Scott and Turner 1997; Turner 1984, 1990, 1992). Small-scale population movements, which, by definition are likely to occur between groups with little morphological differentiation, are therefore hard to recognize, and have only relatively recently been studied (e.g., Coppa et al. 2007; Cucina et al. 1999; Hanihara et al. 2008; Hemphill 1999; Hemphill and Mallory 2004; Irish 2006; Schillaci and Stojanowski 2003; Stefan 1999; Stynder 2009; Sutter and Mertz 2004; Zakrzewski 2007).

Although migration is a well-studied aspect of human behavior (Anthony 1990:895), it was de-emphasized in both processual and post-processual archaeology, and has only recently become re-accepted within archaeology as a potential mechanism for social and cultural change (e.g., Anthony 1990; Burmeister 2000). The apparent unpredictability of migrations, and the difficulty of recognizing them archaeologically without falling back upon a culture-historical approach, has meant that they have been viewed as of limited use for interpretation (Anthony 1990). Migration has been avoided as an explanatory construct as archaeologists have found it difficult to incorporate migrations into models of cultural change (Anthony 1990).

Human migration may take a variety of forms; short distance, long distance, direct, or leapfrogging in nature (Anthony 1990). Most long-distance migration streams, when a series of people follow the same well-defined route over a prolonged period of time (as opposed to a wave model of migration), are associated with return migration, a counter stream of people returning to the migrants' place of origin (Anthony 1990). This, almost by definition, may be especially hard to recognize in the archaeological record. Most short-distance migration, and some long-distance moves, may further be broken down into local, circular, chain, and career migration (Tilly 1978). Recognizing and differentiating these different forms may not be possible archaeologically, but recognizing their differing existences permits different potential biological effects to be considered. Recent research has relied on specialized (primarily isotopic) techniques to identify migration and the use of multiple and varied forms of archaeological evidence to recognize potential migratory ethnic or other identity subgroups or enclaves (e.g., Price et al. 2001; Santley et al. 1987; Stein 2005). By identifying the constructs and patterning of such differences between groups, it may be possible to develop models of skeletal population boundaries and borders.

It is clear that human populations are highly unlikely to evolve morphologically in a discrete and discontinuous manner within a very short time frame, and thus distinct morphological changes or clear discontinuity in morphological patterns may indicate the influx of and (at least partial) replacement of a group with new peoples, or at least new genes (Weiss 1988). By contrast, a rapid change in material culture cannot be used as definitive evidence that there has been an influx of its bearers as the producers, the style and/or the technology may migrate and, as noted above, the match between material culture groupings and social groupings is not direct and causal. Thus it follows that bioarchaeology could provide some of the best evidence of population movement and migration, and therefore of boundaries and borders between population groupings in the archaeological record. The biological differences between groups might be very small and consist of a complex of morphological differences (rather than the highly simplistic description of "narrower" heads given by Ewing (1950)), or may entail variation in skeletal or dental isotopic signature. The two most common isotopic methods used to identify archaeological residential mobility measure strontium and oxygen isotope ratios in dental and skeletal elements formed during an individual's life time and compare them with the signatures of the local bedrock and water (e.g., Bentley et al. 2002; Montgomery et al. 2005; Price et al. 2001). These methods, however, are both destructive of the skeletal and dental material and relatively costly. Other potentially more simple methods may also be applied that are both nondestructive in nature and relatively cheap to undertake, and it is these methods that will be emphasized. This chapter will therefore focus upon morphological aspects of population variation and hence of group identification. As a result, skeletal groups and populations might best be considered in terms of *tribal* or group *microdifferentiation* (Weiss 1988:140). As the term tribe may be laden with racial overtones, the phrase "group microdifferentiation" is preferred here.

Population Variation and Affinity

Humans exhibit patterns of biological variation and diversity. This patterning of variation affects many parts of the body, including soft tissues and skeletal elements. Certain traits, such as blood group B allele frequency in Europe, exhibit a clinal distribution (Mourant 1953). Traits, however, that can be recognized bioarchaeologically, tend to have a mosaic pattern of clinal and discontinuous distribution as the skeletal traits interact with each other and vary in their expression within individuals (Larsen 1997; Saunders 1989). In addition to these biological differences, cultural practices and activities, such as mortuary practices, can elucidate the socio-cultural construction of group or ethnic identity. Human groups are variable in terms of cranial, dental, and postcranial morphology. This variation takes both metric and nonmetric forms. In this chapter, the primary focus will be placed upon craniometric variation, osseous and dental nonmetric variation, and postcranial metric diversity. Examples of such skeletal nonmetric traits (metopic suture and unilateral os japonicum) are shown in Figures 7.1 and 7.2.

Studies of craniometric variation of other populations have shown distinct groupings and clusters existing within global patterns of morphology (e.g., Howells 1973; Lahr 1996). Population groupings have been recognized within African (Morris and Ribot 2006; Zakrzewski 2007), European (Brothwell and Krzanowski 1974; Lalueza Fox et al. 1996; van Vark et al. 1992), American (González-José et al. 2001; Schillaci and Stojanowski 2003), Asian (Hanihara 1992; Hanihara et al. 2008; Hemphill 1999; Hemphill and Mallory 2004), and Australasian (Birdsell 1993; Stefan 1999) samples. Population groupings and variability have also been recognized from the study of discrete dental and skeletal oral traits within African (Irish 2005, 2006), European (Coppa et al. 2007; Turner 1984), American (Scott et al. 1983; Turner 1992), Asian (Hanihara 1992; Lukacs et al. 1998; Turner 1992), and Australasian (Turner 1992) samples. Postcranial metric diversity has

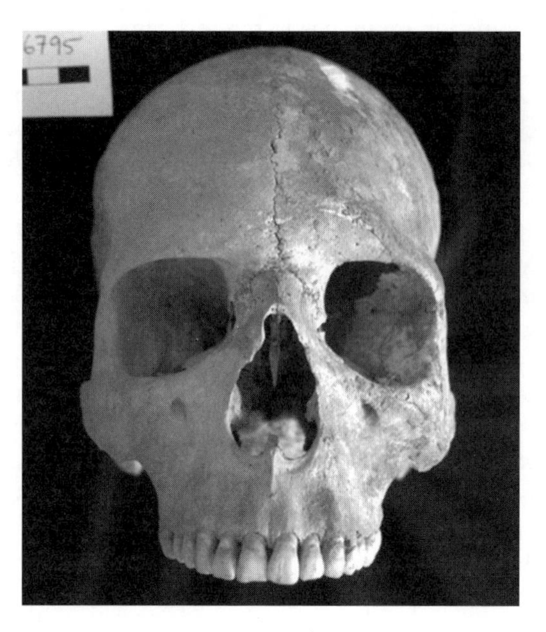

Figure 7.1 Metopic suture in individual 2775 from Écija.
Source: Author's photograph

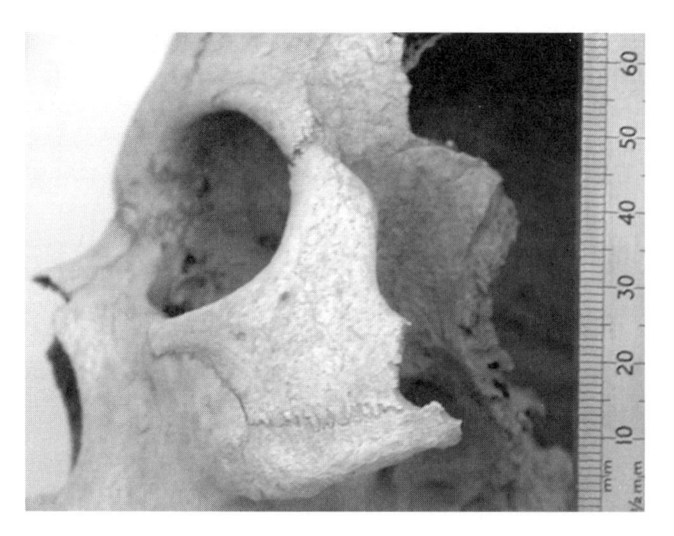

Figure 7.2 Os japonicum (i.e. a bipartite zygomatic bone) in the left malar of skeleton 11830 from Écija. This trait is unilaterally expressed in this individual.
Source: Author's photograph

also been noted within distinct groups and samples (Ruff 2002), including African (Sealy and Pfeiffer 2000; Zakrzewski 2003), European (Holliday 1997; Pomeroy and Zakrzewski 2009), American (Wanner et al. 2007), and Asian (Temple et al. 2008) groups. Given the global nature of skeletal variation, it is suggested that these methods can be applied irrespective of the sample being considered.

Islam and Medieval Iberia

This chapter will employ a case study from a medieval Islamic assemblage from Iberia to demonstrate the ability to identify the biological and social variability and hence aspects of ethnicity and identity within a skeletal sample.

The case study itself is unusual as it comprises an Islamic skeletal sample from Andalucía in southern Spain. There are very few large assemblages of human skeletons from Islamic cemeteries, and only some of these, such as this assemblage from Écija, have been excavated with the support of the local Muslim population. The skeletons described here date from the 8th to 11th centuries A.D.

By early in the 8th century, the Islamic empire controlled much of the Mediterranean region, including the entire North African coast (Fletcher 1992; Jotischky and Hull 2005). Many of the local populations of the conquered areas converted to Islam, although, at the time of the Islamic expansion into Iberia, pockets of tribes and lands still existed where the religion had not yet fully taken hold (Fletcher 2004). The expansion of Islam continued into the remaining uncontrolled areas of the Mediterranean, conquering the Iberian Peninsula in A.D. 711–712, thus greatly expanding the trading routes that the Islamic empire then controlled (Esposito and Voll 2001; Insoll 1999).

The Islamic conquest of Iberia can be considered as an extension of that of North Africa. This conquest initially comprised a raid into Spain crossing the Straits of Gibraltar, followed by a rapid and relatively effortless conquest of the majority of the Iberian Peninsula (Fletcher 1992; Lowney 2005). The conquest was achieved by an alliance of migrant Arabs and indigenous Berbers. The initial invasions were conducted by Berbers including newly converted tribes from Morocco, Algeria, and Tunisia (Fletcher 1992; Glick 1979, 1995; Reilly 1993; Ruggles 2004; Savage 1992). Although eventual control of al-Andalus was by the Arabs, Berbers formed the first invaders, but their invasion involved only small numbers of men along with their families and slaves (Brett and Forman 1980; Fletcher 1992; Glick 1995; Reilly 1993; Ruggles 2004). As the Muslim conquest of the Mediterranean region progressed towards Iberia, religious conversion of the Berbers continued (Fletcher 2004; Glick 1995). This is crucial to understanding the composition of the Andalucian cemeteries associated with the earliest invasions. The Berbers were converts to Islam, but maintained a material culture, including burial rites, which was different from Arab Muslims who had converted much earlier. The two cultures were never fully integrated in al-Andalus (Chejne 1974; Glick 1979; Payne 1973), but maintained somewhat separate communities. After the initial invasions and migrations by Berbers, followed closely by Arabs and their families, other groups also migrated into al-Andalus, including Syrian soldiers. These were followed by an increased flow of slaves from around the Islamic empire, associated with more Berbers and Arabs (Hourani 1991; Payne 1973). The Berbers continued to migrate into al-Andalus throughout the Islamic period in Spain, whereas Arab groups are believed to have arrived mainly during the initial conquest of the Iberian Peninsula (Casas et al. 2006).

As Islam continued to spread, there was expansion of Islamic-controlled trading routes. Slaves comprised a major part of the traded merchandise (Fletcher 1992), and this has a direct impact on this study. These slaves are thought to have derived not only from all areas under Islamic control, but probably also from a number of external locations including Ireland and Eastern Europe (Fletcher 1992).

These different groups of people are important to the history of the expansion into Andalucía as they represent the incoming populations that influenced the Iberian Peninsula and created the extended Islamic empire, thereby increasing trade exchanges and wealth. These groups exemplify distinct ethnic and cultural ways of life being carried into al-Andalus (Glick 1995). The early Muslim residents settled in tribal or subtribal groups (Kennedy 1996). Furthermore, important figures, especially among the Berbers, traveled with their tribal entourage and reconstituted their clans upon settlement in al-Andalus (Ruggles 2004). The Arabs and Berbers did not intermingle once settled and distinct subgroupings remain archaeologically visible (Brett and Forman 1980). These groupings are not only represented in different ways of life, but also in burial rites and hence in cemetery representation. Even within the strict rules of Islam, visible contrasts in burial rites, artistic styles, and other cultural materials existed between different ethnic and cultural groups constituting the *Ummah* or Muslim community (Insoll 1999).

Islamic Écija

The remains of medieval Écija lie beneath the modern town, situated 80 kilometers east of Sevilla in Andalucía (Figure 7.3). Due to its location in the Guadalquivir valley between Córdoba and Sevilla, Écija was of local importance during the

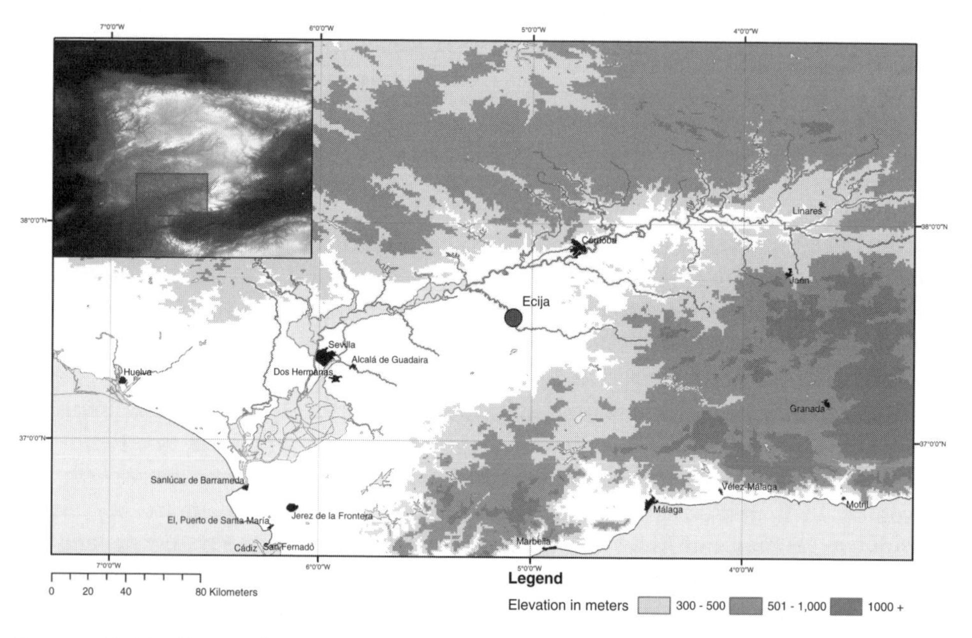

Figure 7.3 Map locating Écija relative to Sevilla and Córdoba in Andalucía (created with the help of Dr. Fraser Sturt).
Source: Author's drawing

medieval period as it was able to exercise some control over the olive oil trade (Fletcher 1992). It was also the site of a major battle in A.D. 711 during the early phase of the Muslim conquest (Jotischky and Hull 2005). The medieval Muslim *maqbara* (cemetery) was located beneath the modern Plaza de España, in the center of the town, and was excavated between 1997 and 2002, yielding in excess of 4,500 inhumed individuals (Jiménez n.d.; Ortega n.d.; Román n.d.). The individuals were identified as Muslim based on characteristics of the burials, which include grave orientation, body position, and a lack of grave goods, as outlined in Insoll (1999:169,172). The *maqbara* was in use from the period immediately following the Muslim conquest in the 8th century A.D. until the 11th century A.D. (Jiménez n.d.; Ortega n.d.; Román n.d.), although the major Christian re-conquest of the region did not take place until the 13th century A.D. The individuals included in the case study sample date primarily to the early period of occupation (8th–9th centuries).

Identity and Ethnicity in Islamic Ecija

The population composition of the Écija *maqbara* is likely to be biologically diverse, potentially comprising both migrating (invading) Berbers and Arabs, associated with some early converts from the indigenous Spanish population. The necropolis group might thus comprise soldiers, their wives and families, slaves, and free men and women; therefore it is expected to be heterogeneous and variable throughout the use of the necropolis. This should result in variation in skeletal patterning and expression, and hence permit the bioarchaeological recognition of aspects of identity including ethnicity.

Funerary Aspects of Identity

Although some geographic variation in burial tradition exists (Abu-Lughod 1993; Jonker 1996), Islamic burials are characterized by their simple nature (Insoll 1999). Islamic burials must occur in a cemetery, or part of a cemetery, especially designated for Muslims only (Al-Kaysi 1999). Multiple burials are uncommon, occurring only during periods of stress and high numbers of deaths, such as periods of war or disease (Sabiq 1991). Children and infants are also usually buried separately from adults as these individuals are considered by Islam to be without sin (Simpson 1995). Burial itself, usually without a coffin, occurs rapidly after death. The body is placed into the grave so that the body lies upon the right-hand side, with the head facing in the direction of Mecca. The head may be supported to ensure that it remains facing in this direction. Graves may be marked but ornamentation is considered ostentatious (Leisten 1990). Graves tend to be relatively shallow as Islam considers it important that the deceased is able to hear the calls to prayer from the *muezzin* (Insoll 1999).

Burials at Écija did vary from these normal and idealized patterns. Some individuals were buried with coffins, although these formed a small minority of the inhumations. Some multiple burials and ossuaries were found within the *maqbara*.

Bodies were generally found lying upon their right sides, leading to bilateral patterning in the degree of skeletal preservation (Inskip et al. in press). This basic funerary analysis indicates that the burials within the necropolis follow the established Islamic norms, but does not provide more detailed elaboration of the identities of the individuals buried.

Migration and Mobility in the Écija Population

Due to the development of the permanent dentition in childhood, the chemical composition of dental enamel from teeth provides a clear isotopic signature of the geographic location of childhood residence. Differences between the isotopic composition of the teeth and the local soil and bedrock indicate the movement of individuals during their life time (i.e., *life-time migration*). Preliminary strontium isotope results (following standard methods using thermal ionization mass spectrometry) have shown that most of the sample from the early Islamic population of Écija have very similar isotope ratios (Figure 7.4), suggesting that these individuals grew up in the same geographic locale (or, at least in regions with the same geological composition). Individual 11333, a young male deriving from the first period of Islamic occupation of Écija, has a strontium signal which is two standard deviations from the mean value for the human sample. A cut-off of two standard deviations from the mean has been used by some researchers (e.g., Bentley et al. 2002, 2008; Price et al. 2001) as a marker of geographic separation, suggesting that this individual grew up in a region different from the rest of the small test sample. As this is the only such individual (of the 20 sampled), it is, however, also possible that this is simply a statistical artifact of sampling and that this individual has a different strontium ratio by chance. The contrast with the results from local medi-

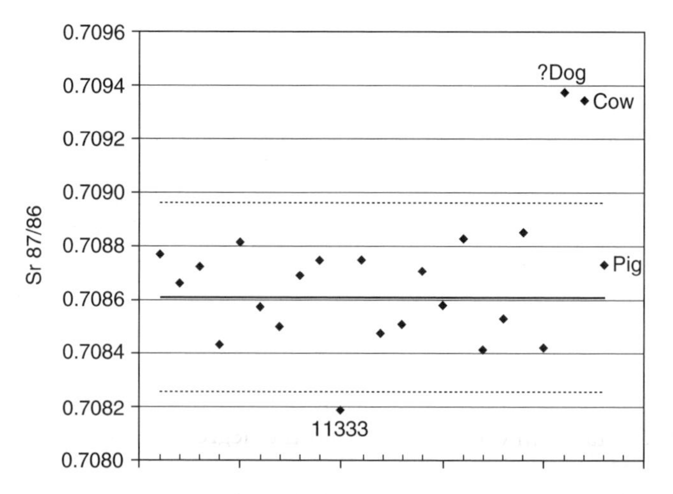

Figure 7.4 Strontium isotope ratios (Sr 87/86) by individual. Solid line represents mean value for the human samples. Dashed lines are two standard deviations from this sample mean.
Source: Author's drawing

eval dog and cow dental samples suggest that the locale in which most individuals spent their childhood was, however, somewhat distinct from the area in which these contemporary animals were kept. This implies that the individuals tested, buried in the earliest period of occupation of the Écija necropolis, did not, during their childhood, live in the same area as those animals tested and thus may have been immigrants to this region.

The rest of this chapter will demonstrate the preliminary results of assessments of other skeletal aspects of ethnicity and migration and hence of identity. As noted above, certain skeletal markers, such as facial characteristics, are inherited (with relatively little influence of local environment). Gross morphological analyses of these portions of the skeleton can thus provide an indication of the biological affinities of a sample. These analyses must be undertaken in a time–space context. In this sense, genetic identity by descent (IBD) is extended to include the effects of time and space. Within modern genetics, studies are undertaken to compute the space–time probabilities of IBD as a measure of the degree of shared ancestry of either a series of populations or genes separated both in time and space (Epperson 1999). The same inferences apply to these studies in that the potential effects of both time and space upon the population structure and hence composition must be considered prior to analysis. Given this premise, this case study considers initially the identity and ethnicity evidence offered by craniometric signals, and then considers the indications provided by the various nonmetric trait data.

Craniometric Ethnicity

Multivariate analysis of standard cranial measurements (and, more importantly, the relationships between such measurements) can provide a guide to the population history of the sample under consideration (e.g., Howells 1973, 1989). These can demonstrate and demarcate patterns of morphological variation and ethnicity both within and between populations (e.g., Stynder 2009; Zakrzewski 2007).

Relethford (1994) has indicated that the level of intergeographic variation in human craniofacial form is low relative to intrageographic variation. Despite this, measurements of the facial skeleton are recognized as being reliable skeletal indicators of population affinities in modern humans (Giles and Elliot 1962; Howells 1973, 1989, 1995; Krogman and Iscan 1986). Furthermore, the majority of studies of craniometric variation have found some clear and highly replicable patterns of such significant differences in facial size and shape between modern human groups (e.g., Hanihara 1996; Howells 1973, 1989; Guglielmino-Matessi et al. 1977; O'Higgins and Strand Viðarsdóttir 1999; Ross et al. 1999).

The analyses presented here do not attempt to apportion the crania into any one particular geographic or race group, but rather attempt initially to delineate the intra-sample variation, and then, to compare the sample with the Howells global cranial data set in order to ascertain the degree of similarity to other cranial samples. The Howells global cranial data set comprises complete measurements (i.e. at least 57 measurements) for 2,524 crania, deriving from approximately 30 different population groups from around the world. All the groups are large enough to provide statistical reliability and yet, at the same time, represent a single locale or population, and can be employed to assess global patterning of variation and similarity. The data is freely available for download from http://konig.la.utk.edu/howells.htm (for an explanation of the data, see Howells 1996). This approach, therefore, does not consider cranial forms to be fixed racial morphs, but instead assumes that variation occurs in a mostly clinal morphological manner, but that distinct differences may also be seen. In addition, the sample comprises individuals from a past historical period who derive from a larger overall grouping of individuals whose ancestry and pattern of morphological variation is unknown. As a result, the anthroposcopic "race" estimation methods utilized in forensic anthropology are completely inappropriate.

A series of 122 crania (comprising 73 males, 48 females, and one individual of unknown sex) were studied. These crania have varying degrees of preservation with some bilateral variation in taphonomic changes potentially from the Islamic burial tradition (Inskip et al. in press). Using the bones of the cranium and the landmarks upon them, a series of standard measurements were taken for each individual following the techniques described by Howells (1973, 1989). As noted above, employing the measures derived by Howells (1973) allowed the comparison of the data with his large global data set in order to identify patterns of craniometric similarity with other groups.

As it is the pattern of morphological variation that is important in aiding in the assessment of ethnic variation, multivariate analyses (comprising both facial and vault measurements) were undertaken. In the early 20th century, scholars used methods such as the calculations of indices (e.g., the cranial index) to assign ethnic or racial groupings to individual crania (e.g., Morant 1928; Pearson and Tippett 1924; Woo and Morant 1932). These methods are obviously highly simplistic and include the inherent assumption that this form of cranial variation is fixed in nature rather than being biologically plastic in response to a variety of both genetic factors and environmental stresses.

As noted above, many of the crania were fragmentary and some were crushed due to their burial position lying on their right sides, and hence, for these individuals, few measurements could be taken. As a result, intra-population analysis focused upon assessment of the expression of cranial sexual dimorphism. Little research has been undertaken upon assessing the global rates of cranial sexual dimorphism in modern humans (e.g., Johnson et al. 1989; O'Higgins et al. 1990) and therefore its use in understanding of ethnicity is necessarily somewhat limited. It does, however, provide an indication of the intra-sample morphological variation. Most variables did not exhibit any sexual dimorphism within the Écija *maqbara* sample, although some variables describing vault length, mastoid size, brow ridge morphology, and aspects of facial prognathism did demonstrate sexual dimorphism. As would be expected, whenever significant sexual dimorphism was noted in measurements, the male sample had larger mean values than the female sample. Postcranially, sexual dimorphism has been noted in terms of both long bone lengths and cross-sections within a sample of the *maqbara* population (Pomeroy and Zakrzewski 2009).

Comparison with cranial samples from neighboring populations or from potential source populations for the Écija sample might enable bioarchaeological affinity to be ascertained. Ideally, analyses would comprise comparisons with potentially related skeletal populations. The majority of potential immigrants to Écija with the Islamic conquest are likely to have originated from the Middle East (Arabs) or from across North and Saharan Africa (Berbers), and this potential genetic and ethnic diversity may be reflected in the morphological heterogeneity noted in later medieval Spanish skulls (Ubelaker et al. 2002). Ideally comparison would be made with potential ethnic groups from which the sample might derive (such as Berbers, Arabs, and Spanish Visigoths), but little published data of suitable form exists. As little published craniometric data exists either for medieval Spain or medieval North and Saharan Africa, analyses were undertaken employing Howells' data set. In order to make these comparisons as valid as possible, the data set was condensed to include only skeletal samples from reasonably proximate locations (i.e. the European samples, comprising the Berg from Austria, the medieval Norse and the Zalavar from Hungary, and the North and Central African samples, comprising the Dogon from Mali, the Egyptians, and the Teita from Kenya).

In order to maximize the number of crania from Écija included in the analysis, a minimum number of variables were selected that provide a basic description of the cranium and that have a relatively large sample size. These variables describe the length of the vault (glabello-occipital length, GOL), the breadth of the vault (biasterionic breadth, ASB), craniofacial breadth (bifrontal breadth, XFB), midfacial prognathism (nasal subtense, NAS), upper facial height (nasal height, NLH and malar height, WMH), and browridge morphology (supraorbital projection, SOS). Principal components and discriminant function analysis were then employed. Principal components analysis is a form of factor analysis that aims to identify the underlying factors (variables) explaining the pattern of correlations within the set of observed variables, and so the variables that are of greatest importance in explaining the variance seen within the ellipse of data points in multidimensional space can be identified.

The principal components analysis derived two significant components that cumulatively explained 61 percent of the variance seen within the crania. These two components are shown in Figure 7.5. This figure demonstrates the

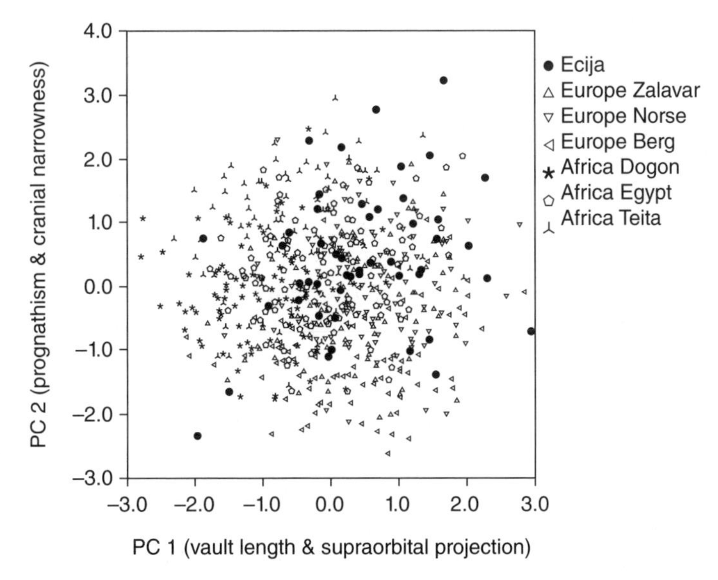

Figure 7.5 Plot of first two principal components (sexes pooled).
Source: Author's drawing

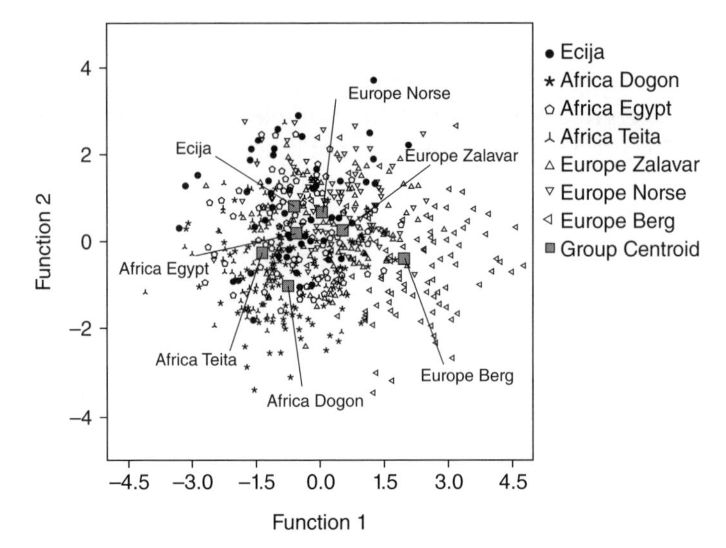

Figure 7.6 Plot of first two discriminant functions (sexes pooled).
Source: Author's drawing

morphological heterogeneity of the cranial samples studied, but also shows that there are trends that may distinguish the European and African samples. The African Dogon and Teita tend to have broader crania than the European samples, while the latter groups tend to have longer cranial vaults. The sample from Écija appears morphologically diverse, with individuals appearing in all quadrants of the plot.

Discriminant function analysis was then applied in order to assign group membership from these same craniometric predictor variables. The main aim is, therefore, to find the dimension or dimensions by which the cranial samples differ and then derive classification functions from this to predict group membership. Discriminant function analysis forms a string of functions and judges whether the groups it predicts from these functions match the imposed groups within the data. As the number of discriminant functions obtained is always one fewer than the number of groups imposed on the data, in this preliminary study there are seven groups and hence six functions derived. The results are shown in Figure 7.6 and Table 7.1. The results demonstrate that, on the basis of these seven variables (selected simply for their abilities to describe basic patterning in craniofacial morphology), there is similarity in craniofacial morphology between the Écija sample and the six comparative groups. Table 7.1, however, indicates that, as only 14 percent of the Écija crania included in the analysis were correctly classified as deriving from the Écija grouping, the Écija sample is harder to describe mathematically on the basis of these particular craniofacial variables and exhibits greater morphological heterogeneity in these variables than the comparators. This suggests that the Écija sample includes individuals deriving from a variety of ethnic or other cranio-morphological patterns, and implies that the individuals buried within the maqbara may be relatively craniometrically, and by implication, genetically heterogeneous.

Table 7.1 Percentage of crania classified into each geographic group, sexes pooled

Original group	*Predicted group membership*						
	Écija	*Afr. Dogon*	*Afr. Egypt*	*Afr. Teita*	*Eur. Zalavar*	*Eur. Norse*	*Eur. Berg*
Écija	**14.0**	10.0	18.0	14.0	18.0	26.0	0.0
Africa Dogon	1.0	**62.6**	11.1	12.1	5.1	6.1	2.0
Africa Egypt	2.7	12.6	**46.8**	10.8	7.2	17.1	2.7
Africa Teita	3.6	20.5	25.3	**45.8**	0.0	4.8	0.0
Europe Zalavar	3.1	11.2	9.2	2.0	**31.6**	24.5	18.4
Europe Norse	1.8	6.4	20.9	4.5	17.3	**42.7**	6.4
Europe Berg	0.0	2.8	5.5	0.0	11.9	9.2	**70.6**

Where Afr. refers to African origin and Eur. refers to European origin. Correct classifications are shown in bold.

Nonmetric Traits and Ethnicity

Analysis of epigenetic nonmetric traits, such as those in the dentition, also provides signals of population affinity. These traits are considered to be genetically inherited, and thus provide a guide to the relative genetic distinctiveness and affinities of the population (Larsen 1997). Recognition of patterns of ethnic affinity may be possible by comparison of a suite of such nonmetric traits with data from neighboring populations. In addition, nonmetric data may be recorded from fragmented or damaged skeletons, and, as noted above, the Écija assemblage has differences in preservation between the left and right sides resulting from either the Islamic burial custom or other taphonomic or excavation processes (Inskip et al. in press).

Morphological variation in the permanent dentition was assessed by use of the Arizona State University dental anthropology system (ASUDAS) (Turner et al. 1991). This method has a series of well-established criteria for scoring using reference plaques (Turner et al. 1991), and has been used in many studies of comparative dental diversity (e.g., Irish 1997, 2005, 2006; Irish and Turner 1990). As noted by Irish (2005), these dental traits are usually observable despite moderate dental attrition, are relatively easy to identify, provide high recording replicability, possess a high genetic component in their expression (Larsen 1997; Rightmire 1999; Scott and Turner 1997), and are conservative in nature (Larsen 1997). Although there is some comparative data available, such as from North Africa (Irish 1997, 2005, 2006), there have been no detailed studies of European group dental microdifferentiation.

A series of 102 adults (comprising 61 males and 41 females) were studied. Because there is a documented lack of trait sexual dimorphism (Bermúdez de Castro 1989; Hanihara 1992; Scott 1980; Turner et al. 1991), the sexes have been pooled, following standard ASUDAS procedures (Irish 1997). Both the dental crowns and roots exhibited variation in expression of the traits suitable for biodistance analyses. A lack of a hypoconulid was noted in 28 percent of lower first molars and in 91 percent of lower second molars. These frequencies compare with 8 percent and 71 percent of Western Europeans and 10 percent and 66 percent of North Africans, respectively (Scott and Turner 1997). Interruption grooves were noted in 39 percent of upper lateral incisors, compared with 42 percent in Western Europeans and 32 percent of North Africans (Scott and Turner 1997), with more specific values of 64 percent in prehistoric Italians (Cucina et al. 1999) to a range between 9 and 33 percent of ancient Egyptians (Irish 2006). Extra upper premolar roots were only noted in 33 percent of cases. Much higher comparative values are found in ancient Egyptians, ranging from 60 to 90 percent (Irish 2006), and prehistoric Italians, at 64 percent (Cucina et al. 1999). More similar values were found in the broader range groups, at 41 percent in samples from Western Europe and 47 percent in samples from North Africa (Scott and Turner 1997). The Écija sample thus appears to exhibit some similarities in dental morphology with both North Africans and other Western Europeans, but insufficient detailed comparative Spanish studies exist to permit calculations of biodistance such as the mean measure of divergence.

Standard skeletal nonmetric traits were recorded for both the crania and postcrania for 124 adults (73 males, 50 females, and 1 unsexed). Nonmetric cranial traits have been commonly used in biodistance analyses (Berry 1963; Berry and Berry 1967; Hauser and De Stefano 1989; Ishida and Dodo 1993; Laughlin and Jorgensen 1956; Pardoe 1991; Saunders 1989; Sjøvold 1973; Sutter and Mertz 2004). As these traits develop during the growth and development of an individual, their degree of expression varies, and hence these traits do not follow simple Mendelian patterns of inheritance, but rather may have quasi-continuous expression (Grüneberg 1963). It is generally thought that multiple genes with small and additive effects are responsible for the development of such traits with the presence of a specific trait being determined by some threshold (Saunders 1989; Sjøvold 1973). There is, however, as with the dental nonmetric traits described above, the implicit assumption that much of the variation is genetic (Berry and Berry 1967; Saunders 1989; Sjøvold 1973), and indeed high heritability has been noted for some cranial traits (Cheverud and Buikstra 1981a, 1981b, 1982; Richtsmeier et al. 1984; Richtsmeier and McGrath 1986).

The early period occupants of the necropolis were found to have high levels of certain of these epigenetic nonmetric skeletal traits. The metopic suture was retained in 15 percent of the adults (n = 107). This is a high incidence, but is similar to frequencies noted in some recent European groups and much higher than in African samples (Hauser and de Stefano 1989). Extra ossicles along the lambdoid sutures of the cranium were present in 30 percent of left sides (n = 63) and 30 percent of right sides (n = 102), with a cranial incidence of 34 percent (n = 102). This is lower than the cranial incidence in contemporary northern European samples (Czarnetzki 1975), suggesting a certain distinctive nature to the Écija Islamic population. It is, however, higher than the frequency in Late Roman/Paleo-Christian individuals from southeastern Portugal (11 percent of left sides, 13 percent of right sides) (McMillan and Boone 1999). The contemporaneous Islamic sample from southeastern Portugal studied by McMillan and Boone (1999) was very small, comprising only four right sides (two of which had lambdoid ossicles) and two left sides (neither of which had ossicles). Ossicles were noted in the sagittal suture in 5 percent of crania (n = 75). Within the small samples from southeastern Portugal, these were noted in only one Late Roman/Paleo-Christian individual (*n* = 7) (McMillan and Boone 1999). Unfortunately only one Islamic skull from the comparator Portuguese site had a preserved sagittal suture, and this lacked Wormian bones. The cranial incidence for sagittal ossicles is similar to that noted in contemporary northern Europeans (Czarnetzki 1975). In addition, ossicles were noted at asterion in 31 percent of left sides (*n* = 49) and 40 percent of right sides (*n* = 101). This is extremely high relative to European and African samples (Hauser and de Stefano 1989). Although Wormian bones are noted to be particularly influenced by aspects of the environment, such as cranial deformation (Del Papa and Perez 2007), the similar degrees of prevalence of such additional ossicles in the sutures within these undeformed crania can imply some genetic similarity, and hence affinity, between the Écija cranial sample and other medieval European groups.

Postcranially, preliminary studies of the Écija skeletons have shown high frequencies of both vastus notches and emarginate patellae. These traits may be genetically controlled, but may potentially be associated with chronic knee flexion, as in squatting (Messeri 1961). The frequency of septal aperture presence in either arm is 30 percent (*n* = 92 for left humerus, *n* = 102 for right humerus), with 44 percent of individuals affected. The frequency in populations of European descent is usually 5–10 percent, but increases to approximately 60 percent in those of North or West African descent (Glanville 1967; Hrdlička 1932; Mays 2008), implying an extra-European affinity of the early Islamic sample studied.

Without the use of the relatively gross anthroposcopic methods used within forensic anthropology for attribution of ancestry (e.g., Bass 1995; Byers 2002), these preliminary studies demonstrate that the burial population of the Islamic *maqbara* at Écija does exhibit both skeletal and dental features that suggest some degree of group microdifferentiation associated with biological affinities to certain European and African groups. Given both their location within southern Andalucía and the historical context of the site, this is not unexpected. Further research and more comparative data are required to develop detailed biodistance measures, such as the mean measure of divergence, in order to place this sample within a broader context.

Iberian Genetic Identity and Ethnicity

The genetic diversity of the modern populations of Europe has been greatly scrutinized in order to identify large-scale migrations, such as those associated with the spread of agriculture from the Near East (e.g., Chikhi et al. 1998; Richards et al. 2000; Simoni et al. 2000). Some studies have also focused, through analyses of modern populations, upon both the expansions of populations into and out of Iberia and the Canary Islands relative to North Africa (Adams et al. 2008; Cherni et al. 2009; Côrte-Real et al. 1996; Gérard et al. 2006; Lucotte et al. 2001; Pereira et al. 2005; Pinto et al. 1996). These studies generally note both the geographic and genetic proximity between Iberia and the Maghreb. High frequencies of both North African Y and mtDNA haplotypes have been noted. The former suggests over 10 percent North African ancestry (Adams et al. 2008), whereas North (and sub-Saharan) African mtDNA sequences suggest a genetic input of over 8 percent

(Pereira et al. 2005) into modern Iberian populations. Archaeological mtDNA has also been extracted from neighboring medieval Islamic sites in Córdoba (Figure 7.3), and indicates that this medieval Islamic population demonstrated greater genetic affinities to modern North African than to modern Iberian samples (Casas et al. 2006).

Bioarchaeological Identity within Islamic Écija

As noted above, identity comprises a variety of divergent yet superimposed aspects, including religion, ethnicity, age, and gender. Although the current chapter has focused upon the former two, other aspects of identity must also be considered. For example, sex and gender have been noted as of importance in defining social ideologies within Andalucía (Brandes 1981) and skeletal markers of such gender differences in activity have been noted within this sample (Pomeroy and Zakrzewski 2009).

Religion may define a group as it can affect many aspects of life such as diet, and even death through burial practice, and therefore religion may be expressed skeletally. Ethnicity can easily be misinterpreted as racial identification or ancestry, but this view ignores the implicit personal and social aspects of ethnic identity. Ethnicity is, however, intrinsically linked to population affinity and hence to mobility and migration within a population sample. Within the current study, these aspects of identity are superimposed and thus they mediate the bioarchaeological expression of the other. This is important as religion, demonstrated here through Islamic funerary practices, can impinge upon the skeleton through differential taphonomic processes (Inskip et al. in press). The preliminary analyses of the Écija *maqbara* population indicate that the skeletons do exhibit traits that suggest both some degree of European and African biological ancestry. This is unsurprising given the complex historical situation within southern Iberia and the expected high levels of religious conversion within the local medieval population.

Migration into Iberia is noted as a direct result of the Islamic conquest of the region, and may be expected to be a mix of both long and short distance. The strontium isotope analyses above do not, however, indicate whether any specific individuals buried in the *maqbara* moved to Écija from other areas, and hence are unable to identify differences in locale of childhood origin within the pilot study sample. The strontium isotope ratios do, however, demonstrate geographic differences between the areas in which the domesticated animals were raised and the location of childhood residence of the individual buried in the oldest portion of the cemetery.

The Écija *maqbara* population has been shown to exhibit morphological, and hence biological, heterogeneity relative to other European and African samples. This implies that craniometric methods are not able to identify group microdifferentiation within this cemetery sample. By contrast, the nonmetric trait analyses demonstrate similarities between the individual buried within the *maqbara* and certain African groups, thereby suggesting extra-European aspects to the ethnicity of the sample. This supports both the modern (Adams et al. 2008; Pereira et al. 2005) and ancient (Casas et al. 2006) genetic evidence which note the relatively great genetic affinity between Andalucíans and North Africans. The case study presented here indicates the multiplicity of identity and the complex interactions between these different strands that may affect the bioarchaeological signal available for interpretation.

ACKNOWLEDGMENTS

The author would like to thank Antonio Fernández Ugalde, José-Manuel Rodriguez Hidalgo, Ana Romo, Sergio Garcia Dils, and the Museo Histórico Municipal de Écija for access to the skeletal material and for support during data collection at Écija. Skeletal data was collected by the author and Lisa Cashmore, Sarah Inskip, Emma Pomeroy, Jolene Twomey, and Jennifer Wainwright. The isotopic analysis was undertaken by Kristi Grinde with support and advice from Clive Trueman and Tina Hayes (the latter now very sadly deceased). Fraser Sturt is thanked for the production of the map. This work was funded by the British Academy (SG-42094).

REFERENCES

AAPA (American Association of Physical Anthropologists) 1996. AAPA Statement on Biological Aspects of Race. American Journal of Physical Anthropology 101:569–570.

Abu-Lughod, L. 1993 Islam and the Gendered Discourses of Death. International Journal of Middle East Studies 25:187–205.

Adams, S. M., with E. Bosch, P. L. Balaresque, S. J. Ballereau, A. C. Lee, E. Arroyo, A. M. López-Parra et al. 2008 The Genetic Legacy of Religious Diversity and Intolerance: Paternal Lineages of Christians, Jews, and Muslims in the Iberian Peninsula. American Journal of Human Genetics 83:725–736.

Al-Kaysi, M. I. 1999 Morals and Manners in Islam. Leicester: The Islamic Foundation.

Anthony, D. W. 1990 Migration in Archeology: The Baby and the Bathwater. American Anthropologist 92:895–914.

Asad, T. 1986 The Idea of an Anthropology of Islam. Center for Contemporary Arab Studies Pamphlet. Georgetown: Georgetown University Press.

Babić, S. 2005 Status Identity and Archaeology. *In* The Archaeology of Identity. M. Díaz-Andreu, S. Lucy, S. Babić, and D. N. Edwards, eds. Pp. 67–85. London: Routledge.

Bass W. M. 1995 Human Osteology: A Laboratory and Field Manual. Columbia: Missouri Archaeological Society.

Beck, L. A. 1995 Regional Cults and Ethnic Boundaries in "Southern Hopewell". *In* Regional Approaches to Mortuary Analysis. L. A. Beck, ed. Pp. 167–187. New York: Plenum Press.

Bentley A., with T. D. Price, J. Luning, D. Gronenborn, J. Wahl, and P. D. Fullagar 2002 Prehistoric Migration in Europe: Strontium Isotope Analysis of Early Neolithic Skeletons. Current Anthropology 43: 799–804.

Bentley R. A., with J. Wahl, T. D. Price, and T. C. Atkinson 2008 Isotopic Signatures and Hereditary Traits: Snapshot of a Neolithic Community in Germany. Antiquity 82:290–304.

Bermúdez de Castro, J. M. 1989 The Carabelli Trait in Human Prehistoric Populations of the Canary Islands. Human Biology 61: 117–131.

Berry, A. C., and R. J. Berry 1967 Epigenetic Variation in the Human Cranium. Journal of Anatomy 101:361–379.

Berry, R. J. 1963 Epigenetic Polymorphism in Wild Populations of *Mus musculus*. Genetical Research 4:193–220.

Birdsell, J. B. 1993 Microevolutionary Patterns in Aboriginal Australia. Oxford: Oxford University Press.

Boas, F. 1912 Changes in Bodily Form of Descendants of Immigrants. American Anthropologist 14:530–562.

Brandes, S. 1981 Like Wounded Stags: Male Sexual Ideology in an Andalusian Town. *In* Sexual Meanings: The Cultural Construction of Gender and Sexuality. S. B. Ortner, and H. Whitehead, eds. Pp. 216–239. Cambridge: Cambridge University Press.

Brett, M., and W. Forman, 1980 The Moors: Islam in the West. London: Orbis.

Brothwell D. R., and W. Krzanowski 1974 Evidence of Biological Differences Between Early British Populations from Neolithic to Medieval Times, as Revealed by Eleven Commonly Available Cranial Vault Measurements. Journal of Archaeological Science 1:249–260.

Buikstra, J. E. 2006a A Historical Introduction. *In* Bioarchaeology: The Contextual Analysis of Human Remains. J. E. Buikstra, and L. A. Beck, eds. Pp. 7–25. London: Academic Press.

Buikstra, J. E. 2006b Repatriation and Bioarchaeology: Challenges and Opportunities. *In* Bioarchaeology: The Contextual Analysis of Human Remains. J. E. Buikstra, and L. A. Beck, eds. Pp. 389–415. London: Academic Press.

Burmeister, S. 2000 Archaeology and Migration. Current Anthropology 41:539–567.

Byers, S. N. 2002 Introduction to Forensic Anthropology. Boston: Allyn and Bacon.

Cartmill, M. 1998 The Status of the Race Concept in Physical Anthropology. American Anthropologist 100:651–660.

Casas, M. J., with E. Hagelberg, R. Fregel, J. M. Larruga, and A. M. González 2006 Human Mitochondrial DNA Diversity in an Archaeological Site in al-Andalus: Genetic Impact of Migrations from North Africa in Medieval Spain. American Journal of Physical Anthropology 131:539–551.

Cavalli-Sforza L. L., with P. Menozzi and A. Piazza 1994 The History and Geography of Human Genes. Princeton: Princeton University Press.

Chejne, A. G. 1974 Muslim Spain: Its History and Culture. Minneapolis: University of Minnesota Press.

Cherni, L., with V. Fernandes, J. B. Pereira, M. D. Costa, A. Goios, S. Frigi, B. Yacoubi-Loueslati et al. 2009 Post-Last Glacial Maximum Expansion from Iberia to North Africa Revealed by Fine Characterization of mtDNA H Haplogroup in Tunisia. American Journal of Physical Anthropology 139:253–260.

Cheverud, J. M., with J. E. Buikstra 1981a Quantitative Genetics of Skeletal Nonmetric Traits in the Rhesus Macaques on Cayo Santiago. I. Single Trait Heritabilities. American Journal of Physical Anthropology 54:43–49.

Cheverud, J. M., with J. E. Buikstra 1981b Quantitative Genetics of Skeletal Nonmetric Traits in the Rhesus Macaques on Cayo Santiago. II. Phenotypic, Genetic, and Environmental Correlations Between Traits. American Journal of Physical Anthropology 54:51–58.

Cheverud, J. M., with J. E. Buikstra 1982 Quantitative Genetics of Skeletal Nonmetric Traits in the Rhesus Macaques on Cayo Santiago. III. Relative Heritability of Skeletal Nonmetric and Metric Traits. American Journal of Physical Anthropology 59:151–155.

Chikhi, L., with G. Destro-Bisol, G. Bertorelle, V. Pascali, and G. Barbujani 1998 Clines of Nuclear DNA Markers Suggest a Largely Neolithic Ancestry of the European Gene Pool. Proceedings of the National Academy of Sciences USA 95:9053–9058.

Collins Cook, Della. 2006. The Old Physical Anthropology and the New World: A Look at the Accomplishments of an Antiquated Paradigm. In: Bioarchaeology: The Contextual Study of Human Remains. Jane E. Buikstra and Lane A Beck, eds. Elsevier: Amsterdam. Pp. 27–72.

Coppa A., with A. Cucina, M. Lucci, D. Mancinelli, and R. Vargiu 2007 Origins and Spread of Agriculture in Italy: A Nonmetric Dental Analysis. American Journal of Physical Anthropology 133:918–930.

Côrte-Real, H. B., with V. A. Macaulay, M. B. Richards, G. Hariti, M. S. Issad, A. Cambon-Thomsen, S. Papiha, J. Bertranpetit, and B. C. Sykes 1996 Genetic Diversity in the Iberian Peninsula Determined from Mitochondrial Sequence Analysis. Annals of Human Genetics 60:331–350.

Cucina, A., with M. Lucci, R. Vargiu, and A. Coppa 1999 Dental Evidence of Biological Affinity and Environmental Conditions in Prehistoric Trentino (Italy) Samples from the Neolithic to the Early Bronze Age. International Journal of Osteoarchaeology 9:404–416.

Czarnetzki, A. 1975 On the Question of Correlation Between the Size of the Epigenetic Distance and the Degree of Allopatry in Different Populations. Journal of Human Evolution 4:483–489.

Del Papa, M. C., and S. I. Perez 2007 The Influence of Artificial Cranial Vault Deformation on the Expression of Cranial Nonmetric Traits: Its Importance in the Study of Evolutionary Relationships. American Journal of Physical Anthropology 134:251–262.

Deniker, J. 1913 The Races of Man. London: Walter Scott Publishing.

Díaz-Andreu, M. 2005 Gender Identity. *In* The Archaeology of Identity. M. Díaz-Andreu, S. Lucy, S. Babić, and D. N. Edwards, eds. Pp. 13–42. London: Routledge.

Edwards, D. N. 2005 The Archaeology of Religion. *In* The Archaeology of Identity. M. Díaz-Andreu, S. Lucy, S. Babić, and D. N. Edwards, eds. Pp. 110–128. London: Routledge.

Epperson, B. K. 1999 Gene Genealogies in Geographically Structured Populations. Genetics 152: 797–806.

Eriksen, T. H. 1991 The Cultural Contexts of Ethnic Differences. Man 26:127–144.

Esposito, J. L., and J. Voll 2001 Makers of Contemporary Islam. Oxford: Oxford University Press.

Ewing, J. F. 1950 Hyperbrachycephaly as Influenced by Cultural Conditioning. Papers of the Peabody Museum of American Archæology and Ethnology, Harvard University 23(2).

Fforde, C., with J. Hubert and P. Turnbull eds. 2002 The Dead and Their Possessions: Repatriation in Principle, Policy and Practice. London: Routledge.

Fletcher, R. 1992 Moorish Spain. Berkeley: University of California Press.

Fletcher, R. 2004 The Cross and the Crescent. London: Penguin.

Gellner, E. 1981 Muslim Society. Cambridge: Cambridge University Press.

Gérard, N., with S. Berriche, A. Aouizérate, F. Diéterlen and G. Lucotte 2006 North African Berber and Arab Influences in the Western Mediterranean Revealed by Y-Chromosome DNA Haplotypes. Human Biology 78:307–316.

Giles, E., and O. Elliot 1963 Sex Determination by Discriminant Function Analysis of Crania. American Journal of Physical Anthropology 21:53–61.

Glanville, E. V. 1967 Perforation of the Coronoid-Olecranon Septum. American Journal of Physical Anthropology 26:85–92.

Glick, T. F. 1979 Islamic and Christian Spain in the Early Middle Ages. Princeton: Princeton University Press.

Glick, T. F. 1995 From Muslim Fortress to Christian Castle. Manchester: Manchester University Press.

González-José, R., with S. L. Dahinten, M. A. Luis, M. Hernández and H. M. Pucciarelli 2001 Craniometric Variation and the Settlement of the Americas: Testing Hypotheses by Means of R-Matrix and Matrix Correlation Analyses. American Journal of Physical Anthropology 116:154–165.

Gowland, R., with C. Knüsel eds. 2006 Social Archaeology of Funerary Remains. Oxford: Oxbow.

Grauer, A. L, with P. Stuart-Macadam eds. 1998 Sex and Gender in Paleopathological Perspective. Cambridge: Cambridge University Press.

Gravlee, C. C., with H. R. Bernard and W. R. Leonard 2003a Heredity, Environment, and Cranial Form: A Reanalysis of Boas's Immigrant Data. American Anthropologist 105: 125–138.

Gravlee, C. C., with H. R. Bernard and W. R. Leonard 2003b Boas's Changes in Bodily Form: The Immigrant Study, Cranial Plasticity, and Boas's Physical Anthropology. American Anthropologist 105:326–332.

Grüneberg, H. 1963 The Pathology of Development: A Study of Inherited Skeletal Disorders in Animals. Oxford: Blackwell.

Guglielmino-Matessi, C. R., with P. Gluckman and L. L. Cavalli-Sforza 1977 Climate and the Evolution of Skull Metrics in Man. American Journal of Physical Anthropology 50:549–564.

Hanihara, T. 1992 Dental and Cranial Affinities Among Populations of East Asia and the Pacific: The Basic Populations in East Asia, IV. American Journal of Physical Anthropology 88:163–182.

Hanihara, T. 1996 Comparison of Craniofacial Features of Major Human Groups. American Journal of Physical Anthropology 99:389–412.

Hanihara, T., with Yoshida and H. Ishida 2008 Craniometric Variation of the Ainu: An Assessment of Differential Gene Flow from Northeast Asia into Northern Japan, Hokkaido. American Journal of Physical Anthropology 137:283–293.

Hauser, G., and G. F. De Stefano, 1989 Epigenetic Variants of the Human Skull. Stuttgart: Schweizerbart.

Hemphill, B. E. 1999 Foreign Elites from the Oxus Civilization? A Craniometric Study of Anomalous Burials from Bronze Age Tepe Hissar. American Journal of Physical Anthropology 110:421–434.

Hemphill, B. E, with J. P. Mallory 2004 Horse-Mounted Invaders from the Russo-Kazakh Steppe or Agricultural Colonists from Western Central Asia? A Craniometric Investigation of the Bronze Age Settlement of Xinjiang. American Journal of Physical Anthropology 124:199–222.

Hodder, I. 1982 Symbols in Action. Cambridge: Cambridge University Press.

Holliday, T. W. 1997 Postcranial Evidence of Cold Adaptation in European Neandertals. American Journal of Physical Anthropology 104:245–258.

Hooton, E. A. 1918 On Certain Eskimoid Characters in Icelandic Skulls. American Journal of Physical Anthropology 1:53–76.

Hourani, A. 1991 A History of the Arab Peoples. Cambridge: Harvard University Press.

Howells, W. W. 1973 Cranial Variation in Man: A Study by Multivariate Analysis of Patterns of Difference Among Recent Human Populations. Papers of the Peabody Museum of Archaeology and Ethnology, Harvard University 67.

Howells, W. W. 1989 Skull Shapes and the Map: Craniometric Analyses in the Dispersion of Modern Homo. Peabody Museum Papers, Harvard University 79.

Howells, W. W. 1995 Who's Who in Skulls: Ethnic Identification of Crania from Measurements. Papers of the Peabody Museum of Archaeology and Ethnology, Harvard University 82.

Howells, W. W. 1996 Howells' Craniometric Data on the Internet. American Journal of Physical Anthropology 101:441–442.

Hrdlička, A. 1932 The Humerus: Septal Aperture. Anthropologie (Prague) 10:31–96.

Inskip, S. A., with S. R. Zakrzewski and A. S. Romo Salas In press Taphonomy of the Islamic Burials from Plaza de España. Astigi Vetus.

Insoll, T. 1999 The Archaeology of Islam. Oxford: Blackwell.

Insoll, T. 2004 Archaeology, Ritual, Religion. London: Routledge.

Insoll, T. 2005 Changing Identities in the Arabian Gulf. *In* The Archaeology of Plural and Changing Identities. E. Conlin Casella, and C. Fowler, eds. Pp. 191–209. New York: Kluwer/Plenum.

Irish, J. D. 1997 Characteristic High- and Low-Frequency Dental Traits in Sub-Saharan African Populations. American Journal of Physical Anthropology 102:455–467.

Irish, J. D. 2005 Population Continuity vs. Discontinuity Revisited: Dental Affinities Among Late Paleolithic Through Christian Era Nubians. American Journal of Physical Anthropology 128:520–535.

Irish, J. D. 2006. Who Were the Ancient Egyptians? Dental Affinities Among Neolithic Through Postdynastic Peoples. American Journal of Physical Anthropology 129: 529–543.

Irish, J. D., and C. G. Turner II 1990 West African Dental Affinity of Late Pleistocene Nubians: Peopling of the Eurafrican-South Asian Triangle II. Homo 41:42–53.

Ishida, H., and Y. Dodo 1993 Nonmetric Cranial Variation and the Population Affinities of the Pacific Peoples. American Journal of Physical Anthropology 90:49–57.

Jiménez, A., n.d. El Sector Noroeste. *In* Intervención Arqueológica en la Plaza de España, Ecija. Memoria Final. Volumen 1: Memoria 1. Ana Romo, ed. Pp. 183–193. Unpublished manuscript.

Johnson, D. R., with P. O'Higgins, W. Moore, and T. J. McAndrew 1989 Determination of Race and Sex of the Human Skull by Discriminant Function Analysis of Linear and Angular Dimensions. Forensic Science International 41:41–53.

Johnson, M. 1999 Archaeological Theory: An Introduction. Oxford: Blackwell.

Jones, S. 1997 The Archaeology of Ethnicity. London: Routledge.

Jonker, G. 1996 The Knife's Edge: Muslim Burial in the Diaspora. Mortality 1:27–43.

Jotischky, A., and C. Hull 2005 The Penguin Historical Atlas of the Medieval World. London: Penguin Books.

Kennedy, H. 1996 Muslim Spain and Portugal: A Political History of al-Andalus. London: Longman.

Knudson, K. J., and C. M. Stojanowski eds. 2009 Bioarchaeology and Identity in the Americas. Gainesville: University Press of Florida.

Krogman, W. M., and M. Y. Iscan 1986 The Human Skeleton in Forensic Medicine. Springfield: Charles C. Thomas.

Lahr, M. M. 1996 The Evolution of Modern Human Diversity. Cambridge: Cambridge University Press

Lalueza Fox, C., with A. González Martín and S. V. Civit 1996 Cranial Variation in the Iberian Peninsula and The Balearic Islands: Inferences about the History of the Population. American Journal of Physical Anthropology 99:413–428.

Larsen, C. S. 1997 Bioarchaeology. Cambridge: Cambridge University Press.

Laughlin, W. S., and J. B. Jorgensen 1956 Isolate Variation in Greenlandic Eskimo Crania. Acta Genetica et Statistica Medica 6:3–12.

Leisten, T. 1990 Between Orthodoxy and Exegesis: Some Aspects of Attitudes in the Shari'a Toward Funerary Architecture. Muquarnas 7: 12–22.

Lowney, C. 2005 A Vanished World: Muslims, Christians, and Jews in Medieval Spain. Oxford: Oxford University Press.

Lucas Powell, M., with P. S. Bridges and A. M. Wagner Mires eds. 1991 What Mean These Bones? Studies in Southeastern Bioarchaeology. Tuscaloosa: University of Alabama Press.

Lucotte, G., with N. Gérard and G. Mercier 2001 North African Genes in Iberia Studied by Y-Chromosome DNA Haplotype H. Human Biology 73:763–769.

Lucy, S. 2005a Ethnic and Cultural Identities. *In* The Archaeology of Identity. M. Díaz-Andreu, S. Lucy, S. Babić, and D. N. Edwards, eds. Pp. 86–109. London: Routledge.

Lucy, S. 2005b The Archaeology of Age. *In* The Archaeology of Identity. M. Díaz-Andreu, S. Lucy, S. Babić, and D. N. Edwards, eds. Pp. 43–66. London: Routledge.

Lukacs, J. R., with B. E. Hemphill, and S. R. Walimbe 1998 Are Mahars Autochthonous Inhabitants of Maharashtra? A Study of Dental Morphology and Population History in South Asia. *In* Human Dental Development, Morphology, and Pathology. J. R. Lukacs, ed. pp.119–153. University of Oregon Anthropological Papers 54. Eugene: University of Oregon Press.

Mays, S. 2008 Septal Aperture of the Humerus in a Mediaeval Human Skeletal Population. American Journal of Physical Anthropology 136:432–440.

McMillan, G. P., and J. L. Boone 1999 Population History and the Islamization of the Iberian Peninsula: Skeletal Evidence from the Lower Alentejo of Portugal. Current Anthropology 40:719–726.

Meskell, L. 2001 Archaeologies of Identity. *In* Archaeological Theory Today. I. Hodder, ed. Pp. 187–213. Cambridge: Polity.

Messeri P. 1961 Morfologia Della Rotula nei Neolitici Della Liguria. Archivio per l'Antropologia e l'Etnologia 91:1–11.

Mirza, M. N., and D. B. Dungworth 1995 The Potential Misuse of Genetic Analyses and the Social Construction of "Race" and "Ethnicity". Oxford Journal of Archaeology 14:345–354.

Montgomery, J., with J. A. Evans, D. Powlesland, and C. A. Roberts 2005 Continuity or Colonization in Anglo-Saxon England? Isotope Evidence for Mobility, Subsistence Practice, and Status at West Heslerton. American Journal of Physical Anthropology 126:123–138.

Moore, J., and E. Scott eds. 1997 Invisible People and Processes. London: Leicester University Press.

Morant, G. M. 1928 A Preliminary Classification of European Races Based on Cranial Measurements. Biometrika 20B:301–375.

Morris, A. G., and I. Ribot 2006 Morphometric Cranial Identity of Prehistoric Malawians in the Light of Sub-Saharan African Diversity. American Journal of Physical Anthropology 130:10–25.

Mourant, A. E. 1953 The ABO Blood Groups: Comprehensive Tables and Maps of World Distribution. Oxford: Blackwell.

O'Brien, O. 1994. Ethnic Identity, Gender and Life Cycle in North Catalonia. *In* The Anthropology of Europe: Identities and Boundaries in Conflict. V. A. Goddard, J. R. Llobera, and C. Shore, eds. Pp. 191–207. Oxford: Berg.

O'Higgins P., with W. Moore, D. Johnson, T. McAndrew, and R. Flinn 1990 Patterns of Cranial Sexual Dimorphism in Certain Groups of Extant Hominoids. Journal of Zoology 222:399–420.

O'Higgins, P., and U. Strand Viðarsdóttir 1999 New Approaches to the Quantitative Analysis of Craniofacial Growth and Variation. *In* Human Growth in the Past. R. D. Hoppa and C. Fitzgerald, eds. Pp. 128–160. Cambridge: Cambridge University Press.

Ortega, M. n.d. El Sector Noreste. *In* Intervención Arqueológica en la Plaza de España, Ecija. Memoria Final. Volumen 1: Memoria 1. Ana Romo, ed. Pp. 117–182. Unpublished manuscript.

Pardoe, C. 1991 Isolation and Evolution in Tasmania. Current Anthropology 32:1–21.

Payne, S. G. 1973 A History of Spain and Portugal. Madison: University of Wisconsin Press.
Pearson, K., and L. H. C. Tippett 1924 On Stability of the Cephalic Indices Within the Race. Biometrika 16:118–138.
Pereira, L., with C. Cunha, C. Alves, and A. Amorim 2005 African Female Heritage in Iberia: A Reassessment of mtDNA Lineage Distribution in Present Times. Human Biology 77:213–229.
Pinto, F., with A. M. González, M. Hernández, J. M. Larruga, and V. M. Cabrera 1996 Genetic Relationship Between the Canary Islanders and Their African and Spanish Ancestors Inferred from Mitochondrial DNA Sequences. Annals of Human Genetics 60:321–330.
Pomeroy, E., and S. R. Zakrzewski 2009 Sexual Dimorphism in Diaphyseal Cross-Sectional Shape in the Medieval Muslim Population of Écija, Spain and Anglo-Saxon Great Chesterford, UK. International Journal of Osteoarchaeology 19:50–65.
Price, T. D., with R. A. Bentley, J. Lüning, D. Gronenborn, and J. Wahl 2001 Human Migration in the Linearbandkeramik of Central Europe. Antiquity 75:593–603.
Reilly, B. F. 1993 The Medieval Spains. Cambridge: Cambridge University Press.
Relethford, J. H. 1994 Craniometric Variation Among Modern Human Populations. American Journal of Physical Anthropology 95:53–62.
Relethford, J. H. 2004 Boas and Beyond: Migration and Craniometric Variation. American Journal of Human Biology 16:379–386.
Richards, M., with V. Macaulay, E. Hickey, E. Vega, B. Sykes, V. Guida, C. Rengo et al. 2000 Tracing European Founder Lineages in the Near Eastern mtDNA Pool. American Journal of Human Genetics 67:1251–1276.
Richtsmeier, J. T., with J. M. Cheverud, and J. E. Buikstra 1984 The Relationship Between Cranial Metric and Nonmetric Traits in the Rhesus Macaques from Cayo Santiago. American Journal of Physical Anthropology 64:213–222.
Richtsmeier, J. T., and J. W. McGrath 1986 Quantitative Genetics of Cranial Nonmetric Traits in Randombred Mice: Heritability and Etiology. American Journal of Physical Anthropology 69:51–58.
Ridley, M. 1993 Evolution. Oxford: Wiley Blackwell.
Rightmire G. P. 1999 Dental Variation and Human History. The Review of Archaeology 20:1–3.
Ripley, W. Z. 1899 Deniker's Classification of the Races of Europe. Journal of the Anthropological Institute of Great Britain and Ireland 28:166–173.
Román, L. n.d. El Sector Suroestse. *In* Intervención Arqueológica en la Plaza de España, Ecija. Memoria Final. Volumen 1: Memoria 1. Ana Romo, ed. Pp. 195–233. Unpublished manuscript.
Ross, A. H., with A. H. McKeown, and L. W. Konigsberg 1999 Allocation of Crania to Groups Via the "New Morphometry'. Journal of Forensic Science 44:584–587.
Ruff, C. B. 2002 Variation in Human Body Size and Shape. Annual Review of Anthropology 31:211–232.
Ruggles, D. F. 2004 Mothers of a Hybrid Dynasty. Journal of Medieval and Early Modern Studies 34:65–94.
Santley, R. S., with C. Yarborough and B. Hall 1987 Enclaves, Ethnicity, and the Archaeological Record at Matacapan. *In* Ethnicity and Culture. R. Auger, M. Glass, S. MacEachern, and P. H. McCartney, eds. Pp. 85–100. Calgary: The University of Calgary.
Sabiq, A.-S. 1991 Fiqh Us-Sunnah: Funerals and Diggers. Indianapolis: American Trust.
Saunders, S. 1989 Nonmetric Skeletal Variation. *In* Reconstruction of Life from the Skeleton. M. Y. Iscan and K. A. R. Kennedy, eds. Pp. 95–108. New York: Liss.
Savage, E. 1992 Berbers and Blacks: Ibadi Slave Traffic in Eighth-Century North Africa. Journal of African History 33:351–368.

Schillaci, M. A., and C. M. Stojanowski 2003 Postmarital Residence and Biological Variation at Pueblo Bonito. American Journal of Physical Anthropology 120:1–15.

Schmidt R. A., and B. L. Voss eds. 2000 Archaeologies of Sexuality. London: Routledge.

Scott, G. R. 1980 Population Variation of Carabelli's Trait. Human Biology 52:63–78.

Scott, G. R, and C G. Turner 1997 The Anthropology of Modern Human Teeth: Dental Morphology and its Variation in Recent Human Populations. Cambridge: Cambridge University Press.

Scott, G. R., with R. H. Yap Potter, J. F. Noss, A. A. Dahlberg, and T. Dahlberg 1983 The Dental Morphology of Pima Indians. American Journal of Physical Anthropology 61:13–31.

Sealy, J. C., and S. Pfeiffer 2000 Diet, Body Size and Landscape Use among Holocene People in the Southern Cape, South Africa. Current Anthropology 41:642–655.

Simoni, L., with F. Calafell, D. Pettener, J. Bertranpetit, and G. Barbujani 2000 Geographic Patterns of mtDNA Diversity in Europe. American Journal of Human Genetics 66:262–278.

Simpson, St. J. 1995 Death and Burial in the Late Islamic Near East: Some Insights from Archaeology and Ethnography. *In* The Archaeology of Death in the Ancient Near East. S. Campbell, and A. Green, eds. Pp. 240–251. Oxford: Oxbow.

Sjøvold, T. 1973 The Occurrence of Minor Non-Metrical Variants in the Skeleton and Their Quantitative Treatment for Population Comparison. Homo 24:204–233.

Sofaer, J. 2006 The Body as Material Culture. Cambridge: Cambridge University Press.

Sokal, R. R., with G. M. Jacquez, N. L. Oden, D. DiGiovanni, A. B. Falsetti, E. McGee, and B. A. Thomson 1993 Genetic Relationships of European Populations Reflect Their Ethnohistorical Affinities. American Journal of Physical Anthropology 91:55–70.

Sparks, C. S., and R. L. Jantz 2002 A Reassessment of Human Cranial Plasticity: Boas Revisited. Proceedings of the National Academy of Sciences 99:14636–14639.

Steckel, R. H., and J. C. Rose eds. 2002 The Backbone of History. Cambridge: Cambridge University Press.

Stefan, V. H. 1999 Craniometric Variation and Homogeneity in Prehistoric/Protohistoric Rapa Nui (Easter Island) Regional Populations. American Journal of Physical Anthropology 110:407–419.

Stein, G. J. ed. 2005 The Archaeology of Colonial Encounters. Oxford: James Currey.

Stocking, G. W. 1968 Race, Culture, and Evolution: Essays in the History of Anthropology. New York: Free Press.

Stynder, D. D. 2009 Craniometric Evidence for South African Later Stone Age Herders and Hunter-Gatherers being a Single Biological Population. Journal of Archaeological Science 36:798–806.

Sutter, R. C., and L. Mertz 2004 Nonmetric Cranial Trait Variation and Prehistoric Biocultural Change in the Azapa Valley, Chile. American Journal of Physical Anthropology 123:130–145.

Temple, D. H., with B. M. Auerbach, M. Nakatsukasa, P. W. Sciulli, and C. S. Larsen 2008 Variation in Limb Proportions Between Jomon Foragers and Yayoi Agriculturalists from Prehistoric Japan. American Journal of Physical Anthropology 137:164–174.

Templeton, A. R. 1998 Human Races: A Genetic and Evolutionary Perspective. American Anthropologist 100:632–650.

Tilly, C. 1978 Migration in Modern European History. *In* Human Migration. W. H. McNeill, and R. S. Adams, eds. Pp. 48–72. Bloomington: Indiana University Press.

Trigger, B. G. 1989 A History of Archaeological Thought. Cambridge: Cambridge University Press.

Turner, C. G. II 1984 Advances in the Dental Search for Native American Origins. Acta Anthropogenetica 8:23–78.

Turner, C. G. II 1987 Late Pleistocene and Holocene Population History of East Asia Based on Dental Variation. American Journal of Physical Anthropology 73:305–322.

Turner, C. G. II 1990 Major Features of Sundadonty and Sinodonty, Including Suggestions About East Asian Microevolution, Population History, and Late Pleistocene Relationships with Australian Aboriginals. American Journal of Physical Anthropology 82:295–318.

Turner, C. G. II 1992 The Dental Bridge Between Australia and Asia: Following Macintosh into the East Asian Hearth of Humanity. Perspectives in Human Biology 2 / Archaeologia Oceania 27:120–127.

Turner, C. G. II, with C. R. Nichol, and G. R. Scott 1991 Scoring Procedures for Key Morphological Traits of the Permanent Dentition: The Arizona State University Dental Anthropology System. *In* Advances in Dental Anthropology. M. A. Kelley, and C. S. Larsen, eds. Pp. 13–32. New York: Wiley-Liss.

Ubelaker, D. H., with A. H. Ross, S. M. Graver 2002 Application of Forensic Discriminant Functions to a Spanish Cranial Sample. Forensic Science Communications. www.fbi.gov/hq/lab/fsc/backissu/july2002/ubelaker1.htm.

van Vark, G. N., with A. Bilsborough and W. Henke 1992 Affinities of European Upper Palaeolithic *Homo sapiens* and Later Human Evolution. Journal of Human Evolution 23:401–417.

Walde, I. D., and N. D. Willows eds. 1991 The Archaeology of Gender. Calgary: The University of Calgary.

Wanner, I. S., with T. S. Sosa, K. W. Alt, and V. T. Blos 2007 Lifestyle, Occupation, and Whole Bone Morphology of the Pre-Hispanic Maya Coastal Population from Xcambó, Yucatan, Mexico. International Journal of Osteoarchaeology 17:253–268.

Wason, P. K. 1994 The Archaeology of Rank. Cambridge: Cambridge University Press.

Weiss, K. M. 1988 In Search of Times Past: Gene Flow and Invasion in the Generation of Human Diversity. *In* Biological Aspects of Human Migration. C. G. N. Mascie-Taylor, and G. W. Lasker, eds. Pp. 130–166. Cambridge: Cambridge University Press.

Woo, T. L., and G. M. Morant 1932 A Preliminary Classification of Asiatic Races Based on Cranial Measurements. Biometrika 24:108–134.

Zakrzewski, S. R. 2003 Variation in Ancient Egyptian Stature and Body Proportions. American Journal of Physical Anthropology 121:219–229.

Zakrzewski, S. R. 2007 Population Continuity or Population Change: Formation of the Ancient Egyptian State. American Journal of Physical Anthropology 132:501–509.

8

Life Histories of Enslaved Africans in Colonial New York

A Bioarchaeological Study of the New York African Burial Ground

Autumn R. Barrett and
Michael L. Blakey

"A man is worked upon by what *he* works on. He may carve out his circumstances, but his circumstances will carve him out as well" (Frederick Douglass, Western Reserve College, July 1854; Douglass 1854:29–30, emphasis in the original).

Introduction

Frederick Douglass' critique of 19th-century physical anthropology proffers an alternative analysis of social inequality emphasizing the dialectical relationship between the human body and a socially enacted environment in which a man, woman, or child lives. Human environments are acted upon and transformed by living people as their lives and bodies are transformed in the acts of living.

A biocultural analysis of past lives seeks to understand the dynamic relationship between biology and the culturally patterned environment in which life is lived. Biocultural approaches to bioarchaeology are multidisciplinary to better understand the complexity of lived experience and the record that is inscribed within the remains of our forebears. As scholars, we are embedded within the cultural, political, and social contexts of our time and place as we seek to understand the

Social Bioarchaeology. Edited by Sabrina C. Agarwal and Bonnie A. Glencross
© 2011 Blackwell Publishing Ltd

complexity of the past. We are guided by questions in the present in our search to understand the past. Taking this into consideration, scholarship may more effectively engage in a critique of the present through understanding the processes that informed and transformed the lived experiences of past peoples.

Bioarchaeology of enslaved populations, forced to provide labor throughout the colonial and early nation-building projects in the Americas, enables a critique of the construction of inequality and the lived interpenetration of sociocultural, economic, political, and biological realities. European colonial expansion involved movement of people, voluntary and involuntary, and shifts in political, economic, and social relationships where trade relations and settlements were established. Population shifts through voluntary and forced migrations, violent conflict, and increased exposure to new environments involved introduction to greater varieties of illness, compromised health, and, for many, mortality (Blakey 2001; Blakey et al. 2004b).

The remains of 419 individuals within the New York African Burial Ground represent lives that spanned approximately a hundred years, primarily within the 18th century. In this study, we employ multiple lines of evidence, including the skeletal, archaeological, and documentary records. While our primary analysis utilizes dental indicators of childhood health, we compare and combine our findings with skeletal indicators of physiological stress and evidence from the documentary and archaeological records. Through these multiple lines of evidence, we attempt to gain insight into the lives of African men, woman, and children who negotiated "the working" and "being worked upon" during the weeks, months or years they lived, before dying in colonial New York. The multidisciplinary scholarship of the New York African Burial Ground Project enables a biocultural analysis that relates statistical patterns of morbidity and mortality to the historic social conditions that individual people were likely to have experienced at different ages in the course of their lives.

African American Bioarchaeology and the New York African Burial Ground Project

Blakey (2001) provides a history of the intellectual traditions of Diasporan scholarship and disciplinary developments of Diasporic studies and African American bioarchaeology, delineating the distinctions between forensic and biocultural approaches. Biocultural approaches in bioarchaeology allow the construction of a more human history of African Diaspora communities, and more importantly have begun to include the involvement of African Diaspora communities in their scholarship (Blakey 2001). The intellectual history of bioarchaeology and how the discipline has historically related (and not related) to Diasporic studies, and Diasporan scholarship is essential to understanding the significance of the New York African Burial Ground Project to bioarchaeology as a discipline.

The New York African Burial Ground site "rediscovery" in 1989 and the resulting research project were an historic joining of Diasporan scholarly trends of political and social activism, "vindicationist" revision (Drake 1980) of eurocentric exclusion, interdisciplinary research and collaboration, and community engaged and biocultural models that complimented these Diasporan traditions (Blakey 2001; see also Harris 1993; Harrison and Harrison 1999). The descendant community's leadership in demanding federal construction be halted, that the analysis of their ancestral remains be conducted by experts on Africa and the Diaspora, and the engagement of descendants in forming the research design, are all aspects of the historical significance of the project, but also characteristic of Diasporan and biocultural approaches to research (Blakey 2001). The following is a brief overview of African American bioarchaeology, which contextualizes the present study within the biocultural and Diasporan scholarly traditions that historically culminated in the New York African Burial Ground Project (for an extensive and complete review, see Blakey 2004a).

The earliest developments within African American bioarchaeology did not incorporate or rely on African American scholarship, nor did the field of study originate from an interest in the African Diaspora. Rather, the first African American bioarchaeological studies resulted from accidental encounters with human remains deemed of interest due to their bearing on studies of race by physical anthropologists (Blakey 2001: Blakey 2004b; Buxton et al. 1938; Stewart 1939). The primary focus was on using skull morphology towards racial identification in order to settle temporal designation for excavated burials, and analyses lacked any depth of cultural and/or historical considerations. A later analysis of two skeletons of enslaved African American men accidentally encountered during excavation of an indigenous site in St. Catherine's island off the coast of Georgia, by Thomas and coworkers (1977), is "less forensic" (Blakey 2001) in combining historical and cultural contexts with paleopathology. The researchers were "examining people, not a race, and probing the conditions of slavery" (Blakey 2001:398). Furthermore, rather than curating the remains, the individuals were reburied, and the researchers provided guidelines for analysis, preservation, and interpretation of future historic cemeteries with consideration for contemporary descendant communities and toward a more thorough analysis (Blakey 2001; Thomas et al. 1977).

African American archaeology and bioarchaeology began to emerge as a more central area of interest due, in large measure, to African American activism and the U.S. National Historic Preservation Act of 1966 (Blakey 2001; Ferguson 1992). The 1966 Act required funding for archaeological investigations where federal development projects threatened historic preservation and cultural heritage sites. Firms for Cultural Resources Management (CRM) became the leading source of employment for archaeologists in the United States (Blakey 2001). However, transformations in research designs and interpretations were yet to occur. African American history became a lucrative commodity for CRM firms, archaeologists and bioarchaeologists. However, the euro-centric exclusionary patterns of scholarship continued. The extensive body of literature on Africans in the Americas, produced by Diasporan scholars, was rarely consulted or cited. Archaeologists and bioarchaeologists did not seek educational training within increasingly prevalent African American studies departments. Archaeologists and bioarchaeologists were not members of academic associations established by African Americans for studies of the Africana world. Collaborations did not occur between archaeologists (and bioarchaeologists) with specialists on the African Diaspora when working on Diasporic sites (Blakey 2001).

The divide between archaeology and African American studies was noted and addressed by plantation archaeologist, Theresa Singleton, and African Americanist, Ronald Bailey. Singleton and Bailey coordinated meetings at the University of Mississippi (1989) between specialists within their respective fields, attempting to facilitate dialogue between the disciplines. Singleton was, at the time, the only black Ph.D. archaeologist studying plantation sites and has since been referred to by Merrick Posnansky as the "mother of African American archaeology" (Blakey 2004a:94). Assessing archaeology of the African Diaspora from 1960s to the 1990s, Singleton and Bograd (1995) found the scope of African American archaeology had expanded in terms of site diversity and the inclusion of discussions on race, ethnicity, acculturation, social inequality, and resistance (23), while analyses, and inclusion of African American perspectives were lacking and continued to demonstrate a "largely descriptive" and distorted, euro-centric objectification of African American ethnicity, characterized by an "archaeology of the 'other'" (Blakey 2001:400; Singleton and Bograd 1995:23–31).

The import of interpretations of past peoples on present realities is explicitly recognized and incorporated within the analyses from Diasporan scholarly traditions (Blakey 2004a; see also Douglass 1854; Fanon 1994[1952]; Firmin 2002 [1885]). The omission of an emic perspective from archaeologies and bioarchaeologies of the African Diaspora serves to obscure the researchers' subjective biases, their positions within larger political-economic systems, and the implications of such academic productions as mechanisms for social control and reproduction of "white privilege" within academic settings and the broader society (Blakey 1998; Blakey 2001). Singleton and Bograd (1995) identify studies of ethnicity as "problematic" partly due to their propensity toward the etic perspective, "how archaeologists and others define ethnics or cultural groups, rather than how ethnics define themselves" (23–24, quoted in Blakey 2001:401).

Physical anthropologists in the 1980s were drawn to the historical, archaeological, and bioarchaeological discussions of quality of life for enslaved Africans and African Americans that included data on demography, nutrition, and health. A culmination of political, social, and academic movements provided the context that enabled the discipline of bioarchaeology to fully emerge in the 1980s. The Black Consciousness, Black Studies, and Civil Rights movements created an interest and marketability for studies of African American history and the history of racism in the United States, to which European American scholars responded during the late 1960s, 1970s, and 1980s (Blakey 2001). The apologetic proposals of the Moynihan report (1965) and Fogel and Engerman's (1974) *Time on the Cross*, though proven to be based on faulty scholarship in terms of historic and economic validity (see David et al. 1976 and Gutman 1975), sparked academic debate on the conditions of slavery in the Americas. The trend for quantifying the impact of slavery on African populations and their descendants in the Americas is seen in the classic debate between Curtin (1969), Lovejoy (1982) and Inikori and Engerman (1992) regarding the exportation and death estimates for Africans forced to migrate to the Americas. Richard Steckel's (1986) documentary analysis of health, nutrition, and labor for enslaved laborers on plantations continued to address the question of how slavery affected the health and demography of African Americans.

According to Blakey (2001), these data and the opportunities to apply evolutionary models to questions posed by biohistorians proved evocative for physical anthropologists, "who were poised to enter the discussion with the bones and teeth of the enslaved people themselves" (401). Assessing the state of research on African American biohistories, Rankin-Hill (1997) states, "little has been accomplished in expanding the conceptual limits of the field. In fact, much of this emphasis has been on the intricacies of quantification, and data manipulation, and not on different approaches of interpreting and/or examining the data generated" (Blakey 2001:402; Rankin-Hill 1997:12). The apologist lens through which research was being focused (characterized by Blakey (2001) as "did whites do anything particularly bad towards blacks during slavery that caused their current condition?" [402]) is not a question of technique, per se. As Singleton and Bograd (1995), Rankin-Hill (1997) and Blakey (2001; 2004) noted, there remained a problem of ideological approaches to research design and interpretation of data. The problem of eurocentric biases is a critique of the larger discipline of anthropology, not solely of archaeology and physical anthropology (Douglass 1854; Drake 1980; Firmin 2002[1885]; Hsu 1979; Willis 1999).

Biocultural analyses of social inequality became more prevalent beginning in the 1970s. Influential biocultural models developed in the 1970s and 1980s at the University of Massachusetts were inter- and intra-disciplinary. Biocultural approaches to paleopathology developed by George Armelagos, human adaptation models by R. Brooke Thomas, the historical demographic research of Alan Swedlund, and collaborations among specialists in the fields of anthropology and African American studies created a uniquely innovative research environment that shaped bioarchaeology of the African Diaspora (Blakey 2001). Integral to this synergy was the inclusion of African American faculty, such as Johnetta Cole, and students, such as Lesley Rankin-Hill and Michael Blakey, a keen awareness of the racist genealogies of physical anthropology, and an emphasis on theoretical transformation of the discipline. Notably, the University of Massachusetts is also the graduate institution of Alan Goodman, Ann Magennis, Debra Martin, Robert Paynter, and Jerome Rose (Blakey 2001).

Indeed, in 1982 Jerome Rose and coworkers excavated Cedar Grove cemetery in Lafayette, Arkansas – the first large-scale African American bioarchaeological excavation (Blakey 2004a; Rose 1985) and possibly the first African American archaeological site to which the National Historic Preservation Act of 1966 was applied (Blakey 2001). In 1985, Rose collaborated with Ted Rathbun to organize the first symposium to present the growing research on African Americans by physical anthropologists, entitled "Afro-American Biohistory: The Physical Evidence" at the Annual Meeting of the American Association of Physical Anthropologists. The symposium included a range of methods and population contexts (see Angel et al. 1987; Blakey 1988; Corruccini et al. 1987; Hutchinson 1987; Martin et al. 1987; Rathbun 1987; Wienker 1987). However, Rankin-Hill (1997) produced the first African American bioarchaeological work to fully synthesize paleopathology, demography, and documentary analyses within a biocultural model that centralizes culture as a source and buffer of stress to understand the lives of African Americans in 19th-century Philadelphia (Blakey 2004a).

The First African Baptist Church (FABC) cemetery site, located in downtown Philadelphia, was in use by African Americans between 1821 and 1843. Most of the African Americans interred in the cemetery were free, though some had been enslaved during earlier years of their lives (Rankin-Hill 1997). Human remains were encountered during construction of a subway tunnel in 1980. Once the site was identified, and the contemporary church congregation was consulted, the FABC site was excavated in 1984 and 1985 by John Milner and Associates (Rankin-Hill 1997).The140 individuals excavated were transferred to the Smithsonian Institution for analysis led by J. Lawrence Angel. The FABC site was unique as a northern, urban site used primarily by free African Americans and the largest archaeological population of African Americans to have been excavated at that time (Blakey 2001; Rankin-Hill 1997). In 1987, the contemporary church congregation reburied the remains of the FABC individuals in Philadelphia's Eden cemetery. The Smithsonian Institution had considered announcing the FABC re-interment ceremony. However, in light of the political pressure being placed on the Institution by Native Americans demanding the reburial of 18,000 remains curated by the Smithsonian, the FABC announcement was cancelled (Blakey 2001).

The prestige of the Smithsonian and Angel in association with the FABC analysis served to further augment disciplinary interest in African American bioarchaeology. The combination of political, academic, and social movements in the 1970s and 1980s that led to the contemporary prominence of African American bioarchaeology, was further strengthened by the success of Native Americans in regaining control over the ancestral remains and grave goods of their sacred sites through the 1990 Native American Graves Protection and Repatriation Act (NAGPRA) (Thomas 2000). African American bioarchaeology became an ideal alternative for American physical anthropologists, who would no longer have access to Native American remains as the primary source for professional production (Blakey 2001). By the 1990s, the biocultural and forensic models were distinctly developed and divergent within African American bioarchaeology.

The New York African Burial Ground Site

The New York African Burial Ground, located in downtown New York City, is the largest excavated colonial archaeological population in the Americas. The life histories embodied within the remains of 419 individuals excavated during 1991 to 1992 (Blakey and Rankin-Hill 2004), provides an opportunity to question the past inherited through narratives embedded within culturally informed networks of power. While no lives are static, the transatlantic experiences of enslaved Africans transported from within the continent of Africa to various points in the Americas, and forced to live and work in an estranged and physically arduous environment is a particular challenge and an important element in interpreting the lives represented within the New York African Burial Ground. Many of the individuals within this population did not experience the entirety of their lives enslaved and, therefore, had to adapt to new social, political, and cultural constraints.

Understanding the trajectory of life experiences for enslaved Africans and their descendants in colonial New York includes exploring how life histories intersected with historical process and how these life histories may be interpreted within the skeletal record. The research presented below addresses how place of birth, childhood environment, social and political power relationships, and changes in the quality of life due to enslavement variously affected people within different stages of life. Through this approach it will often not be possible to test hypotheses. The record of stresses at any particular age may be associated with alternative or multiple stressors whose varied physiological influences are often indiscernible from one another. These health and mortality data thus shadow historical events. However, they provide a voice in a conversation between biology, culture, and history that raises questions, alternative answers, and useful connections among variables.

Brief historical overview

The historical record indicates that West Central Africa and later primarily West Africa were the regions from which Africans were enslaved and brought to the colony of New York where the African Burial Ground served the enslaved community from the late 17th century to 1794 (Blakey 2004; Medford 2004). Enslaved Africans were transported within Africa before being forced to voyage the middle passage and disembark at various points in the Americas where their lives as enslaved workers began in estranged and physically arduous environments. The transatlantic experiences of these individuals are a particular challenge and important element in interpreting the lives represented within the New York African Burial Ground.

After the British took control of colonial New York in 1664, enslaved laborers were imported primarily from West Africa, specifically from the regions of Senegambia, the Sierra Leone–Liberia area, the Gold Coast, the Bight of Benin, and the Niger Delta (Medford 2004). Medford (2004) characterizes these regions as politically fragmented during the 17th and 18th centuries. She states that, "certain polities were vulnerable to attack and their people enslaved. The political instability thus created also led to the kinds of lawlessness that produced banditry and raids conducted for the explicit purpose of acquiring captives for the slave trade" (Medford 2004:60–61). Between 1664 and 1741 specifically, many laborers had been enslaved in the Caribbean prior to being shipped to New York (Goode-Null 2002:37–38).

There remains one outstanding chronicle by an African in Old New York. John Jea was born in "Old Callabar" in 1773 where, in his words, he was "stolen" at the age of two years (Jea 1998:369). Jea and his mother, father, brothers, and sisters were taken to North America and, ultimately, New York. Jea worked as an enslaved laborer into his teen years until he gained his freedom to become a traveling minister. Jea describes his early memories as an enslaved child of only eight or nine years, recounting multiple incidents of violence, poor nutrition, and hunger that he experienced. He also describes strong relationships with his family, and how he visited his parents and brothers upon returning to New York in between his travels. Jea toured throughout the northeast of the United States to England, Ireland, the East Indies, West Indies, and France. Jea was married three times, and had "several children," though, unfortunately, none of them survived (Jea 1998:369–439).

We provide summary points from Jea's manuscript on his life history because Jea was a contemporary of the Africans interred in the New York African Burial Ground, and to highlight the scope and complexity of his experiences. Jea was born in Africa, enslaved, freed, at points imprisoned, and almost enslaved again. Jea's life defies any simple description or analysis.

Childhood in colonial New York

Child mortality rates (see Figure 8.1) in the African Burial Ground population indicate that the youngest were the least likely to survive environmental stress. However, an interpretation of the skeletal record of childhood health and death also informs us about the construction and experience of childhood for enslaved laborers in New York who survived to become adults during the 18th–19th centuries. In 1731 an 11-year-old enslaved African was categorized as an "adult" but earlier in the century the age of 15 marked adulthood (Blakey et al. 2004b:524). Preference for importing young Africans, while reflected in the fluctuating status of "child" or "adult", was further justified by European slaveholders who feared repeated revolt. Susan Goode-Null's (2002) study of childhood health and development in the New York African Burial Ground, drawing on McManus (1966), explains that between 1741 to 1770, due to the cessation of slave trading between the British and Spanish colonies, and due to the fear that an aborted slave revolt in 1741 may have repeated the events of the 1712 slave revolt in New York, enslavers shifted patterns of demand for captive humans. Adult-enslaved men from the Caribbean had been considered the strategists behind the earlier successful and aborted revolts

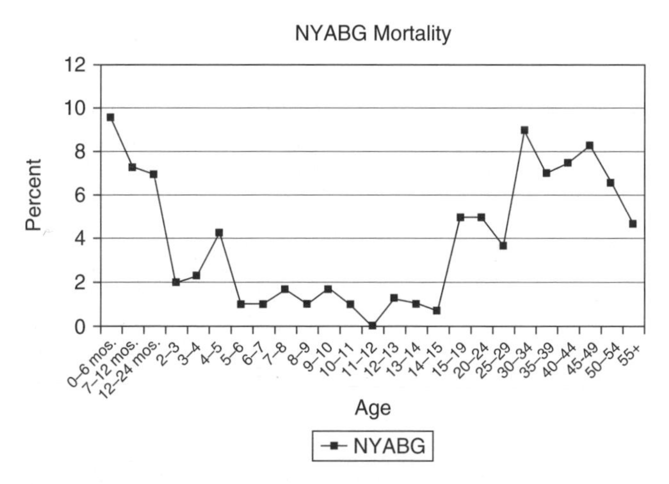

Figure 8.1 New York African Burial Ground mortality.
Source: Rankin-Hill, L. M., M. L. Blakey, J. E. Howson, E. Brown, S. H. H. Carrington, and K. Shujaa 2004 Demographic Overview of the African Burial Ground, and Colonial Africans of New York. *In* The African Burial Ground Project. Skeletal Biology Final Report, vol. 1. M. L. Blakey, and L. M. Rankin-Hill, eds. Pp. 266–304. Washington, DC: The African Burial Ground Project, Howard University for the United States General Services Administration, Northeast and Caribbean Region

(Goode-Null 2002:28). As a result, preference was placed on Africans imported directly from Africa rather than via the Caribbean, and importations shifted to being largely young women aged 13–40 years, and children of preferably 9–10 years of age, rather than adult males. Jea's childhood was consistent with these patterns, though he was only two when enslaved along with his siblings.

Enslaved children in New York were frequently sold by the age of six years, and advertisements indicate domestic skills promoting the marketability of enslaved children (Blakey et al. 2004b; Goode-Null 2002:37–38; Medford 2004:118). We incorporate these historical data in our interpretations of childhood health within the New York African Burial Ground population, addressing the meaning of these justifications and fluctuating definitions of "child" and "adult" in terms of labor demanded of enslaved children. For example, what was the significance of nearing the ages of 6, 9, or 11 for enslaved children in colonial New York? Were these ages, while salient within the documentary record significant within the life histories of the New York African Burial Ground population? Are there indications of the developmental conditions or experience of childhood in Africa? We first and principally focus on the dental record to explore these questions through dental enamel defects. The analyses presented below are a culmination of the New York African Burial Ground project research and results in Blakey and Rankin-Hill (2004) and Medford (2004). These analyses are expanded upon by incorporating the temporal groupings developed by Perry et al. (2006) and Harris Line data.

Dental Enamel Development and Disruption: Hypoplasia

Dental enamel hypoplasia are defects in crown development that appear as a transverse groove or series of pits, partially or entirely around the tooth's circumference (Figure 8.2). The defects occurring on different teeth and in different locations on

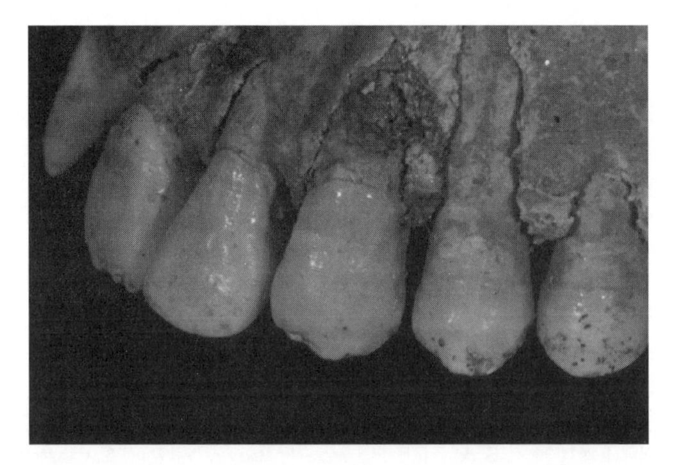

Figure 8.2 New York African Burial Ground, Burial 1, a woman who died between the ages of 20 and 25 years, displaying linear enamel hypoplastic lesions in the anterior maxillary permanent dentition.
Source: Author's photograph

teeth represent stresses at differing ages during childhood and adolescent growth, similar to the analysis of tree rings for a record of droughts during the lifetime of a tree. The evidence of these early stresses remains observable in adult skeletons in which dentition has been retained.

Hypoplastic defects, while manifesting in teeth, result from metabolic disturbances of malnutrition (Blakey and Armelagos 1985; see also Infant 1974; Levy et al. 1970; Shaw and Sweeney 1973) and disease (Blakey and Armelagos 1985; see also Nikiforuk and Fraser 1981; Sarnat and Schour 1941; Sweeney et al. 1969) elsewhere in the body. With rare exception, dental enamel hypoplasia is a result of systemic metabolic stress associated with infectious disease, insufficient calcium, protein or carbohydrates, and low birth weight, characterized together as "general stress" (Blakey et al. 1994; Goodman et al. 1988). Enamel hypoplasia thus provides evidence of physiological stress that may have been brought about by a diverse range of stressors.

Since enamel hypoplasia can develop in childhood and adolescence when both the deciduous and permanent teeth are formed, it is of particular value for the interpretation of developmental health. Deciduous dental enamel begins to develop during the fifth month *in utero*, completing development by the end of the first year of postnatal life (Blakey and Armelagos 1985:371,374; see also Shaw and Sweeney 1973). Permanent dental enamel begins formation at birth and continues into the 16th year of age (Blakey and Armelagos 1985: 371: see also Shaw and Sweeney 1973). In addition to a generalized identification of stress frequency, the rate of enamel matrix formation provides a mechanism for estimating the developmental stage at which the growth arrest occurred (Berti and Mahaney 1992; Blakey and Armelagos 1985; Blakey et al. 1994:372; Goodman and Armelagos 1985a,b; Goodman et al. 1988).

These defects are observed in archaeological and living populations, representing a very broad range of human experiences, from those of early hominids to industrial nations. Included among these are a number of studies of African American and Afro-Caribbean archaeological populations (Blakey and Armelagos 1985; Blakey et al. 1994; Clarke 1982; Condon and Rose 1992; Corruccini et al. 1985; Goodman and Armelagos 1985a,b; Goodman et al. 1984). Like other "physiological stress indicators" such as life expectancy, infant mortality, or rates of growth retardation, frequencies of hypoplastic defects can be compared among different populations as a gross index of physical well-being and the adequacy of resources, upon which the physical quality of life may depend.

The present study builds upon the analysis of deciduous and permanent dentition by Blakey and coworkers (2004a) by incorporating and further synthesizing the archaeological (Perry et al. 2006), historical (Medford 2004), and skeletal biology (Blakey and Rankin-Hill 2004) components of the New York African Burial Ground Project. African revolt and resistance in the first half of the 18th century incited fear in slaveholders and shifted demographic demands toward women and children, considered less threatening than adult males (Blakey et al. 2004b: Medford 2004). The drastic geographic and social transitions of forced migration and enslavement took place at different points in the life course. The present study builds upon Blakey et al. (2004a) by incorporating a temporal analysis to trace these late 17th and 18th century social, political, and historical shifts as they are evident in patterns of biological indicators of stress and developmental disruption within the New York African Burial Ground. While dental enamel defects remain the primary indicator of stress utilized, the presence and chronological age distribution of Harris Lines are also presented and discussed.

Overview of Materials and Method

In accordance with the methodology of Blakey et al. (2004a), "presence" of hypoplasia within an individual was defined by presence of linear or nonlinear hypoplasia in one of the teeth selected for analysis. "Absence" of hypoplasia was defined by absence of hypoplasia in all teeth selected for analysis. According to the research of Goodman and coworkers (1980), secondary canines and incisors display 95 percent of enamel hypoplasia observed when all available dentition are represented. Blakey et al. (2004a) employed this "best tooth" method by selecting individuals with the presence of a permanent left or right maxillary central incisor and a left or right mandibular canine.

Permanent teeth were identified according to codes 2 and 7 in "Standards for Data Collection from Human Skeletal Remains" indicating that teeth are fully developed, in occlusion, and observable (Buikstra and Ubelaker 1994:49). A total of 65 individuals were selected for analysis of permanent dentition, which represents the developmental period between birth and 6.5 years of age. A separate selection was conducted for individuals with permanent third molars, left or right, mandibular or maxillary. There are 111 individuals included within this third molar analysis, which represents the developmental period in life from 9–16 years of age.

Hypoplasia in deciduous dentition was studied by selecting individuals older than one year with one left or right central maxillary incisor, one left or right mandibular canine, and one second molar. Deciduous teeth were identified according to codes 1, 2, and 7 in "Standards for Data Collection from Human Skeletal Remains" indicating that teeth are fully developed and observable (Buikstra and Ubelaker 1994:49). The dentition of 34 subadults are included within this study. The combined studies of deciduous and permanent dentitions span the developmental stages from approximately the fifth month *in utero* to 16 or 17 years of life (Table 8.1).

Controlling for dental attrition

A separate subsample, selected from the permanent canine and incisor study and from the third molar study, was used to control for age- and/or sex-related differences in dental attrition that might affect hypoplasia frequencies. Individuals with moderate to severe dental wear and individuals for whom dental wear could not be scored (including inability to score due to cultural modifications such as filing and pipe notches) were not included in the canine and incisor sample and the third molar sample. Individuals with a dental wear score of 5 or greater, according to Smith (1984), were not included in the permanent incisor and canine sample, resulting in 48 observable individuals. Individuals with a dental wear score of 7 or

Table 8.1 Summary of study samples

Study description	*Dentition*	*Developmental age represented*	*Sample size*	
Hypoplasia	Canines and incisors – Deciduous	5 months *in utero* to 1 year old	34	99
Hypoplasia	Canines and incisors – Permanent	Birth to 6.5 years old	65	
Hypoplasia Controlled for attrition	Canines and incisors – Permanent	9 to approximately 16 years old	48	
Hypoplasia Controlled for attrition	Third molars	9 to approximately 16 years old	97	
Canine chronology for hypoplasia	Canines – Permanent	6 months to 6.5 years old	23	
Hypoplasia	Third molars	9 to approximately 16 years old	111	

Source: Modified from Blakey, M. L., M. Mack, A. R. Barrett, S. S. Mahoney, and A. H. Goodman 2004a Childhood Health, and Dental Development. *In* The African Burial Ground Project. Skeletal Biology Final Report, vols. 1, and 2. M. L. Blakey, and L. M. Rankin-Hill, eds. Pp. 306–331. Washington, DC: The African Burial Ground Project, Howard University for the United States General Services Administration, Northeast and Caribbean Region.

greater, according to Scott (1979), were eliminated from the third molar sample, resulting in 97 observable individuals. SPSS software version 11.5 was employed in the statistical analyses associated with each study.

Canine chronology

Twenty-three individuals were also assessed for the chronology of physiological stress episodes evident in the hypoplastic lesions using the method of measurement and mathematical calculation presented in Blakey et al. (1985). Chronology was determined for defects in the left permanent mandibular canines (Figure 8.3) – right mandibular canines are used when the left are absent or unobservable. Measurements for the hypoplastic lesions were recorded by members of the African Burial Ground Project in the late 1990s. Measurements of the total crown height allows for the estimation of incremental growth, and the estimation of the age of occurrence of the observed incident (Blakey and Armelagos 1985).

Results

The majority of this population experienced physiological stress in their childhood years. Individuals with permanent dentition (n = 65) (representing birth to 6.5 years of age) have hypoplasia in 70.8 percent (n = 46) of the cases, overall. Frequencies for hypoplasia in permanent dentition are higher in the African Burial

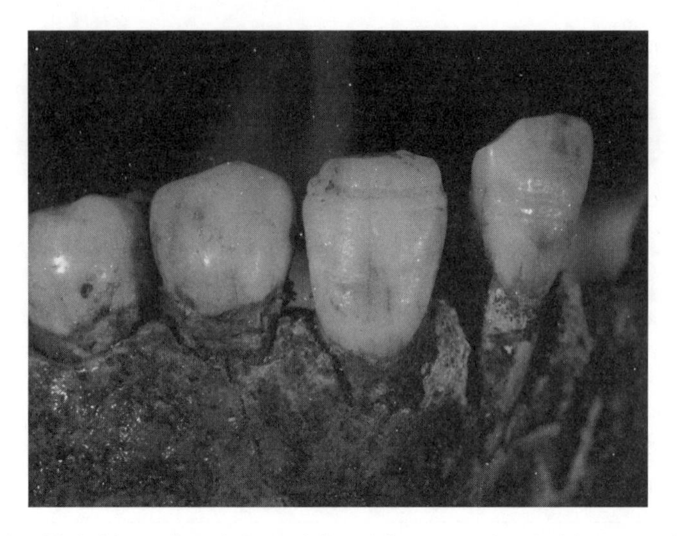

Figure 8.3 New York African Burial Ground, Burial 9, a man who died between the ages of 35 and 45 years, displaying linear enamel hypoplasia in the permanent mandibular canine and lateral incisor.
Source: Author's photograph

Ground population than frequencies observed in the enslaved populations of Catoctin Furnace, Maryland (Kelley and Angel 1987) and Newton Plantation in Barbados (Corruccini et al. 1985), though the difference was only statistically significant in the case of Newton Plantation (Pearson chi-square = 4.494, 1df, $p < .05$). Notably, this frequency is almost 20 percent lower than in the largely free and freed 19th-century First African Baptist Church (FABC) cemetery in Philadelphia (Blakey et al. 1994) and enslaved African Americans buried in 19th-century Charleston, South Carolina 38CH778 (Rathbun 1987) (Table 8.2). Differences in hypoplasia frequencies may reflect variation in the temporal and geographic contexts of these populations. A greater number of people were likely born in Africa within the African Burial Ground (and Barbados) populations than among the 19th-century African American population in Philadelphia and the south. The latter spent their entire lives within the conditions of slavery or as free people living under conditions of economic and social inequality. The New York population, with higher numbers of adults born in Africa, has relatively lower frequencies of hypoplasia, indicating lower levels of childhood stress.

Among children with deciduous dentition, 85.3 percent of the children (29 of 34) have hypoplasia, representing disrupted development between the fifth month *in utero* through the end of the first year of life. In contrast with permanent dentition, this frequency is slightly more than 30 percent higher than in the FABC. Again, the historical contexts for these populations have significant bearing on understanding the levels of childhood stress. The higher frequencies of childhood hypoplasia in children dying in New York reflects, we believe, the extremely high levels of stress experienced by children born into the condition of slavery. The FABC children were relatively less stressed.

Table 8.2 Comparison of frequencies reported in skeletal populations

Site/Location	*Region*	*Rural/ Urban*	*Historical period*	*Hypoplasia frequency/ secondary dentition (%)*	*Hypoplasia in females (%)*	*Hypoplasia in males (%)*	*Hypoplasia in subadults/ deciduous dentition (%)*
African Burial Ground, New York	Northeast, North America	Urban	17th and 18th centuries	70.8 (*n* = 46)	62.5 (*n* = 15)	74.3 (*n* = 26)	85.3 (*n* = 29)
Newton Plantation, Barbados	Barbados, West Indies	Rural	1650s–1834	54.5 (*n* = 56)	*	*	*
FABC, Pennsylvania	Northeast, North America	Urban	1800–1850	89 (*n* = 54)	86 (*n* = 29)	92 (*n* = 25)	55 (*n* = 11)
Catoctin Furnace, Maryland	Southeast, North America	Urban	1790–1820	46 (*n* = 7)	43 Slight 79.3* (*n* = 23) Mod–Sev 37.9* (*n* = 11)	71 Slight 68* (*n* = 17) Mod–Sev 68 (*n* = 17)	*
Charleston, S. Carolina (38CH778)	Southeast, North America	Rural	1840–1870	85 (*n* = 23)	71 (*n* = 10)	100 (*n* = 13)	*

*Frequencies not reported for this category.

Sources: Newton Plantation site frequencies from Corruccini et al. (1985), First African Baptist Church frequencies reported from Blakey et al. (1994) with frequencies in children cited from Rankin-Hill (1997). Catoctin site frequencies reported from Kelley and Angel (1987) for overall frequencies. Frequencies by sex for Catoctin Furnace are from Angel et al. (1987) and Blakey et al. (1994). Frequencies reported by Blakey et al. (1994) have an asterisk (*) and represent frequencies of slight hypoplasia or moderate to severe hypoplasia within the Catoctin Furnace site. Frequencies for males and females in the South Carolina 38CH778 site from Rathbun (1987). Combined secondary dentition frequency calculated from male and female frequencies reported by Rathbun (1987).

The difference in hypoplasia frequencies among men and women in the New York African Burial Ground is not statistically significant (62.5 percent of the women (n = 15) and 74.3 percent of the men (n = 26)). This finding suggests that male and female children experienced similar frequencies of stress episodes from birth to the age of 6.5 years. However, the New York African Burial Ground population does fall into the general pattern established by previous bioarchaeological studies within the Diaspora (mentioned above and below) indicating that men have consistently higher percentages of hypoplasia than females (Khudabux 1991; Owsley et al. 1987; Rathbun 1987). Blakey and coworkers (1994) report 86 percent hypoplasia in women and 92 percent in men among 54 individuals from the FABC population. Angel and coworkers (1987) report 71 percent of men and 43 percent of women at Catoctin Furnace, Maryland, have hypoplasia. The Blakey et al. (1994) study of the Catoctin site indicates that women have higher frequencies of slight linear enamel hypoplasia; however, men have greater frequencies of moderate to severe hypoplasia (68 percent males (n = 17) and 37.9 percent females (n = 11)). Among the populations compared within this study, Rathbun (1987) reports the highest frequencies in men and women at the Charleston, South Carolina, site (71 percent in women and 100 percent for men).

Among the 99 New York African Burial Ground individuals within the deciduous and permanent canine and incisor studies, 37.4 percent (n = 37) died before the age of 15 years, of whom 86.5 percent (n = 32) had hypoplasia. Young adults who died between the ages of 15 and 24 years of age represent 17.2 percent of the population, 76.5 percent of whom had hypoplasia. A total of 45.5 percent of the people died over the age of 25 years (n = 45), of whom 66.7 percent (n = 30) had hypoplasia. The frequency of childhood enamel growth disruption is lowest in the oldest age-at-death groups (Table 8.3).

The data suggest that enslaved children in colonial New York experienced high levels of stress, and that the individuals who experienced early stress episodes resulting in enamel hypoplasia were more likely to have died in childhood. The brisk importation, low fertility, and high child mortality of 18th-century New York meant that an African who lived in New York as an adult was more likely to have been born in Africa (or possibly the Caribbean) than to have been born in New York and survived to adulthood there (Blakey et al. 2004b:530–531). The lower frequency of individuals with hypoplasia among those who were older than 25 when they died

Table 8.3 New York African Burial Ground frequency of hypoplasia by age group

Age group	*Within age group*		
		Men (n = 35)	*Women (n = 24)*
1–14* (n = 37)	86.5% (n = 32)		
1–24 (n = 17)	76.5% (n = 13)	83.3% (n = 5)	75.0% (n = 6)
25–55+ (n = 45)	66.7% (n = 30)	72.4% (n = 21)	56.3% (n = 9)

*Three children within this age category had permanent dentition.

Source: Blakey, M. L., M. Mack, A. R. Barrett, S. S. Mahoney, and A. H. Goodman 2004a Childhood Health, and Dental Development. *In* The African Burial Ground Project. Skeletal Biology Final Report, vols. 1, and 2. M. L. Blakey, and L. M. Rankin-Hill, eds. Pp. 306–331. Washington, DC: The African Burial Ground Project, Howard University for the United States General Services Administration, Northeast and Caribbean Region.

may reflect their forced migration. Based on the documentary record (Blakey et al. 2004b:519; Medford 2004:101–107) these individuals are more likely to have experienced childhood in Africa than in New York and their lower defect frequencies might reflect childhoods elsewhere. Although children were imported, those who died as children in New York seem more likely to have been born there than those who died as adults (see Goodman et al. 2004). Hypoplasia frequencies in the dead children, therefore, seem most likely to reflect the conditions of New York. A temporal analysis of these samples, discussed later, enables us to look more closely at shifting importation patterns and their implications for interpreting the skeletal record. Our focus will now turn to the analysis of third molars, representing the developmental period between 9 and 16 years old.

Third molar analysis

Young enslaved laborers who died between 15–24 years of age have intermediate frequencies (Table 8.3) of defects in the teeth that developed during early childhood (compared with the younger and older age groups). We also examine frequencies of hypoplasia in third molars that developed between 9 and approximately 16 years of age. The late childhood and adolescence stress represented by hypoplastic third molars is present in 44.4 percent ($n = 12$) of those who died between 15–24 years and only 10.7 percent ($n = 9$) of those who died at 25 years of age and older (Figure 8.4). These differences were statistically significant (Pearson chi-square with Yates continuity correction = 13.035, 1df, $p < .0005$).

An interesting point for analysis and interpretation, the 15–24 year olds would have died quite close to the time when these late stresses were occurring. Third molars are less sensitive to hypoplasia than are the anterior teeth so the hypoplastic

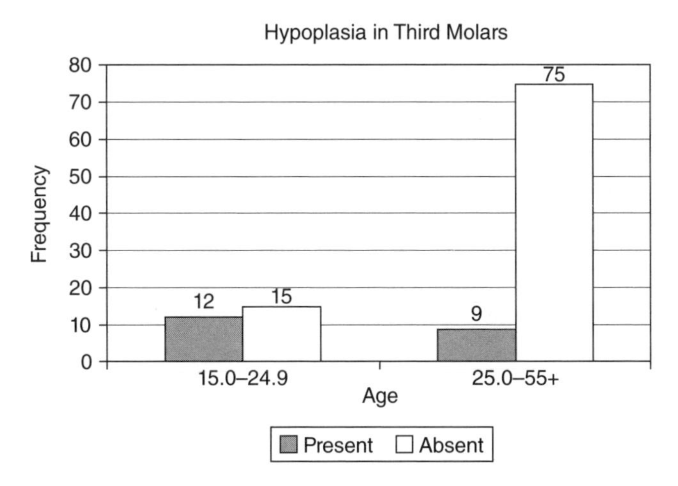

Figure 8.4 New York African Burial Ground hypoplasia in third molars ($n = 111$).
Source: Blakey, M. L., M. Mack, A. R. Barrett, S. S. Mahoney, and A. H. Goodman 2004a Childhood Health, and Dental Development. *In* The African Burial Ground Project. Skeletal Biology Final Report, vols. 1, and 2. M. L. Blakey, and L. M. Rankin-Hill, eds. Pp. 306–331. Washington, DC: The African Burial Ground Project, Howard University for the United States General Services Administration, Northeast and Caribbean Region

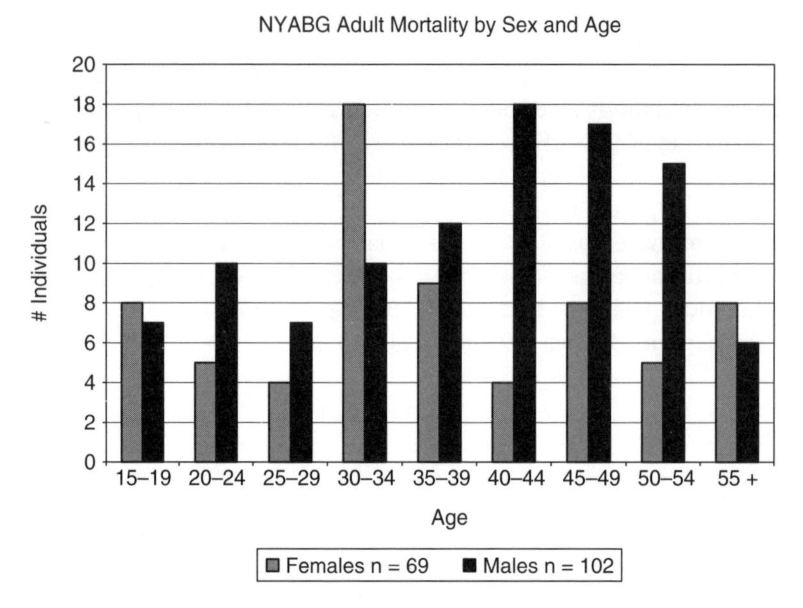

Figure 8.5 New York African Burial Ground mortality by sex and age.
Source: Rankin-Hill, L. M., M. L. Blakey, J. E. Howson, E. Brown, S. H. H. Carrington, and K. Shujaa 2004 Demographic Overview of the African Burial Ground, and Colonial Africans of New York. *In* The African Burial Ground Project. Skeletal Biology Final Report, vol. I. M. L. Blakey, and L. M. Rankin-Hill, eds. Pp. 266–304. Washington, DC: The African Burial Ground Project, Howard University for the United States General Services Administration, Northeast and Caribbean Region

lesions may represent more severe episodes of stress (Goodman and Armelagos 1985a,b). We suspect, based on historical documentation of importation ages (Blakey et al. 2004b:519; Foote 1991:82; see also Medford 2004), that many of the 15–24 year olds are likely to have been new arrivals in New York and that the middle passage within the Atlantic trade in human captives constitutes one plausible stressor in their case. Fifteen years of age was also the beginning of adulthood in most 18th-century censuses in New York (though, at certain points, as discussed earlier, a younger age was used as the criterion for adulthood) (Blakey et al. 2004b:524). Studies of active periosteal lesions in this population show more new infection in the 15–24 year age group than among the older individuals who exhibit a preponderance of sclerotic and healed lesions (Null et al. 2004). Mortality is also very high among the 15–24 year old males, and females, (see Figure 8.5). Changing conditions of life either through forced migration and/or adult status may be involved.

Considering age and dental attrition

Since much of this interpretation relies on the relation of hypoplasia frequency to age, we should examine the extent to which age-related occlusal wear might play a role in reducing our ability to observe hypoplasia, thus reducing the count of defects in older individuals. Subsets of the permanent dentition samples are used to control for the possible effect of dental attrition on hypoplasia frequencies between age and sex groups due to loss of observable data through tooth wear. The incisor and canine study, as well as the third molar study, maintained the previously reported pattern of hypoplasia frequencies when attrition was controlled.

The highest frequencies are found in individuals aged 15–24 years, and lower frequencies in individuals who lived to be 25 years of age and older. These differences are statistically significant in the third molar analysis only (Pearson chi-square with Yates continuity correction = 10.678, 1df, $p < .002$). Men continue to have higher frequencies of hypoplasia than women in both age groups for the canine and incisor study. These gender differences are not statistically significant.

While the gendered patterning of labor for men and women may have provided varying access to nutritional resources within domestic contexts, the data do not support a significant health advantage for women (Medford 2004:103-110; Null et al. 2004; Rankin-Hill et al. 2004). Tables 8.4 and 8.5 provide summaries of hypoplasia frequencies within each study. These findings show that the observed decrease in hypoplasia frequencies for older age groups and the differential frequencies between men and women are not a result of lost data due to tooth attrition.

Table 8.4 New York African Burial Ground frequency of hypoplasia by age and sex in canines and incisors (controlling for attrition), ($n = 48$)

Age group	*Frequency within age group*		
		Men (n = 24)	*Women (n = 21)*
15–24 ($n = 16$)*	81.3% ($n = 13$)	100% ($n = 5$)	75.0% ($n = 6$)
25–55+ ($n = 32$)	71.9% ($n = 23$	65.2% ($n = 15$)	34.8% ($n = 8$)

*Three individuals with adult dentition were too young to determine sex. Therefore, these individuals are not represented in the total number of males and females.

Source: Blakey, M. L., M. Mack, A. R. Barrett, S. S. Mahoney, and A. H. Goodman 2004a Childhood Health, and Dental Development. *In* The African Burial Ground Project. Skeletal Biology Final Report, vols. 1, and 2. M. L. Blakey, and L. M. Rankin-Hill, eds. Pp. 306–331. Washington, DC: The African Burial Ground Project, Howard University for the United States General Services Administration, Northeast and Caribbean Region.

Table 8.5 New York African Burial Ground frequency of hypoplasia by age group in third molars (controlling for attrition), ($n = 97$)

Age group	*Frequency within age group*
15–24 ($n = 26$)	46.2% ($n = 12$)
25–55+ ($n = 71$)	12.7% ($n = 9$)

Source: Blakey, M. L., M. Mack, A. R. Barrett, S. S. Mahoney, and A. H. Goodman 2004a Childhood Health, and Dental Development. *In* The African Burial Ground Project. Skeletal Biology Final Report, vols. 1, and 2. M. L. Blakey, and L. M. Rankin-Hill, eds. Pp. 306–331. Washington, DC: The African Burial Ground Project, Howard University for the United States General Services Administration, Northeast and Caribbean Region.

Canine chronology: mapping childhood periods of stress

Maxillary central incisors are intrinsically the most sensitive to developing hypoplasia, among all teeth, followed by mandibular canines (Goodman and Armelagos 1985a,b). Within this study, we compared hypoplasia frequencies in the permanent maxillary central incisors and the mandibular canines. Among the 65 individuals, 26 (40 percent) evinced hypoplasia defects in the maxillary central incisor versus 41 (63.1 percent) in the mandibular canines. Utilizing the overlap in developmental periods represented by these teeth (birth to 3.5 years in the central incisors and 0.5 to 6.5 years in the mandibular canine) while taking into analytical consideration the intrinsic sensitivity of incisors to hypoplasia in comparison with canines, we sought to assess the periods most stressful in early childhood between birth and 6.5 years for the New York African Burial Ground population.

Hypoplasia chronologies were calculated for the mandibular canines in 23 individuals (Table 8.6). Among the 37 recorded incidences of hypoplasia for these individuals, 73 percent occurred between the ages of 3.51 and 6.5 years (n = 27). Analyzed by individual (n = 23) and age, hypoplasia defects developed between the ages of 3.51 and 6.5 years in 95.7 percent of the cases (n=22). The maxillary incisor frequency may be compared with the mandibular canine chronology frequencies by individual, for an analysis of hypoplasia within the most sensitive teeth, by age range – between birth and 3.5 years (evinced by the most sensitive tooth, the maxillary central incisor) and between 3.51 and 6.5 years of age (evinced by the canine, the most sensitive tooth for this developmental period). The difference between these two hypoplasia frequencies – 40 percent (maxillary central incisors) and 95.7 percent (mandibular canines, between the ages of 3.51 and 6.5) – is, we believe, substantial when utilizing these data to understand stress episodes frequency and quality of life in early childhood (Table 8.7).

Variability of susceptibility to hypoplasia within tooth types is another factor that must be considered in the interpretation of the canine chronology data. Goodman and Armelagos (1985b:485), studying the Dickson Mound population, found mandibular canines to be most sensitive to enamel disruption between ages 3.5 and 4 years of age. Among the 23 New York African Burial Ground individuals in this canine chronology study, only 13.5 percent (n = 5) of the stress episodes occurred during this peak period of enamel susceptibility. However, 59.5 percent (n = 22)

Table 8.6 New York African Burial Ground frequency of hypoplasia by age intervals in mandibular canines

Age (years)	*Frequency*
.5–1	0
1.01–2	0
2.01–3	16.2% (n = 6)
3.01–4	18.9% (n = 7)
4.01–5	46.0% (n = 17)
5.01–6	18.9% (n = 7)
6.01–6.5	0

Table 8.7 New York African Burial Ground frequency comparison of hypoplasia in incisors and canines, $n = 37$ (hypoplasias)

Tooth	*Developmental period/Age (years)*	*Frequency*
Maxillary central incisor	0–3.5	40% (n = 26 of 65)
Mandibular canines	.5–6.5	63.1% (n = 41 of 65)
Mandibular canine chronology	0–3.5	26.2% (n = 6 of 23)
	3.51–6.5	95.7% (n = 22 of 23)

Note: Five individuals within the mandibular canine chronology study had multiple hypoplasias, and are represented in both the 0–3.5 and the 3.5–6.5 developmental period/age category frequencies.

of the hypoplasias occurred between 4.1 years and 6.5 years of age. These patterns are not consistent with Goodman and Armelagos (1985b). Thus, our findings that 95.7 percent of individuals developed hypoplasias in the mandibular canine between 3.51 and 6.5 years of age is likely a reflection of real age-related differences in stress frequencies and not simply an artifact of enamel sensitivity.

The individuals in the age category of 1–14 years are more likely to have been born in New York than individuals who were older at the time of death. Their early death and high levels of stress indicators such as hypoplasia, support an interpretation that these children were born into the arduous conditions of enslavement and therefore experienced overall patterns of greater levels of malnutrition, disease, illness, and possible entry into the work force at young ages.

While the peak frequencies of hypoplasia between the ages of 3–4 years in secondary dentitions observed by Corruccini et al. (1985) are attributed to weaning ages of 2–3 years old, Blakey et al. (1994) tested the weaning hypothesis within African American enslaved populations. The authors argue that enslaved children experience physiological stress from multiple sources and weaning does not account for the peak in hypoplasia frequencies. The test of the weaning hypothesis by Blakey and coworkers further establishes the import of historical and cultural context in the biocultural interpretation of patterns during this age.

The high frequencies of hypoplasia associated with the fifth year in the New York African Burial Ground population demonstrate that this was a vulnerable and stressful age for children who survived early infancy and who usually died as adolescents and young adults. This window on childhood appears to be most pertinent to those who were born in Africa, while childhoods in the Caribbean, New York, and other locations are doubtlessly mixed into our adult sample. How much more stressful the fifth year of age is compared to earlier ages remains difficult to assess using enamel defects due to variation in hypoplastic sensitivity across different parts of the crown (Goodman and Armelagos 1985a,b). Moreover, these data represent the experiences of survivors, while the high death toll of infants clearly represents vulnerability and stress among those who did not survive to exhibit developmental defects in secondary teeth. Those deaths clearly resulted from conditions in New York City, albeit precipitated partly by the poor health of captured mothers whose own experiences of childhood stress were relatively less frequent (Blakey et al. 2004b; Rankin-Hill et al. 2004).

The historical record sheds further light on the fear of African revolt expressed by European enslavers, shifting importation patterns, and changing definitions of child and adult to accommodate European labor needs, as well as their sense of security and psychological comfort (Blakey et al. 2004b; Medford 2004). These historical data further suggest at least two interpretations. One explanation assumes that many children experiencing stress episodes within the ages of 3.5–6.5 years and who lived to adulthood, were born within the colony of New York. Enslaved children in New York were frequently sold by the age of 6 years and advertisements often drew attention to marketable domestic skills in enslaved children (Goode-Null 2002:37–38; Medford 2004).

Therefore, it is possible that children approaching the age of 6 years would experience trauma related to separation from their parents, nutritional decline from inadequate provisions by nonparental custodians or slaveholders, and/or stresses related to an increased exposure to disease corresponding to their introduction into domestic or other labor duties. While children under the age of 15 were highly stressed, approaching the age of 6 years may have been a significant stage within the life histories of children born with the legal status of "slave" in colonial New York. Legal definitions of "adult" were applied to children over the age of 10 years in the 1731, and 1737/8 censuses, and at 16 years in the census data prior to 1731 and after 1737/8 including the 1810 census (Blakey et al. 2004b; Goode-Null 2002). This legal status as "adult" would most likely have affected the character of labor expected of young enslaved Africans under the age of 15 and within the age group of 15 to 24. These data further suggest that a child approaching the age of 9 or 10 may have been charged with tasks associated with adult labor practices or, engaged in training for entry into the labor market. Substantial third molar defect frequencies, especially for those who died between 15 and 24 years of age, represent stresses of older children and adolescents whether or not they were born in New York.

A second interpretation assumes the inclusion of children imported from Africa to New York as enslaved laborers. These children may have experienced high levels of physiological stress during their earlier childhood in relation to shifts in political power and socioeconomic upheaval surrounding the networks of the Atlantic slave trade that may have factored into their enslavement (Medford 2004). Also, these children under the age of 15 years would have experienced the notorious middle passage prior to their arrival in New York. Any of a host of other possible inadequacies of the large, stratified agrarian societies from which they derived may have contributed to moderately high hypoplastic frequencies in the childhoods of those who died as adults in New York. Medford (2004) characterizes the point of origin within West Africa as particularly fragmented, in part due to the European-driven Atlantic trade in human captives. Consistent with other findings of this study, most of the developmental stresses of childhood shown by older adult teeth were likely produced within African contexts while a minority of the adults' teeth developed during childhoods in New York. The high third molar frequencies for those who died between 15 and 24 years of age (44 percent of individuals affected) may reflect stress experienced by enslaved persons who were enslaved between 9 and 16 years of age, and died soon after their arrival in New York. Those who live to old age show far less stress during 9 to 16 years of age.

Our observation that those who lived the longest also had the lowest evidence of childhood stressors may suggest that higher chances of survival to adulthood are associated with having lower stress in childhood, irrespective of where the childhood took place. An attrition of hypoplastic individuals with age has been postulated elsewhere (Blakey and Armelagos 1985). These are not mutually exclusive propositions: those born in Africa may have had fewer childhood stressors and survived to older ages at death in New York than those who were born in New York City. Late adolescence heightened risk of stress or death for those entering the adult work regime irrespective of place of birth and for those recently captured and forced to migrate to New York.

A Temporal Analysis Utilizing the Archaeological Record

Hypoplasia in deciduous and permanent dentitions was also analyzed in relation to archaeological temporal groupings (Figure 8.6; Perry et al. 2006). Canines, incisors, and third molars were selected according to the methodology described above and included in this analysis when temporal designation is also possible (Table 8.8).

The temporal groupings assigned by the New York African Burial Ground archaeologists are based on context and stratigraphy, particularly in relation to features (such as fences), grave-related artifacts, coffin style, and where dating is

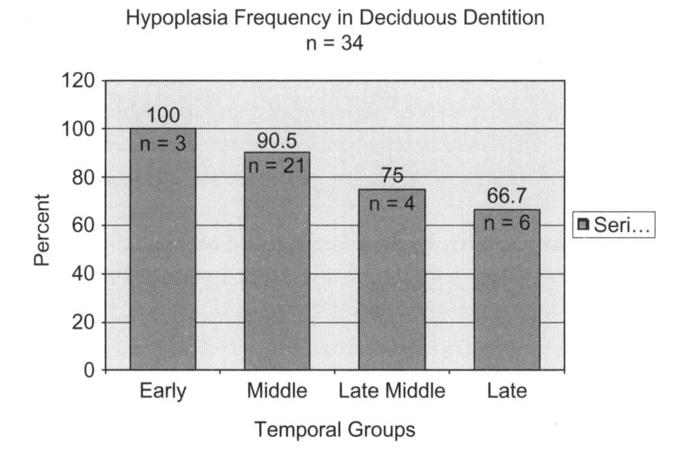

Figure 8.6 New York African Burial Ground hypoplasia frequencies in deciduous dentition across temporal groups.
Source: Author's drawing

Table 8.8 New York African Burial Ground archaeological temporal groups

Early	Late 17th century to approximately 1735
Middle	Approx. 1735–1760
Late Middle	Approx. 1760–1776
Late	Approx. 1776–1794

Table 8.9 New York African Burial Ground temporal analysis subsamples

Sample size	*Age at death*	*Observable*	*Developmental age represented*
n = 34	Subadults 1–7 years	Deciduous incisors, canines, and second molar	5 months *in utero* – 1 year old
n = 51	9–55+ years	Permanent incisors and canines	Birth to 6.5 years old
n = 98	15–55+ years	Third molars	9 to approximately 16 years old

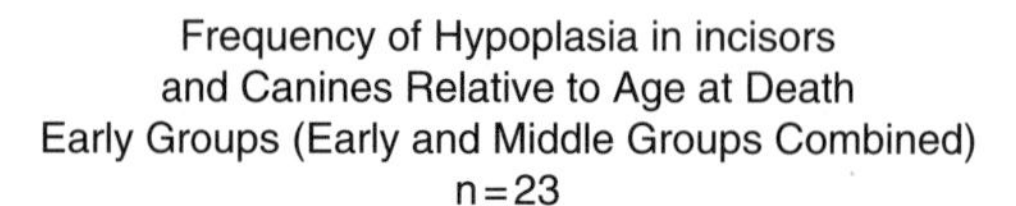

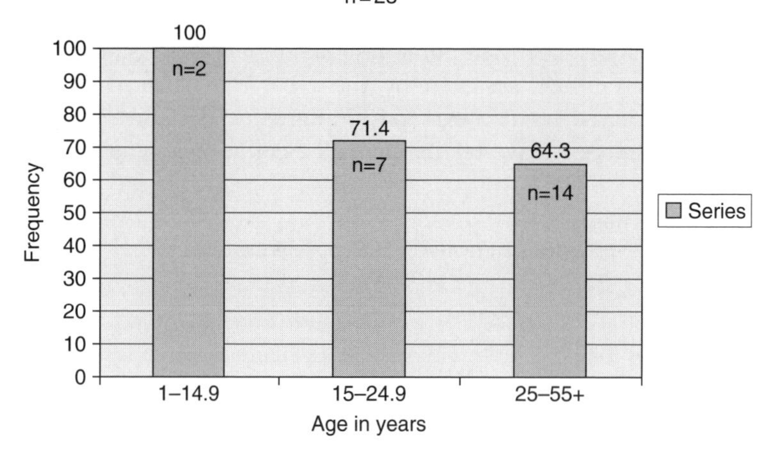

Figure 8.7 New York African Burial Ground frequency of hypoplasia: Early groups (Early and Middle groups combined, representing the late 17th century to approximately 1760).
Source: Author's drawing

more difficult, the burial's relationship to more clearly dated interments (Howson et al. 2006a:105–129). The temporal groups are summarized in Table 8.9. See also Figure 8.7.

Within this group, the frequency of hypoplasia in children with deciduous teeth was 85.3 percent (*n* = 34). A temporal analysis demonstrated a decrease from 100 percent in the Early group to 66.7 percent in the Late group (Figure 8.8). Children with deciduous dentition died between the ages of one and seven years. While the very youngest of the children who died in colonial New York were most likely born there, the documentary record shows increased importation of adolescents, young women, and children, particularly in the latter half of the 1700s (Goode-Null 2002; McManus 1966; Medford 2004:118). It is possible that these children represented by deciduous teeth always include children born in New York yet over time include

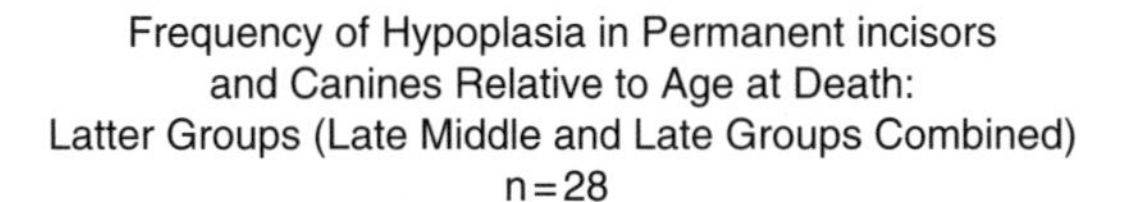

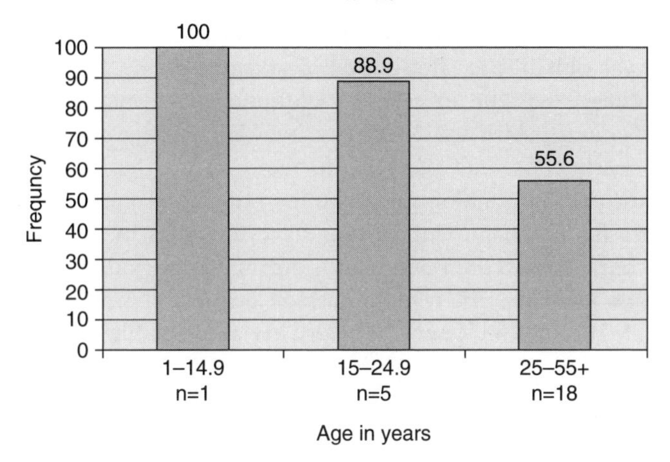

Figure 8.8 New York African Burial Ground frequency of hypoplasia: Latter groups (Late Middle and Late groups combined, representing approximately 1760 to 1794).
Source: Author's drawing

an increasing percentage of African-born children as well. The shift in importation patterns would have increased the likelihood that some of these young children may have spent their first years in Africa and died shortly after arrival in New York. The decrease in hypoplasia within the deciduous dentitions of young children (representing their development between five months *in utero* through the first year), may represent an increase of children entering and dying in colonial New York who spent this early period within Africa, but were enslaved at young ages and forced to migrate to the Americas where they died between the ages of one and seven years.

Free and enslaved Africans in 18th-century New York built independent networks for selling goods and hiring themselves out for wage labor, thus increasing access to much-needed resources (Medford 2004:57,119–122). While enslaved Africans continued to create ways to provide for their children and families, the revolt of 1712 and "conspiracy" of 1741 heightened the anxiety among slaveholders over the continued threat of revolt (Blakey et al. 2004b; Epperson 1999; Foote 1991; Kruger 1985; Lydon 1978; Medford 2004). Legislation was passed to restrict the movement and independence of enslaved and free Africans from the end of the 17th century, continuing into the 18th century (Medford 2004; Perry et al. 2006). As whites sought to secure exclusive access to labor positions for themselves, Africans faced decreasing opportunities for hire (Medford 2004:119–121). The decreasing frequency of hypoplasia during the earliest developmental ages of children who died in New York more likely reflects the increased demand for domestic labor and the importation of young women and children in response to the revolts.

A temporal analysis of the New York African Burial Ground canine and incisor study enabled a more careful investigation of childhood health combined with indications of origin through importation patterns and age at death. In order to maintain the subsample numbers high enough to interpret patterns, the four temporal groupings were combined into Early and Latter groups.

The overall trend is reflected in both early and late 18th-century groups. However, there is an increase in hypoplasia frequencies in the 15–24 year olds, and a decrease in the 25–55 year olds in the later temporal groups. Again, this shift is consistent with the hypothesis that the increased childhood stress may reflect the shift in importation of younger Africans directly to New York. While the canines and incisors reflect the early childhood years, these data indicate that young children were increasingly exposed to physiological stressors in the latter temporal periods. Again, the young years for these individuals may have included the political unrest and upheaval that characterized their homelands during this period, the internal network of the slave trade in Africa, the middle passage prior to arrival, and the possibility of a considerable number of New York births, are all plausible developmental stressors being reflected in this age group.

Eighteenth-century New York newspapers announced the arrival of enslaved Africans between the ages of 9 and 12 years. Merchant preferences were for young laborers, as John Watts stated in the 1760s, "the younger the better if not quite children" (Medford 2004:118; Perry et al. 2006). Enslaved children between the ages of 6 and 12 were advertised for sale and solicited by slaveholders for purchase (Blakey et al. 2004b; Goode-Null 2002; Medford 2004:118; Perry et al. 2006). These ages, represented in the New York African Burial Ground data, were likely periods of increased exposure to health risks, but not necessarily immediate death. These are among the heartiest years for human children in all but peculiar circumstances (including a disease epidemic, Alfredo Coppa personal communication 2008). The evidence throughout the paleodemographic record that one "cannot kill a ten year-old" is reinforced in the age-specific mortality for the New York African Burial Ground as it relates more specifically to the 11-years age group. Mortality between 9–12 years of age is low, while both historical evidence (Blakey et al. 2004b; Goode-Null 2002; Medford 2004:118; Perry et al. 2006) and hypoplasia data (below) provide evidence of high stress experienced during those years.

Harris Lines

Transverse radiopaque lines, also called "Harris Lines," are striations identifiable macroscopically in radiographic images of long bones. The interpretability of these striations in archaeological populations is questioned due to the varying and sometimes contradictory results of analyses performed clinically in human and animal populations, as well as studies concerned with archaeological populations (Alfonso Durruty 2008; Magennis 1990).

However, the cause of transverse radiopaque line formation is most often attributed to growth disruption at the epiphyses in the long bones (due to physiological stress episodes such as illness, malnutrition, and trauma) and subsequent growth resumption (Blanco et al. 1974; Garn et al. 1968; Harris 1931; Magennis 1990). Recent clinical studies of Harris Lines in rabbits resulted in a greater number of Harris Lines during periods of growth acceleration, but with varying and nonlinear results due to differential nutrition levels and rates of resorption (Alfonso Durruty 2008). Regardless of the various suggested causes of opaque line formation, all are consistent with rapid changes in growth. Radiopaque line frequencies in archaeological populations are traditionally measured, recorded, and interpreted as indicators of periodic and/or chronic episodes of growth disruption due to physiological stressors (Khudabux 1991; Owsley 1987; Rathbun 1987). However, without careful consideration, and multiple lines of evidence to indicate the health contexts of a population, Harris Line data in archaeological populations can lead to misinterpretation of health status (Alfonso Durruty 2008). We present the New York African Burial Ground Harris Line data as evidence of rapid change in growth, and as potential indicators of stress. We interpret this data in combination with multiple lines of evidence for the historic, social, and physical stresses experienced by these individuals, as presented in this present study, and most extensively in the social historical, archaeological, and skeletal biology research of the New York African Burial Ground Project (Blakey and Rankin-Hill 2004; Medford 2004; Perry et al. 2006).

Comparison of Harris Lines and enamel hypoplasia

Among the New York African Burial Ground population, 34 individuals were assessed for presence and absence of transverse radiopaque lines and compared with hypoplasia occurrence. The definition of what constituted a transverse radiopaque line was made according to Magennis (1990) and the chronology to determine age of occurrence for each Harris Line was calculated according to Maat (1984). Radiographs were viewed using a light table and radiopaque striations in tibiae recorded as present or absent per individual. Epiphyseal fusion of the tibiae occurs between the approximate ages of 14 to 19 years distally, and 15 to 21.5 years proximally. Therefore, these growth rate change indicators represent the developmental period from birth until the 22nd year of life (Buikstra and Ubelaker 1994).

Among the 34 New York African Burial Ground individuals, 76.5 percent had Harris Lines. There was no statistically significant difference between the frequency of Harris Lines observed among males (77.3 percent) and females (75 percent) presence. Changes in growth rate occur at younger mean ages over the 18th century (Figure 8.9). The Late group has the earliest and latest incidence, and both the Middle and the Late groups have a broader age range of occurrence. These findings support the hypoplasia pattern showing more stress at younger ages in the later 18th-century groups than in the earlier 18th-century groups that coincides with increased importation and reliance on African women and children.

Discussion: Life Histories in Context

The evidence of health and quality of life in the New York African Burial Ground within this study has focused thus far on the period of childhood development. The following is a summative discussion of adulthood with bioarchaeological descriptions of individuals who were analyzed as part of the dental and skeletal subsamples of this study. The developmental years for these individuals who died as adults are

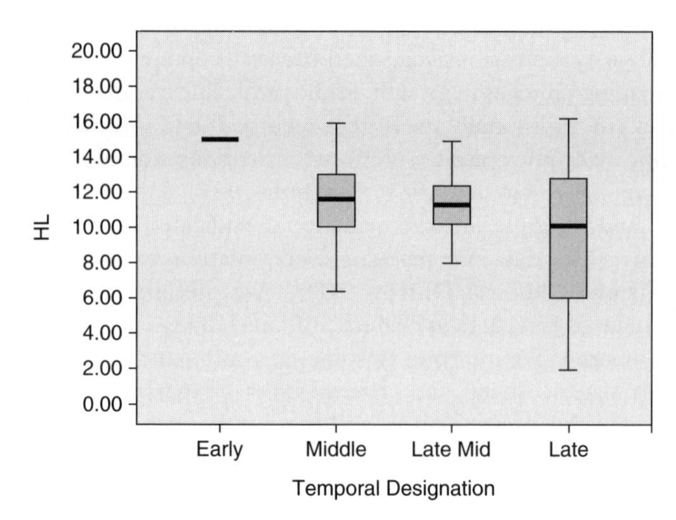

Figure 8.9 New York African Burial Ground Harris Line frequencies by temporal group. Source: Author's drawing

characterized by the statistical patterns of childhood health and development previously presented.

Adulthood is punctuated here to complete the life cycle with an examination of the ends of lives for the Africans who survived childhood to die as men and women in colonial New York. Rather than simply presenting descriptive profiles for a few individuals within the New York African Burial Ground Project, our intent is to bring the research focus back to the men, women, and children who make up the New York African Burial Ground population subsamples selected. We emphasize these individuals within biocultural contexts of social, political, historical, and biological processes that shaped the manner in which they lived, died, and are remembered.

These men and women were most likely born in Africa or the Caribbean, according to documentary sources (Blakey et al. 2004a,b; Medford 2004). Lead and strontium analyses of a dental sample of adults whose teeth are cosmetically filed produce diverse results, including a majority of chemical signatures consistent with African birth (Goodman et al. 2004) and DNA analysis indicates origins from West and West Central Africa (Jackson et al. 2004). While we focus on men and women separately, we weave aspects of West and Central African, Caribbean, and New York historical contexts that variously apply to the men and women buried in the New York African Burial Ground.

Labor in context: Africa, the Caribbean, and colonial New York

West African men and women in the 18th century engaged in agricultural occupations as well as cattle husbandry, mining, textiles, logging, and metal working. Crops included maize, millet, peanuts, and rice (Medford 2004:97). Transhumance in the Senegambia region was a specialty of the Fulbe, whereas the Serer raised and cared for cattle in pens (Medford 2004:97). Along the Gold Coast of West Africa, the Akan mined gold and farmed, which first required removal of dense forests. Fishing and boat building were common occupations along the coast of the Bight of Benin. (Medford 2004:97). West Central Africans in the 18th century engaged in agricultural labor, animal husbandry, fishing, and specialized occupations such as procuring ivory and beeswax, iron mining, and metalworking. As in the West African region, certain ethnic groups specialized in particular occupations, for example, the Imbundus were primarily known for raising cattle and the Matamba for fishing (Medford 2004:97–99).

Africans who died in New York during the late 17th and 18th centuries may also have spent varying amounts of time in the Caribbean islands, spanning from days to months to years. While domestic labor would have been required of enslaved Africans in the 18th century, sugarcane plantations were the primary reason slaveholders demanded enslaved labor. Sugarcane labor is considered by historians of slavery in the Americas to be one of the harshest and dangerous labor systems due to the physical demand required of long days planting and harvesting, as well as the risk of losing digits and limbs during the processing of sugarcane (Medford 2004; see also Goveia 1965; Higman 1995; Moitt 2004). The sugar plantation employed enslaved men, women, and children in the various aspects of planting, harvesting, and processing. Poor nutrition and compromised health exacerbated the effects of such strenuous labor conditions. During the harvesting season, laborers were often required to work without rest until the harvesting was complete (Goveia 1965; Higman 1995; Moitt 2004; Medford 2004).

African laborers in 18th-century colonial New York engaged in a very diverse range of work as tailors, butchers, coopers, sawyers, sail makers, goldsmiths, bricklayers, farm hands, mariners, rope makers, and domestic laborers. Enslaved Africans worked in sugar refineries, textile factories, and tanneries (Medford 2004:103–105). The mixture of urban and rural labor needs within 18th-century New York often required enslaved Africans to perform different types of labor within a single day, and from one day to the next (Medford 2004: 35,103–105; see also Berlin 1998). Medford (2004) states,

> Not least of the factors affecting the health of enslaved people was the toll taken by strenuous and repetitive labor … Dock work, porterage, farming, and other manual jobs that involved heavy lifting and the bearing of weighty objects on the head and shoulders would have produced stress to the body and compromised health. Even the daily routines of domestics (such as carrying water into the house) and the repetitive motions of those women who worked at spinning and weaving exposed them to injury. Such injuries as are found in the African Burial Ground population provide the physical evidence of the consequences of exposure to hard labor (171).

The remnant indicators of labor and mechanical stress in the skeleton are interpreted from patterns of hypertrophic, or raised, muscle attachments, enthesopathies, or stress lesions, and degenerative processes in the joints. Degenerative processes, though generally related to aging, also indicate injury and/or heavy, repetitive use, particularly when present in young people (see Capasso et al. 1999; Wilczak et al. 2004). Within the New York African Burial Ground population, moderate, and severe degenerative joint changes are observed in young men, and women between 15 and 24 years of age (Wilczak et al. 2004). Evidence of musculoskeletal work stress in New York African Burial Ground women and was often pronounced (Wilczak et al. 2004).

Adulthood: women

Women born in West and West Central Africa experienced a range of livelihoods depending on their home of origin, including agricultural labor, textile work, and child rearing (Medford 2004:97). Many African women enslaved in colonial New York, worked as domestic laborers, launderers, and seamstresses.

Slaveholders discouraged women of childbearing years from becoming pregnant, and if a woman became pregnant, she was at risk of being sold (Medford 2004:141). Eighteenth-century census data shows an absence of natural increase due, in obvious part, to high infant death rates and low fertility (Blakey et al. 2004b). The African population in New York gradually increased over the course of the 18th century due to importation patterns rather than increased birth rate (Blakey et al. 2004b). There was a decline in the number of children per woman of childbearing age after importation of enslaved labor ended (Blakey et al. 2004b). While many children represented in the New York African Burial Ground were likely born in Africa and experienced forced migration at young ages, the majority of the youngest of those who died are most likely to represent children conceived by enslaved women and born in New York. High lead levels in teeth of young children suggest exposure to European colonial lead-based technologies at very early ages (Goodman et al. 2004). These children would also experience prenatal and postnatal stress due to the compromised health of their mothers.

Burial 122 (Figure 8.10) exemplifies many of the larger trends of poor health, harsh labor, and early death present in the New York African Burial Ground population. This young woman died between 18 and 20 years of age and is assigned to the Middle temporal group, approximately 1735–1760 (Perry et al. 2006). Though no grave goods were discovered with her skeleton, Burial 122 was buried in a hexagonal, wooden coffin (Perry et al. 2006). She experienced physiological stress between the ages of 9 and 16 years, as indicated by the presence of hypoplasia in her third molars. Consistent with high levels of active and healed infection in young women in the New York African Burial Ground (Null et al. 2004), this woman has evidence of infection in her cranium and in her upper and lower limbs. Burial 122 has enlarged muscle attachments throughout her skeleton and mechanical stress lesions on both humeri and on her left clavicle, indicating heavy and/or repetitive physical labor. Although a young woman, Burial 122 has mild to severe osteoarthritis in her axial and appendicular joints. Bilateral bowing of her leg bones indicates she had rickets, associated with a lack of vitamin D (Blakey and Rankin-Hill 2004). The temporal context of her burial places her life and death after the 1712 revolt, and her death near or after the 1741 revolt. Following the revolts of 1712 and 1741, women were forced to perform labor for which men would otherwise have been preferred (Blakey et al. 2004b; Medford 2004). This young woman was engaged in heavy and/or

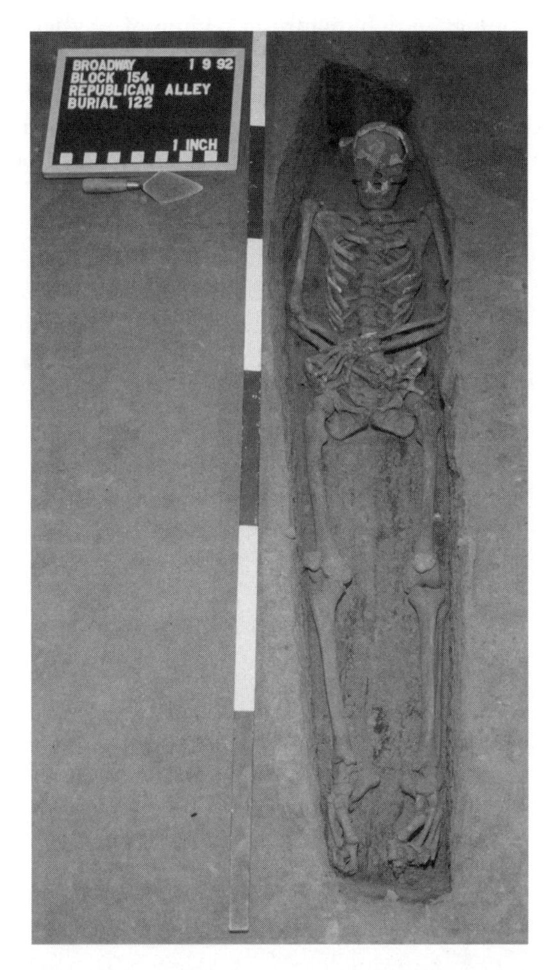

Figure 8.10 New York African Burial Ground, Burial 122, a woman who died between the ages of 18 to 20 years. Her burial dates to the Middle temporal group (approximately 1735–1760). Source: Author's photograph

repetitive labor and likely experienced the increased labor demands placed on enslaved women following the uprising of 1712 and the conspiracy of 1741.

Adulthood: men

Men from West and West Central Africa likely engaged in a wide variety of occupations within their homelands, including animal husbandry, cloth, and iron working in Senegambia, logging in Sierra Leone, mining in the Gold Coast, and cattle rearing in Angola (Medford 2004: 96–100). Due to political upheaval associated with African empire building and the transatlantic slave trade, 18th-century African men, particularly in the West Central region, often were required to go to war (Medford 2004:99). Men were more likely than women to be brought to the

Caribbean for "seasoning," a dehumanizing regime of physical and psychological methods, directed by colonial Europeans with the intent of preventing resistance and fostering labor compliance prior to sale in New York (Medford 2004; see also Mullin 1995). Africans imported from the Caribbean, as well as newly arrived African men, were considered by Europeans in New York to be more of a threat for conspiring revolts (Medford 2004).

African men in colonial New York during the 18th century also labored in a diverse range of work, such as tailors, butchers, coopers, sawyers, sail makers, goldsmiths, bricklayers, farm hands, mariners, and domestic laborers. The shipping industry required men to engage in frequent heavy lifting when unloading and loading ships (Medford 2004:103–110). Burial 47 (Figure 8.11) is a man who

Figure 8.11 New York African Burial Ground, Burial 47, a man who died between the ages of 35 and 45 years. His burial dates to the Middle temporal group (approximately 1735–1760). Source: Author's photograph

died between the ages of 35 and 45 years. The archaeological context of his burial places him within the Middle temporal group, indicating that he died, approximately, between 1735 and 1760. Burial 47 was buried in a wooden coffin, made of pine, and his grave was marked with a granite slab. He may have shared a grave with Burial 31, an adolescent girl or boy between 14 and 16 years of age. Trace elemental signature analysis is not clear in determining this man's place of birth. However, barium and lead concentrations of his third molar indicate the period of his life between approximately 9 and 16 years were spent in Africa. Strontium isotope values indicate possible birth in the Caribbean (see Goodman et al. 2004). Mitochondrial DNA analysis indicates Benin, West Africa as a place of origin (see Jackson et al. 2004). Burial 47 has culturally modified teeth, which strongly supports that he was born and lived part of his life in Africa prior to being enslaved.

Though included in the hypoplasia studies, Burial 47 does not have hypoplasia and, in general, does not have indicators of childhood stress. This man does, however, have infection in his cranium and legs, and he bears the evidence of hard labor in having many mechanical stress lesions throughout his body, consistent with the multifaceted, yet extremely strenuous labor required of African men in colonial New York (Medford 2004; Wilczak et al. 2004). Depending on the age at which Burial 47 was enslaved, and the age at which he entered New York, his death context possibly places his adult life in New York after the 1712 uprising but during the year prior to and/or following the 1741 conspiracy. As mentioned previously, the years during which this man's death is estimated was a period of heightened European fear of African men. This man's culturally modified teeth would have signified an African birth to colonial New York slaveholders. While having survived to adulthood, this man did not survive to his older years.

Superannuated men and women

Superannuated, or elderly, enslaved men and women were considered financial burdens to slaveholders in the same manner that infants were considered a financial loss (Blakey et al. 2004b; Medford 2004). Approximately 5 percent of those buried in the New York African Burial Ground lived to be 55 years of age or older. Archives of the nearby English Trinity Church, a protestant church that banned the burial of Africans in their cemetery in 1697, record 7–9 times as many interments for elderly adults of similar age (Blakey et al. 2004b; Medford 2004). Colonial legislation passed in 1773 cites, "repeated instances in which the owners of slaves have obliged them after they are grown aged and decrepit'" to beg for food, and clothing (Blakey et al. 2004b). The Act of 1773 reports "collusive bargains" where slaveholders paid a person to feign purchase of an elderly enslaved person in order to be rid of the burden of caring for him or her (Blakey et al. 2004b). In 1785 a slaveholder had to procure a certificate from the overseer of the poor in order to free an enslaved person. However, enslaved men and women over 50 were not eligible, due to the tendency of slaveholders to attempt to rid themselves of elderly enslaved men and women who were no longer able to work (Blakey et al. 2004b; Medford 2004).

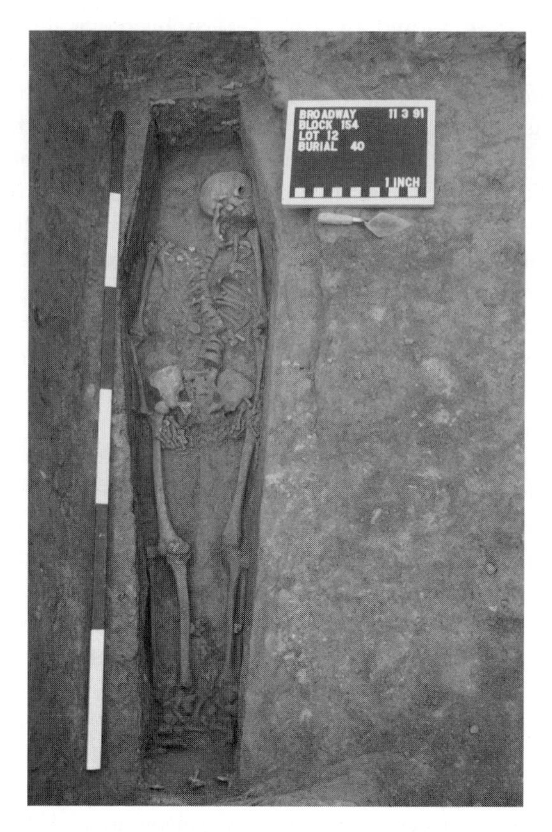

Figure 8.12 New York African Burial Ground, Burial 40, a woman who died between the ages of 50 and 60 years of age. Her burial was assigned to the Late temporal group (approximately 1776–1794).
Source: Author's photograph

Due to the preference of slaveholders for young laborers, older individuals usually came from Africa to New York through the West Indies, where women and men engaged in the harsh labor of sugarcane planting, harvesting, and processing. These men and women performed all manner of labor within colonial New York, though many of the elderly were employed in domestic labor (Medford 2004). Men and women who survived until the ages of 50, and above, generally demonstrate high levels of mechanical stress indicators and age-related, as well as labor-related, joint degeneration (see Wilczak et al. 2004).

Burial 40 (Figure 8.12) was a woman who died between the ages of 50 and 60 years. The archaeological context of her grave places her within the Late group, dying approximately between the years of 1776 and 1794. Burial 40 was buried in a hexagonal Eastern White Pine coffin, and was found with fragments of straight pins, that may have been related to shrouding practices or, likely, to secure a strap around her chin (Howson et al. 2006b; Perry et al. 2006). In common with approximately 14 percent of adults in the New York African Burial Ground population (Null et al. 2004), this woman experienced nutritional deficiency, associated with a lack of vitamin D, during her childhood, which caused bowing of her leg bones.

However, Burial 40 does not have hypoplasia. She has evidence of infection in both legs. Myositis ossificans on her tibiae, as well as her ribs, indicate earlier injuries and a lifetime of hard work is evident in the severe muscle attachment hypertrophies throughout her skeleton. A combination of age and hard work are seen in the moderate to severe osteoarthritis in the joints throughout her body, and the pronounced osteophytosis in her lumbar vertebrae (Blakey and Rankin-Hill 2004; Wilczak et al. 2004).

Most of the oldest individuals in the New York African Burial Ground were born into African societies in which the elderly were revered, cared for, and learned from. In the context of enslavement, high mortality, the increasing reliance on newly imported, young, enslaved labor – the aged were largely discarded (Blakey et al. 2004b). In an economy based on enslaved labor, the youngest and the oldest Africans held no value for slaveholders. This is in contrast with the cosmologies of much of West and Central Africa where a cross-shaped symbol is used to, among other meanings, represent the cycle of life; one of the branches is for the status and role "ancestor," which is how the geriatric remains are now considered in this mortuary context of the New York African Burial Ground.

Conclusion

This study demonstrates the biocultural analysis of paleopathological data with known historic contexts in order to construct a model of the stress-related experiences over the life course of enslaved Africans in 18th-century New York City. Infancy and early childhood for those who were born in the city presented the greatest risks of morbidity and mortality, the latter is demonstrated by the number of individuals who did not survive to become 4 years of age. Dental enamel defects in deciduous dentition give evidence for morbidity during infancy, and likely reflect the health of mothers as well as their newborns. Of those who did survive to show hypoplasia in the permanent teeth indicates that from the ages of 3.5 to 6.5 years proved to be a highly stressful period. Furthermore, third molar data shows evidence of high physiological stress between 9 and 16 years, particularly among individuals who died between 15 and 25 years of age. These findings are associated with the ages at which children were imported and/or introduced to the work regimes and other environmental exposures of slavery. The skeletons that best tell of these events are those who died immediately or within perhaps 10 years of the compounding of childhood stress with early adult life and labor. Third molar data as well as Harris Lines indicative of stresses experienced during late childhood and adolescence show a tendency to increase in the latter part of the 18th century .

Those who survived to adulthood (25 to 55+ years) show far less early childhood hypoplastic stress than did those who died by the time they were young adults. Our most plausible explanation of this trend rests on the historical and chemical data, which indicates that the largest number of persons that survived to and died in adulthood were African born. Thus, their childhood dental indicators of stress frequently reflect life as free persons in their African homelands. As a hypothesis, it seems worth considering whether or not those most stressed in childhood, American born, were at greater risk of morbidity and mortality throughout their lives than were the African born, Those who died as older men and women clearly experienced arduous work regimes and nutritional compromise, very rarely surviving beyond the age of 54 years (Blakey and Rankin-Hill 2004).

This study attempts to understand the complex life histories of enslaved Africans, many of whom were transported across the Atlantic and ultimately lived and died in New York during the 18th century. By integrating historical, archaeological, and skeletal biological research data we use multiple indicators of health to assess the quality of life for men, women, and children over the life cycle. The age at which a man, woman, or child was forced to migrate to the Americas as an enslaved laborer, as well as the historical context in which this migration occurred, shaped the timing of morbidity and mortality of Africans in New York. Children born into slavery were the most vulnerable to poor health and early death. Individuals who died as elders show evidence of high levels of mechanical stress and varying levels of childhood and adolescent health associated with pre- and post-enslavement histories. Children and adolescents were increasingly valued for their labor and perceived ability to be socialized toward compliance in the wake of the "uprising" of 1712 and the conspiracy of 1741 (Blakey et al. 2004b; Medford 2004). Corresponding with increased reliance on the labor of younger Africans, Africans who worked and died in colonial New York were exposed to nutritional and health stressors at younger ages over the course of the 18th century. Whether born in colonial New York or forced to migrate, the quality of life for children, men, and women was greatly compromised upon entry into the environment of enslavement.

NOTE

During the final preparation of this article a revised version of the technical reports of the African Burial Ground Project were published in 2009 by Howard University Press as academic volumes: The Skeletal Biology of the New York African Burial Ground (Michael L. Blakey, and Lesley M. Rankin-Hill, eds.), The Archaeology of the New York African Burial Ground (Warren R. Perry, Jean Howson, and Barbara A. Bianco, eds.), and Historical Perspectives of the African Burial Ground: New York Blacks and the Diaspora (Edna Greene Medford, ed.).

REFERENCES

Alfonso Durruty, M. P. 2008 Biosignificance of Harris Lines as Stress Markers in Relation to Moderate Undernutrition and Bone Growth Velocity: A New Zealand White Rabbit Model for the Study of Bone Growth. Dissertation. Department of Anthropology, State University of New York, Binghamton.

Angel, J. L., J. O. Kelley, M. Parrington, and S. Pinter 1987 Life Stresses of the Free Black Community as Represented by the First African Baptist Church, Philadelphia, 1823–1841. American Journal of Physical Anthropology 74:213–229.

Berlin, I. 1998 Many Thousands Gone: The First Two Centuries of Slavery in North America. Cambridge, MA: Harvard University Press.

Berti, P. R., and M. C. Mahaney 1992 Qantification of the Confidence Interval of Linear Enamel Hypoplasia Chronologies. *In* Recent Contributions to the Study of Enamel Developmental Defects. A. H. Goodman, and L. L. Capasso, eds. Pp. 19–30.Teramo, Italy: Journal of Paleopathology, Monograph Publications, No. 2.

Blakey, M. L. 1988 Social Policy, Economics, and Demographic Change in Nanticoke-Moor Ethnohistory. American Journal of Physical Anthropology 75:493–502.

Blakey, M. L. 1998 The New York African Burial Ground Project: An Examination of Enslaved Lives, a Construction of Ancestral Ties. Transforming Anthropology 7: 53–589.

Blakey, M. L. 2001 Bioarchaeology of the African Diaspora in the Americas: Its Origins and Scope. Annual Review of Anthropology 30:387–422.

Blakey, M. L. 2004 Introduction. *In* The African Burial Ground Project. Skeletal Biology Final Report, Vols. 1 and 2. M. L. Blakey, and L. M. Rankin-Hill, eds. Pp. 306–331. Washington, DC: The African Burial Ground Project, Howard University for The United States General Services Administration, Northeast, and Caribbean Region.

Blakey, M. L., and G. J. Armelagos 1985 Deciduous Enamel Defects in Prehistoric Americans from Dickson Mounds: Prenatal, and Postnatal Stress. American Journal of Physical Anthropology 66(4):371–380.

Blakey, M. L., T. E. Leslie, and J. P. Reidy 1994 Frequency and Chronological Distribution of Dental Enamel Hypoplasia in Enslaved African Americans: A Test of the Weaning Hypothesis. American Journal of Physical Anthropology 95:371–383.

Blakey, M. L., M. Mack, A. R. Barrett, S. S. Mahoney, and A. H. Goodman 2004a Childhood Health, and Dental Development. *In* The African Burial Ground Project. Skeletal Biology Final Report, vols. 1, and 2. M. L. Blakey, and L. M. Rankin-Hill, eds. Pp. 306–331. Washington, DC: The African Burial Ground Project, Howard University for the United States General Services Administration, Northeast and Caribbean Region.

Blakey, M. L., and L. M. Rankin-Hill eds. 2004 The African Burial Ground Project. Skeletal Biology Final Report, Vols. 1 and 2. Washington, DC: The African Burial Ground Project, Howard University for the United States General Services Administration, Northeast and Caribbean Region.

Blakey, M. L., L. M. Rankin-Hill, J. E. Howson, and S. H. H. Carrington 2004b The Political Economy of Forced Migration: Sex Ratios, Mortality, Population, Growth, and Fertility among Africans in Colonial New York. *In* The African Burial Ground Project. Skeletal Biology Final Report, Vols. 1 and 2. M. L. Blakey, and L. M. Rankin-Hill, eds. Pp. 514–540. Washington, DC: The African Burial Ground Project, Howard University for the United States General Services Administration, Northeast, and Caribbean Region.

Blanco, R. A., R. M. Acheson, C. Canosa, and J. B. Salomon 1974 Height, Weight, and Lines of Arrested Growth in Young Guatemalan Children. American Journal of Physical Anthropology 40:39–48.

Buikstra, J. E., and D. H. Ubelaker eds. 1994 Standards for Data Collection from Human Skeletal Remains. Arkansas Archaeological Survey Research Series No. 44. Fayetteville, Arkansas.

Buxton, L. H. D., J. C. Trevor, and A. H. Julien 1938 Skeletal Remains from the Virgin Islands. Man 38:49–51.

Capasso, L., K. A. R. Kennedy, and C. A. Wilczak 1999 Atlas of Occupational Markers for Human Remains. Journal of Paleopathology Monograph Publication 3. Teramo: Edigrafital SpA.

Clarke, S. K. 1982 The Association of Early Childhood Enamel Hypoplasias and Radiopaque Transverse Lines in a Culturally Diverse Prehistoric Skeletal Population. Human Biology 54:77–84.

Condon, K., and J. C. Rose 1992 Intertooth and Intratooth Variability in the Occurrence of Developmental Enamel Defects. Journal of Paleopathology 2:61–77.

Corruccini, R. S., J. S. Handler, and K. P. Jacobi 1985 Chronological Distribution of Enamel Hypoplasia, and Weaning in a Caribbean Slave Population. Human Biology 57:669–711.

Corruccini, R. S., K. P. Jacobi, J. S. Handler, and A. C. Aufderheide 1987 Implications of Tooth Root Hypercementosis in a Barbados Slave Skeletal Collection. American Journal of Physical Anthropology 74:179–184.

Douglass, F. 1854 The Claims of the Negro Ethnologically Considered: An Address, Before the Literary Society of Western Reserve College at Commencement July 12, 1854. Rochester: Lee, Mann & Co., Daily American Office.

Curtin P. 1969 The Atlantic Slave Trade: A Census. Madison: University of Wisconsin Press.

David, P. A., H. G. Gutman, R. Sutch, P. Temin, and G. Wright 1976 Reckoning with Slavery: A Critical Study in the Quantitative History of American Negro Slavery. New York: Oxford.

Drake, St. C. 1980 Anthropology and the Black Experience. Black Scholar II(7):2–31.

Epperson, T. W. 1999 The Contested Commons: Archaeologies of Race, Repression, and Resistance in New York City. *In* Historical Archaeologies of Capitalism. M. P. Leone, and P. B. Potter, Jr., eds. Pp. 81–110. New York: Plenum.

Fanon, F. 1994 Black Skin, White Masks. New York: Grove Press. Originally published in French under the title Peau Noire, Masques Blancs by Editions du Seuil, Paris, 1952.

Ferguson L. 1992 Uncommon Ground: Archaeology, and Early African America, 1650–1800. Washington, DC: Smithsonian Institution.

Firmin, A. 2002 The Equality of the Human Races. Translated from French by Asselin Charles. Originally published in as *De L'egalite des Races Humaines*, 1885. Introduction by Carolyn Fluehr-Lobban. Urbana: University of Illinois Press.

Fogel, R. W., and Engerman, S. L. 1974 Time on the Cross: The Economics of American Negro Slavery. Boston: Little, Brown.

Foote, T. 1991 Black Life in Colonial Manhattan, 1664–1786. Ph. D. Dissertation, Harvard University, Ann Arbor: University Microfilms.

Garn, S. M., F. N. Silverman, K. P. Hertzog, and V. M. Rhomann 1968 Lines and Bands of Density: Their Implication to Growth and Development. Medical Radiography and Photography 44:58–89.

Goode-Null, S. K. 2002 Slavery's Child: A Study of Growth, and Childhood Sex Ratios in the New York African Burial Ground. Ph.D. Dissertation, Department of Anthropology, University of Massachusetts. Ann Arbor: University Microfilms.

Goodman, A. H., and G. J. Armelagos 1985a Factors Affecting the Distribution of Enamel Hypoplasias Within the Human Permanent Dentition. American Journal of Physical Anthropology 68:479–493.

Goodman, A. H., and G. J. Armelagos 1985b The Chronological Distribution of Enamel Hypoplasia in Human Permanent Incisor, and Canine Teeth. Archives of Oral Biology 30:503–507.

Goodman, A. H., G. J. Armelagos, and J. C. Rose 1980 Enamel Hypoplasias as Indicators of Stress in Three Prehistoric Populations from Illinois. Human Biology 52: 515–528.

Goodman, A. H., J. Jones, J. Reid, M. Mack, M. L. Blakey, D. Amarasiriwardena et al. 2004 Isotopic and Elemental Chemistry of Teeth: Implications for Places of Birth, Forced

Migration Patterns, Nutritional Status, and Pollution. *In* The African Burial Ground Project. Skeletal Biology Final Report, vols. 1 and 2. M. L. Blakey, and L. M. Rankin-Hill, eds. Pp. 216–265. Washington, DC: The African Burial Ground Project, Howard University for the United States General Services Administration, Northeast and Caribbean Region.

Goodman, A. H., D. L. Martin, and G. J. Armelagos 1984 Indications of Stress from Bone and Teeth. *In* Paleopathology at the Origins of Agriculture. M. N. Cohen, and G. J. Armelagos, eds. Pp. 13–49. Orlando: Academic Press.

Goodman, A. H., R. B. Thomas, A. C. J. Swedlund, and G. J. Armelagos 1988 Biocultural Perspective on Stress in Prehistoric, Historical, and Contemporary Population Research. Yearbook of Physical Anthropology 31:169–202.

Goveia, E. V. 1965 Slave Society in the British Leeward Islands at the End of the Eighteenth Century. New Haven: Yale University Press.

Gutman, H. G. 1975 Slavery, and the Numbers Game: A Critique of Time on the Cross. Urbana: University of Illinois Press.

Harris H. A. 1931 Lines of Arrested Growth in the Long Bones in Childhood: The Correlation of Histological, and Radiographic Appearances in Clinical and Experimental Conditions. British Journal of Radiology 4:534–640.

Harris, J. E. 1993 Global Dimensions of the African Diaspora. Washington, D C: Howard University Press. Originally published in 1982.

Harrison, I. E. and, Harrison, F. V. eds. 1999 African American Pioneers in Anthropology. Urbana: University of Illinois Press.

Higman, B. 1995 Slave Populations of the British Caribbean 1807–1834. University of the West Indies, Mona: Canoe Press.

Howson, J., W. R. Perry, A. F. C. Holl, and L. G. Bianchi 2006a Relative Dating. *In* New York African Burial Ground Archaeology Final Report, vol. 1. Pp. 105–129. Washington, DC: Howard University for the United States General Services Administration, Northeast and Caribbean Region.

Howson, J., S. Mahoney, and J. L. Woodruff 2006 Pins and Shrouding. *In* New York African Burial Ground Archaeology Final Report, vol. 1. Pp. 288–305. Washington, DC: Howard University for the United States General Services Administration, Northeast and Caribbean Region.

Hsu, F. L. K. 1979 The Cultural Problem of the Cultural Anthropologists. American Anthropologist 81(3):517–532.

Hutchinson, J. 1987 The Age-Sex Structure of the Slave Population in Harris County, Texas: 1850, and 1860. American Journal of Physical Anthropology 74:231–238.

Infant, P. F. 1974 Enamel Hypoplasia in Apache School Children. Ecology Food and Nutrition 2:155–156.

Inikori, J. E., and S. L. Engerman eds. 1992 The Atlantic Slave Trade. Durham: Duke University Press.

Jackson, F. L. C., A. Mayes, M. E. Mack, A. Froment, S. O. Y. Keita, R. A. Kittles et al. 2004 Origins of the New York African Burial Ground Population: Biological Evidence of Geographical, and Macroethnic Affiliations Using Craniometrics, Dental Morphology, and Preliminary Genetic Analyses. *In* The African Burial Ground Project. Skeletal Biology Final Report, vol. 1. M. L. Blakey, and L. M. Rankin-Hill, eds. Pp. 150–215. Washington, DC: The African Burial Ground Project, Howard University for the United States General Services Administration, Northeast and Caribbean Region.

Jea, J. 1998 The Life, History, and Unparalleled Sufferings of John Jea, the African Preacher, Compiled, and Written by Himself. *In* Pioneers of the Black Atlantic: Five Slave Narratives from the Enlightenment, 1772–1815. H. L. Gates, and W. L. Andrews, eds. Pp. 369–439. Washington, DC: Basic Books.

Kelley, J. O., and J. L. Angel 1987 Life Stresses of Slavery. American Journal of Physical Anthropology 74:199–211.

Khudabux, M. R. 1991 Effects of Life Conditions on the Health of a Negro Slave Community in Suriname. The Netherlands: Rijksuniversiteit te Leiden.

Kruger, V. L. 1985 Born to Run: The Slave Family in Early New York, 1626–1827. Ph.D. Dissertation, Columbia University. Ann Arbor: University Microfilms.

Levy, B. M., L. D. Cagnoue, H. V. Anderson, G. Holst, and C. J. Witkop, Jr. 1970 Metabolic Disorders. In Thomas' Oral Pathology, Vol II. R. J. Gorlin, and H. M. Goldman, eds. St. Louis: C. V. Mosby.

Lovejoy, P. E. 1982 The Volume of the Atlantic Slave Trade: A Synthesis. Journal of African History 23:473–501.

Lydon, J. G. 1978 New York and the Slave Trade, 1700, 1774. The William and Mary Quarterly 35:375–394.

Martin, D. L., Magennis, A. L., and Rose, J. C. 1987 Cortical Bone Maintenance in an Historic Afro-American Cemetery Sample from Cedar Grove, Arkansas. American Journal of Physical Anthropology 74:255–264.

Magennis, A. L. 1990 Growth, and Transverse Line Formation in Contemporary Children. Ph.D. Dissertation. University of Massachusetts at Amherst.

McManus, E. J. 1966 A History of Negro Slavery in New York. Syracuse: Syracuse University Press.

Medford, E. G. ed. 2004 The African Burial Ground Project. History Final Report. Washington, DC: The African Burial Ground Project, Howard University for the United States General Services Administration, Northeast, and Caribbean Region.

Moitt, B. ed. 2004 Sugar, Slavery, and Society: Perspectives on the Caribbean, India, the Mascarenes, and the United States. Gainsville: University Press of Florida.

Moynihan, D. P. 1965 The Negro Family in America: The Case for National Action. Washington, DC: Department of Labor, Office of Planning, and Research (March 1965).

Mullin, M. 1995 Africa in America: Slave Acculturation and Resistance in the American South and the British Caribbean, 1736–1831. Champaign: University of Illinois Press.

Nikiforuk, G., and D. Fraser 1981 The Etiology of Enamel Hypoplasias: A Unifying Concept. Journal of Pediatrics 98:888–893.

Null, C. C., M. L. Blakey, K. J. Shujaa, L. M. Rankin-Hill, and S. H. H. Carrington 2004 Osteological Indicators of Infectious Disease and Nutritional Inadequacy. *In* The African Burial Ground Project. Skeletal Biology Final Report, Vols. 1 and 2. M. L. Blakey, and L. M. Rankin-Hill, eds. Pp. 351–402. Washington, DC: The African Burial Ground Project, Howard University for the United States General Services Administration, Northeast, and Caribbean Region.

Owsley, D. W., C. E. Orser, R. W. Mann, P. H. Moore-Jansen, and R. L. Montgomery 1987 Demography, and Pathology of an Urban Slave Population from New Orleans. American Journal of Physical Anthropology 74:185–197.

Perry, W. R., J. Howson, and B. A. Bianco eds. 2006 New York African Burial Ground Archaeology Final Report, Vols. 1–4. Washington, DC: Howard University for the United States General Services Administration, Northeast and Caribbean Region.

Rankin-Hill, L. M. 1997 A Biohistory of 19th-Century Afro-Americans: The Burial Remains of a Philadelphia Cemetery. Westport: Bergin & Garvey.

Rankin-Hill, L. M., M. L. Blakey, J. E. Howson, E. Brown, S. H. H. Carrington, and K. Shujaa 2004 Demographic Overview of the African Burial Ground, and Colonial Africans of New York. *In* The African Burial Ground Project. Skeletal Biology Final Report, Vol. 1. M. L. Blakey, and L. M. Rankin-Hill, eds. Pp. 266–304. Washington, DC: The African Burial Ground Project, Howard University for the United States General Services Administration, Northeast and Caribbean Region.

Rathbun, T. A. 1987 Health and Disease at a South Carolina Plantation: 1840–1870. American Journal of Physical Anthropology 74:239–253.

Rose, J. C. ed. 1985 Gone to a Better Land: A Biohistory of a Rural Black Cemetery in the Post- Reconstruction South. Arkansas Archaeological Survey Research Series No. 25, Arkansas Archaeological Survey, Fayetteville.

Sarnat, B. G., and I. Schour 1941 Enamel Hypoplasia (Chronic Enamel Aplasia) in Relation to Systemic Disease: A Chronologic Morphologic, and Etiologic Classification. Journal of the American Dental Association 28:1989–2000.

Singleton, T. A., and Bograd, M. 1995 The Archaeology of the African Diaspora in the Americas. Historical Archaeology, Guides to the Archaeological Literature of the Immigrant Experience in America. No. 2. Tucson, AZ: The Society of Historical Archaeology.

Scott, E. C. 1979 Dental Wear Scoring Technique. Journal of Physical Anthropology 51:213–218.

Shaw, J. H., and Sweeney, E. A. 1973 Nutrition in Relation to Dental Medicine. *In* Modern Nutrition in Health, and Disease Dietotherapy. R. S. Goodhard, and M. E. Shils, eds. Pp. 733–768. Philadelphia: Lea & Febiger.

Smith, B. H. 1984 Patterns of Molar Wear in Hunter-Gathers and Agriculturalists. American Journal of Physical Anthropology 63:39–56.

Steckel, R. H. 1986 A Dreadful Childhood: The Excess Mortality of American Slaves. Social Science History 10:427–467.

Stewart, T. D. 1939 Negro Skeletal Remains from Indian Sites in the West Indies. Man 39:49–51.

Sweeney, E. A., J. Cabrera, J. Urrutis, and L. Mata 1969 Factors Associated with Linear Hypoplasia of Human Deciduous Incisors. Journal of Dental Research 68:1275–1279.

Thomas, D. H., 2000 Skull Wars: Kennewick Man, Archaeology, and the Battle for Native American Identity. New York: Basic Books.

Thomas, D. H., S. South, and C. S. Larson 1977 Rich Man, Poor Man: Observations on Three Antebellum Burials from the Georgia Coast. Anthropological Papers of the American Museum of Natural History 54:393–420.

Wienker, C. W. 1987 Admixture in a Biologically African Caste of Black Americans. American Journal of Physical Anthropology 74:265–273.

Wilczak, C., R. Watkins, C. Null, and M. L. Blakey 2004 Skeletal Indicators of Work: Musculoskeletal, Arthritic, and Traumatic Effects. *In* The African Burial Ground Project. Skeletal Biology Final Report, Vols. 1 and 2. M. L. Blakey, and L. M. Rankin-Hill, eds. Pp. 403–460. Washington, DC: The African Burial Ground Project, Howard University for the United States General Services Administration, Northeast and Caribbean Region.

Willis, W. S. Jr. 1999 Skeletons in the Anthropological Closet. *In* Reinventing Anthropology. D. Hymes, ed. Pp. 121–151. Ann Arbor: University of Michigan Press.

9

The Bioarchaeology of Leprosy and Tuberculosis

A Comparative Study of Perceptions, Stigma, Diagnosis, and Treatment

Charlotte Roberts

Introduction: Identity and Health

Identity can be defined as: "the state of having unique identifying characteristics held by no other person or thing … the individual characteristics by which a person or thing is recognized" (Hanks 1979). While, as Insoll (2007:1) states, in the media "many of the stories are concerned with, essentially, the struggle of identity manifestations for a voice or for power, be they ethnic, religious, sexual, or related to disability …", it is a term that only seems to have entered the bioarchaeological literature recently. Sociologists appear to have been concerned with questions about "identity" for a long time (e.g., see Bernard et al. 2004; Caplan 1997) although, as bioarchaeologists, we are constantly exploring questions about people's health from an "identity" point of view. Data for health problems experienced by past human groups is linked (or should be) to their biological sex, gender, age at death, religious attitudes, ethnicity, and socioeconomic status; multiple lines of evidence are used to explore all these variables ranging from human remains, through historical documents and iconography, to ethnographic and archaeological data. These many variables make up the person's and population's collective" identity," and takes into account differences there might be in individual and community identity. It is also recognized that the identity of a person and community is "dynamic,

Social Bioarchaeology. Edited by Sabrina C. Agarwal and Bonnie A. Glencross
© 2011 Blackwell Publishing Ltd

reshaping and configuring" (Larsen 2009:xiii), and that identity will change over time and will likely be different in contrasting regions of the same area of the world or different areas of the world. Further, Knudson and Stojanowski (2009:1,5) suggest that identity refers to how people perceive themselves and their relationship with "larger social phenomena that characterize their existence" and that identities research is focused on "about who they thought they were, how they advertised this identity to others, and how others perceived it, and the resulting repercussions of this matrix of interpersonal and intersocietal relationships". They argue that bioarchaeology has contributed very little to the study of identity. Buikstra and Scott (2009) also emphasize the delay in bioarchaeologists focusing on identity and suggest that, when bioarchaeologists do consider "identity," they only tend to look at one or two aspects, which might be age and/or sex and/or ethnicity, and they feel little work has explored identity and disease experience in the past.

There have also been criticisms that identity is not a consideration for people working the reconstruction of past patterns of health and disease because of the "medical" orientation of many scholars in the field of paleopathology (e.g., see Fay 2006). This has particularly been true for people working in the United Kingdom and, to a certain extent, has been accepted, but not for the last 20 years. Gowland and Knüsel (2006) also wonder why reports on the study of human remains from archaeological sites were marginalized to appendices or microfiches, until recently. It has to be remembered that paleopathology has a very different development in the United Kingdom when compared to other countries, and specifically the United States (e.g., see Roberts 2006 for an overview). As it is mostly in Departments of Archaeology where paleopathology is taught currently (rather than Anthropology Departments), and because specific undergraduate and master's level courses have only been recent in development (mainly from the early 1990s when bioarchaeologists started to be employed within those departments), it is also only recently in the United Kingdom where more contextualized approaches to health and disease have been made because the backgrounds of those working in the field have changed from medicine/anatomy/dentistry to archaeology and/or anthropology. Fay (2006) highlights further the problems she perceives with the medical background of people working in paleopathology. Identifying that disease has two forms, fact (paleopathological lesions) and conceptual (language and discourse, disease culture), she suggests that the latter leaves no "corporeal residue, and is thought to be invisible to the palaeopathologist or funerary archaeologists" (2006:190). She continues with, "This kind of interpretation is thought to be fundamentally different to that practised by other scholars (including paleopathologists) who analyse the *material* manifestations of illness." It is agreed that the "conceptual" part of disease is indeed invisible to a paleopathologist whose primary data are pathological changes to human remains, but the paleopathologist utilizes multiple lines of evidence to reconstruct the "conceptual" (or should do) – see examples from the author's work: Roberts 1991, 2000, 2007, Roberts and Buikstra 2003, Roberts and Cox 2003, Roberts and McKinley 2003. If "paleopathologists en masse have not addressed how disease as a conceptual structure was understood in the past" (Fay 2006:192) then the author feels that it is time to retire! One could be forgiven for thinking that paleopathologists were the only culprits who did not consider other data in reconstructing past community and individual experience of disease; this is far from the case. If one uses the analogy of medical practitioners, one can argue that treating the whole person is often ignored in favor of treating the disease from which the person suffers. However, if one takes the World Health Organization's definition of health (1948), it is not merely the absence of disease or infirmity but "a state of complete physical, mental and social well-being" (www.who.int/about/definition/en/print.html); taking an holistic approach to people with ill health today and in the past is essential in order to understand the impact of disease on individual people and the communities in which they lived. Nevertheless, medical practitioners do indeed admit that dealing with social aspects of disease is "much less comprehensible than cellular and biochemical processes" (Rafferty 2005:120), with Kearns and Gesler (1998:9) maintaining that, "The dominant model [in medicine] has a tendency to (mis) represent the *persons* who experience illness (or seek health) as *patients*." Every ill person today and in the past will experience illness differently to the next person and medical practitioners and bioarchaeologists must appreciate these individual experiences. Along with this, paleopathologists in particular could help the discipline move forward by refraining from using terms such as "case of" and "specimen" and treat the remains they study as those of a once living person rather than a curiosity. This tendency in part has developed because of the, until recently, domination of paleopathology with medically qualified practitioners.

However, that said about the United Kingdom, one only has to read the pages of the *American Journal of Physical Anthropology* or the *International Journal of Osteoarchaeology* to see the many papers on past identity and health; the missing keyword, however, has been "identity". The question is: Do we need to use the word identity to describe what we already consider with respect to past health? Related to this, it is rather telling to see Buikstra and Scott's (2009:43–47) need to define terms that are being used when referring to "identity" in sociological, anthropological, and archaeological contexts ("agency," "embodiment," "person," "personhood," and "selfhood"), which suggests that, while bioarchaeologists do not use these terms necessarily in their published research, particularly with respect to paleopathology, they do actually "deal" with them even though they do not explicitly say so. Cross (2007:190–191) when talking about disability and archaeology, also laments that, "The failure to engage with disability politics suggests a lack of awareness of the disability movement … In archaeology there appears to be no concept of 'disability' according to the social model." One wonders whether the right literature has been accessed.

The preceding discussion about the study of "identity" in the bioarchaeology of ill health is highly relevant to the subject matter of this chapter, infectious diseases in medieval England, i.e., the perceptions of those diseases by individuals experiencing them and communities in which they lived, their associated stigma, and the impact of those variables on diagnosis and treatment. To satisfy the critics of paleopathology, this chapter will indeed utilize the term "identity," and the author will continue to understand that, as Fowler (2004:1) states, "Individuality is (was) extremely important to us." Added to this, how we perceive ourselves within our communities and how communities perceive us are equally important in constructing "identity"; this also applies to the past, and in both contexts identity is inherently socially mediated and may be chosen by, or imposed on, individuals and communities of people.

Identity may be considered relevant to consideration of people with leprosy and tuberculosis (TB) in the past (and present). Leprosy and tuberculosis are closely related mycobacterial infections associated today with many factors, including poverty, poor diets, compromised immune systems, urban living, high population density, poor access to health care, and resistance to drugs (www.who.int/topics/tuberculosis/en/; accessed August 2009). In the past, many of these factors would have been important for these infections to flourish. In England they both start to increase in frequency from the 12th century A.D. (Roberts and Manchester 2005), with tuberculosis continuing to make its mark well into the post-medieval period (Roberts and Cox 2003). With an emphasis on the late medieval period (12th–16th centuries A.D.), this chapter aims to make a comparative study of these two infectious diseases that share the same genus, from the standpoint of people's perceptions of what it meant to have tuberculosis and leprosy in past Britain, what the nature was, if any, of the stigma attached to them, and how they were both diagnosed and treated in relation to perceptions and stigma. Using clinical, bioarchaeological, archaeological, and historical evidence the similar characteristics of the two infections will be discussed.

Leprosy and Tuberculosis: Clinical Overview of Mycobacterial Disease

In order to understand how people and communities experience health problems and how those diseases are perceived within communities, both past and present, an essential starting point is to look at those diseases from a clinical standpoint. The author's nursing background has provided a clear and deep understanding that the same disease in different people and populations will evoke different experiences according to many variables, including sex, age, status, ethnicity, and religion (e.g., see Roberts 2000 on disability). How are these infectious diseases, leprosy and tuberculosis, contracted, how do they affect people today (signs and symptoms), and how might they be recognized in the past through studying human remains from archaeological sites?

The bacterial organisms that cause these two infections are closely related. Caused by *Mycobacteria*, leprosy is the result of infection by *Mycobacterium leprae* and tuberculosis (in humans) by *Mycobacterium tuberculosis* (human form) and *bovis* (animal form). As they are both caused by *Mycobacteria*, they are immunologically related; the organisms causing them have the same genus (*Mycobacterium*) but are different species (*leprae* versus *tuberculosis*/*bovis*). The most likely route of transmission for *M. leprae* is suggested to be via inhalation of droplets into the lungs containing the bacteria, that for *M. tuberculosis* being the same, while *M. bovis* is primarily transmitted to humans via meat and milk from infected animals or via swallowed infected sputum (www.who.int/topics/en/).

Leprosy today: effects on the body, stigma, perceptions, frequency, diagnosis, and treatment

> Leprosy occupies a special position among communicable diseases because of the long duration of the disease, the frequency of disabilities and the social and economic consequences it endangers (Dayal et al. 1990:170).

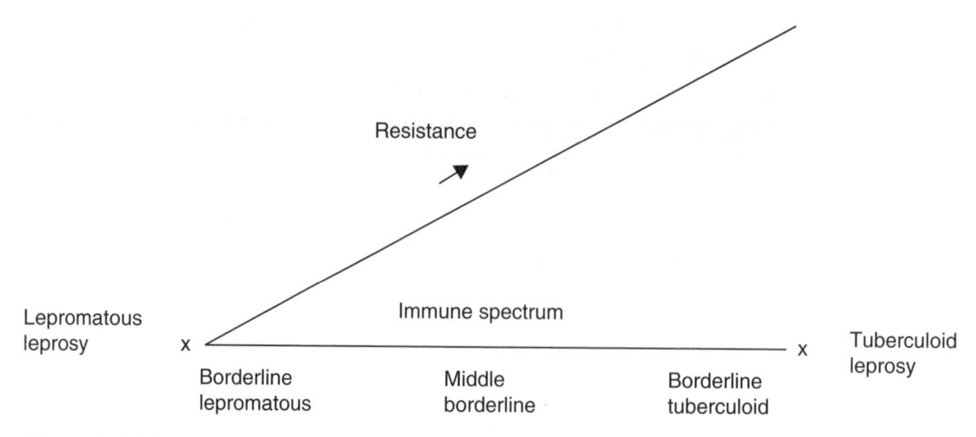

Figure 9.1 The immune spectrum of leprosy after Ridley and Jopling (1966).
Source: Redrawn by Yvonne Beadnell, Durham University

For leprosy, the bacteria mainly affect the peripheral nerves (motor, sensory, and autonomic), leading to paralysis of muscle groups in the limbs and loss of sensation in the hands and feet, particularly, but the skin, eyes, skeleton, kidneys, liver, adrenal glands, and testes can also be involved (Jopling 1982). Depending on the patient's cell-mediated immunity, there could be edema of the lower legs (abnormal accumulation of fluid due to increased permeability of skin capillaries caused by the bacilli in them, damage to autonomic fibers in the skin nerves, and gravity), thinning of the eyebrows, broadening of the nose, deepening of the forehead lines, possible loss of eyelashes, nasal collapse, loss of the incisor teeth, hoarseness of the voice, hardness and ulceration of the legs, ulcers on the soles of the feet, and shortening of the fingers and toes due to bone damage (Jopling 1982). The "type" of leprosy a person develops depends on their immune system strength (Figure 9.1); at each end of the spectrum there is low resistance or lepromatous leprosy (LL) or the type mainly seen in bioarchaeology in skeletal remains, and high resistance or tuberculoid leprosy (TL) which may or may not affect the skeleton (see below).

People with leprosy today may become impaired and disabled, experiencing loss of what is expected to be the normal function of the body. This is particularly seen in the hands and feet where soft tissue and bone damage can lead to deformity and disability, and even handicap (Figure 9.2). Disability can be defined as, "the condition of being unable to perform a task or function because of a physical or mental impairment," whilst disable is defined as "to make ineffective, unfit or incapable ...," and impair can be defined as, "to reduce or weaken in strength, quality" (Hanks 1979); finally, a handicap is, "the social disadvantage that results from an impairment or disability" (Covey 1988:3). Impairments can lead to disabilities and handicaps are the final result. For example, Srinivasan (1993), talking about prevention of disability in leprosy patients, described unemployment, economic and physical dependence, and lack of social integration as handicaps experienced by disability caused by leprosy. However, as Kopparty states (1995:240), "The meaning of handicap changes as per given social situation," and illustrates the impact of the caste system in India on social inequality and acceptance of people with leprosy.

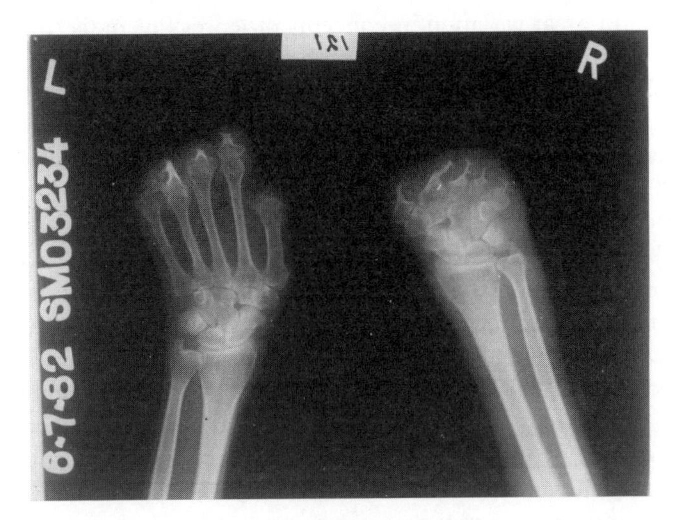

Figure 9.2 Radiograph of a person's hands affected by the bone changes of leprosy.
Source: Johs Anderson, with permission of Keith Manchester

Interestingly, the majority of people with leprosy-associated deformity but no handicap are accepted by the majority of their families regardless of caste. However, of those with a handicap, the higher caste groups were more accepting than the lower caste groups. The definitions described above are highly relevant to what we, as bioarchaeologists, try to impose on the past when trying to recognize and interpret the evidence for impairment in skeletal remains and possible disability.

It is also common for leprosy patients to develop complexes, neurotic symptoms, and psychotic reactions. The most probable reason is social stigma. Ranjit and Verghese (1980), in their study of two households in southern India, where a leprosy control program had been implemented, assessed mental health status and found that mental health was worse in those who had physical deformity of limbs as a result of leprosy. Tsutsumi et al. (2007), in their study in Bangladesh, also found that there was significantly worse mental health for leprous people and that 50 percent of leprosy patients in their study group had perceived stigma, which contributed to a worse "Quality of Life Score." When sex differences in leprosy experience are considered, Cakiner et al. (1993) found that women in Turkey bore not only social, cultural, and economic problems, but those with leprosy also had physical and social consequences of the infection to bear. Morrison (2000) also found in her study of African and Indian women with leprosy that, because they are deemed of lower status, they are less aware of the disease, do not access health services much, and are more stigmatized. With regard to stigma and leprosy today, stigma is often related to people's concepts of causation, and this varies around the world. For example, El-Hassan et al. (2002) studied sociocultural aspects of leprosy in two communities, the Masalit and the Hawsa, in the Eastern Sudan. The Masalit believed that leprosy was caused by eating the meat of wild pig and certain fish, and the Hawsa thought it was caused by eating two types of fish. Both groups did not know leprosy was treatable and turned to spiritual healers for help. Kakar

(1995) also found great variation in concepts of leprosy in India. In some villages in Tamil Nadu, for example, contact with an earthworm is believed to cause leprosy, the blunt ends of a worm likened to stumps of fingers in leprosy. It is clear that stigma, along with traditional medical cures for leprosy, are two of the factors that are affecting leprosy control in some parts of the world (Jacob and Franco-Paredes 2008). For example, in India the traditional medical systems are very popular partly because of the stigma linked to government-run leprosy clinics.

Compared to tuberculosis, leprosy is seen as very much a declining infectious disease. According to official reports received by the World Health Organization during 2008 from 118 countries/territories, the registered prevalence of leprosy in the world at the beginning of 2008 was 212,802 cases, and the number of new cases detected during 2007 was 254,525 (excluding the small number of cases in Europe). This still represents a considerable burden, but the number of new cases detected globally fell by 11,100 cases (a 4 percent decrease) during 2007 compared with 2006 (www.who.int/lep/en/). Most previously highly endemic countries have now reached elimination (a registered prevalence rate of <1 case/10,000 population). The few countries that remain affected are very close to eliminating the disease, but there are areas of high endemicity in Angola, Brazil, Central African Republic, Democratic Republic of Congo, India, Madagascar, Mozambique, Nepal, and the United Republic of Tanzania. Frequency rates have declined because public health workers have worked hard to find, diagnose, educate, and treat people with leprosy, and also educate the general population with the facts of this still very stigmatized, infection. Risk factors for leprosy are many (Table 9.1 and Barkataki et al. 2006). Diagnosis is usually via biopsy of skin lesions and slit skin smears to identify bacilli, and clinical examination, for example to identify skin lesions and enlarged peripheral nerves. Antibiotics are used to treat the infection, preferably at an early stage in the infection to prevent disability, usually as a "cocktail" (MDT or multi-drug therapy) so that if resistance to one antibiotic develops then the others will hopefully be effective; dapsone, rifampicin, and clofazimine are administered for six months for TL and for 12 months for LL (Scollard et al. 2006). Vaccines have also been used for preventing leprosy, particularly the BCG vaccine for tuberculosis, but its efficacy varies around the world (Scollard et al. 2006). In Britain today few people are diagnosed with leprosy. For example Jopling (1982) notes that 1,229 people were treated in England for leprosy between 1951 and 1989, and Gill et al. (2005) record 50 patients presenting at the Liverpool School of Tropical Medicine between 1946 and 2003, with 64 percent of people having been born in the Indian subcontinent.

Table 9.1 Risk factors in leprosy

Risk factors
Poverty
Poor diet
Overcrowding and high population density
Lack of education
Lack of access to health care
Resistance to drugs
Poor compliance with treatment

However, even if leprosy is decreasing, there are still thousands of people who continue to suffer the debilitating physical, social, and psychological consequences of leprosy. In 2005, the Centers for Disease Control and Prevention (CDC) reported that 1–2 million people worldwide were permanently disabled as a result of leprosy, and the World Health Organization (2002) estimated that 177,000 DALYs (disability adjusted life years) were lost due to leprosy in 2001. Meima et al. (2004) suggest that the consequences of leprosy will be around for many years into the future.

Tuberculosis today: effects on the body, stigma, perceptions, frequency, diagnosis, and treatment

> It must be a matter of concern that 20 years after it was realized that tuberculosis was out of control across much of the developing world, the tide of tuberculosis shows no sign of being controlled. As a race, human kind is losing the fight against tuberculosis (Davies 2008:xv).

The causative bacteria for tuberculosis will either enter the lungs or the gastrointestinal tract of the human population. Called primary tuberculosis, at this point, today the infection would normally be diagnosed and treated if it is diagnosed; this will of course be determined by access to health care facilities and this will vary according to where a person lives in the world. However, primary tuberculosis can be harbored for many years with no external signs, and then it can be "reactivated" if a person's health status changes; for example if they experience a poor diet or other disease that compromises their immune system strength, they may be more susceptible to reactivation of their primary tuberculosis, or re-infection with tuberculosis to that which they are exposed (Gordon and Mwandumba 2008).

Signs and symptoms for tuberculosis will of course vary according to where the infection first manifests itself. In pulmonary tuberculosis shortness of breath, cough, chest pain, and coughing up infected phlegm or even blood, loss of weight, and fever are common (Roberts and Buikstra 2003:20). Alternatively, in gastrointestinal tuberculosis, the signs and symptoms include abdominal pain and distension of the abdomen, loss of blood from the tract, fever, and weight loss (Ormerod 2008). However, other general signs and symptoms for people with tuberculosis include anemia and pallor, skin involvement (*lupus vulgaris*), weakness, and loss of appetite (Roberts and Buikstra 2003). If not treated then the bacteria will spread through the body via the blood and lymphatic systems and potentially lodge themselves in the bones of the skeleton with subsequent destruction of bone tissue; this is termed secondary or post-primary tuberculosis (Gordon and Mwandumba 2008).

Today stigma can be associated with tuberculosis in some societies (e.g., see Dodor et al. 2008 on Ghana). Of course the presence of disease-related stigma depends on that particular community's concepts of disease occurrence, although Mak et al. (2006) suggest, for their study of 3,011 Hong Kong Chinese adults, that knowledge of infectious diseases such as tuberculosis, HIV/AIDS, and SARS (severe acute respiratory syndrome) did not significantly affect stigma. There is for many cultures today a link between tuberculosis and stigma and this may be correlated

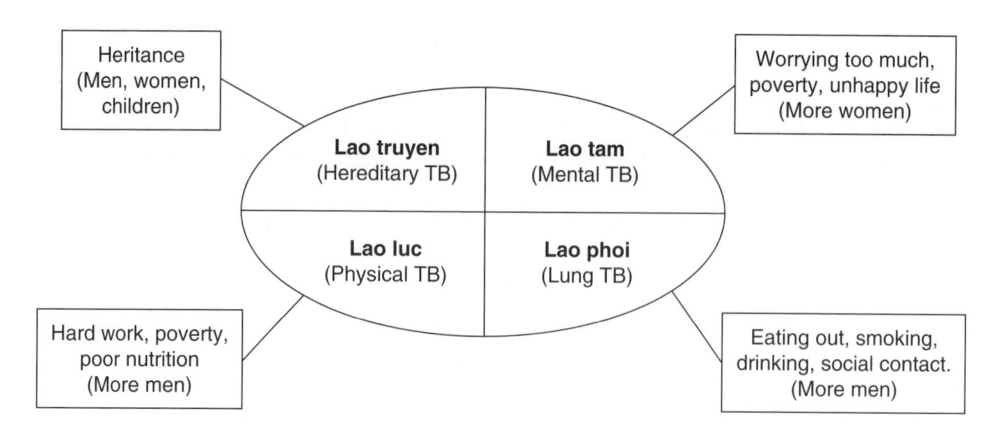

Figure 9.3 Vietnamese beliefs about tuberculosis.
Source: Redrawn after Long, N. H., E. Johanasson, V. K. Diwan, and A. Winkvist 1999 Different Tuberculosis in Men and Women: Beliefs from Focus Groups in Vietnam. Social Science and Medicine 49:815–822

with socially or morally unacceptable behavior (Rangan and Uplekar 1999). It is therefore the case that, for some, an alternative diagnosis is sought, and some doctors avoid diagnosing tuberculosis because of its consequences, such as loss of employment and decreased prospects of marriage (Snider et al. 1994). A study of people with tuberculosis in Pakistan found that people with tuberculosis kept their diagnoses to themselves because they thought that it was not a curable disease, and that the attached stigma would affect their future prospects (Khan et al. 2000). Indeed, even following treatment for tuberculosis, patients may still feel socially stigmatized, as seen in a study of 610 people with tuberculosis in southern India (Rajeswari et al. 2004). Stigma will affect people differently, and females in some countries are stigmatized more often than men with tuberculosis (Hudelson 1999). This is compounded by the stigma attached to tuberculosis and the human immunodeficiency virus (HIV) in many countries, but especially sub-Saharan Africa (Harries and Zachariah 2008). Thus, a population's perceptions of tuberculosis in general will affect the experiences of those people who have a specific disease. For example, Carey et al. (1997), studying New York State Vietnamese refugees' beliefs about the causes of tuberculosis, found that they listed a variety of possible factors, ranging from hard manual labor to lack of sleep. In another study of perceptions and beliefs of tuberculosis, Vietnamese people in Vietnam were studied to explore tuberculosis and its risk factors, and the differences between males and females (Long et al. 1999). While they had a good knowledge of tuberculosis being infectious and contagious, there were traditional beliefs intermingled with this good knowledge, with four types of tuberculosis being recognized (Figure 9.3).

However, because a person with tuberculosis may not, like those suffering from leprosy, display disfiguring external signs, they may not necessarily be stigmatized as much. Nevertheless, if they suffered skin tuberculosis as part of their infection, this might make them more "visible" to their community. Furthermore, skeletal damage to the spine (Figure 9.4) and/or major weight-bearing joints of the body

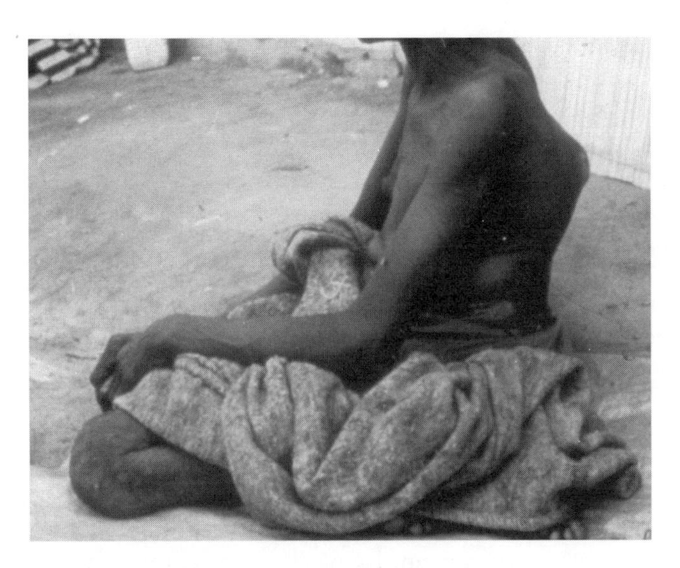

Figure 9.4 Woman with tuberculosis affecting the spine.
Source: Peter Davies

will both make them appear "different" to their peers, and affect their normal mobility. One could argue that if people with tuberculosis live within a community then they may not be singled out for special treatment if they do not look "different" or are not behaving contrary to the rest of the group. In more advanced tuberculosis, people might be more restricted in what they can do, and thus might be more stigmatized later in the course of the infection. Thus, for those diseases that are more invisible, "labeling" the disease and development of associated stigma may not be an issue, although suspicion of a disease without hard evidence can equally be a real problem. In the case of a person with the HIV and AIDS, there may not be any external signs of their condition and yet, once diagnosed and known to their social group, stigma may develop. The same can be said for tuberculosis and therefore the recognition and stigmatization of somebody with this infection may not be as forthcoming or easy as for somebody with a disease like leprosy.

Tuberculosis is a re-emerging infection today and kills around 2 million people each year. In 2007, there were an estimated 9.27 million incident cases of tuberculosis, with 55 percent occurring in Asia and 31 percent in sub-Saharan Africa (World Health Organization 2009); 1.3 million people died of tuberculosis without associated HIV, plus another half a million with HIV. The 9.27 million can be compared with 9.24 million in 2006, 8.3 million in 2000, and 6.6 million in 1990. All six World Health Organization areas currently show the annual number of new cases is in decline, or stable. Problems contributing to this picture include a lack of the right staff in public health programs, insufficient laboratories and resources to make the diagnoses rapidly and effectively, and lack of drugs, ultimately meaning that people with extensively drug-resistant tuberculosis (EDR-TB or XDR-TB) are not necessarily diagnosed and treated (www.who.int/tb/challenges/xdr/en/index.html; accessed August 2009). As with leprosy, risk factors are many

Table 9.2 Risk factors in tuberculosis

Poverty
Poor diet
Overcrowding and high population density
TB-contaminated foods
Particular occupations
Migration and travel
HIV
Lack of education
Lack of access to health care
Resistance to drugs

(Table 9.2). Diagnosis is by clinical examination, radiography, and sputum analysis (Schluger 2008), and treatment is with a "cocktail" of antibiotics, as for leprosy. The antibiotics used are rifampicin, isoniazid, pyrazinamide, and ethambutol for two months and then rifampicin and isoniazid for a further four months (Gordon and Mwandumba 2008). The drugs are administered via DOTS (directly observed therapy short course). In Britain Anderson et al. (2007) note that London accounts for 50 percent of the national burden of tuberculosis and its incidence has doubled in the last 15 years; 75 percent of people with tuberculosis in London were born abroad.

The preceding dialogue has now "set the scene" for exploring leprosy and tuberculosis in late medieval England such that their impact on the population can be understood. We will now consider these mycobacterial infections from the standpoint of evidence of leprosy, perceptions, stigma, diagnosis, and treatment.

Leprosy in Late Medieval England

Skeletal and historical data

Leprosy is well documented in late medieval England in the archaeological and historical record. However, this does not mean that it was not a disease that the English population had to deal with before the 11th century A.D. because the first skeletal evidence for leprosy derives from the early medieval period (mid 5th–mid 11th centuries A.D.) (Roberts 2002). This time period was characterized primarily by rural settlements, but later in the period *buhrs* or forts/fortified towns developed to provide protection against Viking inroads. Leprosy today is a rural disease so it would be expected that leprosy might have appeared in England at this time, as it does. The skeletal evidence is mostly located in cemetery sites in the south, east and southwest of the country, and the earliest dates from the 7th century A.D. In the late medieval period (mid 11th–mid 16th centuries A.D.), there is an increase in individuals excavated with bone changes of leprosy, and specific institutions were founded for segregating people with the infection. Increased population density associated with urbanism likely allowed the bacteria to be transmitted by droplet spread more easily, and the cemetery sites derive from both rural and urban environments.

Recognizing leprosy in skeletal remains relies on characteristic bacterial-induced changes to the bones of the face, hands, feet, and lower legs. Described initially by Møller-Christensen (1961) and developed by Andersen and Manchester (1987, 1988, 1992; Andersen et al. 1992, 1994), the diagnostic criteria are based on the *M. leprae*'s direct effects on the facial bones, and its indirect effect on the sensory, motor, and autonomic peripheral nerves. Absorption of the maxillary alveolar process, absorption and remodeling of the anterior nasal spine and nasal aperture, pitting and even perforation of the palate, osteomyelitis and septic arthritis of hand and foot bones, sometimes with subluxation or dislocation, palmar/plantar grooves of the hand and foot phalanges reflecting paralysis of muscles, dorsal tarsal bars ("drop foot"), carpal and tarsal disintegration with concentric diaphyseal remodeling of the metacarpals, metatarsals and phalanges, knife edge remodeling of the metatarsals, and periosteal new bone formation on the tibiae and fibulae are all characteristic bone changes that, in various combinations, are accepted as being diagnostic of leprosy. Bone changes are bilateral and symmetrical in lepromatous leprosy (LL) and minimal or non-existent in tuberculoid leprosy (TL). In the latter, unilateral changes are the norm, or if bilateral they tend to be asymmetrical (Lee and Manchester 2008). The majority of bioarchaeologists use macroscopic methods of diagnosis, although histological methods (e.g., Blondiaux et al. 1994) and ancient pathogen DNA analysis (e.g., Taylor et al. 2006) have been used.

Identification of skeletal remains with bone changes of leprosy in late medieval English cemetery contexts has revealed a wide distribution pattern at this time. These range from cemetery sites in the south (southeast, southwest, and central areas), east (Norfolk, Suffolk, Nottinghamshire, Lincolnshire, and Cambridgeshire), and north (Yorkshire and Cheshire) (Roberts 2002). Of particular interest is that the majority of burials with bone changes of leprosy derive from cemeteries that do not suggest that there was any different treatment of them in death. Cemetery contexts include parish churchyards, a cathedral burial ground, and a priory chapterhouse, although there are some late medieval hospital cemetery sites that have revealed leprous skeletons. These latter sites, however, are by no means all related to leprosy hospitals. For example, to the east of England there are two sites in Norfolk, two sites in Suffolk, and one each in Nottinghamshire, Lincolnshire, and Cambridgeshire where skeletons with bone changes of leprosy have been excavated; five of the seven were hospital sites. Of the hospital sites, at Grantham, Lincolnshire three individuals were affected (Boulter 1992; Hadley 2001), and at Spittal's Link (St Margaret's), Huntingdon, Cambridgeshire there was one skeleton with clear evidence of leprosy with another 13 individuals with periosteal new bone on lower leg bones (Duhig 1993); this could be suggestive of leprosy, although there are many underlying causes to this condition. At South Acre, Norfolk, Wells (1967) describes four skeletons dated to between A.D. 1100 and A.D. 1350 with accepted bone changes of leprosy in a cemetery probably associated with the leprosy hospital of Rachneise. Bishop (1983) and Manchester and Roberts (1986) also note skeletons dated to between A.D. 1133/4 and A.D. 1642 with leprosy at the cemetery of St. Leonard Hospital, Newark, Nottinghamshire. Finally, Anderson (1998) describes a number of skeletons from St. John, Timberhill, Norwich, Norfolk,

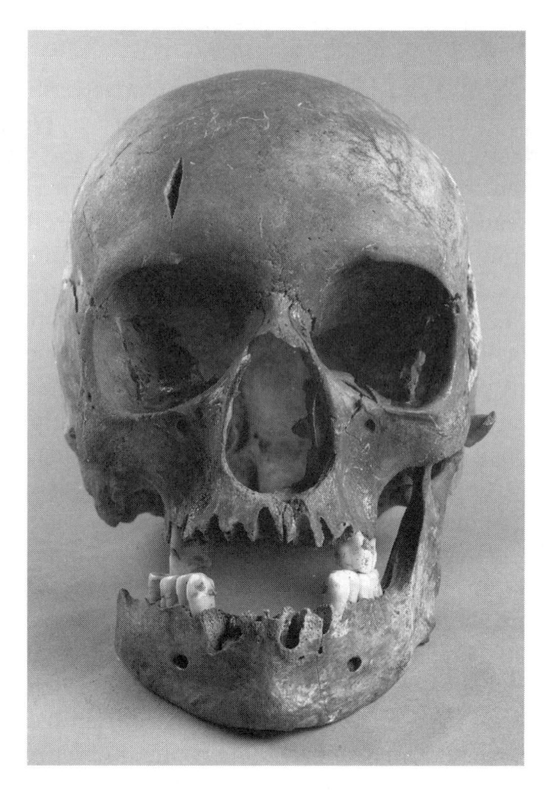

Figure 9.5 Anterior view of the face of a woman who suffered leprosy during life and buried in St. Stephen's churchyard, York.
Source: York Archaeological Trust

including juveniles, with leprosy dated to the 12th–14th/15th centuries A.D.; this cemetery may have served as a leprosy hospital in Norwich.

In the north of England, there is greater evidence for leprosy than in previous periods. Six cemetery sites reveal evidence of leprosy and two are believed to be associated with a hospital (St. Giles by Brompton Bridge in North Yorkshire dated to the late 12th–late 14th centuries A.D. – Chundun and Roberts 1995; and at Doncaster in South Yorkshire – Bayley and Henderson 1989). There are also two other sites in North Yorkshire, in York, where leprosy has been noted in skeletons (the churchyards of Fishergate House (Holst) and St. Stephen's Church, Fishergate); at the latter site a female leprous individual was buried on the margins of the cemetery (McComish, and Figure 9.5). All the sites in the north of England with evidence of leprosy are rural in context apart from those in York, unlike in the south where there are more urban sites. However, at Wharram Percy, North Yorkshire, Mays (2007) describes a juvenile individual, dated to A.D. 960–1000, with leprosy, and confirmed by Taylor et al. (2006) using ancient DNA analysis. Finally, a young male with leprosy was buried in a sandstone coffin under the chapterhouse of Norton Priory, Runcorn, Cheshire in the early 14th century (Manchester and Roberts 1986; Brown and Davis 2008), suggesting he was of high status.

Returning to the south–central part of England and the large late medieval hospital site of St. James and St. Mary Magdalene, Chichester, Sussex (Magilton et al. 2008), 75 adults were identified with leprous bone changes, although many more may have had the infection but show less specific bone changes (Lee and Manchester 2008). The earliest reference to the hospital was a confirmation charter of 1187 which suggests it was founded before A.D. 1118, originally for eight people with leprosy. The last reference to it being a place for people with leprosy was in a will of A.D. 1418; it was then used as a more general hospital/almshouse until the end of the 17th century (post-medieval period), reflecting the function of many (initially) leprosy hospitals at that time. As leprosy declined, hospital function changed. This is the most significant late medieval (definite) leprosy hospital cemetery with leprous skeletons excavated of the four in England known to the author, the rest only containing a few individuals with leprous bone changes.

Overall, late medieval England sees the greatest number of individual skeletons excavated from cemeteries with evidence of leprosy. However, the numbers are relatively low, as now supported by historical research (Rawcliffe 2006:154). Several individuals were associated with hospitals, although most individuals were buried in non-leprosy hospital cemeteries. Dating evidence suggests that most people who died with leprosy in this period are dated prior to the 15th century. Leprosy apparently declines from the 14th century A.D. onwards in England (Manchester 1991), and the skeletal evidence seems to be supportive. It is hypothesized that cross-immunity between leprosy and tuberculosis, increased exposure to tuberculosis with urbanism starting in the 11th–12th centuries led to people becoming tuberculin positive and immune to leprosy. Later in time, people would then be more likely to contract tuberculoid leprosy rather than lepromatous leprosy and decreased infectivity in leprosy led to a decline in the infection.

Nevertheless, any bioarchaeologist is well aware of Wood et al.'s (1992) "osteological paradox"; what the skeletal record for any one disease shows is rather a reflection of the "tip of the iceberg" of the health problem. Many skeletons will not show any bone changes of leprosy even though the person may have been suffering the infection, and the person may have contracted the high resistant form of leprosy (TL) which meant that bone changes may not have developed or may have been more difficult to diagnose as leprous. It should also be noted that leprosy affects the skeleton in only about 3–5 percent of people (Paterson and Rad 1961). Therefore, identifying skeletons with leprosy from archaeological sites is not straightforward. The lack of evidence from post-16th century A.D. in England may reflect that leprosy was in fact declining and not many people contracted it, that people had leprosy but it was TL and did not necessarily lead to clear-cut bone changes, or that the person died without developing any bone changes. Currently, skeletal evidence from the post-medieval period in Britain (A.D. 1550–1850) is restricted to two sites, from the churchyards of St. Augustine the Less in Bristol, Avon, southwest England (O'Connell 1998) and St. Marylebone Church in London (Walker 2009).

Perceptions, stigma, diagnosis, and treatment

That most skeletons with leprosy currently excavated from sites in late medieval England were not buried in leprosy hospital cemeteries (or any differently to the rest of the burials) is surprising when the historical literature for associated stigma in this period is considered. One could argue that this anomaly could be explained by one or more of the following: few leprosy hospitals have been excavated in England and thus the sample is biased; people with TL would not be recognized and thus not institutionalized; perhaps people with leprosy were accepted into society more than has been believed.

It has already been noted for living people with leprosy today that different cultures and countries, and groups within, can view people with leprosy very differently, depending on their medical and other belief systems. As many authors are now appreciating, "leprosy did, indeed, inspire fear and loathing in medieval England. But it was also regarded as a mark of election, akin to a religious calling, and did not automatically lead either to segregation or vilification" (Rawcliffe 2006:43) and "... responses ... were just as varied as the explanations offered for it" (Rawcliffe 2006:103). Much of the view of people with leprosy as being liminal and different has been the responsibility of 19th- and 20th-century writers but these views are now changing as scholars from different disciplines approach their data with more rigor. Additionally, there is a lack of evidence in burials of leprous individuals of the period in England of very much liminality in the disposal of the dead, apart from the skeleton buried on the margins of the cemetery at St. Stephen's Church in York, and of course those skeletons with leprosy excavated from leprosy hospital sites.

Disease in the late medieval period was seen as something that had to be endured and a result of being sinful, including leprosy, and by the start of the late medieval period in England the power of the church was very well established in life and in death (Rawcliffe 2006). Treatment was based on the Graeco-Roman doctrine of the four humors where a disturbance or imbalance in the quantity of any of the humors (black and yellow bile, blood and phlegm) could lead to disease (Rawcliffe 1997). Infectious disease was ascribed to "vapors" that were airborne which the skin absorbed, and the breath of people with leprosy was believed to be a risk to those without the infection. There was later development of a view of a relationship between the body and astrology and astronomy, and the signs of the zodiac were deemed particularly important when the timing of treatment was being considered (Demaitre 2007). Diagnosis of leprosy rested with physicians, priests, and village elders but, as seen today, people with a diagnosis of leprosy could have had a heavy price to pay in terms of their future lives. Examination of the urine, blood, and feces for diagnosis was common, but there were also more unusual diagnostic tests such as putting a raven's egg into a suspect's blood to see if it hardened on cooling (Cule 1970). Physical characteristics of the person's appearance, including the skin, were key to diagnoses along with blood and urine analysis and checking a long list of signs and symptoms; diagnosis could take a long period of time (Demaitre 2007). Rawcliffe (2006:167) notes that diagnosis depended on class, wealth, and where you lived. For example, one could argue that people living in urban situations might have more chance of being diagnosed than those in rural contexts. Was diagnosis accurate? Simpson (1842) stressed that in the early stages of the infection misdiagnosis or non-diagnosis probably occurred. It is possible that some people with leprosy were very accurately diagnosed because of having clear facial lesions of leprosy, but others with early stages of leprosy or TL may have evaded diagnosis.

Management of people with leprosy could involve a number of measures in late medieval England, with women likely responsible for day-to-day care and physicians and surgeons concerned with other procedures (see below). For example, several hundred leprosy hospitals or leprosaria have been identified as being founded in England during the late medieval period, usually by benefactors. Most were founded early in the period and then later became general hospitals/almshouses (e.g., see Magilton et al. 2008; Rawcliffe 2006). Foundations appear to start in the 11th century A.D., peaking in the 13th century and then declining (Roberts 1986). They appear to have been founded on the edge of towns, immediately adjacent to roads (Gilchrist and Sloane 2005:33), which may mark boundaries between towns and parishes, emphasizing the liminal status of those with leprosy. Clearly, these hospitals received people with leprosy and other afflictions, plus the poor who needed charity, but not everybody ended their days in one of these institutions. As we have seen, most skeletons with leprous bone changes in late medieval England have been excavated from cemeteries that were not hospitals and the reasons for this have already been suggested earlier. Leprosaria provided a means to seclude and not to cure and where "patients" could be expelled for bad behavior, could have visitors, could beg outside the hospital and could go to markets (as seen at Sherburn leprosy hospital in Durham, northern England – MacArthur 1953). As Gilchrist suggests (1995:48), if leprosy hospitals were meant to isolate people with the infection then they would have been sited away from main thoroughfares. Rawcliffe (2006:343) also notes that leprosy hospitals appeared to be pleasant places to be being "quieter and less stressful for the newcomer than his or her previous experience of life in an increasingly suspicious and intolerant society." A healthy diet, clothing, and a routine in life would have seemed very beneficial for those on the outside suffering the hardships of life being leprous, although some restrictions on participation in community life might be indicated (Figure 9.6).

Figure 9.6 North side of the "Ripon leprosy chapel," North Yorkshire; note the lower window in the wall allegedly for people with leprosy to view the religious service in the late medieval period. Source: Author's photograph

Treatment of leprosy, apart from segregation, consisted of herbal remedies, laxatives and enemas as purgatives, bathing at specific sites, bloodletting to drain away the infection, cautery of flesh, and ingestion of animal products, minerals, and specific diets (Roberts 1986, see also Rawcliffe 2006). For example, herbs used for leprosy treatment included alder, betony, borage, bugle, dock, elecampne, garlic, honeysuckle, juniper, nettle, rue, sage, scabious and violet and, of interest, many of these herbs were also used for lung and other skin disorders (Potterton 1983). Herbs were also mixed with animal parts and foods, often to make ointments for skin lesions. There is also evidence for some late medieval leprosy hospital sites being associated with wells, for example at Burton Lazars, Leicestershire and Newark, Nottinghamshire (Gilchrist 1995), and Rawcliffe (2006) discusses the evidence for bathing in the treatment of leprosy in late medieval England. However, as Rawcliffe (2006:251) maintains, "the medieval concept of 'a cure' was far less certain than our own. Patients ... could often expect little more than *relative* improvement."

Tuberculosis in Late Medieval England

Skeletal and historical data

Tuberculosis is also well represented in the archaeological and historical record as a disease that affected many in late medieval England although, historically, its frequency rose considerably in the post-medieval period. Prior to the late medieval era, tuberculosis appears to have afflicted Iron Age populations, as seen in the recently identified male from Tarrant Hinton, Dorset, southern England dated to 400–230 B.C. and confirmed with ancient DNA analysis (Mays and Taylor 2003). As with leprosy, tuberculosis only affects 3–5 percent of untreated people at its post-primary or secondary stage (Jaffe 1972), with 30 percent of those with extrapulmonary tuberculosis affected in the skeleton. The spine is the most frequent part of the skeleton observed for bone changes of tuberculosis (Figure 9.7) and is affected in between 25 and 60 percent of people with skeletal tuberculosis (Jaffe 1972). However, there are other parts of the skeleton that are secondary considerations in diagnosis; tuberculosis-induced septic arthritis of hip and knee joints, periosteal new bone formation on visceral surfaces of ribs (Figure 9.8) and calcification of the pleura (lung tuberculosis), new bone formation on long bones (tuberculosis-induced hypertrophic pulmonary osteoarthropathy or HPOA), tuberculous dactylitis of the finger bones, and *lupus vulgaris* (skin tuberculosis) affecting the bone (see Roberts and Buikstra 2008). Despite some individuals from archaeological sites with these nonspecific bone changes being positively analyzed for tuberculous DNA, this still does not prove that tuberculosis caused the bone changes observed (e.g., Mays and Taylor 2002 and HPOA, Taylor et al. 1996 and fused wrist bones). However, a DNA analysis for tuberculosis also can eliminate tuberculosis as a cause (e.g., Mays et al. 2002). None of these skeletal signs in themselves are pathognomonic of tuberculosis because all can be caused by other disease processes and thus, for the purposes of this chapter, skeletons with spinal tuberculosis are only considered even though many skeletons from different sites

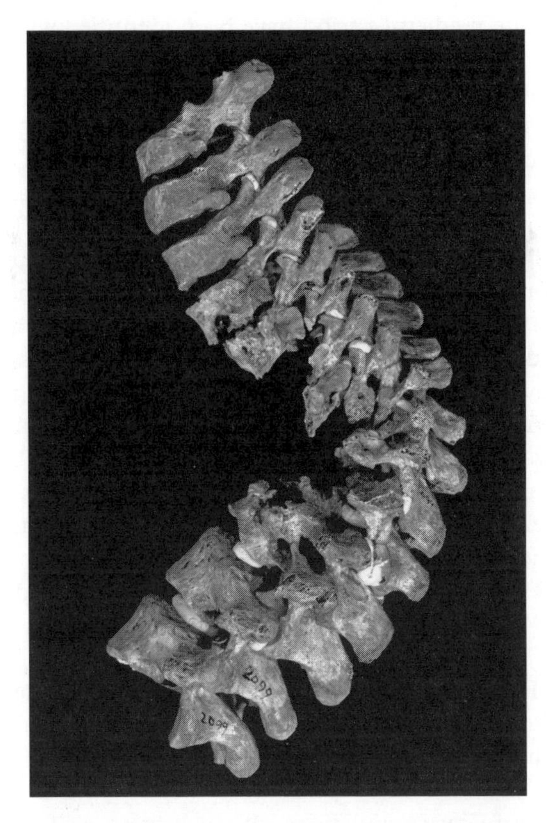

Figure 9.7 Spine of an individual with tuberculosis buried in post-medieval Oxfordshire, England.
Source: Don Ortner

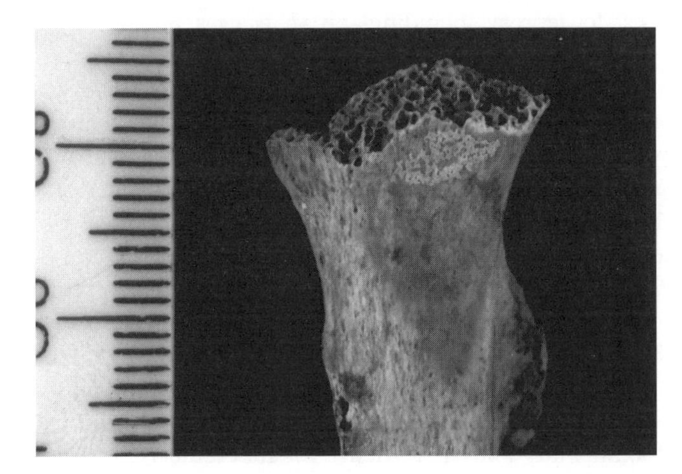

Figure 9.8 Visceral surface of a rib head with a small amount of very subtle new bone formation, reflecting a pulmonary infection.
Source: Author's photograph

in England of the late medieval period have been identified with nonspecific bone changes, especially those on the ribs.

Evidence for tuberculosis afflicting populations in England in the subsequent Roman (southern and eastern areas) and early medieval periods (southern, eastern, southwestern and northern areas) is stronger than for prehistory, but increases quite substantially in the late medieval period, deriving from settled agriculturally based rural or urban communities (Roberts and Buikstra 2003). This likely reflects increases in population density (exacerbated by migration of rural populations to towns and cities for work), poverty and lower standards of hygiene and overall living conditions (Dyer 1998). Of course humans can contract tuberculosis from infected animals, or secondary products they might be working with, such as hides, milk and meat, but they may also be ingesting tuberculous meat or milk through their food intake (see Mays 2005). Animals tend to transmit their tuberculosis between each other in crowded barns via droplet spread (as for humans) but can also spread their tuberculosis to humans in this way. Determining whether a person, through analyzing their skeleton, had suffered pulmonary or gastrointestinal tuberculosis is challenging but ancient pathogen DNA analysis is beginning to answer that question; so far very few individuals have been noted to be affected by *M. bovis* (e.g., see Mays et al. 2001 and the lack of *M. bovis* afflicted people at the rural site of Wharram Percy, Yorkshire, England or Taylor et al. 2007 on *M. bovis* in Siberia 2,000 years ago). Furthermore, there has been little effort made to identify tuberculosis in animal bones from archaeological sites, for various reasons, and therefore it is not known whether there was a significant reservoir of tuberculosis in animal populations in the past.

In late medieval cemetery populations, although there is a further increase in the number of individuals affected by tuberculosis, it is not to the extent reflected in the historical record (see below). The north (Yorkshire), southeast (London), east (Essex, Norfolk, Suffolk), south–central (Oxfordshire and Sussex) and southwest (Gloucestershire) areas see tuberculosis occurring in their communities and most evidence is derived from urban situations. The mortuary contexts cover a variety of types of sites as for leprosy, including abbeys, friaries, priories, a hospital, churchyards, and a plague cemetery. Most individuals affected tend to be located in one of three areas of the country: the southeast where six individuals with tuberculosis were buried at four sites in Suffolk, Essex, and London; the southwest where two individuals were buried at one site in Gloucestershire; and the north where four sites (two in York) had 16 individuals buried with bone changes of tuberculosis. It is only at the hospital site of St. James and St. Mary Magdalene in Chichester where there are larger numbers (14 individuals) seen for one site (Lee and Boylston 2008). Of particular interest here, and relevant to discussions about the frequency of leprosy and tuberculosis through time, most of these individuals were buried in what is suggested to be the latest part of the cemetery where there were fewest people with leprosy. While the late medieval individuals affected seem more evenly spread throughout England, the frequencies are lower than would be expected. If one believes the documentary and art evidence for tuberculosis at that time, especially in London, the figures actually recorded in skeletal evidence are very low indeed, even accounting for the problems in paleopathological diagnosis. If historical data is to be believed, we would expect even more evidence of tuberculosis in post-medieval England, but this is not the case – two sites in London, one in southern–central England, one in the northeast of England. Why is there so little evidence of tuberculosis in skeletal remains for the late medieval period? This could reflect that, as we have seen, only a few percent of people with tuberculosis have skeletal involvement, that people died with the infection before bone changes occurred, that they may have been buried in areas not excavated, that they had subclinical tuberculosis and therefore would not necessarily have ever developed bone damage, or that the historical data is not a good reflection of tuberculosis frequency. This latter explanation might indicate inaccurate diagnosis in many cases.

Perceptions, stigma, diagnosis, and treatment

Today tuberculosis generates associated stigma, as we have noted above. However, in the past there appears to be little evidence for this occurring until the post-medieval period in Europe, and in England specifically. There is also no evidence for people with tuberculosis in the late medieval period being buried apart from the rest of their communities (Roberts and Buikstra 2003); this might be because the signs and symptoms of tuberculosis could have represented any number of pulmonary disorders which would have been particularly commonplace in urban communities at that time. However, when historical data are available for tuberculosis frequency, the numbers given may not be correct for a number of reasons. One of those reasons may, certainly for later periods, relate to the stigma associated with the infection and the possible effect this would have on opportunities for individual people in the future (Bryder 1988; Evans 1998). Some doctors were lobbied not to diagnose tuberculosis, and for some the cause of death from tuberculosis was often not recorded (Hardy 1994). There were many implications for correct and even incorrect diagnoses of tuberculosis for the remaining family, and this may have been the case for earlier periods.

As for leprosy, perceptions of how and why tuberculosis came into their communities varied but were very similar (see above). Diagnosis of tuberculosis would probably have been much more challenging in the late medieval period than leprosy, but perhaps it was not so much of a need because of the difference in social perceptions of tuberculosis when compared to leprosy. The challenge of diagnosis would occur because of the similarities of signs and symptoms for many pulmonary diseases (coughing up blood, shortness of breath, pallor, wheeziness, chest pain, loss of appetite and weight, and fever). However, Pease (1940) felt that diagnosis could be performed by anybody, and he notes that Galen, a 1st–2nd century A.D. Roman writer did link phthisis (tuberculosis) with ulceration of the lungs, chest or throat, cough, fever, and wasting of the body due to malnutrition. Perhaps these symptoms were also recognized by late medieval physicians in their diagnosis, much of their medical knowledge being "inherited" from Graeco-Roman writers.

Management of people with tuberculosis is not as clearly recognized as for leprosy because of the different nature and perception of the two infections. While specific institutions for tuberculosis (sanatoria) were not founded until the 19th and 20th centuries (Roberts and Buikstra 2003 and Figure 9.9), there were many "general" late medieval hospitals established just outside towns, often by rich

Figure 9.9 Sanatorium of the early 20th century in Keighley, West Yorkshire.
Source: Ian Dewhirst

benefactors; several hundred had been founded in England by the end of the period (Orme and Webster 1995; Prescott 1992). These were essentially religious houses as Christian principles instructed nunneries and monasteries to maintain sick houses, but they also cared for travelers and the poor. It is likely that many with tuberculosis were admitted to hospital for care but that many people remained in their own homes while sick. As for leprosy, treatments for tuberculosis revolved around balancing the four humors, with bloodletting (Dormandy 1999), emetics (Daniel 1997), leeches to drain fluid off painful tuberculous joints (Smith 1988), and cautery (Meinecke 1927) being recommended. The following were also recommended, mainly by Graeco-Roman writers, and it is assumed that many of these remedies were used in late medieval England: rest (and exercise), breathing exercises, specific diets (e.g., ingestion of lungs of various animals – Burke 1938), drinking milk (e.g., John of Gaddeson in the 13th–14th century – Daniel 1997) and blood, inhalation of smoke from burning animal dung or infected air from an animal byre (Meinecke 1927), dressings on skin lesions, and herbal (e.g., lungwort and poppy) and mineral remedies. Ingesting or inhaling products of animals may have led to the ultimate development of immunity in a community if the products were infected by tuberculosis. From this period onwards "Touching for the King's Evil" was also practiced in England, the King's Evil being consumption or tuberculosis (Crawfurd 1911). The king or queen apparently had the power of tuberculosis cure by touching the victim on the head and sending them away with a gold piece. This practice started with Edward the Confessor in the 11th century A.D. (Mercer 1964) and then rose and fell in popularity throughout the period; it certainly gave hope and comfort (and money) to many but would of course have been ineffective.

Conclusions: Leprosy and Tuberculosis in Medieval England Compared and Contrasted

The mycobacterial diseases of leprosy and tuberculosis were infections that many had to cope with in late medieval England. From a sociocultural perspective, they have evoked reaction from society in similar and different ways, and people affected have been given a specific "identity," rightly or wrongly, depending on whether a correct or incorrect diagnosis was given. Both caused by the same genus (*Mycobacterium*), but by different species, they have cross-immunity, as we have seen, and rose and fell in frequency in late medieval England at different times. While tuberculosis appears as early as the Iron Age, according to skeletal evidence, leprosy is a later (early medieval) introduction. Both flourish in the early medieval period, with leprosy rising in frequency in the earlier part of the late medieval era. Tuberculosis on the other hand is relatively more frequent through the later late medieval period and into the post-medieval period. This may reflect the cross-immunity hypothesis in that leprosy declines around the 14th century while tuberculosis rises. As Leitman et al (1997) suggest, a species or strain of an organism may competitively exclude another from a host population over a long time period. Both leprosy and tuberculosis also have a long and variable incubation period and have a propensity for subclinical infection. This has implications for its transmission; when people do not know they are infected and are not necessarily showing the external signs of leprosy or tuberculosis, they could be infectious; for late medieval England, any control on transmission would have been affected by this fact. Important too is that they are both today associated with poverty, poor living conditions, and overcrowding, and appear to have been in the past. In the case of tuberculosis, association with infected animals was also a factor in transmission. While both are stigmatized today, it was leprosy that was the focus in late medieval England, and this included opening specific institutions for people diagnosed with leprosy. However, in the bioarchaeological record there is little evidence that people with leprosy or tuberculosis were liminal to society when their burial contexts are considered, apart from those leprous individuals buried in specific leprosy hospital cemeteries. Even when they were segregated into leprosaria, rules were not strict and people could leave the hospital for various reasons. It is suggested that both groups of people were generally accepted into society, as we have seen in living communities from the medical anthropological literature.

However, the specific experience of those with tuberculosis and leprosy in late medieval England will unlikely be appreciated in any great detail because we do not have the personal accounts that we have for later periods (e.g., tuberculosis: MacDonald 1997, Mann 1999, and a more fictional account but based on some facts – leprosy: Hislop 2005). Even so, these accounts may be biased in many ways and for many reasons. As to whether people with tuberculosis and leprosy in late medieval England were disabled in any way, this is a challenging question. As the World Health Organization states, "Disabilities is an umbrella term, covering impairments, activity limitations, and participation restrictions. An impairment is a problem in body function or structure; an activity limitation is a difficulty encountered by an individual in executing a task or action; while a participation restriction is a problem experienced by an individual in involvement in life situations. Thus disability is a complex phenomenon, reflecting an interaction between features of a person's body and features of the society in which he or she lives" (www.who.int/topics/disabilities/en/, accessed June 20, 2009). Clearly, some of the bone lesions for both these infectious diseases appear to be very chronic and possibly disabling, disability meaning the inability to perform a task or function because of impairment. Like leprosy and tuberculosis, disability affects individual people and cultures in different ways, i.e., the same disability might not be so much of a problem in one person when compared to another, and may be more accepted in one community more than another. Disability may also be permanent or temporary, major or minor, visible or invisible and, most importantly, may be something that some people adapt to very well (Roberts 2000). Disability may lead to handicap (social disadvantage resulting from disability), but not necessarily as this depends on how society treats those with disability and how the person themselves deals with their "condition." Today both medical and social models of disability are apparent whereby the medical profession strives to make people with sensory, physical, mental, and psychiatric disorders "normal," not necessarily addressing the social processes and policies that might constrict people's lives (Linton 1998). Whether an "abnormal" person wants to be "made normal" is another matter of debate. All this is in contrast to the social model of disability where society does not meet the needs of the disabled person and thereby restricts their lives and potential. However inferring disability from impairment interpreted in skeletal remains of those with evidence of tuberculosis and leprosy in late medieval England is not easy. Some bioarchaeologists use clinically described signs and symptoms of disease observed to assess the impact on the once living person, bearing in mind that the symptoms and their severity experienced by any one person will vary. This also applies to recovery times from disease and disability; for example a study by Roberts (1985) showed that hospitalized patients recovered at different rates depending on their personalities. It has also been shown that people with the most severe symptoms and possible disability are not necessarily those who have the most bone damage (Rogers *et al.* 1990). The person's potential ability to adapt to a disabling condition both mentally and physically can be impressive, and their contribution to society can be as great as for a nondisabled person (Dettwyler 1991). Thus, it is essential to consider not only impairments, as evidenced in skeletal remains, and the funerary context, but also the social milieu in which people lived in order to infer disability. While bioarchaeologists have been criticized for assuming too much about disability from skeletal remains, historians have also made assumptions about the lives of the disabled ("One can speculate that they did not fare particularly well," Covey 1998:56).

In conclusion, the "identity" of people living with leprosy and tuberculosis in late medieval England can be partly understood, but there are contradictions between bioarchaeological and historical data. It appears that people with both infections were both accepted into their communities to a large extent, but whether diagnosis and "treatment" in its broadest sense was available for all ages, both sexes, and any social status is difficult to assess because the evidence is limited in providing that more nuanced information needed. Here we are dealing with, often, individual burials at many cemeteries. For example, the distribution of people affected by leprosy and tuberculosis by biological sex, status, and age at death is limited because the skeletal samples for study are biased, and it is unknown when a person with the bone changes of either infection developed the infection during their lives as most lesions are chronic, healed, and therefore longstanding. Furthermore, status in late medieval England is difficult to infer from the burial record because grave goods are not common and most people were buried in a coffin; however, burial location within a graveyard may be a possible way of accessing these data as seen in Gilchrist and Sloane's 2005 study. While we know much about these two mycobacterial diseases in late medieval England, there is still much to learn in order that we may fully understand the real impact of these infections on individuals and on communities.

REFERENCES

Andersen, J. G., and K. Manchester 1987 Grooving of the Proximal Phalanx in Leprosy: A Palaeopathological and Radiological Study. Journal of Archaeological Science 14:77–82.

Andersen, J. G., and K. Manchester 1988 Dorsal Tarsal Exostoses in Leprosy: A Palaeopathological and Radiological Study. Journal of Archaeological Science 15: 51–56.

Andersen, J. G., and K. Manchester 1992 The Rhinomaxillary Syndrome in Leprosy: A Clinical, Radiological and Palaeopathological Study. International Journal of Osteoarchaeology 2:121–129.

Andersen, J. G., K. Manchester and R. S. Ali 1992 Diaphyseal Remodelling in Leprosy: A Radiological and Palaeopathological Study. International Journal of Osteoarchaeology 2:211–219.

Andersen, J. G., K. Manchester and C. Roberts 1994 Septic Bone Changes in Leprosy: A Clinical, Radiological and Palaeopathological Review. International Journal of Osteoarchaeology 4:21–30.

Anderson, S. 1998 Leprosy in a Medieval Churchyard in Norwich. *In* Current and Recent Research in Osteoarchaeology. S. Anderson, ed. Pp. 31–37. Proceedings of the 3rd Meeting of the Osteoarchaeology Research Group. Oxford: Oxbow Books.

Anderson, S. R., H. Maguire, and J. Carless 2007 Tuberculosis in London: A Decade and a Half of No Decline. Thorax 62:162–167.

Barkataki, P., S. Kumar, and P. S. Rao 2006 Knowledge and Attitudes to Leprosy Among Patients and Community Members: A Comparative Study in Uttar Pradesh, India. Leprosy Review 77:62–68.

Bayley, J., and J. Henderson 1989 Human Bone from Medieval Doncaster. *In* The Archaeology of Doncaster. 2. The Medieval and Later Town. P. C. Buckland, J. R. Magilton, and C. Hayfield, eds. Pp. 424–431. Oxford: British Archaeological Reports British Series 202(ii).

Bernard, M., B. Bartlam, S. Biggs, and J. Sim 2004 New Lifestyles in Old Age. Health, Identity and Well-Being in Berryhill Retirement Village. Bristol: The Policy Press.

Bishop, M. W. 1993 Burials from the Cemetery of the Hospital of St Leonard, Newark, Nottinghamshire. Transactions of the Thoroton Society of Nottinghamshire 87:23–35.

Blondiaux, J., J.-F. Duvette, S. Vatteoni, and L. E. Eisenberg 1994 Microradiographs of Leprosy from an Osteoarchaeological Context. International Journal of Osteoarchaeology 4:13–20.

Boulter, S. 1992 Death and Disease in Medieval Grantham. Unpublished Undergraduate Dissertation, University of Sheffield.

Brown, F., and C. Howard-Davis eds. 2008 Norton Priory: Monastery to Museum Excavations 1970–87. Lancaster Imprints 16. Oxford: Oxford Archaeology.

Bryder, L. 1988 Below the Magic Mountain. A Social History of Tuberculosis in 20th Century Britain. Oxford: Clarendon Press.

Buikstra, J. E., and R. E. Scott 2009 Key Concepts in Identity Studies. *In* Bioarchaeology and Identity in the Americas. Bioarchaeological Interpretations of the Human Past: Local,

Regional and Global Perspectives. K. J. Knudson, and C. M. Stojanowski, eds. Pp. 24–55. Gainesville: University Press of Florida.

Burke, R. M. 1938 Historical Chronology of Tuberculosis. Springfield, IL: Charles Thomas.

Cakiner, T., A. Yüksel, M. Soydan, T. Saylan, and E. Bahçeci 1993 Women and Leprosy in Turkey. Indian Journal of Leprosy 65:59–67.

Caplan, P. ed. 1997 Food, Health and Identity. London: Routledge.

Carey, J., M. Oxtoby, L. NguyeL, V. Huynh, M. Morgan, and M. Jeffrey 1997 Tuberculosis Beliefs Among Recent Vietnamese Refugees in New York State. Public Health Reports 112:66–72.

Chundun, Z., and C. A. Roberts 1995 Human Skeletal Remains. Archaeological Journal 152:109–245.

Covey, H. 1998 Social Perceptions of People with Disabilities in History, Springfield, IL: Charles C. Thomas.

Crawfurd, R. 1911 The King's Evil. Oxford: Oxford University Press.

Cross, M. 2007 Accessing the Inaccessible. Disability and Archaeology. *In* The Archaeology of Identities. A Reader. T. Insoll, ed. Pp. 179–194. London: Routledge.

Cule, J. 1970 The Diagnosis, Care and Treatment of Leprosy in Wales and the Border in the Middle Ages. Transactions of the British Society for the History of Pharmacy 1:29–50.

Daniel, T. M. 1997 Captain of Death. The Story of Tuberculosis. Rochester, NY: University of Rochester Press.

Davies, P. D. O. 2008 Preface. *In* Clinical Tuberculosis. 4th Edition. P. D. O. Davies, P. F. Barnes, and S. B. Gordon, eds. Pp. xv–xvi. London: Hodder Arnold.

Dayal, R., N. A. Hashmi, P. P. Mathur, and R. Prasad 1990 Leprosy in Childhood. Indian Pediatrics 27:170–180.

Demaitre, L. 2007 Leprosy in Premodern Medicine. A Malady of the Whole Body. Baltimore: Johns Hopkins University Press.

Dettwyler, K. 1991 Can Paleopathology Provide Evidence for "Compassion"? American Journal of Physical Anthropology 84:375–384.

Dodor, E. A., K. Neal, and S. Kelly 2008 An Exploration of the Causes of Tuberculosis Stigma in an Urban District of Ghana. International Journal of Tuberculosis and Lung Disease 12:1048–1054.

Dormandy, T. 1999 The White Death. A History of Tuberculosis. London: Hambledon.

Duhig, C. 1993 Assessment of Skeletons from Pipe Trenches at Huntingdon. *In* A Leper Cemetery at Spittal's Link, Huntingdon. Report No A20. pp. 18–21. D. Mitchell, ed. Cambridge: Archaeological Field Unit of Cambridgeshire County Council (Cambridgeshire Archaeology).

Dyer, C. 1998 Standards of Living in the Later Middle Ages: Social Change c.1200–1520. Cambridge: Cambridge University Press.

El-Hassan, L. A., E. A. Khalil, and A. M. el-Hassan 2002 Socio-Cultural Aspects of Leprosy Among the Masalit and Hawsa Tribes in the Sudan. Leprosy Review 73:20–28.

Evans, C. C. 1998 Historical Background. *In* Clinical Tuberculosis. 2nd Edition. P. D. O. Davies, ed. pp.1–19. London: Chapman and Hall Medical.

Fay, I. 2006 Text, Space and the Evidence of Human Remains in English Late Medieval and Tudor Disease Culture: Some Problems and Possibilities. *In* Social Archaeology of Funerary Remains. R. Gowland, and C. Knüsel, eds. Pp. 90–208. Oxford: Oxbow Books.

Fowler, C. 2004 The Archaeology of Personhood. An Anthropological Approach. London: Routledge.

Gilchrist, R. 1995 Contemplation and Action. The Other Monasticism. London: Leicester University Press.

Gilchrist R., and B. Sloane 2005 Requiem. The Medieval Monastic Cemetery in Britain. London: Museum of London Archaeology Service.

Gill, A. L., D. R. Bell, G. V. Gill, G. B. Wyatt, and N. J. Beeching 2005 Leprosy in Britain: 50 Years Experience in Liverpool. Quarterly Medical Journal 98:505–511.

Gordon, S., and H. Mwandumba 2008 Respiratory Tuberculosis. *In* Clinical Tuberculosis. 4th Edition. P. D. O. Davies, P. F. Barnes, and S. B. Gordon, eds. Pp. 145–162. London: Hodder Arnold.

Gowland, R., and C. Knüsel 2006 Social Archaeology of Funerary Remains. Oxford: Oxbow.

Hadley, D. 2001 Death in Medieval England. Oxford: Tempus.

Hanks, P. ed. 1979 Collins Dictionary of the English Language. London: Collins.

Hardy, A. 1994 Death is the Cure of All Diseases. Using the General Register Office Cause of Death Statistics for 1837–1920. Social History of Medicine 7:472–492.

Harries, A. D., and R. Zachariah 2008 The Association Between HIV and Tuberculosis in the Developing World, with Special Focus on Sub-Saharan Africa. *In* Clinical Tuberculosis. 4th Edition. P. D. O. Davies, P. F. Barnes, and S. B. Gordon, eds. Pp. 315–342. London: Hodder Arnold.

Hislop, V. 2005 The Island. London: Headline.

Holst, M. The Human Bone. In Blue Bridge Lane and Fishergate House, York. Report on Excavations: July 2000–July 2002. C. A. Spall, and N. J. Toop, eds. Archaeological Planning Consultancy Ltd 2003–2006. www.archaeologicalplanningconsultancy.co.uk/mono/001/rep_bone_hum1a.html.

Hudelson, P. 1996 Gender Differences in Tuberculosis: The Role of Socio-Economic and Cultural Factors. Tuberculosis and Lung Disease 77:391–400.

Insoll, T. 2007 Introduction. Configuring Identities in Archaeology. *In* The Archaeology of Identities. A Reader. T. Insoll, ed. Pp. 1–18. London: Routledge.

Jacob, J. T.. and C. Franco-Paredes 2008 The Stigmatization of Leprosy in India and Its Impact on Future Approaches to Elimination and Control. PLoS Neglected Tropical Diseases 2(1):e113. doi:10.1371/journal.pntd.0000113.

Jaffe, H. L. 1972 Degenerative and Inflammatory Diseases of Bones and Joints. Philadelphia: Lea and Febiger.

Jopling, W. H. 1982 Clinical Aspects of Leprosy. Tuberculosis 63:295–305.

Jopling, W. H. 1992 Recollections and Reflections. The Star 51:5–10.

Kakar, S. 1995 Leprosy in India: The Intervention of Oral History. Oral History 23:37–45.

Kearns R. A., and W. M. Gesler 1998 Introduction. *In* Putting Health into Place: Landscape, Identity, and Well-being. R. A. Kearns, and W. M. Gesler, eds. Pp. 1–13. New York: Syracuse University Press.

Khan, A., J. Walley, J. Newell, and N. Imdad 2000 Tuberculosis in Pakistan: Socio-Cultural Constraints and Opportunities in Treatment. Social Science and Medicine 50:247–254.

Knudson, K. J. and, C. M. Stojanowski 2009 Bioarchaeology of Identity. *In* Bioarchaeology and Identity in the Americas. Bioarchaeological Interpretations of the Human Past: Local, Regional and Global Perspectives. K. J. Knudson, and C. M. Stojanowski, eds. Pp. 1–23. Gainesville: University Press of Florida.

Kopparty, S. N. M. 1995 Problems, Acceptance and Social Inequality: A Study of the Deformed Leprosy Patients and Their Families. Leprosy Review 66:239–249.

Larsen, C. S. 2009 Foreword. *In* Bioarchaeology and Identity in the Americas. Bioarchaeological Interpretations of the Human Past: Local, Regional and Global Perspectives. K. J. Knudson, and C. M. Stojanowski, eds. Pp. Xiii–xiv. Gainesville: University Press of Florida.

Lee, F., and A. Boylston 2008 Infection: Tuberculosis and Other Infections. *In* "Lepers Outside the Gate". Excavations at the Cemetery of St James and St Mary Magdalene, Chichester, 1986–87 and 1993. J. R. Magilton, F. Lee, and A. Boylston, eds. Pp. 218–228.

York: Council for British Archaeology Research Report 158 and Chichester Excavations 10.

Lee, F., and K. Manchester 2008 Leprosy: A Review of the Evidence in the Chichester Sample. *In* "Lepers Outside the Gate". Excavations at the Cemetery of St James and St Mary Magdalene, Chichester, 1986–87 and 1993. J. R. Magilton, F. Lee, and A. Boylston, eds. Pp. 208–217. York: Council for British Archaeology Research Report 158 and Chichester Excavations 10.

Leitman, T., T. Porco, and S. Blower 1997 Leprosy and Tuberculosis: The Epidemiological Consequences of Cross-Immunity. American Journal of Public Health 87:1923–1927.

Linton, S. 1998 Claiming Disability. Knowledge and Identity, New York: New York University Press.

Long, N. H., E. Johanasson, V. K. Diwan, and A. Winkvist 1999 Different Tuberculosis in Men and Women: Beliefs from Focus Groups in Vietnam. Social Science and Medicine 49:815–822.

MacArthur, W. P. 1953 Mediaeval Leprosy in the British Isles. Leprosy Review 24:8–19.

MacDonald, B. 1997 The Plague and I. New York: Akadine Press.

Magilton, J. R., F. Lee, and A. Boylston eds. 2008 "Lepers Outside the Gate". Excavations at the Cemetery of St James and St Mary Magdalene, Chichester, 1986–87 and 1993. York: Council for British Archaeology Research Report 158 and Chichester Excavations 10.

Manchester, K. 1991 Tuberculosis and Leprosy: Evidence for Interaction of Disease. *In* Human Paleopathology: Current Syntheses and Future Options. D. J. Ortner, and A. C. Aufderheide, eds. Pp. 23–35. Washington, DC: Smithsonian Institution Press.

Manchester, K., and C. A. Roberts 1986 Palaeopathogical Evidence of Leprosy and Tuberculosis in Britain. SERC Report Grant 337.367.

Mak, W. W.S., P. K. H. Mo, R. Y. M. Cheung, J. Woo, F. M. Cheung, and D. Lee 2006 Comparative Stigma of HIV/AIDS, SARS, and Tuberculosis in Hong Kong. Social Science and Medicine 63:1912–1922.

Mann, T. 1999 The Magic Mountain. London: Vintage.

Mays, S. 2005 Tuberculosis as a Zoonotic Disease in Antiquity. *In* Diet and Health in Past Animal Populations. Current Research and Future Directions. J. Davies, M. Fabiš, I. Mainland, M. Richards, and R. Thomas, eds. Pp. 125–134. Oxford: Oxbow Books.

Mays, S. 2007 The Human Remains. *In* The Churchyard. Wharram. A Study of Settlement on the Yorkshire Wolds X1. S. A. Mays, C. Harding, and C. Heighway, eds. Pp. 77–192. York: University Publications 13.

Mays, S., E. Fysh, and G. M. Taylor 2002 Investigation of the Link Between Visceral Surface Rib Lesions and Tuberculosis in a Medieval Skeletal Series from England Using Ancient DNA. American Journal of Physical Anthropology 119:27–36.

Mays, S., and G. M. Taylor 2002 Osteological and Biomolecular Study of Two Possible Cases of Hypertrophic Osteoarthropathy from Mediaeval England. Journal of Archaeological Science 29:1267–1276.

Mays, S., and G. M. Taylor 2003 A First Prehistoric Case of Tuberculosis from Britain. International Journal of Osteoarchaeology 13:189–196.

Mays, S., G. M. Taylor, A. J. Legge, D. B. Young, and G. Turner-Walker 2001 Paleopathological and Biomolecular Study of Tuberculosis in a Medieval Skeletal Collection from England. American Journal of Physical Anthropology 114:298–311.

McComish, J., Roman, Anglian and Anglo-Scandinavian Activity and a Medieval Cemetery on Land at the Junction of Dixon Lane and George Street, York. York: York Archaeological Trust, Archaeology of York Web Series 9. www.iadb.co.uk/i3/i2_pub.php?PP=47.

Meima, A., W. C. Smith, G. J. van Oortmarssen, J. H. Richardus, and J. D. Hebbema 2004 The Future Incidence of Leprosy: A Scenario Analysis. Bulletin of the World Health Organisation 82:373–380.

Meinecke, B. 1927 Consumption (Tuberculosis) in Classical Antiquity. Annals of Medical History 9:379–402.

Mercer, W. 1964 Then and Now: the History of Skeletal Tuberculosis. Journal of Royal College of Surgeons 9:243–254.

Møller-Christensen, V. 1961 Bone Changes in Leprosy. Copenhagen: Munksgaard.

Morrison, A. 2000 A Woman with Leprosy is in Double Jeopardy. Leprosy Review 71: 128–143.

O'Connell, L. 1998 The Articulated Human Skeletal Remains from St Augustine the Less, Bristol. Bournemouth University, unpublished report.

Orme, N., and W. Webster 1995 The English Hospital 1070–1570. New Haven: Yale University Press.

Ormerod, P. 2008 Non-respiratory Tuberculosis. *In* Clinical Tuberculosis. 4th Edition. P. D. O. Davies, P. F. Barnes, and S. B. Gordon, eds. Pp. 163–188. London: Hodder Arnold.

Paterson, D. E., and M. Rad 1961 Bone Changes in Leprosy: Their Incidence, Progress, Prevention and Arrest. International Journal of Leprosy 29:393.

Pease, A. S. 1940 Some Remarks on the Diagnosis and Treatment of Tuberculosis in Antiquity. Isis 31:380–393.

Potterton, D. ed. 1983 Culpeper's Colour Herbal. London: Foulsham.

Prescott, E. 1992 The English Medieval Hospital 1050–1640. London: Sealy.

Rafferty, J. 2005 Curing the Stigma of Leprosy. Leprosy Review 76:119–126.

Rangan, S., and M. Uplekar 1999 Socio-Cultural Dimensions in Tuberculosis Control. In Tuberculosis. An Interdisciplinary Perspective. J. D. H. Porter, and J. M. Grange, eds. Pp.265–281. London: Imperial College Press.

Rajeswari, R., M. Muniyandi, R. Balasubramanian, and P. R. Narayanan 2004 Perceptions of Tuberculosis Patients About Their Physical, Mental and Social Well-Being: A Field Report from South India. Social Science and Medicine 60:1845–1853.

Ranjit, J. H., and A. Verghese 1980 Psychiatric Disturbances Among Leprosy Patients. An Epidemiological Study. International Journal of Leprosy 48(4):431–434.

Rawcliffe, C. 1997 Medicine and Society in Later Medieval England. Stroud, England: Sutton Publishing.

Rawcliffe, C. 2006 Leprosy in Medieval England. Woodbridge, England: Boydell Press.

Ridley, D. S., and W. H. Jopling 1966 Classification of Leprosy According to Immunity: A Five Group System. International Journal of Leprosy 34:255–273.

Roberts, C. A. 1986 Leprosy and Leprosaria in Medieval Britain. Museum for Applied Science Center for Archaeology Journal 4:15–21.

Roberts, C. A. 1991 Trauma and Treatment in the British Isles in the Historic Period: A Design for Multidisciplinary Research. *In* Human Palaeopathology: Current Syntheses and Future Options. D. J. Ortner, and A. C. Aufderheide, eds. Pp. 225–240. Washington, DC: Smithsonian Institution Press.

Roberts, C. A. 2000 Did They Take Sugar? The Use of Skeletal Evidence in the Study of Disability in Past Populations. *In* Madness, Disability and Social Exclusion. The Archaeology and Anthropology of Difference. J. Hubert, ed. Pp. 46–59. London: Routledge.

Roberts, C. A. 2002 The Antiquity of Leprosy in Britain: The Skeletal Evidence. *In* The Past and Present of Leprosy. Archaeological, Historical, Palaeopathological and Clinical Approaches. C. A. Roberts, M. E. Lewis, and K. Manchester, eds. Pp. 213–222. British Archaeological Reports International Series 1054. Oxford: Archaeopress.

Roberts, C. A. 2006 A View from Afar: Bioarchaeology in Britain. *In* Bioarchaeology. Contextual Analysis of Human Remains. J. E. Buikstra, and L. Beck, eds. Pp. 417–439. New York: Elsevier.

Roberts, C. A. 2007 A Bioarchaeological Study of Maxillary Sinusitis. American Journal of Physical Anthropology 133:792–807.

Roberts, C. A., and J. E. Buikstra 2008 The Bioarchaeology of Tuberculosis: A Global View on a Re-Emerging Disease. Gainesville: University Press of Florida.

Roberts, C. A., and M. Cox 2003 Health and Disease in Britain: From Prehistory to the Present Day. Stroud, England: Sutton Publishing.

Roberts, C. A. and, K. Manchester 2005 The Archaeology of Disease. Stroud, England: Sutton Publishing.

Roberts, C. A., and J. I. McKinley 2003 A Review of Trepanations in British Antiquity Focusing on Funerary Context to Explain their Occurrence. *In* Trepanation. History, Discovery, Theory. R. Arnott, S. Finger, and C. U. M. Smith, eds. Pp. 55–78. Lisse: Swets and Zeitlinger.

Roberts, N. M. 1985 Psychological Aspects of Rehabilitation in Three Orthopaedic Populations. Health Attitudes and Moods Correlating with Disability and Pain in Arthritic Patients, Back Surgery Patients and General Orthopaedic Patients. Unpublished PhD Thesis, University of Bradford.

Rogers, J., I. Watt, and P. Dieppe 1990 Comparison of Visual and Radiographic Detection of Bony Changes at the Knee Joint. British Medical Journal 300:367–368.

Schluger, N. W. 2008 The Diagnosis of Tuberculosis. *In* Clinical Tuberculosis. 4th Edition. P. D. O. Davies, P. F. Barnes, and S. B. Gordon, eds. Pp. 79–103. London: Hodder Arnold.

Scollard, D. M., L. B. Adams, T. P. Gillis, J. L. Krahenbuhl, R. W. Truman, and D. L. Williams 2006 The Continuing Challenge of Leprosy. Clinical Microbiology Reviews 19:338–381.

Simpson, J. Y. 1842 Antiquarian Notices of Leprosy and Leprosy Hospitals in Scotland and England. Edinburgh Medical and Surgical Journal 57:121–156.

Smith, E. R. 1988 The Retreat of Tuberculosis 1850–1950. London: Croom Helm.

Snider, D. E., M. Raviglione, and A. Kochi 1994 Global Burden of Tuberculosis. *In* Tuberculosis: Pathogenesis, Protection and Control. B. R. Bloom, ed. pp. 3–11. Washington, DC: American Society for Microbiology.

Srinivasan, H. 1993 Prevention of Disabilities in Patients with Leprosy. A Practical Guide. Geneva: World Health Organisation.

Taylor, G. M., M. Crossey, J. Saldanha, and T. Waldron 1996 DNA from *M. tuberculosis* Identified in Mediaeval Human Skeletal Remains Using PCR. Journal of Archaeological Science 23:789–798.

Taylor, G. M., C. L. Watson, A. S. Bouwman, D. N. J. Lockwood, and S. A. Mays 2006 Variable Nucleotide Tandem Repeat (VNTR) Typing of Two Palaeopathological Cases of Lepromatous Leprosy from Mediaeval England. Journal of Archaeological Science 33:1569–1579.

Taylor, G. M., E. Murphy, R. Hopkins, P. Rutland, and Y. Christov 2007 First Report of *Mycobacterium bovis* DNA in Human Remains from the Iron Age. Microbiology 153:1243–1249.

Tsutsumi, A., T. Izutsu, A. M. Islam, A. N. Maksuda, H. Kato, and S. Wakai 2007 Quality of Life, Mental Health and Perceived Stigma of Leprosy Patients in Bangladesh. Social Science and Medicine 64:2443–2453.

Walker, D. 2009 The Treatment of Leprosy in 19th Century London: A Case Study From Marylebone Cemetery. International Journal of Osteoarchaeology 19:364–374.

Wells, C. 1967 A Leper Cemetery at South Acre, Norfolk. Medieval Archaeology 9:242–250.

Wood, J. W., G. R. Milner, H. C. Harpending, and K. M. Weiss 1992 The Osteological Paradox: Problems of Inferring Prehistoric Health from Skeletal Samples. Current Anthropology 33:343–370.

World Health Organization 2002 The World Health Report 2002: Reducing Risks, Promoting Healthy Life. Geneva: World Health Organization.

World Health Organization 2009 Global Tuberculosis Control – Epidemiology, Strategy and Financing. Geneva: World Health Organization.

Part III

Growth and Aging: The Life Course of Health and Disease

10

Towards a Social Bioarchaeology of Age

Joanna Sofaer

> Since it is the Other within us who is old, it is natural that the revelation of our age should come to us from outside – from others (Simone de Beauvoir 1972)

Introduction: Age and the Body in Bioarchaeology

Age is a critical axis for the study of human remains. Methods for age estimation, alongside those for the assessment of sex, are among the first suites of techniques learnt by bioarchaeology students. Age estimates act as baseline data on which to hang further interpretations through the analysis of age-related patterns. The emphasis placed on accurate skeletal aging reflects its longstanding importance to a wide range of studies, including demography, human growth, and pathology (Hoppa and Fitzgerald 1999; Hoppa and Vaupel 2002; Lewis 2007).

Bioarchaeology is not, however, alone in its focus on age and the human body. These are key to a range of other disciplines including philosophy, psychology, anthropology, and sociology. As individuals grow, mature, and senesce, changes to the physical body become markers of time passed through, for example, changes in height, the development of secondary sex characteristics, the first gray hairs, and lines of age on the face, or altered physical and mental capabilities. Transformations such as these, and the differences between bodies which they create, have been investigated as signifiers of changing individual social identity, as well as broader cultural responses to these shifts (Featherstone and Wernick 1995; Hockey and James 1993; James 2000).

The need to better integrate knowledge from the natural and social sciences, as well as the humanities, is of increasing importance if we wish to understand the

Social Bioarchaeology. Edited by Sabrina C. Agarwal and Bonnie A. Glencross
© 2011 Blackwell Publishing Ltd

challenges facing human existence in the past, present, and future (Sofaer 2006a; Wilson 1998). Bioarchaeology is uniquely positioned to do this, and there is growing interest in bridging the cultural and biological aspects of the body (Buikstra and Beck 2006; Gowland and Knüsel 2006; Grauer and Stuart-Macadam 1998; Larsen 1997; Robb 2002; Rothschild 2008; Schutkowski 2008a; Sofaer 2006a,b). Yet despite this cross-disciplinary potential, bioarchaeologists have traditionally allied themselves with the sciences alone, seeing little reason to engage with social theory (Sofaer 2006). Since the body is both a biological and a cultural product (Lorentz 2008; Sofaer 2006a,b), there is a need to develop understandings of age and the body that extend "beyond the bone" (Schutkowski 2008b:10). The bioarchaeological study of social identity in relation to age has only recently received attention (Gowland 2006; Sofaer 2004, 2006a, 2006b), and the investigation of what age and aging might have meant on a personal level for the individuals whose skeletons we study has rarely been touched upon (Wilson in press, 2007). Similarly, despite a large number of mortuary studies describing age distributions of artifacts in a range of archaeological contexts, bioarchaeological studies of attitudes to age in past societies are few.

Addressing these gaps requires the development of explicit theoretical foundations for social understandings of age and aging in relation to the study of human remains. This chapter therefore offers suggestions for potentially helpful theoretical directions. It begins by analyzing current understandings of age in bioarchaeology in order to identify areas of both strengths and weaknesses, and moves on to explore approaches to age in four other disciplines that have longstanding interests in age and aging: philosophy, psychology, anthropology, and sociology (Table 10.1). The discussion is necessarily a selective one, focusing on key ideas that may offer relevant or challenging tools for a bioarchaeology of age (although not all of these are necessarily complementary or consistent with each other). It does not aim to provide any definitive answers regarding the theoretical directions in which bioarcheologists should go, rather to offer a series of provocations. By encountering the concept of age differently, bioarchaeologists may be able to expand understandings of age and the body, and thus develop the discipline in fruitful new interpretative directions.

Age and Bioarchaeology

Much bioarchaeological research is implicitly based upon a tripartite model in which the concept of age takes on three distinct but interrelated meanings set within a wider notion of age as a progression along a single linear trajectory from birth to death (Ginn and Arber 1995; Gowland 2006; Lorentz 2008; Sofaer 2004, 2006a). The first meaning of age is physiological or biological age, relating to physical aging of the body, and identified through the sequence of physical changes associated with human growth, maturity, and senescence. The second meaning of age is chronological age. This refers to the amount of time that has elapsed since birth, and is usually quantified in terms of a number of months or years. For bioarchaeologists, construction of a relationship between chronological age and physiological age is critical as it is this axis that enables the categorization of individuals into easily understood age groups and that allows comparisons to be made. The third meaning of age is social age. This is the culturally constructed understanding of what constitutes age-appropriate attitudes and behavior. The link between social and chronological age is an important avenue for archaeologists seeking to interpret data in terms of social identity, allowing exploration of age-related patterns of artifact deposition (e.g., Lucy 2005; Meskell 1994, 2002; Sofaer Derevenski 2000; Welinder 1998, 2001).

This tripartite model is attractive in its apparent simplicity and clarity but closer inspection of each of its three strands reveals a series of assumptions that pose questions regarding its utility for archaeological contexts. With regard to physiological age, although the sequence of age-related biological changes to the human body is well documented and generally understood to be universal, variation occurs in almost all forms of maturation within the body, and the exact timing and speed of such changes, as well as the expression of some traits, may be genetically, population, sex, and life specific (Hoppa and FitzGerald 1999; Kemkes-Grottenthaler 2002; Ríos and Cardoso 2009). Each part of the skeleton reflects a different aspect of aging depending on its location, structure, and function (Kemkes-Grottenthaler 2002), and their use as a proxy for chronological age is constantly under review. For example, despite the reliability of modern reference standards of tooth formation for age estimation of juveniles (e.g. Gustafson and Koch 1974; Moorrees et al. 1963a,b; Schour and Massler 1941), their accuracy for archaeological samples depends upon which standard is used and its fit to past populations (Hoppa and FitzGerald 1999; Liversidge 1994). Confounding factors such as long-term illness and under-nutrition may slow dental eruption and stunt or delay skeletal growth (Arking 1998; Bogin 2001; Cardoso 2006; Goodman and Song 1999; Mays 1999; Olivieri et al. 2008). In adult samples, multifactorial methods are most satisfactory but the accuracy of age estimates deteriorates over the life course and remains a difficult task, particularly for individuals aged about 50 years or more (Angel 1984; Boldsen et al. 2002; Kemkes-Grottenthaler 2002; Mollesson and Cox 1993). Senescent changes in bone are degenerative and are likely to be more variable among individuals and across populations than developmental changes (Boldsen et al. 2002).

To deal with uncertainty in age estimation bioarchaeologists often use discrete age intervals of constant width to categorize individuals. These involve an assumption that all individual age estimates have the same degree of error. However, not all skeletons that are roughly the same age can be assigned with equal confidence to a single age interval since every skeleton has its own degree of error or precision depending on its particular suite of traits and quality of preservation (Boldsen et al. 2002). Thus bioarchaeologists are increasingly turning to complex statistical methods (Boldsen et al. 2002; Gowland and Chamberlain 2002; Lewis and Gowland 2007; Nagaoka and Hirata 2007). Differences between methods can lead to radically different interpretations of archaeological evidence. For example, using traditional regression methods Mays (1993, 2003) argued for infanticide in Roman Britain, whereas Gowland and Chamberlain's (2002) Bayesian analysis of similar data suggested that the distribution of ages at death was similar to a natural mortality profile. Furthermore, with regard to degenerative changes, the observation of these in one body system does not necessarily indicate the same level of

Table 10.1 Concepts of age in bioarchaeology, philosophy, psychology, social anthropology, and sociology

Bioarchaeology	*Philosophy*	*Psychology*
Physiological Age: Sequence of physical changes associated with human growth, maturity, and senescence		
Chronological Age: Time elapsed since birth measured in months or years		Chronological Age: Time elapsed since birth measured in months, years, or cohort
Biological Age: Used in medicine to describe the shortfall between a population cohort average life expectancy and the perceived life expectancy of an individual of the same chronological age	Biological Age: Conception to death	
Social Age: Appropriate role and behavior for chronological age	Social Age: Period of social presence	
	Personal Age: Period of self-awareness requiring sense of social identity, self-consciousness, and agency	
		Behavioral Age: Human development
		Phenomenology of Age: Biocultural behavioral development

Social anthropology	*Sociology*
Chronological Age: Time elapsed since birth measured in months, years or cohort	Chronological Age: Time elapsed since birth measured in months, years, or cohort
Social Age: Timing of major events or rites of passage; appropriate role, and behavior for member of age category linked to status	Social Age: Socially regulated roles linked to social categories and processes of moving through major life course transitions including the development of "habitus"
Functional Age: Capabilities and qualities of the body related to changes in facility	Functional Age: Capacity or incapacity of an individual to perform tasks or roles relative to others of the same chronological age
Historical Age: Time elapsed since birth relative to historical events	
Age as Ontogenetic: Biocultural creation of self	
	Physical Age: Response to personal appearance

degeneration in others (Crews 2003). In addition, a tendency to regard age-related degenerative changes as pathological imposes a particular set of cultural values onto the data (Featherstone and Hepworth 1998). The relationship between physiological and chronological age is not, therefore, straightforward. Physiological age markers are not equivalent to chronological age but are merely an estimate of the physiological status of an individual (Kemkes-Grottenthaler 2002).

In medicine, biological age is a widely used and increasingly validated concept designed to deal with this disjunction between chronological age and physiological status (Jackson et al. 2003; Park et al. 2009; Weale 2008; Wilson in press). It describes the shortfall between a population cohort average life expectancy and the perceived life expectancy of an individual of the same chronological age by quantifying the degree of physical degeneration that an individual has undergone in a range of biomarkers (Jackson et al. 2003). These include physiological functions, such as blood pressure or bone mass, as well as measures of grip strength or mobility. Scores are combined to give the overall biological age, resulting in a general measure of the degree of age-related degeneration while taking into account that not all body systems degenerate at the same rate or in the same order (Jackson et al. 2003; Wilson in press). This approach has not yet had substantial impact in bioarchaeology although it has some potential for modified application within the field (Wilson in press); while some variables, such as bone mass, may be identified from skeletal remains, most employed for determination of biological age require living bodies (Wilson in press).

The complexity of the relationship between physiological and chronological age is often poorly understood by the wider archaeological community who look to bioarchaeologists for absolute answers when they ask the question "How old was s/he when s/he died?" While osteologists constantly strive to improve the precision of age estimation by collecting new baseline data from individuals of known chronological age, improving recording and calibration techniques, and developing sophisticated multivariate and statistical analyses (Belcastro et al. 2008; Boldsen et al. 2002; Cardoso 2008), the methodological challenges facing osteological practice often disappear into what Latour (1987) calls a "black box." This allows other archaeologists to add yet another layer to the existing transformation from physiological to chronological age, by freely converting chronological age to social age (Sofaer 2006a,b). In turn, this has the effect of naturalizing methodologically driven age intervals, turning them into "real" social categories, despite the fact that the relationship between chronological age categories and social age categories are culturally variable (Gowland 2006). It also turns a *process* – aging – into a series of distinct *categories* – age – that say little about the history of the individual and which implicitly separate off different phases of the life course through the labeling of categories such as infant, juvenile, adult, and old age. In other words, archaeological interpretations tend to be akin to photographic stills – snapshots of a particular point in time – rather than moving images (Budden and Sofaer 2009).

Some of these difficulties may be solved by promoting wider appreciation of osteological methods, so as to take aging out of the "black box," along with clearer more careful use of terminology. The most pressing problem, however, is that the notion of social age – what it means and how it is understood – is not well articulated in relation to skeletal remains. A layer-cake approach that simply overlays social identity onto chronological age gains little, as one merely becomes a synonym for each other. Configuring this in terms of what behavior is appropriate for given age categories echoes the contested sex–gender relationship (Sofaer 2006b; Sørensen 2000). In particular, it raises serious issues regarding the relationship between bodies and artifacts, and the primacy accorded to objects when it comes to questions of the social. In other words, the chronological age estimates given to skeletons become the basis for identifying age distributions of objects. These objects are then used to form interpretations about the social age of individuals such as, for example, the age at which Anglo-Saxon children became legal adults (Crawford 1991). Here the body is used as a foundation for archaeological interpretation but is not itself a further source for understanding social age since "the social" is removed from the body and taken to reside in artifacts (Sofaer 2006a).

It is important to note that there is nothing intrinsic about the generation of categories that necessarily leads to this state of affairs. Categories can be useful units of analysis. Indeed, it may be argued that an emphasis on categories is to some extent inevitable in a discipline whose subjects are dead and which is therefore confined to cross-sectional, rather than longitudinal, investigations. The problem lies in the way they are routinely used, and consequently the way that understandings of age as a process tend to be overlooked. What is at stake here are perceptions of the concept of age and its workability within archaeological practice because the notion of social age as it is currently defined in the tripartite model (i.e., age-appropriate attitudes and behaviors) and methodologically executed (through an emphasis on artifacts) hinders the development of bioarchaeological practice in dealing with questions of sociality and age. Furthermore, the osteological emphasis on the relationship between physiological and chronological age that services this model by attempting to fix an individual's age at a point in time as precisely as possible using universal principles, tends to overlook the process of development of the individual in terms of their life experience *as a whole*, including the influence of the specific cultural milieu in which they are situated. This means that the link between biology, age, and social life is weak. To solve this problem, bioarchaeology requires the development of theoretical axes that understand the human body itself as having particular social qualities related to age. This in turn requires a reconceptualization of what the "social" might mean for bioarchaeological analysis.

Outside archaeology there are a number of approaches which develop the idea of the social. Many of these focus on understanding variability and contextuality in human development through the notions of the life course, life history or life span, rather than a notion of age as a single measured point in time or a biologically orientated notion of the life cycle with its cross-cultural connotations (cf. Gilchrist 2000:326). They are therefore concerned with process rather than categories *per se*. In the rest of this chapter I therefore want to explore approaches to social age that may offer inspiration for a social bioarchaeology of age.

Age and Philosophical Enquiry

One of the most influential philosophical discussions of age can be found in the work of Rom Harré (1991). In a discussion of the temporality of the body, Harré

(1991) fragments the traditional view of age as having a linear trajectory with a single origin point and a single end point, and replaces it with three parallel strands: the biological life span, the personal life span, and the social life span. Each has its own beginning and end that need not overlap with those of other strands. The biological life span begins at conception and ends at death. Harré (1991) argues that although conception and death are absolutes in the human time frame, they are external to a sense of self, and therefore stand outside the social and personal. The personal life span, on the other hand, fits within the time span of the physical body. Reflective of consciousness, it begins in late infancy and often ends in old age in senility, before the death of the body; conception and death lie outside the period of self-knowledge. The personal life span is intimately connected to the notion of autobiography and is itself composed of three complementary strands. The first of these is a sense of personal identity through which a person conceives of himself as a singular being with a continuous and unique history. The second is self-consciousness, which involves both knowing what one is experiencing as well as noting that one is experiencing it. This requires the capacity to make some form of self-reference and the ability to identify something, rather than to simply react to it as a stimulus. The third component of personal being is agency. For Harré (1991), to be an agent is to conceive of oneself as being in possession of the power to make decisions and take actions, and to be capable of deciding between alternatives.

The social life span is longer than both the biological and personal life spans. Social identity may begin before bodily identity takes shape (as in the parental definition of a child to come), and may persist after death (c.f. Hockey and Draper 2005). Thus Harré's (1991) notion of social age refuses to conflate the body, self, and social identity, by pointing to situations where people have a social presence, yet lack an active or living body. Harré (1991) points out that the separation of the body from the person is "routinely accomplished" by a series of "separation practices" including anesthetization for surgery and states of drug-induced disembodiment (Harré 1991:15). In archaeological settings, situations where people have a social presence, yet lack a living body, include the case of ancestors (Bloch and Parry 1982; Parker Pearson 1999), a corpse in preparation for disposal (Kligman 1988), or even when a body becomes a focus for archaeological study. Thus, while social identity and the lived experiences of the body are closely related (Featherstone 1982; James 2000; Shilling 1993), for Harré (1991) a concentration on the living body and its embodiment, to the exclusion of its counterpoint disembodiment, leads to the fallacy that the biological death of the body means that the individual has ceased to be (Hallam et al. 1999). Attitudes to death and commemoration also illustrate how meanings and identity can detach themselves from the living body (Tarlow 1999), while Gell (1998) has argued that the biographical careers of people may be prolonged long after death through memories, traces, leavings, and material objects.

Harré's (1991) model of age fractures the layer-cake approach implicit in current bioarchaeological approaches because it identifies human life as being composed of three distinct strands, rather than a single timeline which can be described in terms of one set of unique narrative events. As a result it offers distinctive analytical opportunities since, while the strands are interconnected, each may be described independently. Therefore, unlike current bioarchaeological approaches, investigations of different aspects of age do not rely on transforming one into another. This leads to analytical clarity and the realization that different forms and degrees of access to each strand can be obtained since they need not be equally visible in the archaeological record. Thus it offers a theoretical framework for what aspects of age we may be able to investigate, rather than assuming that we can examine age in its entirety in the past. For example, bioarchaeology may be able to contribute to the study of many (although not all) parts of the biological life span including, in some cases, study of fetal individuals. Bioarchaeological investigation of the personal life span may be more problematic as it is difficult to explore notions of "I as an individual," and human consciousness through skeletal remains. Nonetheless, in adult skeletons, the study of activity patterns through, for example, muscle markings and activity-related skeletal modifications (Capasso et al. 1998) could be understood as investigations of expressions of agency (Sofaer 2006a). With regard to social age, applying Harré's (1991) definition to the ways that the dead body can retain an influential presence in the living, and the consequent extension of social age beyond biological death, indicates that understanding age in social terms cannot be simply achieved by overlaying the social onto chronology defined in terms of years of life. If adopted into bioarchaeology, Harré's (1991) version of social age in archaeological contexts would require close contextual analysis of the deposition and use of human remains in relation to their depositional context.

Harré's (1991) work is useful in offering new definitions of age, but also has lacunae. In particular, he is less concerned with the *means* by which people become social. As the development of sociality involves change over time it is intimately linked to age. For exploration of this it is useful to turn to developmental psychology.

Age and Psychological Enquiry

Developmental psychology is concerned with the scientific understanding of age-related changes in experience and behavior, not only in children but throughout the life span. The aim is to discover, describe, and explain how development occurs, from its earliest origins, into childhood, adulthood, and old age (Butterworth and Harris 1994). Much of developmental psychology revolves around the investigation of the relationship between two notions of age: chronological age and behavioral age. The former is identified as a measure of duration, usually given as years or months but sometimes in terms of a cohort such as school grade. The constitution of behavioral age is more controversial, being part of the so-called "nature vs. nuture" debate (Butterworth and Harris 1994).

Some workers have adopted a largely biological model of behavioral age, focusing on processes considered to be internal to normal human development (Whitehouse 2001). In such studies, chronological age is a causal variable (Green 1984). For example, Chomsky (1975) proposed the concept of modularity, positing the existence of a hard-wired "language acquisition device" in the cognitive architecture of the brain. Piaget (1923; 1928; 1963) identified human development in terms of the natural biological growth and development of the person as organism, documenting its regulation and formulating a fixed scheme for its analysis (Inhelder and Piaget 1958, 1964). He identified a series of stages in the development of cognitive capacities related to chronological age categories which were proposed as constants in child development. Both Chomsky and Piaget have, however, been critiqued for their lack of attention to the role of culture and environment. Thus some researchers see adults and environment as important influences in language acquisition (Hayes *et al* 2001). Further, the universal nature of Piaget's stages has been questioned, and the role of culture in, for example, the development of perceptual skills has been highlighted (Cole and Scribner 1974; Shore 1996; Valsiner 2000, 2007).

Other workers have emphasized behavioral aspects of the social environment and their effects on development, including the ways that children are actively involved in their interpersonal worlds. In these approaches behavior is understood as sociologically grounded. Bowlby (1969, 1973, 1980), for example, famously developed "attachment theory," positing that the mother–child bond was critical to behavioral outcomes for children in later life (Bretherton 1997; Holmes 1993). Another important proponent of a sociocultural approach was Vygotsky (1962), whose work continues to have influence today through the modern field of sociocultural psychology (Greenfield 2000; Valsiner and Rosa 2007; Wertsch 1985). Vygotsy's (1962) "culture-historical" psychology suggests that individual development is guided by culture and interpersonal communication. As children grow, they acquire the use of tools and speech through social interaction, these cultural forms being the product of historical development (Greenfield 2000; Scribner 1985). Thus as societies change over time and space, human development also changes (Greenfield 2000). In other words, human development is culturally mediated.

Despite the longstanding polarization of views regarding the relative roles of culture and biology in psychology, there has been a move toward a middle ground where it is increasingly recognized that human development is as much a process of acquiring culture as it is of biological growth, with many scholars identifying behavioral age as a combination of both (Butterworth and Harris 1994; Durkin 2001). Such a perspective bridges the traditional Cartesian division implicit in the nature vs. nurture debate between understandings of human development as grounded in either the biological body or the cultural mind, paving the way to a phenomenology of age. To explore this further it is useful to briefly turn to the work of Merleau-Ponty (1962).

Merleau-Ponty was both inspired by, and had a significant influence on, modern psychology. He was particularly interested in research into infant motor skills, language acquisition, and vision experiments that demonstrated the importance of interaction with people and objects for human development (Merleau-Ponty 1962). Merleau-Ponty (1962) identified perception as the key to this interaction; perception being defined as sensorimotor behavior through which the world (familiar natural and cultural things as well as people) is constituted as human consciousness prior to any explicit or reflexive thought about it (Edie 1964). Perception takes place through the human body, and is therefore the means through which the world is experienced, but the body also develops as a result of those interactions; particular experiences lead to particular developmental trajectories and in turn to particular behaviors. Hence behavior cannot be understood without reference to the experience of the body and the development of the body cannot be understood without reference to the impact of behavior formed through experience. This phenomenological notion of the body as simultaneously interface and developing self was described by Merleau-Ponty as the "incarnate cogito" (Merleau-Ponty 1964).

Recent work in neuroscience has both adopted and added to this perspective by demonstrating the plasticity of the brain in response to varying experiences over the life course (DeFelipe 2006; Doidge 2007; Ramachandran and Hirstein 1998). Merleau-Ponty's (1964) phenomenology may also have potential as a general theoretical framework for a bioarchaeology of age, particularly given the behavioral emphasis of bioarchaeology in terms of the development of skeletal modifications related to human life ways (Larsen 1997). Although there are methodological difficulties in identifying the chronological age at which behaviors may begin or end from skeletal material, as well as in pinning down skeletal modifications to specific behaviors (Jurmain 1990; Knüsel 2000; Rogers and Waldron 1995), the plasticity of the skeleton means that changes to bones and teeth are formed through a recursive relationship between behavior and the world (Ingold 1998, 2004; Sofaer 2006a). Taking on a phenomenological approach suggests that skeletal modifications are not just abstract markers of activity or reflections of sensitivity to the environment, but are developmental processes arising from interactive human experiences and their impact on bodies. Thus a phenomenological view shifts the emphasis away from age as a singular causal variable in human development, towards a more sophisticated behavioral-developmental conceptualization of age that is complex and multifactorial. To understand potential variation in this complexity it is useful to turn to social anthropological approaches to age.

Age and Social Anthropological Enquiry

Age has traditionally been an important area of investigation within social anthropology. A number of classic, although sometimes controversial studies (e.g. Fortes 1938; Mead 1928, 1930; van Gennep [1908]1960; Whiting 1963) have focused on age, although as a principle of social organization it has never received the full anthropological treatment given to kinship, sex, or ethnicity (Keith 1980). In contrast to psychology's focus on understanding the regularities of child development or explaining deviations from a "norm," anthropological studies tend to explore variability and multi-vocality in experiences of age. Thus anthropological studies raise questions about an approach that assumes a universal progression from childhood to adulthood, from incompetency to competency and immaturity to maturity (Bluebond-Langner and Korbin 2007). This leads to some tension between ethnographic evidence and concepts of "normal" child development in the West (Lancy 2007).

A consequence of this emphasis on plurality is, however, that anthropologists have explicitly struggled with definitions of age, considering issues of chronology, agency, roles, responsibilities, and development with regard to the formulation of definitions (Bluebond-Langner and Korbin 2007). This, in turn, has resulted in several notions of age including chronological age, historical age, social age, and functional age (Counts and Counts 1985), as well as wider theorized understandings of human ontogeny (Toren 1999, 2001). This variety of approaches to age should also be understood in terms of contrasting views within the discipline of the ways that the study of age contributes to anthropological understandings.

Chronological age in years or months has been widely used as a measure of age. For anthropologists interested in social organization, however, chronological age offers limited information about social position and the physical or mental abilities that may contribute to that positioning. Indeed, sweeping statements made in chronological terms regarding the categorization of age can often be misleading (Counts and Counts 1985). For example, the notion that in many ethnographic or historical contexts people might be considered "old" by the time they are 45 or 50 years of age, whereas in modern Western societies old age is thought to begin at 60, 65 or even 70 years of age, begs the question whether we should assume that old people in both contexts have similar functional abilities. Are they in the same place in the life cycle or does the meaning of old age vary between societies (Counts and Counts 1985)? This observation has relevance to bioarchaeologists who frequently compare demographic profiles of chronological age between human groups and ought to prompt some consideration of the interpretations formed from such comparisons. Furthermore, problems of chronological categorizations, particularly in relation to societies who keep alternative measures of time, or for whom records of the passage of time are not kept, have been widely recognized (Counts and Counts 1985; Goody 1976). In such cases, historical age may be more relevant, with people classified as old or young through reference to their birth in relation to significant events during the history of the society or ability to remember such events (Counts and Counts 1985); historical age need not be conceptualized in units of months or years.

Social age involves notions about the appropriate timing of major events or rites of passage in the life cycle and has been the focus of many important anthropological studies. It commonly includes concepts of "age grade" (a social category or status based on a chronological age range), and "age set" (a recognizable social group defined by those who share the same age status). Membership of social age categories may depend on changes to individuals (for example a girl's first menstruation causing her to be viewed as a woman of marriageable age rather than a child), linked to the experiences of a peer group moving through life stages together, or by reference to another person or social unit against which a person is defined (such as becoming a grandparent) (Counts and Counts 1985). The study of social age has frequently been used as a lens through which to understand social life and social relations. Thus age grades, rites of passage, or the roles, status, and treatment of old people have been explained as a societal trait or explained through correlation with other factors (Keith 1980) and age symbolism has also assumed importance. In this kind of approach participants assume a less personal role as part of the systematic study of the conditions and consequences of age differentiation in various cultural settings and at various points in the life course. It has been described as "the study of age *in* anthropology" (Keith 1980), and has had a direct influence on bioarchaeological approaches to the study of age through the investigation of the relationship between grave goods and age (and sex) distributions (cf. Gowland 2006; Lucy 2005). Importantly, however, whereas social anthropologists are able to relate material culture to the social age of living people without knowing their absolute chronological age, bioarchaeologists frequently rely upon an assumed link between chronological age and social age to explore patterns in the archaeological record. In cases where social age is not synchronous with chronological age or maturational stage (Counts and Counts 1985), bioarchaeological methods are clearly problematic.

Less familiar to bioarchaeologists is the notion of functional age. Functional age focuses on the capabilities and qualities of the body related to changes in facility (sensitivity to smell, taste, hearing, sight, pain, and vibration, strength, muscle tone, dental development and tooth loss, mental acuity, ability to walk, and ease of movement), appearance (graying or loss of hair, smoothness of skin or wrinkled skin, dental development or tooth loss, body development), activity (participation in community activities and the ability to independently meet one's own needs), and body action. The latter includes biological functions (basal metabolic rate, oxygen intake in the brain, kidney function, cardiac output, and ability to readily digest food and eliminate waste) but is also frequently experienced in terms of gain or loss (including emotional changes through deaths of parents, spouses, friends and relatives, births of children, marriage, and friendships), and thus is both related, and contributes, to activity levels (Counts and Counts 1985). Functional age has been studied particularly for old age (Counts and Counts 1985) although it is also applicable to other parts of the life course including culturally specific classifications of infancy, childhood, adolescence, and adulthood. In a study of 60 societies from the Human Relations Area Files, functional age was the most common basis for classifying people as aged (Keith 1980). It also allows the division of people of similar chronological age into different categories, for example some elderly people might be regarded as active respected elders, while others are identified as decrepit or in decline. Similarly, where developmental changes take place over a chronological age range, functional age may form the criteria for transition from one social category to another. For instance, linguistic competency is important as criteria for the Kaluli of Papua New Guinea in the acquisition of social identity and the transition from infancy to childhood (Ochs and Schieffelin 2009), while in modern Britain this same shift is defined by walking, a transition expressed in the term "toddler," Anthropological investigations of functional age tend to have an emic emphasis, the focus being on the understanding of personal experiences of changing ages and life stages in relation to culturally determined categories and contextually specific influences on these. They therefore seek to understand "the native's point of view" by listening to what subjects say. This has been called "the anthropology *of* age" (Keith 1980).

The anthropological reliance upon living informants to gain insights into experiences of age means that investigation of functional age may, at first sight, appear difficult for bioarchaeology, particularly since the experience of the body is necessarily individual and the expression of feelings such as pain are culturally specific (Sofaer 1998). Even within a single cultural milieu it is difficult to correlate experience with skeletal changes. For instance, the relationship between degree of pain expressed by patients with osteoarthritis and the severity of degenerative changes detected radiographically is not straightforward (Resnick and Niwayama 1981). Nonetheless, investigation of functional age is arguably highly relevant to bioarchaeology since it should be possible to explore functional changes and impairment to function through skeletal remains and to link these to social categories or life stages. For example, an increase in bone density and changes to trabecular architecture in the lower limbs has been related to the transition from crawling to walking as a result of increased strain related to weight bearing (Gosman and Ketcham 2008). Conversely, as individuals age, the development of osteoporosis means that the amount of bone making up the skeleton diminishes, meaning that bones can fracture in response to relatively minor stresses and this, in turn, can lead to functional impairment (Crews 2003; Wilson in press). Thus, even if oral personal experiences are not available, individual osteobiographies or narratives can be reconstructed (Robb 2002; Saul and Saul 1989). This offers particular possibilities for bioarchaeology. Recent work has begun to explore aging as a variable process that affects people individually and the ways that the capacities and capabilities of the body have a significant effect upon the construction of the person (Wilson in press, 2007). The impact of degenerative joint changes, ante-mortem tooth loss, fractures caused by stresses to diminished bone mass including vertebral compression factures, Colles' fracture of the wrist, femoral neck or hip fractures (the predominant fracture in elderly patients with modern orthopedic trauma) (Lofman et al. 2002) have been investigated in relation to functional capacity including movement, in prehistoric contexts, and also linked to the distribution of grave goods (Wilson 2007, in press). While such stresses are not themselves related to age, their effects may be magnified in older individuals because their bodies are less able to respond to them (Crews 2003; Singer et al. 1998; Wilson in press). The aim of such work is to move away from "floating" ideas of old age in order to allow the embodied nature of aging and the person to be properly considered. This kind of approach relates the biological and the social in a way that does not necessarily require chronological age as the interpretative key. In particular, given the low upper chronological age limits of current skeletal aging techniques (Cox 2000a), it enables old age to be studied in a finer grained manner by distinguishing between different levels of function and categories of elderly people.

In addition to the categorical notions of age described above, recent extensions to ideas about age and aging in social anthropology have explored them in the wider context of human ontogeny (Ingold 1999, 2001; Toren 1994, 1998). This work follows in the French tradition of Mauss (1935, 1947), and his vision of the *homme total* in which people are understood to literally create themselves through culturally specific interactions with the world, including encounters with material culture and other people (Budden and Sofaer 2009; Ingold 2001; Schlanger 2006; Sofaer 2006a; Toren 1998). As in Merleau-Ponty's (1964) phenomenology of perception, physical and mental development are conjoined as material changes to the body involve changes to mind (and vice versa), although in the anthropological view perception is not given the primacy that it is in psychology and broader cultural influences are understood to be at play on the body. For anthropologists, the body is a biological organism but is also a historical phenomenon that develops in specific ways through its interaction with the world (Toren 2001); process and structure are inseparable (Toren 2002). Age thus becomes a measure of this historicity and its specific contextualized articulation.

This emphasis on the body as a biohistorical phenomenon places the investigation of ontogeny squarely within a bioarchaeological purview. It also moves away from traditional categorical classificatory approaches to age based on regular predictable links between physical and chronological age familiar to many bioarchaeologists (such as the sequence of epiphyseal fusion), toward a more synthetic exploration of the relationship between biology and culture over the human life course (Sofaer 2006b). Key to this approach is the idea that humans grow in social worlds so the response of the body (including the skeleton) – what people do, how they act, respond, eat, work, play or learn – reflects this social world. This emphasis shifts the focus of bioarchaeological study toward the study of human plasticity and the analysis of factors such as activity-related skeletal modifications, posture, and diet (Sofaer 2006a).

Age and Sociological Enquiry

Sociology also has a longstanding interest in the study of age. Like social anthropology, sociological concepts of age have tended to focus on the variability of experiences of age and aging using chronological, social, and functional age (Featherstone and Wernick 1995; Hockey and James 2003). Nonetheless, there are important differences between the disciplines in their approaches to age. A key distinction lies in a persistent sociological focus on the human body as the primary site for interrogation of sociality, individual and group identity (Bourdieu 1977; Elias 1978; Goffman 1959; Hepworth and Featherstone 1982; Hockey and James 1993, 2003; Prout 2000; Shilling 1993; Turner 1996; Wacquant 2004), as opposed to an emphasis on culture through the identification of socio-symbolic dimensions of age, their cultural ramifications, and the influence of culture on the body. This means that importance is frequently placed on another type of age – physical age – understood in terms of personal appearance. Each of the different concepts of age in sociology is used to explore the relationship between diachronic changes in social structures and aging over the life course. Age is thus simultaneously identified as a social process as well as a category.

In sociological studies, chronological age may be used as a matter of convenience, providing a commonly understood unit of measurement in years or months. Other chronological measures such as school grade or age of retirement may be used as a proxy for age in years in research looking at the impact of a range of socioeconomic variables such as education, wealth, or class on achievement at different points of the life course. It is commonly recognized, however, that the social significance of chronological age may be highly variable. For example, age of retirement varies greatly between countries (Gorman 1999; Roebuck 1979; Thane 1978). Reliance upon chronological age can also lead to a common methodological error in as much as differences between age groups identified in cross-sectional studies may be interpreted as being *caused* by the process of aging, when in fact they are related to the different experiences of different generations. For example, the young of today will age differently to people who lived through two world wars and the great depression (Riley 1987). Differences in level of education between older less educated people and younger more educated ones can lead to similar interpretive problems (Riley 1987). In addition, use of a single comparative group (frequently white Western males) as the baseline data against which other groups are compared can cause difficulties since the assumption that they will age in the same way may not hold. These observations have obvious relevance for bioarchaeologists who have also recognized the difficulties of using similar baseline samples for comparative purposes when working with cross-sectional data.

Social age is regarded as a structural feature of societies and groups, with both people and roles differentiated by age, but it is also a dynamic process that takes place over the life course (Bury 1995; Riley 1987). Structure and process are linked by the succession of cohorts as people move through the life course. They also provide a link between society and the individual agent since the aging of people and societal change are interdependent, each transforming the other (Bury 1995; O'Rand and Elder 2008; Riley 1987). Age structures are most commonly understood in terms of socially regulated roles linked to social categories and major life course transitions such as marriage, parenthood, or retirement. These may be linked to chronological age to different degrees. Thus a woman's experience of motherhood and her taking on the role of mother both begin with the birth of her baby whether she is 15 or 40 years old. The chronological age of motherhood may be highly variable between societies (Kramer 2008), as is the degree to which it is regulated within specific social contexts (illustrated by recent debates regarding the cut-off age for IVF in the UK). Other sociological studies of age as a structural feature include investigations of the influences of major role transitions on the behavior of individuals for example, marriage, parenthood, education, employment, or the consequences of changes in family structure (Correll et al. 2007; Giordano et al. 2007; Glauber 2007; Laub and Sampson 2003; Leicht 2008). Although many such transitions do not have skeletal implications, a few – such as motherhood – do. Osteological methods for the assessment of parturition have been described (Cox 2000b), while lactation places demands upon the mother may be detected in lowered bone density and increased bone resorption (Agarwal 2001; Agarwal and Stuart-Macadam 2003; Argarwal et al. 2004; Cox 2000b); part of the mineral requirements of lactation are met by bone resorption irrespective of dietary calcium and then replaced after weaning (Ellinger et al. 1952). The investigation of these could be reconfigured to address issues of social transition, offering interesting possibilities for a social bioarchaeology of age, particularly when combined with historical and ethnographic data. In addition, the so-called "grandmother effect" (Hawkes and Smith 2009) offers a ready example of the ways that theorized interpretations of human demography can be used to consider social age based on skeletal samples. Investigation of age-related risks run through social behaviors such as interpersonal violence or life ways might also be osteologically accessible by comparing the prevalence of specific ante-mortem fracture patterns between age groups (Judd 2000; Jurmain 1999; Lovell 1997), although it is important to try to distinguish between effects of social experience and length of exposure to risk (Djurić et al. 2006). On a more general level, there is a growing literature on how early social experiences become biologically embedded (Dickens 2001). Since growing up involves the social regulation of behaviors and development of "habitus" (Bourdieu 1990; Bourdieu and Passeron 1990), then in theoretical terms, skeletal evidence for life ways can be understood as "ways of growing up," and thus linked to age, even if it is difficult to link life style to chronological age of activity.

Social age as process is frequently understood as a cultural construction that may or may not be tied to roles and life course transitions. This more fluid understanding of age places emphasis on individual agency and the ways that people perceive that they can take control over their lives either irrespective of their social roles or in relation to them, rather than membership of a particular social group (Bury 1995; Mirowsky and Ross 2007): many old and young people choose to act outside the supposed constraints of their age (Hockey and James 2003; Neugarten and Neugarten 1996). This perspective therefore deconstructs notions of the life course that rely upon a fixed set of stages (Bury 1995; Featherstone and Hepworth 1991) focusing on the ways that life stages become blurred or radically reconstructed under the impact of social and demographic change (Bury 1995). For example, shifts in retirement mean that people may exit the labor market well before state pension age or stay in work well beyond it. Similarly, in the United Kingdom and United States, higher education is no longer limited to an elite group of young people but increasingly accessed by people of a range of ages and backgrounds (Archer et al. 2003; Trow 1995).

Such observations have led some workers to question, and in some cases sever, the relationship between chronological age and social age, particularly in the sociology of old age, and to move toward functional age as the key analytical axis (Neugarten and Neugarten 1996). Functional age is the age-related capacity or incapacity of an individual to perform certain tasks or roles relative to others of the same chronological age, such as an inability to dress oneself as a result of degenerative joint disease in old age (Greene 2000). Neugarten and Neugarten (1996) have argued that social policy should be related to functional age, rather than chronological age, since the latter does not provide an accurate measure of disability or need for social care. Consequently, services currently linked to state age of retirement should be offered on the basis of need alone; not all people of retirement age will be physically or mentally in need of these services while some much chronologically younger individuals may require them. Just as this radical decoupling of functional and chronological age offers new provocations for policy makers, it also invites bioarchaeologists to consider the importance of chronological age estimates for social interpretation. Indeed, it has recently been suggested that by focusing on chronological age osteologists have been looking at the wrong thing (Wilson in press). An emphasis on functional age presents interesting possibilities for a bioarchaeological focus on physical disability. What is at stake here, however, is the *strength* of the relationship between chronological age and functional age, and thus the extent to which the removal of a link between chronological age and functional age is useful. While a total decoupling might prove interesting, a concern may be that such studies would simply become investigations of disability alone which, while useful, and interesting, could lose the dimension of age entirely. The relationship between chronological and functional age need not, however, be "all or nothing," and different degrees of linkage may be useful for different contexts or types of research.

Physical age differs from functional age in as much as physical age relates to the ways that the visible appearance of a body is responded to in face-to-face encounters by others. For example, age of menarche and growth can vary considerably (Ohsawa et al. 1998), and a girl who has entered puberty and has developed breasts may elicit different social responses to a girl of the same chronological age who has not. Nonetheless, there may be some overlap between physical and functional age, particularly when physical appearance is linked to functional abilities or impairments; both physical and functional age emphasize the body as a site for social interpretation with body differences the basis for policies, practices, and perceptions of identity which create inequalities between people of different ages (Craib 1998; Hockey and James 2003). In this sense the body is a resource for making and breaking identity, unstable bodily forms being paradigmatic of social transition (Prout 2000). Physical age provides a sociological response to purely biological models of age, emphasizing that aging cannot be simply regarded as a biological process of cellular growth, maturity, and decay (Hockey, and James 2003), and to social constructionism through its emphasis on the materiality of age (Prout 2000). It is most frequently investigated through social responses, repercussions, and influences on the process of physical aging. Studies include explorations of attitudes to beauty and body modification (Hepworth and Featherstone 1982; Sennett 1998), childhood and growth of the body (Prout 2000), as well as the influence of societal changes on the aging process mediated through diverse structures and processes in the social system (such as the influence of changes in health care) (Riley 1987).

These topics echo themes in bioarchaeology. Thus, although physical age may initially seem a rather difficult notion for bioarchaeologists to employ other than in the most general terms because of its emphasis on external physical appearance, it should be of particular interest because of these thematic links. Furthermore, it represents an attempt to debate the relationship between the biological and the social by understanding the changing materiality of the body in relation to social life. Indeed, physical age has particular potential for understanding human skeletal remains in terms of the social identity of the living body (Lorentz 2005, 2008). Recent work has examined specific, highly visual, deliberate body practices that may be carried out at particular points in the life course; some of these require constant maintenance although others are one-off events (Lorentz 2008). These include head-shaping, carried out between birth and the second year when the plasticity of the cranium is greatest, and which, once instigated, is fixed (Lorentz 2005, 2008), Chinese foot-binding begun between the ages of three to ten years old, before the bones of a young girl's feet were fully developed but requiring constant attention to restrict foot growth and maintain its shape (Gates 2001; Jackson 1997), intentional tooth removal (Robb 1997), intentional modification of the dentition (Milner and Larsen 1991; Tiesler 2002) or severing of fingers (Lorentz 2008). In addition, some soft tissue modifications, such those arising from labrets, may lead to skeletal lesions (Torres-Rouff 2003). Again, these may also be inserted at specific points in the life course. Such modifications are active components in the construction of social difference in terms of gender, status, and ethnicity (Lorentz 2004). They act as displays of physical capital (Lorentz 2007, 2008; Shilling 1993) accumulated over the life course of a "becoming body" (Prendergast 2000:124); a body in constant process (Lorentz 2004). Interestingly, the practice of cradle-boarding, which may also lead to skull deformation, has considerable effects upon the behavior and physiological status of infants (Chisholm 1978), suggesting that there may be room to link physical appearance, age, patterns of caretaking, and infant behaviors.

Conclusion

Age and aging are not straightforward concepts. They are subject to a wide range of definitions, forming the basis of several different approaches. Readers may feel that not all of those described here are equally useful to bioarchaeology. Nonetheless, it is clear that the existing tripartite division of age in bioarchaeology is somewhat limiting and more use could be made of other notions of age to describe and interpret human remains. The challenges facing bioarchaeologists are how to explore the relationship between biology and society through the skeleton without falling back entirely on artifacts as interpretative keys, and how to understand age as both category and process. These require more sophisticated theoretical understandings of what is meant by the "social," and how this is related to the human body.

The body is the nexus of biology and culture (Sofaer 2006a). Bioarchaeological studies of age and aging need to take biology into account but this has to be situated within an understanding of human development as a whole, including social relations, culturally specific life experiences, and local attitudes to age and aging. Other disciplines have much to offer us in providing ways to appreciate these, although it will always be necessary for bioarchaeologists to tailor approaches to their specific interests and data. The suggestions and examples presented here are by no means exhaustive and are intended only to act as prompts for future work. By opening up a range of possibilities, some complementary, some conflicting, we can move forward into an exciting new social bioarchaeology of age.

REFERENCES

Agarwal, S. C. 2001 The Effects of Pregnancy and Lactation on the Maternal Skeleton. A Historical Perspective. American Journal of Physical Anthropology 32:30.

Agarwal, S. C., M. Dimitriu, and M. D. Grynpas 2004 Medieval Trabecular Bone Architecture: The Influence of Age, Sex, and Lifestyle. American Journal of Physical Anthropology 124:33–44.

Agarwal, S. C., and P. Stuart-Macadam 2003 An Evolutionary and Biocultural Approach to Understanding the Effects of Reproductive Factors on the Female Skeleton. *In* Bone Loss and Osteoporosis: An Anthropological Perspective. S. C. Agarwal, and S. D. Stout, eds. Pp. 105–120. New York: Kluwer Academic/Plenum Publishers.

Angel, J. L. 1984 Variation in Estimating Age at Death of Skeletons. Collegium Anthropologicum 8:163–168.

Archer, L., M. Hutchings, and A. Ross 2003 Higher Education and Social Class: Issues of Exclusion and Inclusion. London: Routledge.

Arking, R. 1998 Biology of Aging. Observations and Principles. Sunderland, MA: Sinauer.

Belcastro, M. G, E. Rastelli, and V. Mariotti 2008 Variation of the Degree of Sacral Vertebral Body Fusion in Adulthood in Two European Modern Skeletal Collections. American Journal of Physical Anthropology 135:149–160.

Bloch, M., and J. Parry eds. 1982 Death and the Regeneration of Life. Cambridge: Cambridge University Press.

Bluebond-Langner, M., and J. Korbin 2007 Challenges, and Opportunities in the Anthropology of Childhoods. An Introduction to "Children, Childhoods, and Childhood Studies." American Anthropologist 109:241–246.

Bogin, B. 2001 The Growth of Humanity. New York: Wiley-Liss.

Boldsen, J. L., G. R. Milner, L. W. Konigsberg, and J. W. Wood 2002 Transition Analysis: A New Method for Estimating Age from Skeletons. In Paleodemography. Age Distributions from Skeletal Samples. R. D. Hoppa, and J. W. Vaupel eds. Pp. 73–106. Cambridge: Cambridge University Press.

Bourdieu, P. 1977 Outline of a Theory of Practice. Cambridge: Cambridge University Press.

Bourdieu, P. 1990 The Logic of Practice. Cambridge: Polity Press.

Bourdieu, P., and J. C. Passeron 1990 Reproduction in Education, Society, and Culture. London: Sage.

Bowlby, J. 1969 Attachment and Loss (Vol. 1). New York: Basic Books.

Bowlby, J. 1973 Separation: Anxiety and Anger (Vol. 2). London: Hogarth Press.

Bowlby, J. 1980 Loss: Sadness and Depression (Vol. 3). London: Hogarth Press.

Bretherton, I. 1997 Bowlby's Legacy to Developmental Psychology. Child Psychiatry and Human Development 28:33–43.

Budden, S., and J. Sofaer 2009 Non-discursive Knowledge, and the Construction of Identity Potters, Potting, and Performance at the Bronze Age Tell of Százhalombatta, Hungary. Cambridge Archaeological Journal 19:203–220.

Buikstra, J. E., and L. A. Beck eds. 2006 Bioarchaeology: The Contextual Analysis of Human Remains. Burlington, MA: Academic Press.

Bury, M. 1995 Aging, Gender, and Sociological Theory. *In* Connecting Gender and Aging: A Sociological Approach. S. Arber, and J. Ginn, eds. Pp. 15–29. Milton Keynes: Open University Press.

Butterworth, G., and M. Harris 1994 Principles of Developmental Psychology: An Introduction. Hove: Psychology Press.

Capasso, L., K. A. R. Kennedy, and C. Wilczak 1998 Atlas of Occupational Markers on Human Remains. Teramo: Edigrafital SPA.

Cardoso, H. 2006 Environmental Effects on Skeletal Versus Dental Development: Using a Documented Subadult Skeletal Sample to Test a Basic Assumption in Human Osteological Research. American Journal of Physical Anthropology 132:223–233.

Cardoso, H. 2008 Age Estimation of Adolescent and Young Adult Male, and Female Skeletons II, Epiphyseal Union at the Upper Limb, and Scapular Girdle in a Modern Portuguese Skeletal Sample. American Journal of Physical Anthropology 137:97–105.

Chisholm, J. 1978 Swaddling, Cradleboards, and the Development of Children. Early Human Development 2/3:255–275.

Chomsky, N. 1975 Reflections on Language. New York: Pantheon Books.

Cole, M., and S. Scribner 1974 Culture and Thought: A Psychological Introduction. New York: Wiley.

Correll, S. J., S. Benard, and I. Paik 2007 Getting a Job: Is There a Motherhood Penalty? American Journal of Sociology 112:1297–1338.

Counts, D. A., and D. R. Counts 1985 Aging and its Transformations. Moving Toward Death in Pacific Societies. Pittsburgh: University of Pittsburgh Press.

Cox, M. 2000a Aging Adults from the Skeleton. *In* Human Osteology in Archaeology and Forensic Science. M. Cox, and S. Mays, eds. Pp. 61–81. Cambridge: Cambridge University Press.

Cox, M. 2000b Assessment of Parturition. *In* Human Osteology in Archaeology and Forensic Science. M. Cox, and S. Mays, eds. Pp. 131–142. Cambridge: Cambridge University Press.

Craib, I. 1998 Experiencing Identity. London: Sage.

Crawford, S. 1991 When Do Anglo-Saxon Children Count? Journal of Theoretical Archaeology 2:17–24.

Crews, D. E. 2003 Human Senescence: Evolutionary and Biocultural Perspectives. Cambridge: Cambridge University Press.

de Beauvoir, S. 1972 The Coming of Age. New York: G. P. Putnam.

DeFelipe, J. 2006 Brain Plasticity, and Mental Processes: Cajal Again. Nature Reviews Neuroscience 7:811–817.

Dickens, P. 2001 Linking the Social and Natural Sciences: Is Capital Modifying Human Biology in Its Own Image? Sociology 35:93–110.

Djurić, M. P., C. A. Roberts, Z. B. Rakočvić, D. D. Djonić, and A. R. Lešić 2006 Fractures in Late Medieval Skeletal Populations from Serbia. American Journal of Physical Anthropology 130:167–178.

Doidge, N. 2007 The Brain That Changes Itself. Stories of Personal Triumph from the Frontiers of Brain Science. New York: Penguin.

Durkin, K. 2001 Developmental Social Psychology. From Infancy to Old Age. Oxford: Blackwell.

Edie, J. 1964. Introduction. In M. Merleau-Ponty The Primacy of Perception, and Other Essays on Phenomenological Psychology. Evanston: Northwestern University Press.

Elias, N. 1978. The Civilizing Process, New York: Urizen.

Ellinger, G., J. Duckworth, A. Dalgarno, and M. Quenouille 1952 Skeletal Changes in Pregnancy and Lactation in the Rate: Effect of Different Levels of Dietary Calcium. British Journal of Nutrition 6:235–253.

Featherstone, M. 1982 The Body in Consumer Culture. Theory, Culture and Society 1:18–33.

Featherstone, M., and M. Hepworth 1991 The Mask of Aging, and the Postmodern Life Course. *In* The Body: Social Process and Cultural Theory. M. Featherstone, M. Hepworth, and B. Turner, eds. Pp. 371–388. London: Sage.

Featherstone, M., and Wernick, A. eds. 1995 Images of Aging: Cultural Representations of Later Life. London: Routledge.

Fortes, M. 1938 Social and Psychological Aspects of Education in Taleland. Supplement to Africa 11(4). London: Oxford University Press.

Gates, H. 2001 Footloose in Fujian: Economic Correlates of Footbinding. Comparative Studies of Society and History 43:130–148.

Gell, A. 1998 Art and Agency: Towards a New Anthropological Theory. Oxford: Clarendon Press.

Gilchrist, R. 2000 Archaeological Biographies: Realizing Human Lifecycles, -courses, -histories. World Archaeology 31:325–328.

Ginn, J., and S. Arber 1995 Only Connect: Gender Relations and Aging. *In* Connecting Gender and Aging: A Sociological Approach. S. Arber, and J. Ginn eds. Pp. 1–14. Buckingham: Open University Press.

Giordano, P. C., R. D. Schroeder, and S. A. Cernkovich 2007 Emotions, and Crime Over the Life Course: A Neo-Meadian Perspective on Criminal Continuity and Change. American Journal of Sociology 112:1603–1661.

Glauber, R. 2007 Marriage, and Motherhood. Wage Penalty Among African Americans, Hispanics, and Whites. Journal of Marriage and Family 69:951–961.

Goffman, E. 1959 (1990 edition). The Presentation of Self in Everyday Life. London: Penguin

Goodman, A. H., and R.-J. Song 1999 Sources of Variation in Estimated Ages at Formation of Linear Enamel Hypoplasias. *In* Human Growth in the Past. Studies from Bones and Teeth. R. D. Hoppa, and C. M. FitzGerald, eds. Pp. 210–240. Cambridge: Cambridge University Press.

Goody, J. 1976 Aging in Non-industrial Societies. *In* Handbook of Aging and the Social Sciences. R. Binstock, and E. Shanas, eds. Pp. 117–129. New York: van Nostrand-Reinhold.

Gorman, M. 1999 Development and the Rights of Older People. *In* The Aging and Development Report: Poverty, Independence, and the World's Older People. J. Randel, T. German, and D. Ewing, eds. Pp. 3–21. London: Earthscan.

Gosman, J., and R. Ketcham 2008 Patterns in Ontogeny of Human Trabecular Bone from SunWatch Village in the Prehistoric Ohio Valley: General Features of Microarchitectural Change. American Journal of Physical Anthropology 138:318–332.

Gowland, R. 2006 Aging the Past: Examining Age Identity from Funerary Evidence. *In* Social Bioarchaeology of Funerary Remains. R. Gowland, and C. Knüsel, eds. Pp. 143–155. Oxford: Oxbow.

Gowland, R., and A. Chamberlain 2002 A Bayesian Approach to Aging Perinatal Skeletal Material from Archaeological Sites: Implications for the Evidence for Infanticide in Roman Britain. Journal of Archaeological Science 29:677–685.

Gowland, R., and C. Knüsel, eds 2006 Social Bioarchaeology of Funerary Remains. Oxford: Oxbow.

Grauer, A. L., and P. Stuart-Macadam, eds. 1998 Sex and Gender in Palaeopathological Perspective. Cambridge: Cambridge University Press.

Green, S. K. 1984 Age as a Causal Variable. Senility Versus Wisdom: The Meaning of Old Age as a Cause for Behavior. Basic and Applied Social Psychology 5:105–110.

Greene, R. R. 2000 Social Work with the Aged and Their Families. New Brunswick: Aldine Transaction.

Greenfield, P. 2000 Children, Material Culture, and Weaving. Historical Change and Developmental Change. In Children and Material Culture. J. Sofaer ed. Pp. 72–86. London: Routledge.

Gustafson, G., and G. Koch 1974 Age Estimation Up to 16 Years of Age Based on Dental Development. Ondontologisk Revy 25:295–306.

Hallam, E., J. Hockey, and G. Howarth 1999 Beyond the Body: Death and Social Identity. London: Routledge.

Harré, R. 1991 Physical Being: A Theory for a Corporeal Psychology. Oxford: Blackwell.

Hawkes, K., and K. Smith 2009 Evaluating Grandmother Effects. American Journal of Physical Anthropology 10. 1002/ajpa. 21061.

Hayes, S. C., D. Barnes-Holmes, and B. Roche eds. 2001 Relational Frame Theory: A Post-Skinnerian Account of Human Language and Cognition. New York: Kluwer Academic/Plenum Press.

Hepworth, M., and M. Featherstone 1982 Surviving Middle Age. Oxford: Blackwell.

Hockey, J., and J. Draper 2005 Beyond the Womb, and the Tomb: Identity, (Dis)embodiment, and the Life Course. Body and Society 11:41–57.

Hockey, J., and A. James 1993 Growing Up, and Growing Old: Aging, and Dependency in the Life Course. London: Sage.

Hockey, J., and A. James 2003 Social Identities Across the Life Course. New York: Palgrave Macmillan.

Holmes, J. 1993 John Bowlby and Attachment Theory. London, Routledge.

Hoppa, R. D., and C. M. FitzGerald 1999 From Head to Toe: Integrating Studies from Bones and Teeth. *In* Human Growth in the Past. Studies from Bones and Teeth. R. D. Hoppa, and C. M. FitzGerald, eds. Pp. 1–31. Cambridge: Cambridge University Press.

Hoppa, R. D., and Vaupel, J. W. eds. 2002 Paleodemography. Age Distributions from Skeletal Samples. Cambridge: Cambridge University Press.

Ingold, T. 1998 From Complementarity to Obviation: On Dissolving the Boundaries Between Social and Biological Anthropology, Archaeology, and Psychology. Zeitschrift für Ethnologie 123:21–52.

Ingold, T. 1999 Foreword. *In* The Social Dynamics of Technology. M. A. Dobres and C. R. Hoffman, eds. Washington, DC: Smithsonian Institute.

Ingold, T. 2001 Beyond Art and Technology: The Anthropology of Skill. *In* Anthropological Perspectives on Technology. M. B. Schiffer ed., Pp. 17–31. Albuquerque: University of New Mexico Press.

Ingold, T. 2004 Situating Action VI: A Comment on the Distinction Between the Material, and the Social. Ecological Psychology 8:183–187.

Inhelder, B., and J. Piaget 1958 The Growth of Logical Thinking from Childhood to Adolescence. New York: Basic Books.

Inhelder, B., and J. Piaget 1964 The Early Growth of Logic in the Child: Classification and Seriation. London: Routledge and Kegan Paul.

Jackson, B. 1997 Splendid Slippers. A Thousand Years of an Erotic Tradition. Berkley: Tenspeed Press.

Jackson, S. H., M. R. Weale, and R. A. Weale 2003 Biological Age – What Is It and Can It Be Measured? Archives of Gerontology and Geriatrics 36:103–115.

James, A. 2000 Embodied Being(s): Understanding the Self and the Body in Childhood. *In* The Body, Childhood, and Society. A. Prout ed. Pp. 19–37. London: Macmillan.

Judd, M. 2000 Trauma and Interpersonal Violence in Ancient Nubia During the Kerma Period (ca. 2500–1500 BC). Unpublished Ph.D. Thesis, University of Alberta, Edmonton, Alberta, Canada.

Jurmain, R. 1990 Paleoepidemiology of a Central California Prehistoric Population from CA–ALA–329: II. Degenerative Disease. American Journal of Physical Anthropology 83:83–94.

Jurmain R. 1999 Stories from the Skeleton. Behavioral Reconstruction in Human Osteology. Amsterdam: Gordon, and Breach Publishers.

Keith, J. 1980 The Best is Yet To Be: Toward an Anthropology of Age. Annual Review of Anthropology 9:339–364.

Kemkes-Grottenthaler, A. 2002 Aging Through the Ages: Historical Perspectives on Age Indicator Methods. In Paleodemography. Age Distributions from Skeletal Samples. R. D. Hoppa, and J. W. Vaupel eds. Pp. 48–72. Cambridge: Cambridge University Press.

Kligman, G. 1988 The Wedding of the Dead: Ritual, Poetics, and Popular Culture in Transylvania. Berkeley: University of California Press.

Knüsel, C. 2000 Bone Adaptation and its Relationship to Physical Activity in the Past. *In* Human Osteology in Archaeology and Forensic Science. M. Cox, and S. Mays, eds. Pp. 381– 402. Cambridge: Cambridge University Press.

Kramer, K. 2008 Early Sexual Maturity Among Pumé Foragers of Venezuela: Fitness Implications of Teen Motherhood. American Journal of Physical Anthropology 136:338–350.

Lancy. D. 2007 The Anthropology of Childhood. Cambridge: Cambridge University Press.

Larsen, C. S. 1997 Bioarchaeology. Interpreting Behaviour from the Human Skeleton. Cambridge: Cambridge University Press.

Latour, B. 1987 Science in Action: How to Follow Scientists, and Engineers through Society. Milton Keynes: Open University Press.

Laub, J. H., and R. J. Sampson 2003 Shared Beginnings, Divergent Lives: Delinquent Boys to Age 70. Cambridge, MA: Harvard University Press.

Leicht, K. 2008 Broken Down by Race and Gender? Sociological Explanations of New Sources of Earnings Inequality. Annual Review of Sociology 34:237–255.

Lewis, M. 2007 The Bioarchaeology of Children. Perspectives from Biological and Forensic Anthropology. Cambridge: Cambridge University Press.

Lewis, M., and R. Gowland 2007 Brief and Precarious Lives: Infant Mortality in Contrasting Sites from Medieval and Post–Medieval England (AD 850–1859). American Journal of Physical Anthropology 134:117–129.

Liversidge, H. 1994 Accuracy of Age Estimation from Developing Teeth in a Population of Known Age (0–5.4 years). International Journal of Osteoarchaeology 4:37–45.

Lofman, O., K. Berglund, L. Larsson, and G. Toss 2002 Change in Hip Epidemiology, Redistribution Between Ages, Gender, and Fracture Type. Osteoporosis International 13:18–25.

Lorentz, K. 2004 Age and Gender in Eastern Mediterranean Prehistory: Depictions, Burials and Skeletal Evidence. Ethnographisch-Archäologische Zeitschrift 45:297–315.

Lorentz, K. 2005 Late Bronze Age Burial Practices: Age as a Form of Social Difference. *In* Cyprus: Religion, and Society from the Late Bronze Age to the End of the Archaic Period. V. Karageorghis ed. Pp. 41–55. Erlangen, Mohnesee-Wamel: Bibliopolis.

Lorentz, K. 2007 From Life Course to Longue Durée: Headshaping as Gendered Capital? *In* Gender Through Time in the Ancient Near East. D. Bolger ed. Pp. 281–312. Lanham: AltaMira.

Lorentz, K. 2008 From Bodies to Bones, and Back: Theory, and Human Bioarchaeology. *In* Between Biology and Culture. H. Schutkowski ed. Pp. 273–303. Cambridge: Cambridge University Press.

Lovell, N. 1997 Trauma Analysis in Paleopathology. Yearbook of Physical Anthropology 40:139–170.

Lucy, S. 2005 The Archaeology of Age. *In* The Archaeology of Identity: Approaches to Gender, Age, Status, Ethnicity, and Religion. M. Díaz-Andreu, and S. Lucy eds. Pp. 43–66. London: Routledge.

Mauss, M. 1935 Les Techniques du Corps. *In* Marcel Mauss: Techniques, Technology, and Civilisation. N. Schlanger, ed. Pp. 77–95 2006. Oxford: Durkheim Press. Translation by B. Brewster previously published in Economy and Society 1973.

Mauss, M. 1947 Technology Manuel d'ethnographie. *In* Marcel Mauss: Techniques, Technology, and Civilisation. N. Schlanger, ed. Pp. 97–140 2006. Oxford: Durkheim Press. Translation by Dominique Lussier previously published in Paris: Paypout 2nd edition 1967.

Mays, S. 1993 Infanticide in Roman Britain. Antiquity 67:883–888.

Mays, S. 1999 Linear and Appositional Long Bone Growth in Earlier Human Populations: A Case Study from Mediaeval England. *In* Human Growth in the Past. Studies from Bones and Teeth. R. D. Hoppa, and C. M. FitzGerald, eds. Pp. 290–312. Cambridge: Cambridge University Press.

Mays, S. 2003 Comment on "A Bayesian Approach to Aging Perinatal Skeletal Material from Archaeological Sites: Implications for the Evidence for Infanticide in Roman Britain" by R. L. Gowland, and A. T. Chamberlain. Journal of Archaeological Science 30:1695–1700.

Mead, M. 1928 Coming of Age in Samoa. New York: Morrow.

Mead, M. 1930 Growing Up in New Guinea: A Study of Adolescence and Sex in Primitive Societies. New York: Morrow.

Merleau-Ponty, M. 1962 Phenomenology of Perception. London: Routledge and Kegan Paul.

Merleau-Ponty, M. 1964 The Primacy of Perception and Other Essays on Phenomenological Psychology. J. M. Edie, ed. Evanston: Northwestern University Press.

Meskell, L. 1994 Dying Young: The Experience of Death at Deir el Medina. Archaeological Review from Cambridge 13:35–45.

Meskell, L. 2002 Private Life in New Kingdom Egypt. Princeton, NJ: Princeton University Press.

Milner, G. R., and C. S. Larsen 1991 Teeth as Artefacts of Human Behaviour: Intentional Mutilation and Accidental Modification. *In* Advances in Dental Anthropology. M. A. Kelley, and C. S. Larsen, eds. Pp. 357–378. New York: Wiley-Liss.

Mirowsky, J., and C. E. Ross 2007 Life Course Trajectories of Perceived Control, and Their Relationship to Education. American Journal of Sociology 112:1339–1382.

Mollesson, T., and M. Cox 1993 The Spitalfields Project. Vol. 2. CBA Research Report 86. London: Council for British Archaeology.

Moorrees, C. F. A., E. A. Fanning, and E. E. Hunt 1963a Age Variation of Formation Stages for Ten Permanent Teeth. Journal of Dental Research 42:1490–1502.

Moorrees, C. F. A., E. A. Fanning, and E. E. Hunt 1963b Age Variation of Formation and Reabsorption of Three Deciduous Teeth in Children. American Journal of Physical Anthropology 21:205–213.

Nagaoka, T., and K. Hirata 2007 Reconstruction of Paleodemographic Characteristics from Skeletal Age at Death Distributions: Perspectives from Hitotsubashi, Japan. American Journal of Physical Anthropology 134:301–311.

Neugarten, B. L., and D. A. Neugarten 1996 The Meanings of Age: Selected Papers of Bernice L. Neugarten. Chicago: University of Chicago Press.

O'Rand, A., and G. Elder 2008 Changing Societies and Changing Lives. *In* Within the Social World. Essays in Social Psychology. J. Chin, and J. Jacobsen eds. Pp. 202–212. Boston: Longman.

Ochs, E., and B. Schieffelin 2009 Language Acquisition, and Socialization: Three Developmental Stories, and Their Implications. *In* Linguistic Anthropology. A Reader. 2nd Edition. A. Duranti, ed. Pp. 263–301. Oxford: Wiley-Blackwell.

Ohsawa, S., C.-Y. Ji, and N. Kasai 1998 Age at Menarche and Comparison of the Growth and Performance of Pre- and Post-Menarcheal Girls in China. American Journal of Human Biology 9:205–212.

Olivieri, F., S. Semproli, D. Pettener, and S. Toselli, 2008 Growth and Malnutrition of Rural Zimbabwean Children (6–17 Years of Age). American Journal of Physical Anthropology 136:214–222.

Park, J. H., B. L. Cho, H. T. Kwon, and C. M. Lee 2009 Developing a Biological Age Assessment Equation Using Principal Component Analysis, and Clinical Biomarkers of Aging in Korean Men. Archives of Gerontology and Geriatrics 49:7–12.

Parker Pearson, M. 1999 Fearing and Celebrating the Dead in Southern Madagascar. *In* The Loved Body's Corruption: Archaeological Contributions to the Study of Human Mortality. J. Downes, and T. Pollard eds. Pp. 9–18. Glasgow: Cruithne Press.

Piaget, J. 1923 Le Langage et la Pensée Chez L'Enfant. Paris: Delachaux et Niestlé.

Piaget, J. 1928 The Child's Conception of the World. London: Routledge and Kegan Paul.

Piaget, J. 1963 Intellectual Operations, and Their Development. Reprinted in The Essential Piaget: An Interpretative Reference and Guide. H. E. Griber, and J. J. Voneche eds. Pp. 342–358 (1977). New York: Basic Books.

Prendergast, S. 2000 To Become Dizzy in Our Turning: Girls, Body-Maps and Gender as Childhood Ends. *In* The Body, Childhood and Society, A. Prout, ed. Pp. 101–124. London: Macmillan.

Prout, A. 2000 Childhood Bodies. *In* The Body, Childhood, and Society. A. Prout, ed. London: Macmillan.

Ramachandran, V. S., and W. Hirstein 1998 The Perception of Phantom Limbs: The D. O. Hebb Lecture. Brain 121:1603–1630.

Resnick, D., and G. Niwayama 1981 Diagnosis of Bone, and Joint Disorders. London: W. B. Saunders.

Riley, M. W. 1987 On the Significance of Age in Sociology. American Sociological Review 52:1–14.

Ríos, L., and H. Cardoso 2009 Age Estimation from Stages of Union of the Vertebral Epiphyses of the Ribs. American Journal of Physical Anthropology. 10.

Robb, J. 1997 Intentional Tooth Removal in Italian Neolithic Women. Antiquity 71:659–669.

Robb, J. 2002 Time, and Biography: Osteobiography of the Italian Neolithic life-span. *In* Thinking Through the Body: Archaeologies of Corporeality. Y. Hamilakis, M. Pluciennik, and S. Tarlow, eds. Pp. 153–171. New York: Kluwer Academic/Plenum Publishers.

Roebuck, J. 1979 When Does Old Age Begin? The Evolution of the English Definition. Journal of Social History 12:416–428.

Rogers, J., and T. Waldron 1995 A Field Guide to Joint Disease in Archaeology. Chichester: Wiley.

Rothschild, N. A. 2008 Colonised Bodies, Personal, and Social. *In* Past Bodies. Body-Centred Research in Archaeology. D. Borić, and J. Robb eds. Pp. 135–144. Oxford: Oxbow.

Saul, F. P., and J. M. Saul 1989 Osteobiography: A Maya Example. *In* Reconstruction of Life from the Skeleton. M. Y. Iscan, and A. R. Kennedy, eds. Pp. 287–301. New York: Liss.

Schlanger, N. 2006 Marcel Mauss: Techniques, Technology, and Civilisation. Oxford: Durkheim Press.

Schour, I., and M. Massler 1941 The Development of the Human Dentition. Journal of the American Dental Association 28:1153–1160.

Schutkowski, H. ed. 2008a Between Biology and Culture. Cambridge: Cambridge University Press.

Schutkowski, H. 2008b Introduction. *In* Between Biology and Culture. H. Schutkowski ed. Pp. 1–11. Cambridge: Cambridge University Press.

Scribner, S. 1985 Vygotsky's Use of History. In Culture, Communication, and Cognition: Vygotskian Perspectives. J. V. Wertsch, ed. Pp. 119–145. Cambridge: Cambridge University Press.

Sennett, R. 1998 The Corrosion of Character. New York: Norton.

Shilling, C. 1993 The Body and Social Theory. London: Sage.

Shore, B. 1996 Culture in Mind. Cognition, Culture, and the Problem of Meaning. Oxford: Oxford University Press.

Singer, B., G. McLaughlan, C. Robinson, and J. Christie 1998 Epidemiology of Fractures in 15,000 Adults: The Influence of Age and Gender. Journal of Bone and Joint Surgery 80:243–248.

Sofaer, B. 1998 Pain. Principles, Practice, and Patients. Cheltenham: Stanley Thornes.

Sofaer, J. 2004 The Materiality of Age: Osteoarchaeology, Objects, and the Contingency of Human Development. Ethnographisch-Archäologischen Zeitschrift. Heft 2–3:165–180.

Sofaer, J. 2006a The Body as Material Culture. A Theoretical Osteoarchaeology. Cambridge: Cambridge University Press.

Sofaer, J. 2006b Gender, Bioarchaeology, and Human Ontogeny. *In* Social Bioarchaeology of Funerary Remains. R. Gowland, and C. Knüsel, eds. Pp. 155–167. Oxford: Oxbow.

Sofaer Derevenski, J. 2000 Rings of Life: The Role of Early Metalwork in Mediating the Gendered Life Course. World Archaeology 31:389–406.

Sørensen, M. L. S. 2000 Gender Archaeology. Cambridge: Polity Press.

Tarlow, S. 1999 Bereavement and Commemoration: An Archaeology of Mortality. Oxford: Blackwell.

Thane, P. 1978 The Muddled History of Retiring at 60 and 65. New Society 45:234–236.

Tiesler, V. 2002 New Cases of an African Tooth Decoration from Colonial Campeche, Mexico. Homo 52:277–282.

Toren, C. 1994 On Childhood Cognition and Social Institutions. Man 29:979–981.

Toren, C. 1999 Mind, Materiality, and History: Explorations in Fijian Ethnography. London: Routledge.

Toren, C. 2001 The Child in Mind. *In* The Debated Mind: Evolutionary Psychology Versus Ethnography. H. Whitehouse ed. Pp. 155–179. Oxford: Berg.

Toren, C. 2002 Anthropology as the Whole Science of What it Means to Be Human. *In* Anthropology Beyond Culture. R. Fox, and B. King, eds. Pp. 105–124. Oxford: Berg.

Trow, M. 2005 Reflections on the Transition from Elite to Mass To Universal Access: Forms and Phases of Higher Education in Modern Societies Since WWII. *In* International

Handbook of Higher Education. J. J. F. Forest, and P. G. Altbach, eds. Pp. 243–280. Dordrecht: Springer.

Turner, B. S. 1996 The Body and Society. London: Sage.

Valsiner, J. 2000 Culture and Human Development. London: Sage.

Valsiner, J. 2007 Culture in Minds and Societies. New Delhi: Sage.

Valsiner, J., and A. Rosa eds. 2007 The Cambridge Handbook of Sociocultural Psychology. Cambridge: Cambridge University Press.

van Gennep, A. [1908] 1960 The Rites of Passage. Translated by M. B. Vizedom, and G. L. Caffee. Chicago: Chicago University Press.

Vygotsky, L. S. 1962 Thought and Language. Cambridge, MA: MIT Press.

Wacquant, L. 2004 Body and Soul: Ethnographic Notebooks of an Apprentice Boxer. New York: Oxford University Press.

Weale, R. 2008 Biomarkers by Gender. Archives of Gerontology and Geriatrics doi:10. 1016/j. archger. 2008 07. 013.

Welinder, S. 1998 The Cultural Construction of Childhood in Scandinavia, 3500 BC–1350 AD. Current Swedish Archaeology 8:185–205.

Welinder, S. 2001 The Archaeology of Old Age. Current Swedish Archaeology 9:163–178.

Wertsch, J. V. ed. 1985 Culture, Communication, and Cognition: Vygotskian Perspectives. Cambridge: Cambridge University Press.

Whitehouse, H. 2001 Introduction. *In* The Debated Mind. Evolutionary Psychology Versus Ethnography. H. Whitehouse ed. Oxford: Berg.

Whiting, B. ed. 1963 Six Cultures. New York: Wiley.

Wilson, E. O. 1998 Consilience: The Unity of Knowledge. New York: Alfred Knopf.

Wilson, J. 2007 The Social Role of the Elderly in the Early Bronze Age of Central Europe. Unpublished Ph.D. Thesis. University of Cambridge.

Wilson, J. In press Why We Need an Archaeology of Aging, and a Suggested Approach. Norwegian Archaeological Review.

11

It is Not Carved in Bone

Development and Plasticity of the Aged Skeleton

Sabrina C. Agarwal and
Patrick Beauchesne

Introduction

The human skeleton is popularly characterized as a dry and inert material that acts primarily as a soft tissue scaffold and protector of the vital body. While the skeleton does indeed perform these roles dutifully, it is also a dynamic and living tissue that has the ability to grow, mold, and maintain itself over the life course. The dynamic nature of the skeleton resides in its basic biology – at its cellular level, bone tissue is able to respond to the physiological and biomechanical needs of the body. The fact that the skeleton can respond and adapt to the biological and cultural environment in which it resides forms the basis for the central tenets of bioarchaeology. The well-established biocultural approach in bioarchaeology emphasizes the importance of the interaction between humans and their larger social, cultural, and physical environments, recognizing that the skeleton is influenced by environmental variables (see Zuckerman and Armelagos this volume). This approach has been the cornerstone of bioarchaeology in investigating patterns of skeletal health and disease, and is particularly utilized in studies that seek to sort out the influences that may have affected bone aging and bone loss in past populations (Agarwal 2008; Agarwal and Grynpas 1996).

However, even within biocultural models, environmental and cultural effects on skeletal maintenance and bone loss are often viewed as only potential modifiers that are still tightly constrained by biology. For example, while lifestyle factors such as reproductive behavior (parity and/or breastfeeding) (Mays et al. 2006; Poulsen

Social Bioarchaeology. Edited by Sabrina C. Agarwal and Bonnie A. Glencross
© 2011 Blackwell Publishing Ltd

et al. 2001; Turner-Walker et al. 2001) or diet (Martin 1981; Martin and Armelagos 1979, 1985) are considered to influence bone maintenance in the past, they are still only considered as isolated agents that exacerbate inevitable biological (hormonal or genetic) changes to bone loss. As such, indications of bone loss or osteoporosis in the past are generally regarded to reflect the irreversible course of menopause and aging (Macho et al. 2005; Mays 1996; Mays et al. 1998). Further, bioarchaeologists often hypothesize about the influence of environmental factors on bone morphology over a short period of time during the life of an individual(s) or during a distinct phase of the life cycle (typically the adult and post-menopause phase). This is in part due to the nature of archaeological samples that obviously do not permit looking at changes in morphology longitudinally over a given individual's life cycle. Skeletal samples permit only cross-sectional studies of bone loss and fragility and generally attract focus on individuals with unusual pathology, rather than lend themselves to life course approaches in the study of bone health. The result, however, is that while bone loss and fragility fracture have been widely reported and studied in bioarchaeology, they are regarded primarily as the result of skeletal degeneration that reflects senescence of the body (Agarwal 2008). In bioarchaeological studies the focus on bone maintenance and loss is at the end of the life cycle, particularly in females. The a priori assumption is that it is inevitable that women will lose bone and have more fragile skeletons (Agarwal 2008).

Where does this assumption about bone loss and maintenance in bioarchaeology come from, and is it really inevitable that women age into fragile skeletons? The assumption that bone maintenance and bone loss is tied entirely to menopause and old age is well perpetuated in popular biomedicine. While the level of sex steroids plays a vital role in bone maintenance across the life cycle in both sexes, particularly in old age, it is increasingly well known in clinical and epidemiological studies that there are many other biological and environmental influences on bone health that can change the outcome of bone loss and fragility. For example, biomechanical influences (physical activity), reproductive behaviors, diet and nutrition are just some of the factors now known to interact and potentially change the course of adult bone maintenance and loss (Sowers and Galuska 1993; Stevenson et al. 1989; Ward et al. 1995). While bioarchaeologists have strived to investigate environmental influences on bone health in past populations, it seems they are tied to the notion that the biological influences of menopause and senescence are primary. This may be related in part to the fact that bioarchaeological approaches to bone maintenance and aging are also shaped within, and struggle against, the larger framework of biological anthropology that gives primacy to biology and the gene in explaining bone morphology. In these developmental biological frameworks the morphology of the skeleton is seen as limited by regulatory mechanisms and a set range of possible responses in human tissue (Lovejoy et al. 2003). While insights from development biology have been revolutionary in our analyses of the evolution of the human primate skeleton, they should not overshadow the importance of postnatal influences on bone morphology during growth and aging. These nonpredetermined influences do not act in isolation, and often act synergistically with one another and with biological (genetic, hormonal) influences on bone morphology. More importantly, these influences act throughout the life course, beginning even in utero, to shape the skeleton (e.g. Cooper et al. 2006). The adult-aged skeleton, in both its strength and frailty, is the creation of life history and trajectories taken during growth.

In this chapter we examine alternative perspectives on human morphology as the result of development and plasticity, and the specific history of these theories as applied to understanding growth and aging of the human skeleton. We then review some of the applications of developmental approaches in bioarchaeological studies of bone maintenance and loss in past populations. Finally, we explore the new directions in the study of maintenance and aging of the skeleton that are possible with the integration of ideas in both biological and social theory on the role of ontogenetic process and embodied lived experience in the construction of skeletal form.

Theoretical Understandings of Development and Plasticity

Definitions

Plasticity, growth, and development are essential concepts in anthropology as they form the foundation to understanding patterns of phenotypic variability. However, their meaning and use varies within the biological and archaeological literature. *Growth* is generally understood to reflect stepwise or progressive changes in size and morphology during the development of an individual (Scheuer and Black 2004). Growth is generally correlated with chronological age, however, differences in rates of growth are still common between individuals due to divergent life-history trajectories (Scheuer and Black 2004). As such, while growth in size is correlated with biological maturity, they diverge enough so that "individuals reach developmental milestones, or biological ages, along the maturity continuum at different chronological ages" (Scheuer and Black 2004:4). Growth can then be seen as the enlargement and differentiation of tissues advancing with chronological age, while development comprises the pathways of biological milestones along the life course, including for example embryogenesis and puberty. Rates of growth and timing of developmental changes differ between individuals, leading to considerable debate over normal growth and development trajectories (Bogin 1999; Worthman and Kuzara 2005), and the health (Clark et al. 1986; Klaus and Tam 2009; Mays et al. 2008), adaptation (Lasker 1969; Lewis 2007; Roberts 1995; Schell 1995; Worthman and Kuzara 2005) and evolutionary significance of growth rates (Ellison 2005; Nelson and Thompson 1999; Ruff et al. 1994). The complexity and debate on the role of growth and development is exciting as it allows us to explore how gene–environment relationships operate to produce a wide range of phenotypes at different stages of the life course.

Plasticity is a broader utilized concept that is much more difficult to grasp as there are inconsistencies in how the term is used to describe its role in the formation of the adult phenotype through developmental processes. Most of the confusion with the concept of plasticity resides in its conceptual link to adaptation (Lasker 1969; Roberts 1995; Schell 1995). Prior to the 1950s and 1960s the working definition of plasticity was simply an understanding that human morphology appeared to be malleable during growth and development (Bogin 1995). Yet this vague conceptualization of plasticity was purely descriptive and was not amenable to hypothesis testing. Dobzhansky (1957) was one of the first to view plasticity as a form of adaptation. In this view natural selection produces genotypes "that permit their possessors to adjust themselves to a spectrum of environments by homeostatic modification of the phenotype" (Roberts 1995:2). Lasker (1976) is considered to have truly merged plasticity with adaptation and in the process redefine the plasticity concept altogether. Lasker's (1969) view of plasticity operated within three modes of adaptation. The first of these was natural selection itself, where the selection of genotypes directly influences the genetic spectrum of the population (Roberts 1995). The second form of adaptive plasticity, acclimatization, is a nonpermanent physiological and behavioral response that adapts an individual to the immediate environment (Roberts 1995). The third and most important mode (in this discussion) is developmental or ontogenetic adaptation (Roberts 1995). The key features of ontogenetic modifications are that plastic responses operate through growth and development, and that the changes are not reversible and also not heritable (Schell 1995). Numerous others have studied variation in human phenotypes through the lens of plasticity and while they all have their own definition of plasticity, adaptation and a concern with trade-offs are central to most (see Worthman and Kuzara 2005 for an excellent review). While a concern for adaptation has been helpful in trying to fit plasticity into the framework of modern Darwinian thought (e.g. McDade et al. 2008), we believe a broadening of focus would offer a better understanding of the process of plasticity and its role in development.

Plasticity in development

Understanding plasticity in the developmental context beyond a strictly adaptationist model has been put forward by a number of researchers (Cooper et al. 2006; Lewontin 2001; Oyama 2000a; Sofaer 2006; Worthman and Kuzara 2005) often using terms such as developmental plasticity, developmental systems theory or approach (DST/DSA), and developmental dynamics. All of these approaches share a general concern with the developmental processes in embryogenesis, fetal growth, early postnatal growth, and adolescence that give rise to variation through plastic responses. While these areas of research have much common ground, there are differences in nomenclature and conceptual divides about the limits of plasticity. Plasticity studies working primarily in fetal development (e.g. Hallgrimsson et al. 2002) are conceptualized differently than research that extends plasticity to include infancy, childhood, and adolescence (Fausto-Sterling 2005). In essence, this mirrors the larger tension between the two most prominent approaches, evolutionary developmental biology (EDB) and developmental systems theory (DST) (Table 11.1). Both are concerned with understanding how plasticity operates rather than solely looking at the products and evaluating their adaptive fitness and both give an alternative to reductionist approaches. However, the EDB perspective is limited primarily to embryology/fetal development and is less concerned with postbirth plastic and developmental changes (Hallgrimsson et al. 2002; Robert et al. 2001). Further, in EDB genes are given primacy during development as they are seen to supply the material needs of development (Robert et al. 2001); genes can exist without

Table 11.1 Developmental approaches in biological and social theory that can be used specifically in the study of bone morphology, maintenance, and loss

Theoretical approach	*Primary concepts and interests*	*Some key references*
Evolutionary developmental biology (EDB)	– Interested in the evolution of development (ontogeny)– Focus is on the role of development (particularly embryonic/fetal) in phenotypic evolution– Interested in how developmental modifications effect evolutionary change– The gene is given primacy, and considered the primary unit of inheritance	Hall 1999, 2005 Roberts et al. 2001 Lovejoy et al. 2003
Developmental systems theory (DST)	– Development is considered contingent on context (broadly environment) and can extend well into postnatal growth– Interaction of developmental influences is key (and can include molecular, cellular, organismal, ecological, social, and biogeographical influences)– Developmental information is thought to reside in the interaction of genes and environment– Inheritance is extended to include non-gene factors such as ecological and social resources, and other epigenetic processes	Oyama 2000a, 2000b Oyama et al. 2001 Gray, 1992, 2001 Griffiths and Gray 1994
Life course approaches	– Emphasize the role of physical and social exposures during gestation, childhood, adolescence, young adulthood, and later adult life (e.g., the development and physical manifestations of disease risk)– Focus is on biological, social, and psychosocial pathways that operate over the life course, as well as across generations	Bengston and Allen 1993 Ben-Shlomo and Kuh 2002 Elder et al. 2003 Fausto-Sterling 2005
Embodiment	– As a concept, can be taken to refer to how beings literally biologically incorporate the world in which they exist, including social and ecological variables– Emphasizes the process of creation or transformation of beings and organisms over time as the product engagement with their world	Ingold 1998 Joyce 2005 Krieger 2001, 2009 Sofaer 2006

development, but there is no development without genes. EDB does emphasize the importance of variation, with the goal to observe patterns of variability to better understand underlying developmental systems that can ultimately be linked to how development intersects with natural selection and evolutionary change (Hallgrimsson et al. 2002). Perhaps most importantly, variation in developmental processes is studied in the context of conservation of form, where "individual variation is minimal and seemingly constrained" (Robert et al. 2001:959). The developmental systems theory (DST) or approach diverges from EDB in many ways. DST contrasts with EDB in that variation is primarily focused on in terms of plasticity rather than conservation of form. Developmental information is believed to reside neither in the genes nor the environment, but rather in the interaction of the two (Robert et al., 2001). As such, genes have no primacy in the DST model and plasticity is the defining feature of the development system that is defined as the interplay of all influences on development including the "molecular, cellular, organismal, ecological, social, and biogeographical" (Robert et al. 2001:954). As such development is seen to extend well into postnatal growth (Robert et al. 2001; Worthman and Kuzara 2005). There are a number of examples of this, including neurological growth (Kamm et al. 1990) and immune functions (Worthman 1995).

Common ground between EDB and DST approaches may be argued in the study of epigenetics (Robert et al. 2001). While there are many definitions of epigenetics, it can be broadly defined as the study of genetic and nongenetic interactions on development (Hallgrimsson et al. 2002; Robert et al. 2001). Robert et al. (2001) suggest that epigenetics may be the "practice of what DST proposes," a place for scientific testing of DST. While both DST and EBD approaches advocate for both acceptance of genetic and nongenetic influence during developmental processes, DST goes one step further in suggesting that inheritance is also epigenetic (Robert et al. 2001). For DST theorists again the gene is not the only player in inheritance, and instead inheritance is extended to include ecological, social resources, or other interactants that influence development (Oyama 2000b). As such, epigenetic processes are seen as heritable and are constructed and reconstructed during each life cycle.

Whether nongenetic influences are heritable, particularly in skeletal morphology, continues to remain uncertain. This uncertainty, of whether or not nongenetic forces significantly shape postnatal and intergenerational skeletal morphology, has limited the theoretical explorations of plasticity and development in bioarchaeology. Moreover, EDB paradigms in biological anthropology essentially greatly minimize the role of environmental and postnatal influence on the plasticity of morphology (Lovejoy et al. 2003). The focus for studies of bone plasticity in biological anthropology has thus been primarily on evolutionary and adaptive change, rather than postnatal development over the life course. One thing that does unite all studies of plasticity is a desire to understand the roots of phenotypic diversity. Pritchard (1995) has noted that plastic responses in a given tissue or tissues not only react to external stimuli but also generate their own effects in other tissues. In this context plasticity during development is a generative force in shaping the body as much as a reactive one and should then be viewed as more than a side-note or byproduct of discussions on gene–environment dynamics. Furthermore it is unclear how plasticity can successfully modulate and affect existing genetic networks in widely different developmental and environmental landscapes rather than relying on the evolution of novel genes or genetic pathways to produce phenotypic variation (Young and Badyaev 2007). The interdependency between genes, development, and environment are at the heart of the matter in understanding plasticity. We now turn to discuss how theoretical approaches to plasticity and development been applied in anthropology and bioarchaeology, and specifically how developmental approaches can help us better understand bone maintenance and aging across the life course.

The Concept of Plasticity in Skeletal Growth and Morphology

The formal history of the study of plasticity in anthropology can arguably be said to have begun with Boas (1912), although earlier studies do exist that similarly observed generational changes in growth in migrants (Baxter 1875; Bowditch 1879). Through detailed anthropometric measurements of body size and shape Boas (1912) observed that the children of new immigrants (of European descent) to the United States displayed different growth patterns than their parents. Moreover, he noted the change was accentuated with each generation (Boas 1912). In an earlier work commissioned by the U.S. Congress, Boas (1910:53) remarked "we must speak of plasticity (as opposed to permanence) of types." Boas's 1912 article was pivotal as it presented solid evidence that environmental changes, which included changing cultural milieus, could produce changes in body size and shape in future generations. Growth and adult stature was seen as more than the sole product of genetic heritability. Boas's work was supported by Shapiro's (1939) often-cited growth study of Japanese children in Japan and Hawaii that also showed significant differences in growth, stature, and development, which he also attributed to environmental triggers. Numerous migrant studies have repeatedly confirmed the correlation of changing environments to changes in growth and development (Baker et al. 1986; Bogin 1995; Bogin and Rios 2003; Goldstein 1943; Kasl and Berkman 1983; Lasker and Evans 1961). Plasticity studies were not limited to migrants only; plasticity was studied within cultures as well to account for the fact that those who stayed behind might have differed in some important ways (e.g. Mascie-Taylor 1984). Plasticity was also studied through observation of so-called natural experiments (Roberts 1995). For example, Roberts and Bainbridge (1963) observed a population of three Nilotic tribes living in the same environment but with slight cultural differences. Somatotype and anthropometric measures demonstrated small but significant differences between the three tribes (Roberts 1977). Roberts (1977) concluded that these differences were environmentally based through ways of life and dietary differences in particular. Similar studies among Polynesian and other traditional cultures have observed similar results in cases of changing or differing socioeconomic conditions between two closely related migrant/sedante groups (Baker et al. 1986; Kasl and Berkman 1983).

Schell (1995) has argued that by 1954 with Kaplan's review of migrant-sedante studies in American Anthropologist that plasticity was firmly established as a recognized phenomenon of growth and development. Research into plasticity has faced numerous challenges since Boas first set out to develop a new model for how phenotypes vary. A dominant challenge, as noted previously, has been defining the concept of plasticity itself. This appears to have become an interpretive problem only after Lasker (1969) permanently tied plasticity to adaptation. There is much we do not know about the adaptiveness or relative benefit of plastic modifications made during growth and development, in part because of the difficulty of interpreting growth patterns (Humphrey 2000; Lewis 2007; Saunders 2000; Schell 1995; Worthman and Kuzara 2005). Two general interpretations have been put forward in attempts to understand variation in growth and development. Both models address the issue of morphological variation and compare stress and health among and between cultures from an adaptationist perspective, but from very different theoretical positions. The first of these interpretations is the "medical model" common in public health policies, pediatrics, and nutritional science (Schell 1995). The medical model views growth as a reflection of health, and with this it literally becomes a measure of health and consequently, of adaptation (Schell 1995). Growth to the full extent of an individual's genetic potential is interpreted as good health while slow or stunted growth signifies ill health (Schell 1995). The implicit assumption is that the body will always reach its full genetic potential if no boundaries are presented. Determining this with archaeological skeletons is difficult given that retarded growth and development may not show clear outward signs (Humphrey 2000). The opposing model is termed the human adaptability paradigm (HAP) (Schell 1995). The HAP views growth and development as the mechanism of plasticity (Schell 1995). In other words, "growth patterns can be a mode of adaptation" and in "this context growth is a means of achieving an adapted state rather than a result of that adaptation" (Schell 1995:223 emphasis in original). The problem here is that the modifications that reduce stress/strain can be seen as adaptive but they cannot be proven so in a strict sense (Bogin 1995; Schell 1995). Further, the medical and HAP models conflict because growth cannot be both a measure and a means of adaptation (Schell 1995). As such, the models are mutually exclusive. Clearly this poses a problem for which model to use in bioarchaeology. To some degree this may depend on what influences or stressors are being considered as causes for the observed plastic changes. Schell (1995) has offered that the medical model may be better suited to interpreting plasticity as a feature of human-made environmental changes, such as slums, where nutrition is poor and disease load high, while the HAP may be beneficial for interpreting plastic responses induced by the physical environment. However, it is unclear how to structure bioarchaeological research questions and analyses when typically both human-made and naturally occurring environmental factors are at play. While both these models have contributed significantly to studies of growth and development, neither works fully when applied to bioarchaeological analyses, particularly to the interpretation of patterns of bone maintenance and loss.

Considering development as a generative force (Oyama 2000a), rather than a "reading out" of genetic material during key periods of growth, may help us better understand how the human body and skeleton is shaped and reformed throughout life. Recent biomedical and epidemiological studies have specifically explored how plasticity during growth and development can influence aspects of lifelong bone health, such as bone mineral density and loss. For example, infant and adolescent growth spurts seem to be highly influential in defining bone quality and quantity at later life stages (Cooper et al. 2001, 2006; Javaid and Cooper 2002; Javaid et al. 2006; Miller 2005). Peak bone mass (the maximal amount of bony tissue accrued during growth) is generally thought to be mostly inherited (Duncan et al. 2003), but Cooper et al. (2002:391) remark that "only a small proportion of the variation in individual bone mass" is accounted by genetic markers. Seeman (1999:91) has also noted that the contribution of heritability in bone health is not a constant proportion, and that statements claiming "80 percent of areal BMD (bone mineral density) is genetically determined leaving only 20 percent to modify" is flawed. Heritability is a complex, fluid measure based on a relationship between population and environment variance (Seeman 1999). As age, height, gender, and body composition vary, so do heritability measures of bone mass or density (Seeman 1999). Cooper et al. (2006) posit that environmental cues early in life interact with the genome to create the boundaries of growth and development for a given individual. It has been hypothesized that these types of developmental boundaries or trajectories may originate in expectation of future environmental conditions and serve as predictive adaptive responses (or PARS, Gluckman and Hanson 2005). For example, fetal programming by maternal under-nutrition is a risk factor for low birth weight (Cooper et al., 2002). Low birth weight is strongly correlated with lower levels of basal level growth hormones, even during adult life, placing the individual at risk for lower peak bone mass, reduced mineralization, and an elevated rate of bone loss later in life (Cooper et al. 2002; Dennison et al. 2005). Numerous epidemiological studies have shown that impaired fetal and childhood growth place individuals at risk for fragility fractures later in life (Cameron and Demerath, 2002; Cooper et al. 1995, 1997, 2001; Dennison et al. 2004; Dennison et al. 2005; Gale et al. 2001). These studies emphasize the dramatic role of environmental influences on phenotypic plasticity in early life, and more importantly underscore how this early exposure can change the trajectory of development and aging of skeletal morphology throughout life.

Studies of Plasticity in Bone Development and Maintenance in Bioarchaeology

The general concept of skeletal plasticity is fundamental in bioarchaeology, particularly in the study of temporal and spatial differences in skeletal morphology as related to influences such as nutrition, activity or disease (Bogin 1999; Hind and Burrows 2007; Knüssel 2000; Larsen 1999; Lewis and Gowland 2007; Lloyd and Cusatis 1999; Mcdade et al. 2008; Prentice et al. 2006; Rauch 2005; Ruff et al. 2006; Saggese et al. 2002; Schwartz et al. 2003; Skerry 2006). However, studies of plasticity in growth and development in past populations have largely followed approaches developed in studies of living human biology. Most notably, patterns of long bone growth in archaeological skeletal samples have been widely used as a proxy for comparing health and stress statuses between and among populations (Humphrey 2000; Kemkes-Grottenthaler 2005; Lewis 2007; Mays et al. 2008; Saunders 2000). In studies of bone maintenance and loss in bioarchaeology the focus has been primarily on the influence of nutrition and levels of physical activity in either encouraging, or protecting against, the onset of age- and sex-related bone loss and fragility (Agarwal 2008; Agarwal and Glencross in press). There has been some study of the affect of early growth and development on the maintenance of the mature skeleton in archaeological samples. For example, the classic studies of bone loss in prehistoric Sudanese Nubia were some of the first studies to consider and compare bone growth and maintenance in both juvenile and adult skeletons. Armelagos et al. (1972) suggested that the significant cortical bone loss in the femur found in young-aged female Nubians, as compared to males, was likely due to early growth disturbance and stress as young adults during pregnancy and lactation. Similarly, a study of cortical bone growth maintenance in prehistoric juvenile Nubians from the Kulubnarti site found that while bone mineral content increases after birth, processes of modeling combined with likely periods of nutritional stress, cause a reduction in percent cortical area during early and late childhood (Van Gerven et al. 1985), although this study does not comment on the role of early bone maintenance on later femoral bone loss. Two recent studies have focused on the structural variation of trabecular bone during ontogeny. Kneissel et al. (1997) examined the ontogeny and aging patterns of vertebral trabecular bone in a juvenile and adult skeletal sample from Medieval Lower Egyptian Nubia. The authors found the largest bone trabecular volume during adolescence when the rod-like trabeculae of childhood begin to change to plate-like structures. In addition, age-related loss of trabecular structure was observed in adults, with changes occurring earlier than those seen in modern populations (Kneissel et al. 1997). Gosman and Ketcham (2009) also examined patterns of ontogeny in trabecular bone in their study of tibial bones from the prehistoric Ohio Valley, particularly noting changes in trabecular structure and connectivity from growth to skeletal maturity and with increasing ambulatory activities.

More recent studies have attempted to more directly correlate growth patterns and developmental stress, with variation in skeletal morphology and bone loss. For example, a study by Rewekant (2001) examined the correlation of adult cortical bone loss with indicators of growth disturbance (specifically compression of the skull base and vertebral stenosis) in two Polish medieval populations with differing socioeconomic status. Rewekant (2001) found greater adult age-related cortical bone loss in the metacarpal in the population that also showed greater disturbance of bone growth during childhood. Interestingly, lower sexual dimorphism in measurements of metacarpal cortical bone and skull base height were also found in the population that appeared to have suffered greater environmental stress during growth. This study suggests a relationship between the disturbance of growth and the achievement of peak bone mass, as well as the age- and sex-related patterns of bone loss later in life. Similarly, McEwan et al. (2005) examined the correlation of bone quantity in the radius to overall growth patterns and indicators of growth disturbances typically attributed to poor nutrition (specifically Harris lines, and cribra orbitalia) in juvenile skeletons in a medieval British sample. The authors found that while bone mineral density (BMD) was well correlated to overall growth, cortical index (a measure of total cortical bone) was not (McEwan et al. 2005). This again suggests that some aspects of bone maintenance such as the overall amount of cortical bone may be compromised during development under the influence of environmental (nutritional) stress with a lasting effect on cortical bone content and morphology well into adulthood (Mays 1999; McEwan et al. 2005).

There has also been focus on influences after childhood, into young adulthood that may play a significant role in later bone fragility. For example, several bioarchaeological studies within and between skeletal populations suggest that physical activity during adulthood can result in a conservation of bone quantity during life and offer protection against the affects of bone loss in old age (Ekenman et al. 1995; Lees et al. 1993). The opposite has also been noted, with observations of decline in bone quantity and strength in more sedentary agricultural populations as compared to physically active hunter-gatherer groups (Larsen 2003; Ruff et al. 1984; Ruff et al. 2006), although this observation is not universal as workloads were likely variable in agriculturalists depending on region and local terrain (Larsen 2003; Nelson et al. 2002). While it is known that bone tissue responds to mechanical loading, the biomedical literature is unclear on what type and level of physical activity or exercise is needed to affect bone mass and more importantly bone strength into adulthood. There may be an ideal "window of opportunity" for physical activity to contribute to the growth and robusticity of the skeleton during the acquisition of peak bone mass (Pearson and Lieberman 2004), but it seems likely that some high strain stress activity may still be effective at older ages (Rittweger 2006). Reproductive behavior is another factor that may influence the trajectory of bone maintenance and loss in older age. Several studies have suggested that young age females in the archaeological record show evidence of bone loss that is result of physiological stress on the skeleton due to pregnancy and/or breastfeeding (Martin and Armelagos 1979, 1985; Martin et al. 1985, 1984; Mays et al. 2006; Poulsen et al. 2001; Turner-Walker et al. 2001). However, it can be argued that the loss of bone in reproductive-age women in the past was transitory, and that bone loss during reproduction would have little or no affect on long-term bone fragility in women who would have survived to old age (Agarwal 2008; Agarwal and Grynpas 2009; Agarwal et al. 2004; Agarwal and Stuart-Macadam 2003). In fact, high parity and prolonged breastfeeding in some past populations would have provided women in the past with a very different hormonal milieu and steroid exposure that could have offered protection against the sudden postmenopausal drop of hormones experienced by modern women (Agarwal et al. 2004; Agarwal 2008; Weaver 1998).

All of these studies take the first step in exploring the role of development in bone morphology and maintenance, and emphasize the importance of earlier life experiences on the strength and fragility of the aged skeleton. While influences such as nutrition, physical activity, and reproduction are critical to understanding bone growth and maintenance, it is increasingly evident that what is really important is how these influences are played out over the life course, and the cumulative effect that they may have on the skeleton at the end of life.

Pushing Beyond Plasticity and Adaptation: The (Re)construction of the Skeleton Through Time

Despite the numerous studies of bone aging and osteoporosis in bioarchaeology, the etiology of bone loss in the past remains unclear (Agarwal 2008; Agarwal and Grynpas 1996, 2009). Paleo-populations of similar temporal or spatial origin show similar patterns of bone loss, while others do not, and most differ from the typical age- and sex-related patterns of bone loss and fragility observed in modern Western populations (Agarwal 2008). For example, bone loss is often seen in young age and equal in both males and females, and there is a low prevalence of fragility fracture in comparison to modern populations (Agarwal 2008; Agarwal and Grynpas 1996, 2009; Brickley and Agarwal 2008). The explanation for these observed patterns in the bioarchaeological record is complex, and the use of often incomplete and biased skeletal samples is an ongoing issue in the analysis of any indicator of health and disease in the past (see also Jackes this volume). However, the variable patterns of bone maintenance fragility in the past are also not surprising given that groups in the past would have had very different biosocial histories from our own. The fact that human skeletons in the archaeological record vary in overall morphology and indicators of skeletal health, such as bone maintenance and loss, perfectly illustrates the plasticity and development of the body. Yet, recent bioarchaeological studies have used familiar patterns of bone loss in the past to ratify traditional paradigms of aging in the female skeleton, while discounting patterns that simply do not fit a *priori* expectations. How then can we begin to make more meaningful interpretations of bone maintenance, loss, and aging in the past?

More comprehensive explanations for the observed patterns of bone loss may be gained through the appreciation of the cumulative nature of bone maintenance over the life cycle. For example, a second look at the patterns of bone loss and fragility in the British medieval skeletal sample, Wharram Percy, discussed earlier, from a life course perspective offers new insights on bone health. The Wharram Percy sample shows evidence for age- and sex-related cortical bone loss at multiple bone sites typical of modern populations, and has been used to illustrate how despite historic lifestyle practices, human females are inevitably subject to menopausal and age-related bone loss (Mays 1996; Mays et al. 1998). However, the same population also shows evidence for stress-related reduction of bone mass during growth (Mays 1999; McEwan et al. 2005) that would have changed the trajectory of bone maintenance later in life regardless of expected changes with senescence or menopause, and little or no evidence for typical fragility fracture (Agarwal and Grynpas 2009; McEwan et al. 2005). Further, study of bone loss in the trabecular bone tissue of the vertebrae shows atypical patterns of bone loss in females that suggests that other factors such as reproductive behavior may have played a role in bone loss in young adulthood. This may have conserved and strengthened bone post-menopause (Agarwal et al. 2004; Agarwal and Grynpas 2009). While there are many hypotheses that can be suggested from the patterns of bone maintenance and loss at Wharram Percy, none support the notion that bone loss in the rural medieval population was solely an outcome of aging and menopause. What is evident is that bone maintenance and loss is the result of ontogenetic processes over the life course, with trajectories of bone maintenance laid out in early growth, refined during adulthood, and played out and modified within the everyday individual and generational choices of behavior and life experience (Figure 11.1). Observing one snapshot of bone maintenance at one scale (such as bone loss only in adulthood; one area of the skeleton, or using one methodology) will give a skewed perspective on the complex and unique path that has created the observed bone morphology.

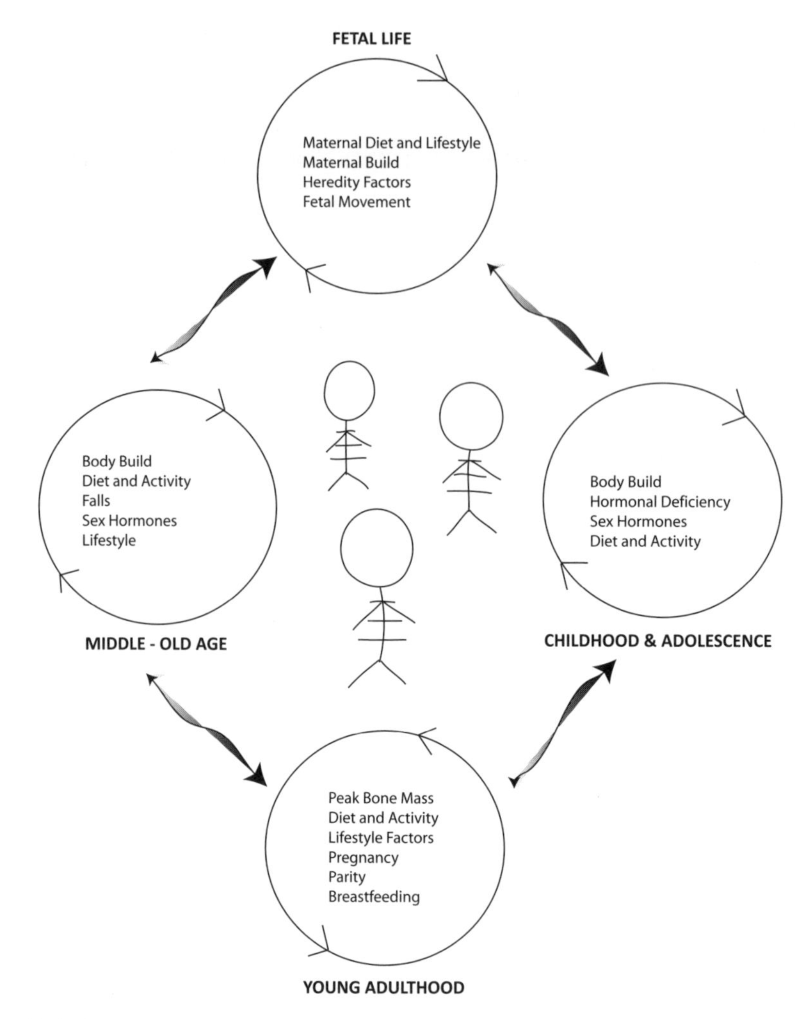

Figure 11.1 Diagrammatic model of the plasticity in development and maintenance of the skeleton over the life course.

Figure 11.1 shows a diagrammatic model of the plasticity in development and maintenance of the skeleton over the life course. Circles represent major periods of the biological life cycle (fetal life, childhood and adolescence, young adulthood, and middle/old age) each containing examples of some of the major influences within each life stage in human skeletal development. Influences within each stage are interdependent (represented with arrows around each circle), and influences in each stage are cumulative and dependent on influences in earlier life stages (represented as arrows between circles). Cumulative influences shape skeletal morphology, and affect bone maintenance and bone loss. These influences account for the variation in skeletal morphology and maintenance observed over an individual's life course, as well as within communities/populations (represented by varied skeletal figures in middle of model). Note, interdependent arrows are shown even between middle/old age and the fetal life stage, as representation of the potential intergenerational effect of individual and population life history to skeletal morphology in subsequent generations.

This life course perspective of plasticity and development of the skeleton is suitably grounded in the DST approach to the development of organisms discussed earlier. DST approaches emphasize the interaction of both environmental and biological influences on the development of the organism that occur over the entire life cycle (Oyama 2000a). Fausto-Sterling (2005) has applied this model specifically to understanding skeletal morphology and osteoporosis in modern humans. Borrowing from life course approaches that have been used in the study of chronic diseases, Fausto-Sterling (2005) highlights the cumulative nature that influences have on bone health, and suggests that prior events during life can alter the trajectory of bone development in later points of the life cycle. Life course approaches extend this model of "critical periods" in fetal development (Ben-Shlomo and Kuh 2002), and suggest there could also be modifiers on bone form and health later in life (Table 11.1).

The concept of the body as a product of developmental context (both biological and social) is not limited to DST and life course approaches, and is also found in archaeological perspectives on embodied life experience (Table 11.1). Ingold (1998) has argued that the body is a developmental system that is contextually dependent, and that more importantly humans grow and are active in their development through engagement with the social world. This engagement with the world in which bodies are situated can be both conscious (with agency) or unconscious (Krieger 2001, 2009), and dilutes the belief that organisms are primarily passively built by their genetic code. In her discussion of skeletal markers of gendered behavior in archaeological skeletons, Sofaer (2006:161) notes that while it may be difficult to directly correlate skeletal markers with distinct activities or lifeways in the past, "plasticity of the body means that the body is never pre-social and is contextually dependent". There is no pristine bodily state that is outside of the environmental and cultural context in which it operates (Oyama 2000a). This is not to say that the plasticity and development of organisms are limitless (Oyama 2000a; Sofaer 2006). Bone's ability to shape itself is bound by, among other things, genetics, environment, age, and sex (Oyama 2000a; Hallgrimsson et al. 2002, Lovejoy et al. 2003; Pearson and Lieberman 2004; Ruff et al. 2006; Sofaer 2006). For example, processes such as canalization and developmental stability tightly control fetal skeletal development (Hallgrimsson et al. 2002). However, novel or stressful environments can reduce the ability of these processes from limiting variation (Young and Badyev 2007). While the traditional view gives the gene formative power as keeper of the plan or code, the developmental perspective sees the gene not as an information-containing device, but as an information-generating device that depends on immediate environment (Oyama 2000a). While bioarchaeologists do not dismiss the notion that genes and the environment interact, it is that the flow of information in development is thought to move outwards in one direction from the genome, which then interacts with the environment (see Oyama 2000b). This leads to the idea that there are "two kinds of developmental processes, one controlled primarily from the inside and another more open to external forces" (Oyama 2000b:21). What this means for our discussion, is that while the coded forces of bone physiology and senescence play vital roles in bone growth and maintenance, they need to be viewed as interwoven in a larger developmental process driven by cumulative life experience. While it may be suggested that the focus on life experience limits the exploration of bone morphology and health to the individual context, these theoretical approaches to the body and development over the life course are inherently intergenerational. For example, epidemiological life course approaches contextualize early life exposure in structures that include the role of parents, grandparents, households, and communities (Ben-Shlomo and Kuh 2002). Here biological and social risk is seen as playing across entire generations. DST approaches go one step further, extending what we traditional think of as heredity. Inheritance is seen as more than the passing of a trait or blueprint, but instead the transmission of entire developmental contexts, which can include genes, cellular machinery as well as social and ecological systems (Oyama et al., 2001; Robert et al. 2001). Social and environment context are seen as potential intergenerational influences on the phenotypic variation of the skeleton. As such, skeletal variation in bone maintenance and loss potentially could be the result of developmental processes that have acted at the level of the individual, generations, or entire communities. This has great relevance for how bioarchaeologists observe variation in not only bone maintenance but all aspects of bone morphology.

Lived experiences over the entire life cycle build the final skeleton observed in the archaeological record. In this sense plasticity is viewed as more than adaptation to specific environmental context. Instead through a developmental process, plasticity constructs and reconstructs the body and skeleton again and again over the life course, and potentially over generations of multiple life cycles. The quantity and quality of bone tissue is an exceptional bony indicator in the analyses of past life as it literally reflects the lived experience of the body crafted at the cellular level through bone remodeling. The trajectory of skeletal development is not "carved in stone," and similarly the fate of degeneration with aging and menopause is not inevitably carved onto the bony tissue at birth. The fact that patterns of bone maintenance, loss, and fragility are different among and between human populations provides an opportunity to go beyond biological reductionism toward hypotheses of biosocial development of whole beings and populations.

REFERENCES

Agarwal, S. C. 2008 Light and Broken Bones: Examining and Interpreting Bone Loss and Osteoporosis in Past Populations. In Biological Anthropology of the Human Skeleton, 2nd Edition. M. A. Katzenburg, and S. R. Saunders eds. Pp. 387–442. New York: Wiley-Liss.

Agarwal, S. C., and M. Grynpas 1996 Bone Quantity and Quality in Past Populations. Anatomical Record 246:423–432.

Agarwal, S. C., and M. Grynpas 2009 Measuring and Interpreting Age-Related Loss of Vertebral Bone Mineral Density in a Medieval Population. American Journal of Physical Anthropology 139:244–252.

Agarwal, S. C., M. Dumitriu, G. Tomlinson, and M. Grynpas 2004 Medieval Trabecular Bone Architecture: The Influence of Age, Sex, and Lifestyle. American Journal of Physical Anthropology 124:33–44.

Agarwal, S. C., and B. A. Glencross in press. Examining Nutritional Aspects of Bone Loss and Fragility Across the Life Course in Bioarchaeology. In Biosocial Perspectives on Human Diet and Nutrition. T. Moffat, and T. Prowse eds. Oxford: Berghahn Press.

Agarwal, S. C., and P. Stuart-Macadam 2003 An Evolutionary and Biocultural Approach to Understanding the Effects of Reproductive Factors on the Female Skeleton. In Bone Loss and Osteoporosis: An Anthropological Perspective. S. C. Agarwal, and S. D. Stout, eds. Pp. 105–120. New York: Kluwer Academic/Plenum Publishers.

Armelagos, G., J. Mielke, K. Owen, D. Van Gerven, J. Dewey, and P. Mahler 1972 Bone Growth and Development in Prehistoric Populations from Sudanese Nubia. Journal of Human Evolution 1:89–119.

Baker, P. T., M. Hannah, J. and T. S. Baker 1986 The Changing Samoans. Oxford: Oxford University Press.

Baxter, J. H. 1875 Statistics, Medical and Anthropological, of Over a Million Recruits. Washington, DC: Government Printing Office.

Bengston, V. L., and K. R. Allen 1993 The Life Course Perspective Applied to Families Over Time. In Sourcebook of Families, Theories, Methods: A Contextual Approach. P. Boss, W. J. Doherty, R. LaRosaa, W. R. Schumm, and S. K. Steimetz, eds. New York: Springer.

Ben-Schlomo, Y., and D. Kuh 2002 A Life Course Approach to Chronic Disease Epidemiology: Conceptual Models, Empirical Challenges, and Interdisciplinary Perspectives. International Journal of Epidemiology 31:1–9.

Boas, F. 1910 Changes in Body Form of the Descendents of Immigrants. Senate Document 208, 61st Congress, 2nd session. Washington, DC.

Boas, F. 1912 Changes in Body Form of the Descendents of Immigrants. New York: Columbia University Press.

Bogin, B. 1995 Plasticity in the Growth of Mayan Refugee Children Living in the United States. In Human Variability and Plasticity. C. G. N. Mascie-Taylor, and B. Bogin, eds. Pp. 46–74. Cambridge: Cambridge University Press.

Bogin, B. 1999 Patterns of Human Growth. Cambridge: Cambridge University Press.

Bogin, B., and L. Rios 2003 Rapid Morphological Change in Living Humans: Implications for Modern Human Origins. Comparative Biochemistry and Physiology, Part A 136:71–84.

Bowditch, H. P. 1879 The Growth of Children, a Supplementary Investigation, pp. 25–62. Boston, State Board of Health of Massachusetts.

Brickley, M., and S. C. Agarwal 2003 Techniques for the Investigation of Age-Related Bone Loss and Osteoporosis in Archaeological Bone. In Bone Loss and Osteoporosis: An Anthropological Perspective. S. C. Agarwal, and S. D. Stout, eds. Pp. 157–172. New York: Kluwer Academic/Plenum Publishers.

Cameron, N., and E. W. Demerath 2002 Critical Periods in Human Growth and Their Relationship to Diseases of Aging. American Journal of Physical Anthropology 119:159–184.

Clark, G., N. Hall, G. Armelagos, G. Borkan, M. Panjabi, and F. Wetzel 1986 Poor Growth Prior to Early Childhood: Decreased Health and Life-Span in the Adult. American Journal of Physical Anthropology 70:145–160.

Cooper, C., M. I. D. Cawley, A. Bhalla 1995 Childhood Growth, Physical Activity and Peak Bone Mass in Women. Journal of Bone and Mineral Research 10:940–947.

Cooper, C., J. G. Eriksson, T. Forsen, C. Osmond, J. Tuomilehto, and D. J. P. Barker 2001 Maternal Height, Childhood Growth and Risk of Hip Fracture in Later Life: A Longitudinal Study. Osteoporosis International 12:623–629.

Cooper, C., C. Fall, P. Egger, R. Hobbs, R. Eastell, and D. Barker 1997 Growth in Infancy and Bone Mass Later in Life. Annals of the Rheumatic Diseases 56:17–21.

Cooper, C., M. K. Javaid, P. Taylor, K. Walker-Bone, E. Dennison, and N. Arden 2002 The Fetal Origins of Osteoporotic Fracture. Calcified Tissue International 70: 391–394.

Cooper, C., S. Westlake, N. Harvey, K. Javaid, E. Dennison, and M. Hanson 2006. Review: Developmental Origins of Osteoporotic Fracture. Osteoporosis International 17:337–347.

Dennison, E. M., H. E. Syddall, S. Rodriguez, A. Voropanov, I. N. Day, and C. Cooper, 2004 Polymorphism in the Growth Hormone Gene, Weight in Infancy, and Adult Bone Mass. Journal of Clinical Endocrinology and Metabolism 89:4898–4903.

Dennison, E. M., H. E. Syddall, A. Sayer, H. Gilbody, and C. Cooper 2005 Birth Weight and Weight at 1 Year are Independent Determinants of Bone Mass in the Seventh Decade: The Hertfordshire Cohort Study. Pediatric Research 5:582–586.

Dobzhansky, T. 1957 Evolution, Genetics and Man. New York: John Wiley.

Duncan, E., L. Cardon, J. Sinsheimer, J. Wass, and M. Brown 2003 Site and Gender Specificity of Inheritance of Bone Mineral Density. Journal of Bone and Mineral Research 18:1531–1538.

Elder, G., M. K. Johnson, and R. Crosnoe 2003 The Emergence and Development of Life Course Theory. In Handbook of the Life Course. B. J. Lee, J. T. Mortimer, and M. J. Shanahan, eds. Pp. 3–19. New York: Plenum Press.

Ellison, P. 2005 Evolutionary Perspectives on the Fetal Origins Hypothesis. American Journal of Human Biology 17:113–118.

Ekenman, I., S. A. Eriksson, and J. U. Lindgren 1995 Bone Density in Medieval Skeletons. Calcified Tissue International 56:355–358.

Fausto-Sterling, A. 2005 The Bare Bones of Sex: Part 1 – Sex and Gender. Signs: Journal of Women in Culture and Society 30:1491–1527.

Gale, C. R., C. N. Martyn, S. Kellingray, R. Eastell, and C. Cooper 2001 Intrauterine Programming of Adult Body Composition. Journal of Clinical Endocrinology and Metabolism 86:267–272.

Goldstein, M. S. 1943 Demographic and Bodily Changes in Descendents of Mexican Immigrants. Austin: University of Texas, Institute of Latin American Studies.

Gosman, J. H., and R. A. Ketcham 2009 Patterns in Ontogeny of Human Trabecular Bone from Sunwatch Village in the Prehistoric Ohio Valley: General Features of Microarchitectural Change. American Journal of Physical Anthropology 138:318–332.

Gluckman, P., and M. Hanson 2005 The Fetal Matrix: Evolution, Development and Disease. Cambridge: Cambridge University Press.

Gray, R. 1992 Death of the Gene: Developmental Systems Strikes Back. *In* Tree of Life: Essays in the Philosophy of Biology. Pp. 165–209. Boston, MA: Kluwer.

Gray, R. 2001 Selfish Genes or Developmental Systems? In Thinking about Evolution: Historical, Philosophical, and Political Perspectives. S. R. Krimbas, D. Paul, and J. Beatty, eds. Cambridge: Cambridge University Press.

Griffiths P. E., and R. Gray 1994 Developmental Systems and Evolutionary Explanations. Journal of Philosophy 91:277–304.

Hall, B. K. 1999 Evolutionary Developmental Biology. 2nd Edition. Amsterdam: Kluwer Academic.

Hall, B. K. 2005 Bones and Cartilage: Developmental and Evolutionary Skeletal Biology. London: Elsevier.

Hallgrimsson, B., K. Willmore, and B. Hall 2002 Canalization, Developmental Stability, and Morphological Integration in Primate Limbs. Dedicated to the Memory of Nancy Hong. American Journal of Physical Anthropology 119:131–158.

Hind, K., and M. Burrows 2007 Weight-Bearing Exercise and Bone Mineral Accrual in Children and Adolescents: A Review of Controlled Trials. Bone 40:14–27.

Humphrey, L. T. 2000 Growth Studies of Past Populations: An Overview and an Example. In Human Osteology in Archaeology and Forensic Science. M. Cox, and S. Mays, eds. Pp. 23–38. Cambridge: Cambridge University Press.

Ingold, T. 1998 From Complementarity to Obviation: On Dissolving the Boundaries Between Social and Biological Anthropology, Archaeology, and Psychology. Zeitschrift für Ethnologie 123:21–52.

Javaid, M., and C. Cooper 2002. Prenatal and Childhood Influences on Osteoporosis. Best Practice and Research Clinical Endocrinology and Metabolism 16:349–367.

Javaid, M., S. Lekamwasam, J. Clark, E. Dennison, H. Syddall, N. Loveridge, J. Reeve, T. Beck, and C. Cooper 2006 Infant Growth Influences Proximal Femoral Geometry in Adulthood. Journal of Bone and Mineral Research 21:508–512.

Joyce, R. A. 2005 Archaeology of the Body. Annual Review of Anthropology. 34:139–158.

Kamm, K., E. Thelen, and J. L. Jensen 1990 A Dynamical Systems Approach to Motor Development. Physical Therapy 70:763–775.

Kasl, S. V., and L. Berkman 1983 Health Consequences of the Experience of Migration. Annual Review of Public Health 4:69–90.

Kemkes-Grottenthaler, A. 2005 The Short Die Young: The Interrelationship Between Stature and Longevity – Evidence from Skeletal Remains. American Journal of Physical Anthropology 128:340–347.

Klaus, H. D., and M. E. Tam 2009 Contact in the Andes: Bioarchaeology of Systemic Stress in Colonial Mórrope, Peru. American Journal of Physical Anthropology 138:356–368.

Kneissel, M., P. Roschger, W. Steiner, D. Schamall, G. Kalchauser, A. Boyde, and M. Teschler-Nicola 1997 Cancellous Bone Structure in the Growing and Aging Lumbar Spine in a Historic Nubian Population. Calcified Tissue International 61:95–100.

Knüssel, C. 2000 Bone Adaptation and Its Relationship to Physical Activity in the Past. In Human Osteology in Archaeology and Forensic Science. M. Cox, and S. Mays, eds. Pp. 381–401. Cambridge: Cambridge University Press.

Krieger, N. 2001 Theories for Social Epidemiology in the 21st Century: An Ecosocial Perspective. International Journal of Epidemiology 30:668–677.

Krieger, N. 2005 Embodiment: A Conceptual Glossary for Epidemiology. Journal of Epidemiology and Community Health 59:350–355.

Larsen, C. 1999 Bioarchaeology: Interpreting Behavior from the Human Skeleton. Cambridge: Cambridge University Press.

Larsen, C. 2003 Animal Source Foods and Human Health During Evolution. Journal of Nutrition 133:3893S–3897S.

Lasker, G. W. 1969 Human Biological Adaptability. Science 166:1480–1486.

Lasker, G. W. 1976 Physical Anthropology, 2nd Edition. New York: Holt Rinehart and Winston.

Lasker, G. W., and Evans, F. G. 1961 Age, Environment and Migration: Further Anthropometric Findings on Migrant and Non-Migrant Mexicans. American Journal of Physical Anthropology 19:203–211.

Lees, B., T. Molleson, T. R. Arnett, and J. C. Stevenson 1993 Differences in Proximal Femur Bone Density Over Two Centuries. Lancet 341:673–675.

Lewis, M. E. 2007 The Bioarchaeology of Children: Perspectives from Biological and Forensic Anthropology. Cambridge: Cambridge University Press.

Lewis, M. E., and R. Gowland 2007 Brief and Precarious Lives: Infant Mortality in Contrasting Sites from Medieval and Post-Medieval England (AD 850–1859). American Journal of Physical Anthropology 134:117–129.

Lewontin, R. C. 2001 Gene, Organism and Environment. In Cycles of Contingency: Developmental Systems and Evolution. S. Oyama, P. E. Griffiths, and R. D. Gray, eds. Pp. 59–66. Cambridge; MIT Press.

Lloyd, T., and D. C. Cusatis 1999 Nutritional Determinants of Peak Bone Mass. In The Aging Skeleton. G. J. Rosen, J. Glowacki, and J. P. Bilezikian, eds. Pp. 95–104. San Diego: Academic Press.

Lovejoy, C., M. McCollum, P. Reno, and B. Rosenman 2003 Developmental Biology and Human Evolution. Annual Review of Anthropology 32:85–109.

Macho, G. A., R. L. Abel, and H. Schutkowski 2005 Age Changes in Bone Microstructure: Do They Occur Uniformly? International Journal of Osteoarchaeology 15:421–430.

Martin, D. L. 1981 Microstructural Examination: Possibilities for Skeletal Analysis. Amherst: Department of Anthropology, University of Massachusetts.

Martin, D. L., and G. J. Armelagos 1979 Morphometrics of Compact Bone: An Example from Sudanese Nubia. American Journal of Physical Anthropology 51:571–577.

Martin, D. L., and G. J. Armelagos 1985 Skeletal Remodeling and Mineralization as Indicators of Health: An Example from Prehistoric Sudanese Nubia. Journal of Human Evolution 14:527–537.

Martin, D. L., G. J. Armelagos, A. H. Goodman, and D. P. Van Gerven 1984 The Effects of Socioeconomic Change in Prehistoric Africa: Sudanese Nubia as a Case Study. In Paleopathology at the Origins of Agriculture. M. N. Cohen, and G. J. Armelagos, eds. Pp. 193–214. New York: Academic Press.

Martin, D. L., A. H. Goodman, and G. J. Armelagos 1985 Sketetal Pathologies as Indicators of Quality and Quantity of Diet. In The Analysis of Prehistoric Diets. R. I. Gilbert, and J. H. Mielke, eds. Pp. 227–279. New York: Academic Press.

Mascie-Taylor, C. G. N. 1984 The Interaction Between Geographical and Social Mobility. In Migrants and Mobility. A. J. Boyce, ed. Pp. 161–178. London: Taylor and Francis.

Mays, S. 1996 Age-Dependent Cortical Bone Loss in a Medieval Population. International Journal of Osteoarchaeology 6:144–154.

Mays, S. 1999 Linear and Appositional Long Bone Growth in Earlier Human Populations: A Case Study from Medieval England. Cambridge Studies in Biological and Evolutionary Anthropology 25:290–312.

Mays, S. 2006 Age-Related Cortical Bone Loss in Women from a 3rd–4th Century AD Population from England. American Journal of Physical Anthropology 129:518–528.

Mays, S., M. Brickley, and R. Ives 2008 Growth in an English Population from the Industrial Revolution. American Journal of Physical Anthropology 136:85–92.

Mays, S., B. Lees, and J. C. Stevenson 1998 Age-Dependent Bone Loss in the Femur in a Medieval Population. International Journal of Osteoarchaeology 8:97–106.

Mays, S., G. Turner-Walker, and U. Syversen 2006 Osteoporosis in a Population from Medieval Norway. American Journal of Physical Anthropology 131:343–351.

Mcdade, T. W., V. Reyes-García, S. Tanner, T. Huanca, and W. R. Leonard 2008 Maintenance Versus Growth: Investigating the Costs of Immune Activation Among Children in Lowland Bolivia. American Journal of Physical Anthropology 136:478–484.

McEwan, J. M., S. Mays, and G. M. Blake 2005 The Relationship of Bone Mineral Density and Other Growth Parameters to Stress Indicators in a Medieval Juvenile Population. International Journal of Osteoarchaeology 15:155–163.

Miller, M. 2005 Hypothesis: Fetal Movement Influences Fetal and Infant Bone Strength. Medical Hypotheses 65:880–886.

Nelson, A. J., and J. L. Thompson 1999 Growth and Development in Neandertals and Other Fossil Hominids: Implications for the Evolution of Hominid Ontogeny. In Human Growth in the Past: Studies from Bones and Teeth. R. D. Hoppa, and C. M. Fitzgerald, eds. Pp. 88–110. Cambridge: Cambridge University Press.

Nelson, D. A., N. Sauer, and S. C. Agarwal 2002 Evolutionary Aspects of Bone Health. Clinical Review in Bone and Mineral Metabolism 1:169–179.

Oyama, S. 2000a Evolution's Eye: A Systems View of the Biology-Culture Divide. Durham: Duke University Press.
Oyama, S. 2000b The Ontogeny of Information: Developmental Systems and Evolution. Durham: Duke University Press.
Oyama, S., P. E. Griffiths and R. Gray eds. 2001 Cycles of Contingency: Developmental Systems and Evolution, Cambridge: MIT Press.
Pearson, O., and D. Lieberman 2004 The Aging of Wolff's "Law": Ontogeny and Responses to Mechanical Loading in Cortical Bone. Yearbook of Physical Anthropology 47:63–99.
Poulsen, L. W., D. Qvesel, K. Brixen, A. Vesterby, and J. L. Boldsen 2001 Low Bone Mineral Density in the Femoral Neck of Medieval Women: A Result of Multiparity? Bone 28:454–458.
Prentice, A., I. Schoenmakers, M. A. Laskey, S. de Bono, F. Ginty, and G. Goldberg 2006 Symposium on "Nutrition and Health in Children and Adolescents" Session 1: Nutrition in Growth and Development. Proceedings of the Nutrition Society 65:348–360.
Pritchard, D. J. 1995 Plasticity in Early Development. In Human Variability and Plasticity. C. G. N. Mascie-Taylor, and B. Bogin, eds. Pp. 18–45. Cambridge: Cambridge University Press.
Rauch, F. 2005 Bone Growth in Length and Width: The Yin and Yang of Bone Stability. Journal of Musculoskeletal and Neuronal Interactions 5:194–201.
Rewekant, A. 2001 Do Environmental Disturbances of an Individual's Growth and Development Influence the Later Bone Involution Processes? A Study of Two Medieval Populations. International Journal of Osteoarchaeology 11:433–443.
Rittweger, J. 2006 Can Exercise Prevent Osteoporosis? Journal of Musculoskeletal and Neuronal Interactions 6:162–166.
Robert, J., B. Hall, and W. Olson 2001 Bridging the Gap Between Developmental Systems Theory and Evolutionary Developmental Biology. Bioessays 23:954–962.
Roberts, D. F. 1977 Physique and Environment in the Northern Nilotes. Mitteilungen der Anthropologischen Gesellschaft in Wien 107:161–168.
Roberts, D. F. 1995 The Pervasiveness of Plasticity. In Human Variability and Plasticity. C. G. N. Mascie-Taylor, and B. Bogin, eds. Pp. 1–17. Cambridge: Cambridge University Press.
Roberts, D. F., and D. Bainbridge 1963 Nilotic Physique. American Journal of Physical Anthropology 21:341–370.
Ruff, C., B. Holt, and E. Trinkaus 2006 Who's Afraid of the Big Bad Wolff?: "Wolff's Law" and Bone Functional Adaptation. American Journal of Physical Anthropology 129:484–498.
Ruff, C., C. S. Larsen, and W. C. Hayes 1984 Structural Changes in the Femur with the Transition to Agriculture on the Georgia Coast. American Journal of Physical Anthropology 64:125–136.
Ruff, C., A. Walker, and E. Trinkaus 1994 Postcranial Robusticity in Homo. III: Ontogeny. American Journal of Physical Anthropology 93:35–54.
Saggese, G., G. Baroncelli, and S. Bertelloni 2002 Puberty and Bone Development. Best Practice and Research Clinical Endocrinology and Metabolism 16:53–64.
Saunders, S. R. 2000 Subadult Skeletons and Growth-Related Studies. In Biological Anthropology of the Human Skeleton. M. A. Katzenburg, and S. R. Saunders, eds. Pp. 135–162. New York: Wiley-Liss.
Schell, L. M. 1995 Human Biological Adaptability with Special Emphasis on Plasticity: History, Development and Problems for Future Research. In Human Variability and Plasticity. C. G. N. Mascie-Taylor, and B. Bogin, eds. Pp. 213–237. Cambridge: Cambridge University Press.
Scheuer, L., and S. Black 2000. Developmental Juvenile Osteology. London: Elsevier.

Schwartz L., H. Maitournam, C. Stolz, J. Steayert, M, Ho Ba Tho, and B. Halphen 2003 Growth and Cellular Differentiation: A Physico-Biochemical Conundrum? The Example of the Hand. Medical Hypotheses 61:45–51.

Seeman, E. 1999 Genetic Determinants of the Population Variance in Bone Mineral Density. In The Aging Skeleton. G. J. Rosen, J. Glowacki, and J. P. Bilezikian, eds. Pp. 77–94. San Diego: Academic Press.

Shapiro, H. L. 1939 Migration and Environment. New York: Oxford University Press.

Skerry, T. 2006 One Mechanostat or Many? Modifications of the Site-Specific Response of Bone to Mechanical Loading by Nature and Nurture. Journal of Musculoskeletal and Neuronal Interactions 6:122–127.

Sofaer, J. 2006 Gender, Bioarchaeology and Human Ontogeny. In The Social Archaeology of Funerary Remains. R. Gowland, and C. Knüsel, eds. Pp. 155–167. Oxford: Oxbow:

Sowers M. R., and D. A. Galuska 1993 Epidemiology of Bone Mass in Premenopausal Women. Epidemiological Reviews 15:374–398.

Stevenson J. C., B. Lees, M. Devenport, M. P. Cust, and K. F. Ganger 1989 Determinants of Bone Density in Normal Women: Risk Factors for Future Osteoporosis? British Medical Journal 298:924–928.

Turner-Walker, G., U. Syverson, and S. Mays 2001 The Archaeology of Osteoporosis. Journal of European Archaeology 4:263–268.

Van Gerven, D. P., J. R. Hummert, and D. B. Burr 1985 Cortical Bone Maintenance and Geometry of the Tibia in Prehistoric Children from Nibia's Batn el Hajar. American Journal of Physical Anthropology 66:272–280.

Ward, J. A., S. R. Lord, P. Williams, K. Anstey, and E. Zivanovic 1995 Physiologic, Health and Lifestyle Factors Associated with Femoral Neck Bone Density in Older Women. Bone 16:373S–378S.

Weaver, D. S. 1998. Osteoporosis in the Bioarchaeology of Women. In Sex and Gender in Paleopathological Perspective. A. Grauer, and P. Stuart-Macadam, eds. Pp. 27–46. Cambridge: Cambridge University Press.

Worthman, C. 1995 Hormones, Sex, and Gender. Annual Review of Anthropology 24:593–617.

Worthman, C., and J. Kuzara 2005 Life History and the Early Origins of Health Differentials. American Journal of Human Biology 17:95–112.

Young, R. L., and A. V. Badyaev 2007 Evolution of Ontogeny: Linking Epigenetic Remodeling and Genetic Adaptation in Skeletal Structures. Integrative and Comparative Biology 47:234–244.

12

The Bioarchaeological Investigation of Children and Childhood

Siân E. Halcrow and Nancy Tayles

Introduction

During the past decade there has been a dramatic increase in the study of children and childhood in archaeology and anthropology, with the publication of many papers, books, and theses on this topic (e.g., Baxter 2005b; Baxter 2008; Benthall 1992; Crawford and Lewis 2008; Gottlieb 2000; Kamp 2001b; Lewis 2007; Moore and Scott 1997; Panter-Brick 1998; Schwartzman 2001; Scott 1999; Sofaer Derevenski 2000; Stearns 2006; Wileman 2005). In past archaeological research, children were marginalized, much like women were, especially because of the perceived close association between children and women (Baxter 2008). Children were seen as being unimportant to social life, and their study was thought to be hindered by lack of preservation and therefore under-representation of children in the archaeological record (Lewis 2007). The initial interest in the archaeology of childhood occurred with the rise of feminist approaches in the 1970s, but with the relatively recent emphases on identity and agency in archaeology, childhood has become recognized as an important focus of study (Baxter 2008). In fact, contrary to the dismissal of children in earlier archaeological work it is now recognized that children are significant social and economic actors (e.g., Kamp 2001b) and overlooking them would ignore an important demographic proportion of past societies (Baxter 2008:160). Recent bioarchaeological work also recognizes that they are central for understanding biological adaptation and health to their social environment (Lewis 2007).

Recently, rising tensions between social archaeologists and bioarchaeologists in their approach to the study of human remains has been noted (Sofaer 2006),

Social Bioarchaeology. Edited by Sabrina C. Agarwal and Bonnie A. Glencross
© 2011 Blackwell Publishing Ltd

although this is perhaps more evident in Britain than in the American school of anthropology. These tensions are also apparent in the study of childhood in past populations. Increasing research interest in aspects of social identity in archaeology, including age, illustrates a conscious move away from biological determinism (Insoll 2007b; Sofaer 2006:119). It has been observed by some that the social view seems to predominate in the study of childhood in the past (Prout 2005; cf. Lally and Ardren 2008). Because of the perceived distinction between biological and social phenomena, and in turn research approaches, Insoll (2007a:4) reminds us that there:

> ... sometimes seems to be within the archaeology of identities an emphasis upon forgetting the prosaic, but equally important foundational rudiments such as biology, in favour of the more popularly perceived social theoretical elements ... the empirical body from which adequate interpretation and theory are generated in pursuing past identities must also not be neglected; otherwise there is a danger that empty shells are created.

Bioarchaeological approaches can ensure that both biological and social factors of childhood are acknowledged. Recent developments involving exploration of social aspects using human bones, including those of children, explicitly attempt to erase the dualistic construction of the biological and social (e.g., Gowland 2006; Lewis 2007; Sofaer 2006).

However, there is little written on the practicalities, problems, and prospects of incorporating social childhood theory into bioarchaeological analysis. This chapter brings together the relevant theoretical information from bioarchaeological methods, childhood social theory, and archaeological mortuary ritual analysis that are vital to consider when adopting this approach. First, the issues of terminology in subadult research are presented. A short historical overview of the development of subadult bioarchaeological health analysis and a review of heath indicators used in this field are then presented, followed by an overview of childhood social research. For more detailed accounts of these historical developments in bioarchaeology and social archaeology, see Lewis (2007) and Kamp (2001b), respectively. Finally, the practical and theoretical issues of applying childhood social theory to bioarchaeological research and the potential of bioarchaeological techniques for understanding childhood are discussed.

Terminology and Age Categories

Various problems exist with the terminology used to define childhood, including age categories. With an increased emphasis on research about children and childhood from the past, questions are being asked about the appropriate terminology and age categories used in subadult bioarchaeological analysis (Kamp 2001b; Perry 2005). Further, few bioarchaeological studies integrate social age categories into the analysis of skeletal populations (Gowland 2006; Perry 2005). This is a particularly important issue as the age categories used have implications for the analysis and interpretation of biological data, including aspects of mortality, fertility and other indicators of health, and also in comparisons among bioarchaeological studies and with the health of living populations. The purpose of this review is not to provide a solution for terminological problems by way of advocating specific terminology, but rather to highlight some of these issues, which have also been discussed elsewhere in detail (e.g., Halcrow and Tayles 2008b; Lally and Ardren 2008; Lewis 2007; Sofaer 2006).

Firstly, it is important to distinguish the different "types" of age (Ginn and Arber 1995; Gowland 2006; Sofaer 2006). These are:

1 physiological or biological age (including skeletal and dental age), estimated from the biological changes in the body;
2 chronological age, the time since birth; and
3 social age, the culturally constructed norms of appropriate behavior and status of individuals within society for a given age.

Unfortunately, in archaeological and bioarchaeological research there are instances where the basis upon which age is estimated is not acknowledged and the "types" of age often not distinguished.

Historically there have been many different ways of dividing the human life span, so that today terminology varies among medical pediatrics, developmental osteology and anthropology, and even within these fields. In anthropological research there is a problem with varying definitions of childhood or subadulthood and the different age categories within this classification. In fact, numerous terms are used for subadults and related age categories. For example, "subadult," "non-adult," "juvenile," and "child" are all used interchangeably for individuals who have not reached adulthood (Hoppa 1992; Saunders 2000; Scheuer and Black 2000a,b; Scheuer and Bowman 1995). These terms are themselves problematic. "Subadult" implies that the individuals are hierarchically inferior to adults (Lewis 2007:2; Sofaer 2006:121). "Non-adult," which is used as an alternative to "subadult" (e.g., Lewis 2007), has the same problem because this categorization is defined in contrast to what they are not: "adult" (Halcrow and Tayles 2008b). In using this dualism, children are defined as a deviant from the norm, similar to the process known as "Othering" (de Beauvoir 1949). The use of the terms "adult" and "child" have also been criticized in that, by juxtaposing these concepts, there is a construction or iteration of modern Western notions of roles and relationships that mask the complexities of social age (Sofaer Derevenski 1994b:7–9, cited in Baxter 2005a:19).

Further, the age categories used in bioarchaeology are defined using definitions in the wider literature, which are not necessarily relevant to biological development or social identity. For example, Lewis (2007:1,5–6) has noted the problem of the different definitions for the "infant" age category used in biological anthropology. Some assign this term to individuals younger than one year of age, based on the medical definition, while others use this term to refer to individuals up to three or five years of age (Lewis 2007:5–6). Lewis (2007:5–6) argues that it is problematic to include individuals up to five years of age in an "infant" age category because it ignores the major physiological and social development that occurs from birth to five years of age.

In this chapter, the term "subadult" is based on biological age, unless described otherwise, and is used for convenience to refer to infants and children. Because the definition of this term varies among bioarchaeologists we are not defining the "cut-off" point between "subadult" and "adult". The term "childhood" refers to a social age category.

Bioarchaeological Approaches to Subadult Health and Disease and Childhood

At the beginning of the last century, subadults from archaeological contexts were often ignored. This can be understood in the context of research interests at that time with a focus on human taxonomy achieved through description and metrics. "Physical" anthropologists were mainly interested in comparative craniometry, which required the analysis of adult crania (Gould 1996). Subadult crania were deemed useless because they were often found disarticulated in archaeological contexts. Hooton (1930:15) typifies the disinterest in the analysis of subadults at the time:

> in the case of infants and immature individuals, the cartilaginous state of epiphyses and the incomplete ossification of sutures, as well as the fragility of the bones themselves usually results in crushing and disarticulation. In any event, the skeletons of young subjects are of comparatively little anthropological value.

Johnston (1961, 1962) was a pioneer of subadult investigations of growth, development, and mortality with his research on the Indian Knoll skeletal sample. It is now widely acknowledged that subadult human remains are particularly useful for the study of patterns of health and disease in prehistory, as they are the most demographically variable and sensitive indicators of biocultural change (Buikstra and Ubelaker 1994:39; Van Gerven and Armelagos 1983:39). Disease, subsistence mode, weaning patterns, and genetic disorders can leave evidence on the dentition and bones of infants and children and thus provide clues to aspects of the health of the community and of the environment in which they live (Buikstra and Ubelaker 1994:39). The merits of subadult health investigation, including the potential to contribute to our understanding of social factors, are succinctly described by Goodman and Armelagos (1989:239):

> monitoring the health of infants and children can provide the prehistorian with a rich variety of information about the health of a community. As this segment of the population is very sensitive to environmentally and culturally produced insults, changes in morbidity ... could provide one of the first signs of changes in environment and culture.

The usefulness of subadult remains in bioarchaeology has resulted in a large number of studies on the mortality, growth, and growth disruption of subadult samples (e.g., Buckley 2000; Halcrow et al. 2008; Humphrey 2000; Lewis 2004; Lovejoy et al. 1990; Mays 1995; Mays 1999; Saunders 2000), and a recent comprehensive book devoted entirely to the study of subadult remains by Lewis (2007).

It is accepted in biological anthropology that health data need to be interpreted in the context of the cultural environment. This interpretive tool was recognized in the 1980s with the adoption of the biocultural general adaptationist model after Goodman et al. (1984b). The underlying premise of this model, that human health and disease occur in an ecological setting and are ultimately affected by human behavior, makes it crucial to understand interactions among people and between people and their environment. In turn, the study of health can contribute to an understanding of the environment in which an individual lives. Bioarchaeologists therefore generally incorporate information on the cultural context of children's roles and activities in society when interpreting their data. For example, Lewis (2002), with reference to high child mortality at a late period site in medieval Britain, argues that many rural and urban children from as young as seven years of age were sent to work as apprentices (Cunningham 1995), in conditions which would have been deleterious to their health. In doing so, children are viewed in the context of their social role in society, which is complementary to this central theme within social childhood theory.

A relatively new approach in the biological sciences, "life history theory" has also been adopted in biological anthropology. Bogin (2003:15) describes this theory as "a field of biology concerned with the strategy an organism uses to allocate its energy toward growth, maintenance, reproduction, raising offspring to independence, and avoiding death." Life history theory assumes that resources are limited and that energy is allocated between three essential life functions: growth, maintenance, and reproduction. McDade (2003:100) notes that over the past couple of decades anthropologists have become increasingly interested in the human immune function and its implications for human health. As the immune system is crucial to an organism's survival and is costly in terms of maintenance, it is important to look at its role, given the theory of limited resource allocation (McDade 2003). McDade (2003) discusses immunological processes and life history events across different human life stages. Each stage in a person's life has a set of adaptive challenges that the immune system has to meet/overcome. This framework is a good interpretive tool to assess the health of individuals in relationship to the specific demands on immune function at particular times throughout life. For example, in infancy the competing demands are high between growth, immune development and maintenance, and are exacerbated if only limited resources are available.

The incorporation of biological life history information with social age can be helpful in interpretation of bioarchaeological data. For example, going back to Lewis's (2002) research, the children in her sample sent to work as young as seven years of age would have had to deal with multiple competing demands. Energy demands would be high since children of this age are growing at a fast rate and developing their immune system. Additional demands include the hard physical labor the children were involved in, as well as probable poor nutrition and sleep deprivation, all lending to compromised growth and/or immune development.

Measures of Health and Disease

Bioarchaeological methods of analysis of mortality, growth, growth disruption, pathology, and trauma (Larsen 1997; Lewis 2007) are used to assess childhood health and therefore, as discussed, aspects of the social environment. Isotopic analyses of bone and teeth and investigations of bone morphology also have potential to give insights on health and social aspects of childhood.

Bone is a dynamic structure that is shaped and remodeled throughout life in response to physiological and biomechanical forces. Cells called osteoblasts and osteoclasts reside within bone and are responsible for constant turnover or renewal of the tissue (Ortner and Turner-Walker 2003). These specialist cells also respond to stressors, for example, pathogens, by the production and/or resorption of bone (Ortner 2003:48–49). Some diseases may leave specific lesions on bone and these can be used in pathological diagnosis, for example, tuberculosis, syphilis, and leprosy (Aufderheide and Rodriguez-Martin 1998). However, many pathogens leave only generalized changes in bone, making a specific diagnosis impossible. Because of this, paleopathologists can usually define only a group of diseases rather than a specific pathogen. However, this is acceptable in research projects interested in general measures of the quality of life in past human populations (Buikstra and Cook 1980:436).

In addition to this nonspecificity of bone change, there are also many pathogens that leave no evidence of pathology on the skeleton, as Aufderheide and Rodriquez-Martin (1998:117–118) state:

> ... acute, often lethal infections were probably the most frequent in antiquity ... yet these usually leave few recognizable skeletal changes. Those infections to which the body has developed at least sufficient immunity to prolong its coexistence as a chronic infection are the most apt to generate obvious skeletal lesions, though quantitatively these chronic infections probably had lesser demographic impact ...

In light of the possibility of acute fatal infections having no associated skeletal changes, it has been acknowledged that skeletal and dental markers of stress may paradoxically indicate good health (Ortner 1991; Wood et al. 1992). Following the lessons of Goodman (1993), the key to making sense from apparent paradoxes in bioarchaeological interpretation, is to employ an analysis of multiple indicators of stress. These multiple indicators, in conjunction with a heightened appreciation of the biological and social or cultural contexts, can aid in painting a more informed picture of health.

Nonspecific stress indicators

The study of nonspecific stress indicators is an approach that has been increasingly used in bioarchaeology, especially since the 1980s (e.g., Goodman et al. 1988; Larsen 1997; Lewis and Roberts 1997). As noted, Goodman et al. (1988), advocating the biocultural theory, served to reorientate bioarchaeological inquiry away from the identification of specific pathologies and toward the analysis of health and disease at a population level. According to this model, health is viewed as an adaptation to a combination of stressors in the environment. These stressors may be in the form of a wide range of factors including infection, weaning, climate, food shortages, and social status. Therefore, health status is often a composite of nutrition, disease, and other factors. The multiple etiologies of skeletal and dental stress indicators are described elsewhere (e.g., Lewis and Roberts 1997). Nonspecific stress indicators used in bioarchaeology include mortality rates from demographic profiles, linear and appositional bone growth, dental enamel defects (including linear enamel hypoplasia), and skeletal pathologies (including cribra orbitalia, subperiosteal new bone growth and endocranial new bone growth) (Lewis 2007).

Paleodemography and mortality patterns

Death represents the ultimate level of ill health. Although the death of an old adult eventually becomes inevitable, the death of an infant, child, or young adult is generally the result of adverse environmental factors. Hence, patterns of infant and child mortality are especially useful in the interpretation of the overall success of a population (Lewis 2007). Mortality patterns are also important in the interpretation of stress indicators. Often patterns in paleopathology that give insight into the etiology of the stress indicator are only confirmed when age categories are incorporated into analysis. For example, Lewis (2002) notes no significant difference in indicators of health between the subadult samples from medieval and post-medieval British sites, but when the severity of lesions and age profiles are compared, differences are identified in the data. Lewis (2002) argued that the occurrence of cribra orbitalia before the first six months of life, and early enamel hypoplasia formation times at Christ Church Spitalfields in London, indicates stress for young infants in this industrial community.

Infant mortality rates are often used by anthropologists as a measure of a population's adaptation to challenges in their environment (Baker and Dutt 1972; Little 1989). Wiley and Pike (1998:316) have argued that moving the focus away from overall infant mortality rates to patterns of mortality during infancy can show the interaction of the biological and social factors contributing to mortality at specific developmental phases.

When assessing infant mortality it is important to understand the environmental factors that can affect the fetus and infant. Compared with postnatal life, the fetus generally experiences a relatively stable environment with adequate nutrition and protection from infection (Chandra 1975; Cole 1998:491; McDade 2003:109). The birth transition from life *in utero* to life independent of the uterine environment can be classed as the first crisis in a human's life where the environment has its first direct impact on the subadult itself (Lewis 2007:81). Unlike the fetus, the newborn is subject to environmental imbalances and therefore to warmth and nutrition deprivation (Bornstein and Lamb 1992:125). Even full-term newborns are biologically immature, including their immune system (McDade 2003) and ability to regulate temperature, so are less able to respond to stresses. These factors can explain why mortality is particularly high during the first year of life in most communities (Mosley 1984). Subadult mortality often follows a general pattern with the majority of deaths occurring in infancy (defined here as before the age of one year), especially around the time of birth, and after that a decrease in mortality with age, which represents the reducing vulnerability of young subadults to stress and the increasing resilience children acquire as they mature (Weiss 1973:26).

At what stage during infancy that death occurs is important in determining the relative contributions of the environment and biology to the event (Wiley and Pike 1998). Demographers and clinicians make an interpretive distinction between neonatal mortality (the first 27 days postpartum), which is seen as a consequence of endogenous causes including low birth weight and trauma and death after the neonatal period caused by exogenous environmental factors (e.g., infectious disease, poor nutrition, and accidents) (Wiley and Pike 1998:318). This distinction has been used as an interpretive tool for assessing infant mortality from archaeological contexts (e.g., Lewis and Gowland 2007). Lewis and Gowland (2007) in an assessment of age-at-death profiles found that at Spitalfields, compared with other British medieval and post-medieval sites, had an underrepresentation of neonates as the result of unbaptized babies being buried in a different area. Spitalfields also had a high number of post-neonatal deaths, which they argue is a result of exogenous factors including early weaning age and infection arising from poor sanitation (Lewis and Gowland 2007).

As burial rites and customs can determine whether an infant or child is included in a cemetery, and thereby influence the age distribution of the infant sample in archaeological contexts (La Fontaine 1986:18; Mays 1993; Scheuer and Black 2000b), the study of subadult age distributions is also a valuable tool to investigate burial practices and provide insights into religious ideologies and treatment of infants and children (Tocheri et al. 2005:328).

Longitudinal and appositional growth

It is well recognized that human growth is a very plastic process that is readily modified by environmental factors.

> A child's growth rate reflects, perhaps better than any other single index, his [*sic*] state of health and nutrition ... a well-designed growth study is a powerful tool with which to monitor the health of a population ... (Eveleth and Tanner 1990:1).

Growth-related studies of archaeological samples assume that the differences observed in growth among and within populations reflect differences in health (Larsen 1997; Saunders et al. 1993). Growth disruption can result from nutritional inadequacy, pathological processes, or a combination of these factors causing physiological stress (Ulijaszek et al. 1998). Although the study of growth itself does not indicate the specific cause of the stress, evidence from other dental and skeletal markers of stress and archaeological information about the natural environmental conditions and cultural factors can aid in the interpretation of growth patterns (Humphrey 2000). In the case of *in utero* development, the pattern of growth is largely genetically determined (Mays 1998:42–43). However, as the baby develops, growth may be constrained by poor maternal health as a result of environmental conditions. If these conditions are severe, this may result in growth retardation of the fetus, with subsequent effects on the health of the individual throughout life (Fowden 2001).

Since the pioneering growth research by Johnston (1961, 1962, 1969) there has been an impressive array of work on linear growth in archaeological populations (Hoppa 1992; Jantz and Owsley 1984; Lewis and Roberts 1997; Merchant and Ubelaker 1977; Ribot and Roberts 1996; Steyn and Henneberg 1996; Walker 1969; Wall 1991). Generally in bioarchaeological research maximum diaphyseal lengths of long bones are measured and plotted against estimated dental age. It has been advocated that cortical bone mass, when compared with linear growth, can produce additional insights to health than either variable in isolation (Mays 1995; Mays 1999; McEwan et al. 2005). Research that has found linear bone growth to be maintained at the expense of cortical thickness in stressed individuals supports this argument (Garn et al. 1964; Hummert and Van Gerven 1983).

Growth disruption of teeth

Like the process of bone growth, the growth of enamel, or amelogenesis, is also subject to disturbance by physiological disruptions, but rather than reducing the size of the tooth, it results in defects in its structure (Goodman and Rose 1990). Hypoplastic dental defects are deficiencies in enamel thickness during enamel secretion that cause a disruption in the contour of crown surfaces (Hillson 1996). Dental enamel defects, particularly linear enamel hypoplasia (LEH), are often used in assessing health in bioarchaeological research (e.g., Goodman and Rose 1990; King et al. 2002). Teeth provide an advantage in investigating aspects of health in archaeological contexts as their robust structure lends to good preservation, and because enamel does not remodel, enamel defects are permanent markers of stress (Goodman and Rose 1990). Deciduous dental enamel defects can give useful insights into maternal and early infant health. Studies have linked enamel hypoplasia to a wide range of etiologies including: hemolytic disease in newborn infants; premature birth or low birth weight; dietary deficiencies of vitamin A, C and D; low social status; fever; starvation; congenital infections; and parasitic infections (Brook et al. 1997; Goodman et al. 1984a; Goodman and Rose 1990; Jontell and Linde 1986; Seow 1992; Whittington 1992). However, the exact etiology of these defects cannot be known so they are therefore nonspecific indicators of stress (e.g., Katzenberg et al. 1996).

Skeletal pathology

Types of nonspecific skeletal pathology often investigated in subadults include cribra orbitalia, periostitis, and endocranial new bone formation (Lewis and Roberts 1997). Cribra orbitalia has been related to chronic blood loss, deficiencies in diet and high pathogen loads, inflammation, rickets, scurvy and nonspecific stressors (Mensforth et al. 1978; Ribot and Roberts 1996; Stuart-Macadam 1992; Walker et al. 2009; Weinberg 1992). Periosteal reactions have been linked to infectious disease and trauma (Lewis 2004; Mensforth et al. 1978; Ortner and Putschar 1985). New bone growth on the endocranial surface has been adopted as a nonspecific indicator of stress, which Lewis (2004) has argued could be the result of trauma, tumors, specific and nonspecific meningitis, syphilis, vitamin deficiencies, and congenital defects.

In summary, the assessment of subadult health through nonspecific stress indicators including mortality, linear, and appositional bone growth, dental enamel defects and various skeletal pathologies are important in assessing population health as a whole. Using a biocultural approach bioarchaeologists acknowledge that human health and disease occur in an ecological setting and therefore the importance of understanding the interactions among people and between people and their environment. Therefore, the study of subadult health is valuable to an understanding of the social environment in which they live.

Subadult diet

Stable isotopes

Breastfeeding and weaning (defined here as the *process* of the supplementation of diet with solid foods and not necessarily a single event marked by the cessation of breastfeeding) are important social stages of child rearing that can be inferred through stable isotope analysis. The timing of these events can have consequences for subadult and maternal health, the roles of adults, both males and females, and children in childcare, as well as fertility through lactational amenorrhea (Bocquet-Appel and Naji 2006; Maher 1992; Sellen and Mace 1997). Bioarchaeological studies have investigated breastfeeding and weaning using stable isotopes, and whether it has a relationship with mortality and morbidity (Fuller et al. 2003; Herring et al. 1998; Katzenberg et al. 1996; Richards et al. 2002; Schurr 1998; Wright and Schwarcz 1998).

Dietary analysis is also useful for looking at social aspects of childhood. For example, an interesting isotopic study of diet in Sudanese Nubia shows variation between the subadults and adults and suggests a set of age-related dietary practices specific to infancy, subadulthood, and adulthood (Turner et al. 2007). Ethnographic analysis has shown that food allocation strategies closely follow cultural definitions of childhood, vulnerability and gender (Messer 1997), emphasizing the potential contribution of isotopic analysis to understanding childhood social age.

Dental health and disease

Another avenue to studying diet and health is the investigation of dental health. Caries is a multifactorial disease with diet as a main cause, but with other influences including infection, saliva flow rate and pH, dental morphology, and environmental factors such as exposure to fluoride (Ferjerskov et al. 1993).

As most research on dental health in bioarchaeological samples has been undertaken on adults, it is important to consider the factors that affect the susceptibility of subadult teeth to developing dental caries. These factors include enamel defects (e.g., Duray 1990), the enamel structure of deciduous dentition (Lussi, et al. 2000; Wilson and Beynon 1989), and other dental pathology (e.g., Halcrow and Tayles

2008a:2219). For example, as there is a relationship between periodontitis and age (Hillson 1996:266–267), the argument for an increase in caries at the cemento-enamel junction with the greater reliance on cereal agriculture, as has been put forward from some adult data (e.g., Moore and Corbett 1971, 1975), is not applicable to subadults. Rather, it is more common for fissure and interproximal caries to become dominant types of caries when sugars are introduced into the diet of subadults (Hillson 1996:283).

Acknowledging that there is not a simple relationship between caries and diet, an analysis of subadult dental health compared with dental health within an archaeological sample has the potential to provide information on social age-related food allocation patterns.

Trauma

Trauma is rarely identified in subadults from archaeological contexts compared with adults (Lewis 2007). However, this does not mean that subadults did not suffer from trauma in the past. Rather, the nature of bone remodeling and repair can mask the evidence of trauma, resulting in its underestimation. The study of trauma patterns in subadults from past populations has potential to identify social information including interpersonal relationships, occupation, subsistence activities, accidents, child abuse, parental care, and the home environment (Lewis 2007:169). While it is acknowledged that there are limitations of assessing trauma from archaeological subadult remains, its merit is still recognized (e.g., Lewis 2007). For example, Lewis (2007:183) states that the age at which fractures occurred and their type and frequency may provide information on the age at which children started to do physical labor, perhaps as apprentices.

Musculoskeletal markers of stress

Another possible avenue, although not previously assessed for its feasibility in subadult samples, is the investigation of activity patterns from musculoskeletal stress markers at muscle insertion sites. Hawkey and Merbs (1995) have justified the exclusion of children from their analysis of activity-induced musculoskeletal stress because muscle insertion sites may not be acting on a localized part of the bone, due to the continual growth of the subadult bone (Enlow 1976). However, the examination of these activity markers could potentially give information regarding physical activities that the subadults were carrying out. For example, the cross-cultural occurrence of individuals generally becoming more active and self-sufficient at around five years of age (Bogin 1997) has been used to explain the apposition of bone in an archaeological population at this time (Mays 1999). Studies of modern children, which have shown that bone mass and density is influenced by mechanical loading and muscle stress, support the potential for this type of analysis (Ruff 2003; Welten et al. 1994). However, in investigating activity-induced musculoskeletal stress markers, factors including nutritional stress and normal biological patterns of linear and appositional bone growth, especially around puberty, would have to be considered and are not currently documented well enough. The inclusion of children in this type of analysis is also potentially valuable in investigating their physical contribution to society in their everyday life and their active role in production.

Sample Bias and Osteological Interpretation

Theoretical and practical issues in bioarchaeology that are particularly relevant to subadult bioarchaeology, include issues of representativeness of biological aging standards, sample representativeness, selective mortality bias, preservation of subadult skeletons, and the "osteological paradox" (Goodman 1993; Lewis 2007; Saunders and Hoppa 1993; Wood et al. 1992).

Skeletal growth and skeletal and dental maturation are used to estimate age-at-death in subadult bioarchaeology. A major assumption made in age estimation is that skeletal (or dental) age is the same as chronological age. However, the biological changes on which aging methods for subadults are based vary at the intra-individual, inter-individual, and inter-population levels (Halcrow et al. 2007; Heuzé and Cardoso 2008; Tompkins 1996).

Sample representativeness (as discussed by Jackes, this volume) is one of the main issues that must be considered in the interpretation of health and disease from skeletal samples. Waldron (1991:17) writes:

> the underlying assumption that is inherent in any attempt to use a death assemblage to predict something about the living is that the dead population is representative – or at least typical – of the live population. Given all the non-random events that surround death and burial, not to mention preservation and recovery, this is at best an approximation, and at worst the two (the live and the dead) bear no epidemiological relation to each other whatsoever.

Skeletal samples are composed of a special subset – the non-survivors of the population. This is particularly relevant to subadult samples, as they are a subset of people who have died prematurely. Although researchers have argued that mortality, except for certain selective circumstances, will operate randomly within the population (Cohen 1994:631), it is generally accepted that there are deterministic elements for which individuals or groups are at a higher risk of death (Wood et al. 1992). As stated by Wood et al. (1992:344) the implication of this issue is that "... the observed frequency of pathological conditions should overestimate the true prevalence of the conditions in the general population" thus giving a false frequency of pathological incidence in the sample under study.

As discussed earlier, Wood et al. (1992) emphasize the apparent paradoxical nature of the interpretation of health and disease in the past, based on the premise that indicators of stress in bone and dentitions may in fact represent survival of a person after a morbidity event. Conversely, a lack of skeletal evidence of pathology may be the result of a person dying without a sufficient immune response to enable survival to a chronic stage of a disease (Goodman 1993; Ortner 1991). In response, Goodman (1993) illustrates that, although evidence from a single indicator may be open to a number of different interpretations, the adoption of multiple indicators can aid in the interpretation of health. Goodman (1993) reviewed a "paradoxical" example presented by Wood et al. (1992) where three imaginary populations are described. Group A experienced no stress, group B experienced moderate stress, and group C experienced heavy stress. Wood et al. (1992) state that the paradox is that group C may, at least in terms of skeletal pathology, appear to be healthier than those in group B. However, Goodman (1993) explains that using mortality data combined with pathology he could correctly identify group A by the combination of low morbidity and low mortality, group B by high morbidity and low mortality, and group C by low morbidity and high mortality. In this instance, an infant who dies around the time of birth and has no evidence of pathology is assumed less healthy in comparison to an older infant who has pathological lesions.

Excavated skeletal samples frequently have proportions of infants and elderly well below expectations based on age-at-death distributions from modern populations (Lovejoy et al. 1977; Weiss 1973). Much research attributes underrepresentation of infants to taphonomic processes (Gordon and Buikstra 1981; Guy et al. 1997; Specker et al. 1987). Young subadults have bones that are high in organic components and correspondingly low in mineral content making them more susceptible to postmortem decay than the bones of older individuals (Guy et al. 1997). A relationship between low soil pH and low bone preservation has been shown with juvenile bone durability declining rapidly with increasing soil acidity (Gordon and Buikstra 1981). Bello et al. (2006) have recently produced a study aiming to assess age and sex biases in preservation and whether preservation factors are more dependent on factors extrinsic or intrinsic to anatomical features of human bones. Assessing over 600 skeletons from archaeological sites in France and the United Kingdom they found that subadult remains are generally poorly preserved and less complete than adult human bones and that the lack of preservation is mainly due to intrinsic features of the bones (e.g., lower mineral density in subadult bone). Extrinsic factors can lead to a further differentiation of preservation between subadults and adults, but these are largely dependent on the physical or intrinsic properties of the bones (Bello et al. 2006). This argument is based on the finding that the frequencies of well-preserved bones were similar in the three archaeological samples investigated.

In addition, cultural reasons are inferred for the underrepresentation of subadult skeletons in archaeological contexts. As stated by Lewis (2000:40–41) "(w)hen using data derived from cemetery samples it is important to remember that we are actually measuring *burial rates* and not *mortality rates*. Cultural practices may dictate if, and where, certain individuals were placed within a cemetery; non-adults are often clustered ...". These practices may include infanticide and cemetery subdivision resulting in babies being buried apart from the main burial area (Jamieson 1995; Mays 1998:23-25; Scrimshaw 1984). For example, it has been shown in several Ghanaian tribes, West Africa, that infants and children were interred away from main community burial places (Ucko 1969:271). Among the Ashanti tribe, when an infant dies under the age of eight days old they are buried in a pot in a latrine, as they are considered to be a "ghost child" who did not have any intention of staying in this world (Ucko 1969:271).

However, although it is reported that there is an underrepresentation of subadults at some sites, there have also been large numbers of subadults excavated from cemetery sites around the world, for example in the United Kingdom (see Lewis 2007:20–21), Southeast Asia (e.g., Halcrow et al. 2008), South Asia (e.g., Robbins 2007; Robbins in press), and North America (e.g., Herring et al. 1998; Johnston 1962). Therefore it seems that infants and children are not always underrepresented and can therefore potentially inform us of a raft of important aspects of past populations including paleodemography, aspects of health and disease, and social aspects as inferred through mortuary ritual.

Childhood Social Theory and Identifying Social Age in Archaeology

Phillipe Ariès (1962) produced the first book on the topic of the history of childhood. This work describes the changes in the concept of childhood from medieval to early modern Europe. Although Ariès' (1962) work has been criticized (Hanawalt 1993; Kuefler 1991), it is important in illustrating that childhood is a socially and culturally constructed category, a central tenet in childhood social theory (Allison and Prout 1997).

The rise of feminist approaches in anthropology during the 1970s sparked an initial interest in the place of children in the archaeological record, although the main emphasis was on the place of women (Scott 1997:6–7; Sofaer Derevenski 1997). The lack of interest in the study of childhood by archaeologists is attributed to the perceived notion of the underrepresentation and fragility of subadults from archaeological contexts (Lewis 2007:20), and to the notion of "children" in modern Western societies as being passive and unimportant in contributing to economic and political life, and dependent on adults (Kamp 2001b; Nieuwenhuys 1996; Sofaer Derevenski 1997).

Lillehammer's (1989) publication: "A Child is Born. The Child's World in an Archaeological Perspective" has been described as the "birth" of archaeological investigations of childhood (Baxter 2005b:16). Lillehammer (1989) acknowledges that children were left out of archaeological investigations and advocates an approach that focuses on the child's relationship with the environment and the adult world. In response to the absence of children in archaeological narratives, Kamp (2001b:1) asks in her evocative paper title: "Where Have All the Children Gone?" Since this time there has been an increasing amount of research on childhood in the past (Finlay 2000; Ingvarsson-Sundström 2003; Kamp 2001b; Moore and Scott 1997; Scott 1999), a focus of interest that is evident throughout the social sciences, particularly sociology (Bowman 2007; Corsaro 2005; Hopkins and Barr 2004; James et al. 1998; James and Prout 1997; Jenks 2005; Prout 2000b; Prout 2005; Qvortrup et al. 1994).

There are currently several main themes in contemporary sociological, anthropological, and archaeological work on childhood. These include childhood as a social and cultural construction, the investigation of children's agency and the role that children play in societies (Baxter 2005a; Bluebond-Langner and Korbin 2007; James and Prout 1997; Politis 2006; Prout 2000b).

Here we address the prospects and problems of the use of childhood social theory in bioarchaeological research. First, the role that children play in society is considered. Following this, theoretical and methodological aspects of the use of social age categories in bioarchaeological research are critically examined. In discussing these issues, the importance of an approach that considers biological factors, and therefore the contribution of bioarchaeology, to understanding childhood in the past is illustrated.

The social child

Most cultures recognize childhood as a period in the human life span (Bogin 1997:63; Stearns 2006:3). What is universal in humans is the period of biological and social immaturity, but there are numerous ways in which cultures negotiate this, making childhood a life stage of variable length and diverse associated social roles (Prout 2005:111).

Without considering the role that children play in society, bioarchaeologists may unwittingly portray children as passive victims of their environment. However, children *do* determine the majority of the day-to-day activities of the family in terms of the care that is provided to them, the contribution that they make to the household and society, as well as their social relationships with their parents, siblings, and extended family.

Human infancy and childhood is a crucial and vulnerable time biologically (Prentice and Prentice 1988; Trevathan 2005), which requires a lot of energy input from adults and often other children (Stearns 2006:1). It is understood that childhood diseases and mortality, and the prevention of these, would have been a preoccupation of parents, caregivers, and their wider community in the past (Stearns 2006:1). It is partly because of the biological immaturity of infants and children, and therefore susceptibility to morbidity and mortality (McDade 2003), that complex social arrangements for their care and well-being exist, and that certain social age identity categories exist in human societies (Halcrow and Tayles 2008b). Bioarchaeologists, by assessing age-specific mortality rates, can contribute to understanding childhood and the treatment of children within societies. Infant and child morbidity and mortality are important factors to consider when understanding the social meanings of childhood and the social relationships between adults and children.

From birth, humans have communication skills that are important in the facilitation of social childcare relationships. Complex sociobiological models of the relationship between maternal behavior, hormones, and the suckling infant have been described in the literature (Winberg 2005). Through close contact after birth, mothers have been shown to regulate the newborn's temperature, respiration, crying, and nursing behaviors (Winberg 2005). Similarly, the baby may regulate and therefore increase their mother's attention through initiation and maintenance of breastfeeding and the efficiency of maternal exploitation of ingested calories by gastrointestinal hormone release (Winberg 2005). It can be argued from an evolutionary standpoint that infants are socially "primed" from birth to increase the nurturing behavior from adults and other children, therefore ensuring their survival. When born, an infant has the ability to perform behaviors including making eye contact, following movements of a caregiver's face, and crying, all of which have been shown to contribute to ensuring the proximity of the caregiver and the development of social relationships (Katz et al. 1973; Winberg 2005).

Based on ethnographic works, historical records, and archaeological evidence, it has been argued that children were a major contributor to economic production in prehistoric societies (Cain 1977; Kamp 2001a,b; Nag et al. 1978; Wileman 2005). For example, Maya boys from Xculoc, Yucatán Peninsula, Mexico between the age of 7 and 15 years produce more than half of what they consume (Kramer 2005). However, industrialization and capitalism could have contributed to an increase in child labor recorded in some ethnographic and historical works. Historical archaeological investigations also show that children are important agents in forming their environment, by creating their own material culture, social spaces, and social networks (Wilkie 2000).

Definition of social age categories in bioarchaeology

Recently, researchers have begun to advocate the use of social age categories in bioarchaeological analyses (Kamp 2001b; Perry 2005). This section addresses the methods and theory important when assessing social age.

> Most of the studies of childhood paleo-health and nutrition are weakened because they fail to use archaeological data to establish age group boundaries. Studies usually start with a definition of groups that seems logical to the investigator, then test for differences between the groups, rather than beginning the exploration by looking for differences that might imply local age definitions. Because the burial remains, which are often the basis for research into children's health and nutrition, are one of the primary sources for establishing age groups archaeologically, this area of investigation should be one of the pioneers in such a process (Kamp 2001b:10).

This quote illustrates existing tensions between biological and archaeological investigations of age in the past, which arise partly from the assumption that "biological" age is directly linked to "social" age (see Sofaer, this volume). Social age categories are difficult to define in past populations, particularly prehistoric societies in the absence of written records of child and adult status, social roles and relationships. Establishment of social age groups is also contingent on skeletal samples large enough to demonstrate recognizable patterns of different social processes. Alternatively, researchers have relied on ethnographic analogy and/or analysis of subadult burial rituals, although both are known to have inherent problems.

Another issue that adds complexity to defining social age is that the definition of childhood can change over time within a society. For example, the age at which children are perceived to be adults as recorded in legal documents in Anglo-Saxon Britain changed from 10 years in the 7th century to 12 years in the 10th century A.D. (Crawford 1991). Also, in contrast to modern Western societies where social age is closely linked to chronological age, in many "traditional" societies, social age relates to a particular stage of maturation (Cox 2000; Fortes 1984). These stages take into account not only the chronological age but also the skills, personality, and capacities of the individual (Kamp 2001b:4). For example, Stoodley (2000) has argued in a study of burial rites in early Anglo-Saxon Britain that "age identity" was only loosely related to biological age or chronological age. Also, it is important to acknowledge that the number of age categories may vary from one population to the next, as well as with gender and other social attributes including class (Kamp 2001b:25; Lesick 1997). There is a tendency when assessing age categories to think in terms of the singular (Insoll 2007a:6). However, while different categories are important, aspects of identity are "multivalent," defined by multiple elements (Insoll 2007a:6; Kealhofer 1999:63). For example, Rega (1997) identified a pattern for the different placement of needles in the graves of women and girls at the cemetery of Mokrin in Hungary. However, there was no overlapping artifact type for males indicating that a different type of transition existed for age categories in boys and men (Rega 1997).

The problem of using biological age as a proxy for social age of a child is becoming more widely acknowledged in the literature (Baxter 2005b:98; Lewis 2007). Some texts are starting to state explicitly the "type" of age used when presenting results and analyses. For example, it is acknowledged by Lewis (2007:2) that the categorical terms used within subadulthood in her book on the bioarchaeological study of children provide a biological basis for discussion, and are not intended to describe the social age of these individuals. Recently, some are even attempting to identify transitions in social age through the analysis of burial ritual (e.g., Gowland 2006).

Mortuary treatment including grave wealth can be considered when assessing the socially important age categories in a population (Saxe 1970; Tainter 1978). It has been shown in ethnographical and archaeological work, for example, that infants and children are often treated differently from adults in terms of burial location, position, and grave goods (Boric and Stefanovic 2004; Jamieson 1995; Kamp 1998; Murail et al. 2004). These can reflect beliefs about personhood and social age categories.

However, it is well known that there are problems with an approach that ascribes differing mortuary treatment to status during life (Parker Pearson 1982; Scott 1993; Ucko 1969:266-268; Wason 1994). Archaeologists have rethought the perspectives employed by the New Archaeologists, where mortuary rituals were seen as a "... passive reflection of abstract concepts of society and social structure" (Parker Pearson 1999:84). It is now acknowledged that there are a variety of reasons for the grave wealth interred with particular burials, other than just material wealth and a society's value system (Parker Pearson 1999:84). It has also been argued that, although the rules of mortuary interpretation using Saxe's (1970) work, for example, may hold true for adults, these may not be applicable to children, as the individuality of the child may be subsumed into that of family status (Crawford 1991:18).

Although these factors of burial treatment complicate interpretation of the subadult mortuary remains, this need not necessarily negate their use in assessment of social age categories. Rather these issues need to be considered to help strengthen analysis and interpretation of childhood in past populations.

To summarize, a number of issues complicate the incorporation of social age categories into bioarchaeological research. Given that societies have different notions of what constitutes a child and definitions of social age categories that may not always match those based on chronological age, bioarchaeological data collected and presented using these social age categories can prevent comparison of these data among different populations. One way to get around this, although work intensive, could be to explicitly present data within age categories traditionally used in bioarchaeological health analysis for comparative purposes in addition to social age categories.

So, where do we start in bioarchaeological investigations defining childhood and its subdivisions? Just as bioarchaeologists incorporate information on the cultural context of children's roles and activities in the interpretation of health and mortality, some are also looking at developing a more integrated biological and social picture of childhood and age (e.g., Crawford 1991; Ingvarsson-Sundström 2003; Sofaer 2006).

Perry (2005) has analyzed the health of subadults from the Byzantine Near East. Historical evidence indicates that the Byzantine empire established legal codes defining the ages at which an individual could marry, and marking a cultural transition to adulthood, which began at around 13–15 years. Perry (2002:269, cited in Perry 2005:97) found that, compared with individuals from a Roman period community and a Byzantine period urban trading center, the Byzantine-period site of Rehovot, Israel contained a high number of older children and adolescents (defined as 7–15 years). Perry (2002:270, cited in Perry 2005:97) argues that the high mortality in these age groups marked a period of "self-sufficiency" possibly being the cause of physiological stress coinciding with increased labor. Perry (2005:97) comments that by adhering to modern Western notions of age categories, these individuals, some of whom were probably married and independent, were incorrectly defined as subadults. This is a good example of the importance of the recognition of both social age and biological age. Interpreting these in terms of McDade's (2003) life history theory of resource allocation to biological processes, these young individuals who were socially self-sufficient and working, probably fared worse than biologically mature individuals because of the trade-off in energy that was needed to carry out these demanding tasks while continuing biological growth and development.

Wiley and Pike (1998) advocate analyzing mortality using developmental stages such as crawling and weaning. In this way children's roles and activities in relation to their environment are the focus (Wiley and Pike 1998), which follows the central theme in childhood social theory of children as social actors. This could be applied to understanding aspects of social age groups in skeletal samples. For example, whether there are any patterns of health and/or mortuary treatment that are related to developmental stages in life including crawling and weaning could be investigated, taking into account that the duration and timing of these stages are also mediated by the social environment and therefore can differ among societies. As discussed, weaning age can be determined through isotopic analysis of teeth.

Ingvarsson-Sundström (2003:170) hypothesizes in her work on skeletal remains from the Middle Helladic period that changing social identities from "subadult" to "adult" is when a girl or boy moved to start their own family. This could perhaps be determined through isotopic analysis of young females and males at given sites to assess whether the enamel chemistry fits with the local signature of the people in this area (Bentley 2006). Of course, this assumes that individuals would have moved to another "geological area" to start their family.

Sofaer (2006, and this volume) provides a theoretical discussion with the aim of reducing the chasm between bioarchaeological and social archaeological approaches to understanding age. Using sociological theory, Sofaer (2006:129) discusses the use of conceptualizing the body as a "hybrid," the notion of the body as being socially and biologically unfinished and therefore the cumulative formulation of a complex entity that develops over time (Prout 2000a; Shilling 1993). Here the body is viewed as both a material and a cultural object. With this approach there is a move beyond the biocultural model, where osseous change is seen as an adaptation to the environment, and away from simply drawing inferences about the relationship between people and objects in mortuary contexts, and therefore beyond the nature–culture dualism, to an analysis of the total milieu in which people are situated and the objects which they use causing changes to the body (Sofaer 2006:141). To illustrate this approach Sofaer (2006:140) refers to her study of individuals from the island of Ensay, Outer Hebrides. This research focuses on the analysis of gendered activities, where she attributes the differences in osseous changes in the spines between adult men and women to load-bearing, where women, primarily, carried creel baskets. Sofaer (2006:140) using the concept of hybridity takes this one step further and looks at the life course in which these activities were carried out. She argues that because these gendered skills were acquired in childhood these were having an effect on the body early in life.

Lorentz (2003) uses a similar approach in her research on modifications of children's bodies from Cyprus. Focusing on the analysis of head-shaping, contextual burial analysis, and anthropomorphic depictions, the young body was viewed either as a depiction (figurine), or physically, as being able to be manipulated and modified by material practices both in life and death. This approach acts to draw together specific entities that are presented as dualistic in the literature including: nature–culture, mind–body, agent–artifact. Further research into age-related growth and malleability of cranial structure could also be useful as this is possibly intertwined with cultural ideas about manipulation and modification.

Bioarchaeologists are aware that the bodies of infants and children are shaped throughout life by both biological, and social and cultural factors. In fact, bioarchaeology is very well situated to assess the contribution of biology and social/cultural factors in shaping children in the past. Lally and Ardren (2008) have recently produced a paper examining approaches to understanding the archaeological infant. While this is helpful in contributing to the examination of social aspects of infancy in the archaeological record, in criticizing biological approaches, it ignores the biology, thereby reinforcing the dualistic view of the infant body. Rather, approaches to understanding infancy and childhood do not have to view biological and social aspects of the body as mutually exclusive.

Conclusion

The past decade has witnessed an increased interest in children and childhood in archaeological research. The social archaeological and bioarchaeological approaches to assess age and childhood demonstrate the dualistic way in which "social" and "biological" aspects of the body are viewed. This chapter has discussed the practical and theoretical issues that need to be considered in the endeavor of incorporating social theory into bioarchaeological analysis. Issues discussed included terminology and age categories used and the problems of identifying social age in past populations. The important contribution of biology in the analysis of childhood and age in past societies is outlined, and various bioarchaeological methods described. A practical solution to the problem of integrating social age into bioarchaeology remains somewhat elusive and we do not presume to suggest that we have provided answers. The new theoretical approaches for understanding the body we have reviewed are useful to place the skeleton into discussions of childhood in the past, and importantly, in turn, to start the process of developing a more integrated and balanced biological and social picture of childhood and age.

REFERENCES

Allison, J., and A. Prout eds. 1997 Constructing and Reconstructing Childhood: Contemporary Issues in the Sociological Study of Childhood. London: Falmer.

Ariès, P. 1962 Centuries of Childhood: A Social History of Family Life. London: Cape.

Aufderheide, A., and C. Rodriguez-Martin 1998 The Cambridge Encyclopedia of Human Paleopathology. Cambridge: Cambridge University Press.

Baker, P. T., and J. S. Dutt 1972 Demographic Variables as Measures of Biological Adaptation: A Case Study of High Altitude Populations. *In* The Structure of Human Populations. G. A. Harrison, and A. J. Boyce, eds. Pp. 352–378. Oxford: Clarendon Press.

Baxter, J. E. ed. 2005a Children in Action: Perspectives on the Archaeology of Childhood. Berkeley: University of California Press.

Baxter, J. E. ed. 2005b The Archaeology of Childhood: Children, Gender, and Material Culture. Volume 10. Walnut Creek: AltaMira Press.

Baxter, J. E. ed. 2008 The Archaeology of Childhood. Annual Review of Anthropology 37:159–175.

Bello, S. M., A. Thomann, M. Signoli, O. Dutour, and P. Andrews 2006 Age and Sex Bias in the Reconstruction of Past Population Structures. American Journal of Physical Anthropology 129:24–38.

Benthall, J. 1992 A Late Developer? The Ethnography of Children. Anthropology Today 8:1.

Bentley, R. A. 2006 Strontium Isotopes from the Earth to the Archaeological Skeleton: A Review. Journal of Archaeological Method and Theory 13:135–187.

Bluebond-Langner, M., and J. E. Korbin 2007 Challenges and Opportunities in the Anthropology of Childhoods: An Introduction to "Children, Childhoods, and Childhood Studies". American Anthropologist 109:241–246.

Bocquet-Appel, J. P., and S. Naji 2006 Testing the Hypothesis of a Worldwide Neolithic Demographic Transition. Current Anthropology 47:341–365.

Bogin, B. 1997 Evolutionary Hypotheses for Human Childhood. Yearbook of Physical Anthropology 40:63–89.

Bogin, B. 2003 The Human Pattern of Growth and Developmenti Paleontological Perspective. *In* Patterns of Growth and Development in the Genus *Homo*. J. L. Thompson, G. E. Krovitz, and A. J. Nelson, eds. Pp. 15–44. Cambridge: Cambridge University Press.

Boric, D., and S. Stefanovic 2004 Birth and Death: Infant Burials from Vlasac and Lepenski Vir. Antiquity 78:526–546.

Bornstein, M. H., and M. E. Lamb 1992 Development in Infancy: An Introduction. New York: McGraw-Hill.

Bowman, V. ed. 2007 Scholarly Resources for Children and Childhood Studies: A Research Guide and Annotated Bibliography. Lanham, MD: The Scarecrow Press.

Brook, A. H., J. M. Fearne, and J. M. Smith 1997 Environmental Causes of Enamel Defects. *In* Dental Enamel. D. J. Chadwick, and G. Cardew, eds. Pp. 212–225. Chichester: John Wiley and Sons.

Buckley, H. R. 2000 Subadult Health and Disease in Prehistoric Tonga, Polynesia. American Journal of Physical Anthropology 113:481–505.

Buikstra, J. E., and D. C. Cook 1980 Paleopathology: An American Account. Annual Review of Anthropology 9:433–470.

Buikstra, J. E., and D. H. Ubelaker 1994 Standards for Data Collection from Human Skeletal Remains. Arkansas Archaeological Survey Report No. 44. Arkansas: Fayetteville.

Cain, M. 1977 The Economic Activities of Children in a Village in Bangladesh. Population and Development Review 3:210–227.

Chandra, R. K. 1975 Fetal Malnutrition and Postnatal Immunocompetence. American Journal of Diseases in Childhood 129:450–454.

Cohen, M. N. 1994 The Osteological Paradox Reconsidered. Current Anthropology 35(5):629–637.

Cole, F. B. 1998 Bacterial Infections of the Newborn. *In* Avery's Diseases of the Newborn. H. M. Taeusch, and R. A. Ballard, eds. Pp. 490–512. Philadelphia: W. B. Saunders Company, A Division of Harcourt Brace and Company.

Corsaro, W. A. 2005 The Sociology of Childhood. Thousand Oaks: Pine Forge Press.

Cox, M. 2000 Ageing Adults from the Skeleton. *In* Human Osteology in Archaeology and Forensic Science. M. Cox, and S. Mays, eds. Pp. 61–81. London: Greenwich Medical Media.

Crawford, S. 1991 When do Anglo-Saxon Children Count? Journal of Theoretical Archaeology 2:17–24.

Crawford, S., and C. Lewis 2008 Childhood Studies and the Society for the Study of Childhood in the Past. Childhood in the Past: An International Journal 1:5–16.

Cunningham, H. 1995 Children and Childhood in Western Society Since 1500. London: Longman.

de Beauvoir, S. 1949 The Second Sex. Harmondsworth: Penguin.

Duray, S. M. 1990 Deciduous Enamel Defects and Caries Susceptibility in a Prehistoric Ohio Population. American Journal of Physical Anthropology 81:27–34.

Enlow, D. H. 1976 The Remodelling of Bone. Yearbook of Physical Anthropology 20:19–34.

Eveleth, P. B., and J. M. Tanner 1990 Worldwide Variation in Human Growth. Cambridge: Cambridge University Press.

Ferjerskov, O., V. Baelum, and E. S. Østergaard 1993 Root Caries in Scandinavia in the 1980s and Future Trends to be Expected in Dental Caries Experience in Adults. Advances in Dental Research 7:4–14.

Finlay, N. 2000 Outside of Life: Traditions of Infant Burial in Ireland from *Cillín* to Cist. World Archaeology 31:407–422.

Fortes, M. 1984 Age, Generation, And Social Structure. *In* Age and Anthropological Theory. D. I. Kertzer, and J. Keith, eds. Pp. 99–122. New York: Cornell University Press.

Fowden, A. B. 2001 Growth and Metabolism. *In* Fetal Growth and Development. R. Harding, and A. D. Bocking, eds. Pp. 44–69. Cambridge: Cambridge University Press.

Fuller, B. T., M. P. Richards, and S. A. Mays 2003 Stable Carbon and Nitrogen Isotope Variations in Tooth Dentine Serial Sections from Wharram Percy. Journal of Archaeological Science 30:1673–1684.

Garn, S. M., C. G. Rohnmann, M. Behar, F. Viteri, and M. A. Guzman 1964 Compact Bone Deficiency in Protein-Calorie Malnutrition. Science 145:144–145.

Ginn, J., and S. Arber 1995 "Only Connect": Gender Relations and Ageing. *In* Connecting Gender and Ageing: A Sociological Approach. S. Arber, and J. Ginn, eds. Pp. 1–14. Buckingham: Open University Press.

Goodman, A. H. 1993 On the Interpretation of Health from Skeletal Remains. Current Anthropology 34:281–288.

Goodman, A. H., and G. J. Armelagos 1989 Infant and Childhood Morbidity and Mortality Risks in Archaeological Populations. World Archaeology 21(2):225–243.

Goodman, A. H., G. J. Armelagos, and J. C. Rose 1984a The Chronological Distribution of Enamel Hypoplasias from Prehistoric Dickson Mounds Populations. American Journal of Physical Anthropology 65:259–266.

Goodman, A. H., J. Lallo, G. J. Armelagos, and J. C. Rose 1984b Health Changes at Dickson Mounds, Illinois (A.D. 950–1300). *In* Paleopathology at the Origins of Agriculture. M. N. Cohen, and G. J. Armelagos, eds. Pp. 271–305. Orlando: Academic Press.

Goodman, A. H., and J. C. Rose 1990 Assessment of Systemic Physiological Perturbations from Dental Enamel Hypoplasias and Associated Histological Structures. Yearbook of Physical Anthropology 33:59–110.

Goodman, A. H., R. B. Thomas, A. C. Swedlund and G. J. Armelagos 1988 Biocultural Perspectives on Stress in Prehistoric, Historical and Contemporary Population Research. Yearbook of Physical Anthropology 31:169–202.

Gordon, C. C., and J. E. Buikstra 1981 Soil pH, Bone Preservation, and Sampling Bias at Mortuary Sites. American Antiquity 46(3):566–571.

Gottlieb, A. 2000 Where Have All the Babies Gone? Towards an Anthropology of Infants (and their Caretakers). Anthropological Quarterly 73:121–132.

Gould, S. J. 1996 The Mismeasure of Man. New York: Norton.

Gowland, R. L. 2006 Ageing the Past: Examining Age Identity from Funerary Evidence. *In* Social Archaeology of Funerary Remains. R. L. Gowland, and C. Knüsel, eds. Pp. 143–154. Oxford: Oxbow.

Guy, H., C. Masset, and C.-A. Baud 1997 Infant Taphonomy. International Journal of Osteoarchaeology 7:221–229.

Halcrow, S. E., and N. Tayles 2008a Stress Near the Start of Life? Localised Enamel Hypoplasia of the Primary Canine in Late Prehistoric Mainland Southeast Asia. Journal of Archaeological Science 35:2215–2222.

Halcrow, S. E., and N. Tayles 2008b The Bioarchaeological Investigation of Children and Childhood: Problems and Prospects. Journal of Archaeological Method and Theory 15:190–215.

Halcrow, S. E., N. Tayles, and H. R. Buckley 2007 Age Estimation of Children in Prehistoric Southeast Asia: Are the Standards Used Appropriate? Journal of Archaeological Science 34:1158–1168.

Halcrow, S. E., N. Tayles, and V. Livingstone 2008 Infant Death in Prehistoric Mainland Southeast Asia. Asian Perspectives 47:371–404.

Hanawalt, B. A. 1993 Growing Up in Medieval London: The Experience of Childhood in History. New York: Oxford University Press.

Hawkey, D. E., and C. F. Merbs 1995 Activity-induced Musculoskeletal Stress Markers (MSM) and Subsistence Strategy Changes Among Ancient Hudson Bay Eskimos. International Journal of Osteoarchaeology 5:324–338.

Herring, D. A., S. R. Saunders, and M. A. Katzenberg 1998 Investigating the Weaning Process in Past Populations. American Journal of Physical Anthropology 105(4):425–439.

Heuzé, Y., and H. F. V. Cardoso 2008 Testing the Quality of Nonadult Bayesian Dental Age: Assessment Methods to Juvenile Skeletal Remains: The Lisbon Collection Children and Secular Trend Effects. American Journal of Physical Anthropology 135:275–283.

Hillson, S. W. 1996 Dental Anthropology. Cambridge: Cambridge University Press.
Hopkins, B., and R. G. Barr eds. 2004 The Cambridge Encyclopedia of Child Development. New York: Cambridge University Press.
Hoppa, R. D. 1992 Evaluating Human Skeletal Growth: An Anglo-Saxon Example. International Journal of Osteoarchaeology 2:275–288.
Hooton, E. A. 1930 The Indians of Pecos Pueblo: A Study of Their Skeletal Remains. New Haven, CT: Yale University Press.
Hummert, J. R., and D. P. Van Gerven 1983 Skeletal Growth in a Medieval Population from Sudanese Nubia. American Journal of Physical Anthropology 60:471–478.
Humphrey, L. T. 2000 Growth Studies of Past Populations: An Overview and an Example. *In* Human Osteology in Archaeology and Forensic Science. M. Cox, and S. Mays, eds. Pp. 23–38. London: Greenwich Medical Media.
Ingvarsson-Sundström, A. 2003 Children Lost and Found: A Bioarchaeological Study of the Middle Helladic Children in Asine with a Comparison to Lerna. Ph.D. Thesis, Uppsala University.
Insoll, T. 2007a Introduction: Configuring Identities in Archaeology. *In* The Archaeology of Identities: A Reader. T. Insoll, ed. London: Routledge.
Insoll, T. ed. 2007b The Archaeology of Identities: A Reader. London: Routledge.
James, A., C. Jenks, and A. Prout 1998 Theorizing Childhood. Cambridge: Polity Press.
James, A., and A. Prout eds. 1997 Constructing and Reconstructing Childhood: Contemporary Issues in the Sociological Study of Childhood. London: RoutledgeFalmer.
Jamieson, R. W. 1995 Material Culture and Social Death: African-American Burial Practices. Historical Archaeology 29:39–58.
Jantz, R. L., and D. W. Owsley 1984 Long Bone Growth Variation Among Arikara Skeletal Populations. American Journal of Physical Anthropology 63:13–20.
Jenks, C. 2005 Childhood. London: Routledge.
Johnston, F. E. 1961 Sequence of Epiphyseal Union in a Prehistoric Kentucky Population from Indian Knoll. Human Biology 33:66–81.
Johnston, F. E. 1962 Growth of the Long Bones of Infants and Young Children at Indian Knoll. American Journal of Physical Anthropology 20:249–254.
Johnston, F. E. 1969 Approaches to the Study of Developmental Variability in Human Skeletal Populations. American Journal of Physical Anthropology 31:335–342.
Jontell, M., and A. Linde 1986 Nutritional Aspects on Tooth Formation. World Review of Nutrition and Dietetics 48:114–136.
Kamp, K. A. 1998 Social Hierarchy and Burial Treatments: A Comparative Assessment. Cross-Cultural Research 32:79–115.
Kamp, K. A. 2001a Prehistoric Children Working and Playing: A Southwestern Case Study in Learning Ceramics. Journal of Anthropological Research 57:427–450.
Kamp, K. A. 2001b Where Have All the Children Gone?: The Archaeology of Childhood. Journal of Archaeological Method and Theory 8:1–34.
Katz, S. H., H. Rivinus, and W. Barker 1973 Physical Anthropology and the Biobehavioural Approach to Child Growth and Development. American Journal of Physical Anthropology 38:105–117.
Katzenberg, M. A., D. A. Herring, and S. R. Saunders 1996 Weaning and Infant Mortality: Evaluating the Skeletal Evidence. Yearbook of Physical Anthropology 39:177–199.
Kealhofer, L. 1999 Creating Social Identity in the Landscape: Tidewater, Virginia, 1600–1750. *In* Archaeologies of Landscape. W. Ashmore, and A. B. Knapp, eds. Pp. 58–82. Oxford: Blackwell.
King, T., S. W. Hillson, and L. T. Humphrey 2002 A Detailed Study of Enamel Hypoplasia in a Post-Medieval Adolescent of Known Age and Sex. Archives of Oral Biology 47:29–39.
Kramer, K. L. 2005 Maya Children: Helpers at the Farm. Cambridge, MA: Harvard University Press.

Kuefler, M. S. 1991 "A Wryed Existence": Attitudes Toward Children in Anglo-Saxon England. Journal of Social History 24:823–830.

La Fontaine, L. 1986 An Anthropological Perspective on Children in Social Worlds. *In* Children of Social Worlds. M. Richards, and P. Light, eds. Pp. 10–30. Cambridge: Polity Press.

Lally, M., and T. Ardren 2008 Little Artefacts: Rethinking the Constitution of the Archaeological Infant. Childhood in the Past: An International Journal 1:62–77.

Larsen, C. S. 1997 Bioarchaeology: Interpreting Behaviour from the Human Skeleton. Cambridge: Cambridge University Press.

Lesick, K. S. 1997 Re-engendering Gender: Some Theoretical and Methodological Concerns on a Burgeoning Archaeological Pursuit. *In* Invisible People and Processes: Writing Gender and Childhood into European Archaeology. J. Moore, and E. C. Scott, eds. Pp. 31–41. London: Leicester University Press.

Lewis, M. E. 2000 Non-adult Palaeopathology: Current Status and Future Potential. *In* Human Osteology in Archaeology and Forensic Science. M. Cox, and S. Mays, eds. Pp. 39–57. London: Greenwich Medical Media.

Lewis, M. E. 2002 Urbanisation and Child Health in Medieval and Post-Medieval England: An Assessment of the Morbidity and Mortality of Non-Adult Skeletons from the Cemeteries of Two Urban and Two Rural Sites in England (AD 850–1859). Oxford: Archaeopress.

Lewis, M. E. 2004 Endocranial Lesions in Non-Adult Skeletons: Understanding Their Aetiology. International Journal of Osteoarchaeology 14:87–97.

Lewis, M. E. 2007 The Bioarchaeology of Children: Perspectives from Biological and Forensic Anthropology. Cambridge: Cambridge University Press.

Lewis, M. E., and R. L. Gowland 2007 Brief and Precarious Lives: Infant Mortality in Contrasting Sites from Medieval and Post-Medieval England (AD 850–1859). American Journal of Physical Anthropology 134:117–129.

Lewis, M. E., and C. Roberts 1997 Growing Pains: The Interpretation of Stress Indicators. International Journal of Osteoarchaeology 7:581–586.

Lillehammer, G. 1989 A Child is Born. The Child's World in an Archaeological Perspective. Norwegian Archaeological Review 22:89–105.

Little, M. A. 1989 Human Biology of African Pastoralists. Yearbook of Physical Anthropology 32:215–247.

Lorentz, K. O. 2003 Cultures and Physical Modifications: Child Bodies in Ancient Cyprus. Stanford Journal of Archaeology 2:1–17.

Lovejoy, C. O., R. S. Meindl, T. R. Pryzbeck, T. S. Barton, K. G. Hemple, and D. Knotting 1977 Paleodemography of the Libben Site, Ottawa County, Ohio. Science 198:291–293.

Lovejoy, C. O., K. F. Russell, and M. L. Harrison 1990 Long Bone Growth Velocity in the Libben Population. American Journal of Human Biology 2(5):533–541.

Lussi, A., N. Kohler, and D. Zero 2000 A Comparison of the Erosive Potential of Different Beverages in Primary and Permanent Teeth Using an In Vitro Model. European Journal of Oral Sciences 108:110–114.

Maher, V. 1992 Breast-feeding and Maternal Depletion: Natural Law or Cultural Arrangements? *In* The Anthropology of Breast-feeding: Natural Law or Social Construct. V. Maher, ed. Pp. 151–180. Oxford: Berg.

Mays, S. A. 1993 Infanticide in Roman Britain. Antiquity 67:883–888.

Mays, S. A. 1995 The Relationship Between Harris Lines and Other Aspects of Skeletal Development in Adults and Juveniles. Journal of Archaeological Science 22:511–520.

Mays, S. A. 1998 The Archaeology of Human Bones. London: Routledge.

Mays, S. A. 1999 Linear and Appositional Long Bone Growth in Earlier Human Populations: A Case Study from Mediaeval England. *In* Human Growth in the Past: Studies from

Bones and Teeth. R. D. Hoppa, and C. M. Fitzgerald, eds. Pp. 290–312. Cambridge: Cambridge University Press.
McDade, T. W. 2003 Life History Theory and the Immune System: Steps Toward a Human Ecological Immunology. Yearbook of Physical Anthropology 46:100–125.
McEwan, J. M., S. Mays, and G. M. Blake 2005 The Relationship of Bone Mineral Density and Other Growth Parameters to Stress Indicators in a Medieval Juvenile Population. International Journal of Osteoarchaeology 15:155–163.
Mensforth, R. P., C. O. Lovejoy, J. W. Lallo, and G. J. Armelagos 1978 Part Two: The Role of Constitutional Factors, Diet, and Infectious Disease in the Etiology of Porotic Hyperostosis and Periosteal Reactions in Infants and Children. Medical Anthropology 2(1):1–59.
Merchant, V. L., and D. H. Ubelaker 1977 Skeletal Growth of the Protohistoric Arikara. American Journal of Physical Anthropology 46:61–72.
Messer, E. 1997 Intra-household Allocation of Food and Health Care: Current Findings and Understandings – Introduction. Social Science and Medicine 44:1675–1684.
Moore, J., and E. C. Scott, eds. 1997 Invisible People and Processes: Writing Gender and Childhood into European Archaeology. London: Leicester University Press.
Moore, W. J., and M. E. Corbett 1971 The Distribution of Dental Caries in Ancient British Populations 1: Anglo-Saxon Period. Caries Research 5:151–168.
Moore, W. J., and M. E. Corbett 1975 Distribution of Dental Caries in Ancient British Populations: III The 17th Century. Caries Research 9:163–175.
Mosley, W. H. 1984 Child Survival: Research and Policy. Population and Development Review Supplement 10:3–23.
Murail, P., B. Maureille, D. Peresinotto, and F. Geus 2004 An Infant Cemetery of the Classic Kerma Period (1750-1500 BC, Island of Sai, Sudan). Antiquity 78: 267–277.
Nag, M., B. N. F. White, and R. C. Peet 1978 An Anthropological Approach to the Study of the Economic Value of Children in Java and Nepal. Current Anthropology 19:293–306.
Nieuwenhuys, O. 1996 The Paradox of Child Labor and Anthropology. Annual Review of Anthropology 25:237–251.
Ortner, D. J. 1991 Theoretical and Methodological Issues in Paleopathology. *In* Human Paleopathology: Current Synthesis and Future Options. D. J. Ortner, and A. C. Aufderheide, eds. Pp. 5–11. Washington, DC: Smithsonian Institution Press.
Ortner, D. J. ed. 2003 Methods Used in the Analysis of Skeletal Lesions. Volume 45-72. San Diego: Academic Press.
Ortner, D. J., and W. G. J. Putschar 1985 Identification of Pathological Conditions in Human Skeletal Remains. Washington DC: Smithsonian Institution Press.
Ortner, D. J., and G. Turner-Walker 2003 The Biology of Skeletal Tissues. *In* Identification of Pathological Conditions in Human Skeletal Remains. D. J. Ortner, ed. Pp. 11–35. San Diego: Academic Press.
Panter-Brick, C. 1998 Biosocial Perspectives on Children. Cambridge: Cambridge University Press.
Parker Pearson, M. 1982 Mortuary Practices, Society and Ideology: An Ethnoarchaeological Study. *In* Symbolic and Structural Archaeology. I. Hodder, ed. Pp. 99–103. Cambridge: Cambridge University Press.
Parker Pearson, M. 1999 The Archaeology of Death and Burial. Phoenix Mill: Sutton Publishing.
Perry, M. A. 2005 Redefining Childhood Through Bioarchaeology: Towards an Archaeological and Biological Understanding Of Children In Antiquity. *In* Children in Action: Perspectives on the Archaeology Of Childhood. J. E. Baxter, ed. Pp. 89–111. Archaeological

Papers of the American Anthropological Association, Number 15. Berkeley: University of California Press.

Politis, G. G. 2006 Children's Activity in the Production of the Archaeological Record of Hunter-Gatherers: An Ethnoarchaeological Approach. *In* Global Archaeological Theory: Contextual Voices and Contemporary Thoughts. P. P. Funari, A. Zarankin, and E. Stovel, eds. Pp. 121–144. New York: Kluwer Academic.

Prentice, A., and A. Prentice 1988 Reproduction Against the Odds. New Scientist 118:42–46.

Prout, A. 2000a Childhood Bodies: Construction, Agency and Hybridity. *In* The Body, Childhood and Society. A. Prout, ed. Pp. 1–18. Basingstoke: Macmillan.

Prout, A. ed. 2000b The Body, Childhood and Society. Basingstoke: Macmillan.

Prout, A. 2005 The Future of Childhood: Towards the Interdisciplinary Study of Children. London: RouledgeFalmer.

Qvortrup, J., M. Bardy, G. Sgritta, and H. Wintersberger eds. 1994 Childhood Matters: Social Theory, Practice and Policy. Avebury: Aldershot.

Rega, E. 1997 Age, Gender and Biological Reality in the Early Bronze Cemetery at Mokrin. *In* Invisible People and Processes: Writing Gender and Childhood into European Archaeology. J. Moore, and E. C. Scott, eds. Pp. 229–247. London: Leicester University Press.

Ribot, I., and C. Roberts 1996 A Study of Nonspecific Stress Indicators and Skeletal Growth in Two Mediaeval Subadult Populations. Journal of Archaeological Science 23: 67–79.

Richards, M. P., S. Mays, and B. T. Fuller 2002 Stable Carbon and Nitrogen Values of Bone and Teeth Reflect Weaning Age at the Medieval Wharram Percy Site, Yorkshire, UK. American Journal of Physical Anthropology 119:205–210.

Robbins, G. 2007 Population Dynamics, Growth and Development in Chalcolithic Sites of the Deccan Plateau, India. Ph.D. Dissertation, University of Oregon.

Robbins, G. in press Saving the Babies and the Bathwater: Expanding Available Methods for Fertility-Centered Demography. Current Anthropology.

Ruff, C. 2003 Growth in Bone Strength, Body Size, and Muscle Size in a Juvenile Longitudinal Sample. Bone 33:317–329.

Saunders, S. R. 2000 Subadult Skeletons and Growth-Related Studies. *In* Biological Anthropology of the Human Skeleton. M. A. Katzenberg, and S. R. Saunders, eds. Pp. 135–162. New York: Wiley-Liss.

Saunders, S. R., and R. D. Hoppa 1993 Growth Deficit in Survivors and Non-Survivors: Biological Mortality Bias in Subadult Skeletal Samples. Yearbook of Physical Anthropology 36:127–151.

Saunders, S. R., R. D. Hoppa, and R. Southern 1993 Diaphyseal Growth in a Nineteenth Century Skeletal Sample of Subadults from St. Thomas' Church, Bellville, Ontario. International Journal of Osteoarchaeology 3:173–188.

Saxe, A. A. 1970 Social Dimensions of Mortuary Practices. Ph.D. Dissertation, Michigan University.

Scheuer, J. L., and S. Black 2000a Development and Ageing of the Juvenile Skeleton. *In* Human Osteology in Archaeology and Forensic Science. M. Cox, and S. Mays, eds. Pp. 9–21. London: Greenwich Medical Media.

Scheuer, J. L., and S. Black 2000b Developmental Juvenile Osteology. San Diego, California: Academic Press.

Scheuer, J. L., and J. E. Bowman 1995 Correlation of Documentary and Skeletal Evidence in the St. Brodes's Crypt Population. *In* Grave Reflections. Portraying the Past Through Cemetery Studies. S. R. Saunders, and D. A. Herring, eds. Pp. 49–70. Toronto: Canadian Scholars Press.

Schurr, M. R. 1998 Using Stable Nitrogen-Isotopes to Study Weaning Behavior in Past Populations. World Archaeology 30(2):327–342.

Schwartzman, H. B., ed. 2001 Children and Anthropology: Perspectives for the 21st Century. Westport, CN: Bergin and Garvey.

Scott, E. C. 1993 Images and Contexts of Infants and Infant Burials: Some Thoughts on Cross-Cultural Evidence. Archaeological Review from Cambridge 11:77–92.

Scott, E. C. 1997 Introduction: On the Incompleteness of Archaeological Narratives. *In* Invisible People and Processes: Writing Gender and Childhood into European Archaeology. J. Moore and E. C. Scott, eds. Pp. 1–12. London: Leicester University Press.

Scott, E. C. 1999 The Archaeology of Infancy and Infant Death. Oxford: Archaeopress.

Scrimshaw, S. C. M. 1984 Infanticide in Human Populations: Societal and Individual Concerns. *In* Infanticide: Comparative and Evolutionary Perspectives. G. Hausfater, and S. B. Hardy, eds. Pp. 439–462. New York: Aldine.

Sellen, D. W., and R. Mace 1997 Fertility and Mode of Subsistence: A Phylogenetic Analysis. Current Anthropology 38:878–889.

Seow, K. W. 1992 Dental Enamel Defects in Low Birthweight Children. *In* Recent Contributions to the Study of Enamel Developmental Defects. A. H. Goodman, and L. Capasso, eds. Pp. 321–330, Vol. 2. Chieti: Journal of Paleopathology Monograph Publications.

Shilling, C. 1993 The Body and Social Theory. London: Sage.

Sofaer Derevenski, J. R. 1997 Engendering Children, Engendering Archaeology. *In* Invisible People and Processes. J. Moore, and E. C. Scott, eds. Pp. 192–202. London: Leicester University Press.

Sofaer Derevenski, J. R. 2000 Children and Material Culture. London: Routledge.

Sofaer, J. R. 2006 The Body as Material Culture: A Theoretical Osteoarchaeology. Cambridge: Cambridge University Press.

Specker, B. L., W. Brazerol, R. C. Tsang, R. Levin, J. Searcy, and J. Steichen 1987 Bone Mineral Content in Children 1 to 6 Years of Age. American Journal of Disease in Children 141:343–344.

Stearns, P. N. 2006 Childhood in World History. New York: Routledge.

Steyn, M., and M. Henneberg 1996 Skeletal Growth of Children from the Iron Age Site K2 (South Africa). American Journal of Physical Anthropology 100:389–396.

Stoodley, N. 2000 From the Cradle to the Grave: Age Organization and the Early Anglo-Saxon Burial Rite. World Archaeology 31:456–472.

Stuart-Macadam, P. L. 1992 Porotic Hyperostosis: A New Perspective. American Journal of Physical Anthropology 87:39–47.

Tainter, J. A. 1978 Mortuary Practices and the Study of Prehistoric Social Systems. Advances in Archaeological Method and Theory 1:105–141.

Tocheri, M. W., T. L. Dupras, P. Sheldrick, and J. E. Molto 2005 Roman Period Fetal Skeletons from the East Cemetery (Kellis 2) from Kellis, Egypt. International Journal of Osteoarchaeology 15:326–341.

Tompkins, R. L. 1996 Human Population Variability in Relative Dental Development. American Journal of Physical Anthropology 99:79–102.

Trevathan, W. R. 2005 The Status of the Human Newborn. *In* The Cambridge Encyclopedia of Child Development. B. Hopkins, ed. Pp. 188–192. Cambridge: Cambridge University Press.

Turner, B. L., J. L. Edwards, E. A. Quinn, J. D. Kingston, and D. P. Van Gerven 2007 Age-related Variation on Isotopic Indicators of Diet at Medieval Kulubnarti, Sudanese Nubia. International Journal of Osteoarchaeology 17:1–25.

Ucko, P. J. 1969 Ethnography and Archaeological Interpretation of Funerary Remains. World Archaeology 1:262–280.

Ulijaszek, S. J., F. E. Johnston, and M. A. Preece eds. 1998 The Cambridge Encyclopedia of Human Growth and Development. Cambridge: Cambridge University Press.

Van Gerven, D. P., and G. J. Armelagos 1983 "Farewell to Paleodemography?" Rumours of Its Death Have Been Greatly Exaggerated. Journal of Human Evolution 12:353–360.

Waldron, T. 1991 Rates for the Job. Measures Of Disease Frequency in Paleopathology. International Journal of Osteoarchaeology 1:17–25.

Walker, P. L. 1969 The Linear Growth of Long Bones in Late Woodland Indian Children. Proceedings of the Indiana Academy of Sciences 78:83–87.

Walker, P. L., R. R. Bathurst, R. Richman, T. Gjerdrum, and V. A. Andrushko 2009 The Causes of Porotic Hyperostosis and Cribra Orbitalia: A Reappraisal of the Iron-Deficiency-Anemia Hypothesis. American Journal of Physical Anthropology 139:109–125.

Wall, C. E. 1991 Evidence of Weaning Stress and Catch-Up Growth in the Long Bones of a Central Californian Amerindian Sample. Annals of Human Biology 18:9–22.

Wason, P. 1994 The Archaeology of Rank. Cambridge: Cambridge University Press.

Weinberg, E. D. 1992 Iron Withholding in Prevention of Disease. *In* Diet, Demography and Disease. P. Stuart-Macadam, and S. Kent, eds. Pp. 105–150. New York: Aldine De Gruyter.

Weiss, K. M. 1973 Demographic Models for Anthropology. American Antiquity 38:1–88.

Welten, D. C., H. C. G. Kemper, G. B. Post, W. Van Mechelen, J. Twisk, P. Lips, and G. J. Teule 1994 Weight-bearing Activity During Youth is a More Important Factor for Peak Bone Mass Than Calcium Intake. Journal of Bone and Mineral Research 9:1089–1233.

Whittington, S. L. 1992 Enamel Hypoplasia in the Low Status Maya Population of Prehispanic Copan, Honduras. *In* Recent Contributions to the Study of Enamel Developmental Defects. A. H. Goodman, and L. L. Capasso, eds. Pp. 185–205. Journal of Paleoplathology, Monographic Publication 2. Chieti (Italy): Associazione Anthropologica Abruzzese.

Wileman, J. 2005 Hide and Seek: The Archaeology of Childhood. Stroud, Gloucestershire: Tempus.

Wiley, A. S., and I. L. Pike 1998 An Alternative Method for Assessing Early Mortality in Contemporary Populations. American Journal of Physical Anthropology 107(3): 315–330.

Wilkie, L. 2000 Not Merely Child's Play: Creating a Historical Archaeology of Children and Childhood. *In* Children and Material Culture. J. R. Sofaer Derevenski, ed. Pp. 100–113. London: Routledge.

Wilson, P. R., and A. D. Beynon 1989 Mineralisation Differences Between Human Deciduous and Permanent Enamel Measured by Quantitative Microradiology. Archives of Oral Biology 34:85–88.

Winberg, J. 2005 Mother and Newborn Baby: Mutual Regulation of Physiology and Behavior – A Selective Review. Developmental Psychobiology 47:217–229.

Wood, J. W., G. R. Milner, H. C. Harpending, and K. M. Weiss 1992 The Osteological Paradox: Problems of Inferring Prehistoric Health from Skeletal Samples. Current Anthropology 33:343–370.

Wright, L. E., and H. P. Schwarcz 1998 Stable Carbon and Oxygen Isotopes in Human Tooth Enamel: Identifying Breastfeeding and Weaning in Prehistory. American Journal of Physical Anthropology 106:1–18.

13

Moving from the Canary in the Coalmine

Modeling Childhood in Bahrain

Judith Littleton

Introduction

One bioarchaeological approach to children can be characterized as the "canary in the coalmine" approach, where the health of children, the most vulnerable segment of society is seen as a measure of the health of the entire community. This approach underlies bioarchaeological studies in which relative measures of "health" such as childhood mortality are used as a basis for comparison between subsistence patterns or within regions (e.g., the studies in Cohen and Armelagos 1984). More recently, there has been a greater focus on contextual analysis (e.g., Rautman 2000), and children (along with women) have emerged as the object of study within themselves (e.g., Baxter 2005). Focusing more exclusively on the subadults alone in an archaeological context, leads to new insights on those who "failed to adapt" (Lewis 2007). But is it possible to more fully interpret the experiences of those who failed to survive along with those who survived childhood; in other words to model childhood experience rather than to reflect only on a subset of society or the health of the entire community? In this chapter I will attempt to do that by analyzing the pattern of skeletal indicators among a group of human remains from Bahrain, the Arabian Gulf, and examining the relationship between the development and survivorship of children and the risks they faced.

Any analysis or attempt to interpret health or survivorship from the dead faces multiple interpretative difficulties including those highlighted by Wood et al. (1992). The extent to which skeletal populations represent stable demographic regimes is questionable – at the very least the record is time-averaged (Wood et al. 1992).

Social Bioarchaeology. Edited by Sabrina C. Agarwal and Bonnie A. Glencross
© 2011 Blackwell Publishing Ltd

Furthermore, a collection of the dead may represent individuals who are heterogeneous not simply because of intrinsic variation between individuals in frailty but also due to variation in exposure (Goodman 1993). Not all members of a society will necessarily be exposed to the same stressors and of those exposed some will be better able than others to withstand the threat. Argument has continued over the extent to which these issues are significant or can even be identified in an archaeological record (Cohen 1994; Goodman 1993; Saunders and Hoppa 1993; Wright and Yoder 2003). Certainly new techniques offer novel perspectives on these conundrums (Fitzgerald et al. 2006; Wright and Yoder 2003). However, we are still beset by the issue of age estimation (Saunders and Hoppa 1993), the inaccuracies and uncertainty of which may serve to mask mortality bias, although least among children. In addition, the uncertainty of the relationship between skeletal indicators and their relationship to stressors remains problematic (e.g., Lewis 2004; Ortner 1991).

In this chapter, I aim to analyze a series of indicators (porotic hyperostosis, child growth, and linear enamel hypoplasia) and their relationship with the distribution of ages at death for children. The analysis, based on a notion of survivorship, seeks to identify what were particularly "risky" periods of childhood for part or all of the society (approaching the issue of heterogeneity), what possible ecological circumstances are implicated, and what this combination might mean in terms of "growing up." Using multiple indicators among children who are the most accurately aged segment of skeletal samples, serves to highlight consistent trends and systemic stressors. While this may sound like a "biological" reconstruction, as Oris et al. (2004:360) point out: "Biological, environmental, economic, social and cultural factors interact to influence the survival chances of the youngest members of society. Diseases and malnutrition no longer are mere causes of death, but the consequence of an interaction between social and biological contexts." Constructing a model of "growing up" requires a biocultural perspective.

Materials

Bahrain

The modern kingdom of Bahrain is a group of small islands lying approximately halfway up the Arabian Gulf off the eastern coast of Saudi Arabia (Figure 13.1). The islands, particularly the main island of Awal, were noted in prehistoric and historic times because, in a largely arid area, the island afforded harborage and supplies of fresh water (Larsen 1983). This water, from underground aquifers, could support a local population engaged in intensive agriculture, pastoralism, and fishing along with successful trading enterprises (Andersen 2007; Larsen 1983; Salles 1999).

The main constraints on settlement and agriculture prior to the modern period were water and arable land. Climatically the island lies within the Saharo-Arabian zone (Rumney 1968). Rainfall is spare, never more than 7 cm per year. The main island was totally dependent upon groundwater supplies for cultivation of arable land at the northern end of the island and down a narrow strip along the west coast (which is where the DS3 site is located). The average annual temperature is 26.4 degree Celsius and at all times humidity is high (average 85 percent) (Vine 1986).

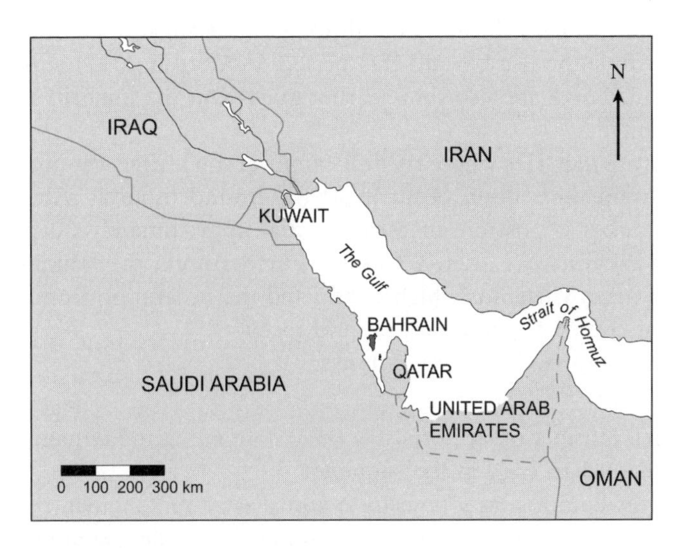

Figure 13.1 Map of Bahrain and the Arabian Gulf.

While intensive agriculture was the major part of subsistence in Bahrain, it was supplemented by pastoralism, primarily of sheep and goats, and fishing. Traditionally cultivation was based on date palm gardens, an oasis crop (Daggy 1959). On Bahrain other fruit and vegetables are intercropped with dates (Hansen 1968) with a gravity-fed water supply from springs. One problem of this form of agriculture is its association with malaria. Historically Bahrain was holoendemic for malaria with *Plasmodium falciparum* (and some *P. vivax*) recorded (Daggy 1959). The breeding of mosquitoes was encouraged by the presence of standing water particularly in irrigation systems and wells. Evidence from Saudi Arabia prior to eradication demonstrates continual malarial attack throughout the year (Daggy 1959). Such conditions also favor other parasites, for example shells of *Bulinus*, a vector of bilharzia (*Shistosomias haemotobium*), have been identified in deposits from the Islamic site of Bilad al-Qadim located just south of modern day Manama on Awal Island (Insoll and Hutchins 2005).

Bioarchaeological work in the Arabian Gulf has been ongoing since the work of Højgaard on dental anthropology in the 1980s (e.g., Højgaard 1980). Many, but by no means all publications of burials, have included a descriptive osteological report (e.g., Blau 1999; Kunter 1983, 1991, 2001; Schutkowski and Hermann 1987). More detailed bioarchaeological analyses have tended to focus upon earlier remains than those discussed here (e.g., Blau 1998; Kieswetter 2006; Macchiarelli 1989).

DS3 cemetery

The human remains analyzed here date to the Tylos (Hellenistic period) traditionally dated from c. 300 B.C. to A.D. 600 (Salles 1999). Precise dating of the cemetery within this period is difficult although finds suggest use from c. 300 B.C. to A.D. 250 with some use beyond this period (Andersen 2007) although the evidence is limited. There are also some earlier graves but the majority fall within this single time frame.

Unlike many cemeteries where the full extent of the burials cannot be discerned, burials in Bahrain most often occur in above-ground mounds. Analysis of aerial photographs makes it possible to estimate that approximately 80 percent of the cemetery was excavated (Littleton 1998b). Furthermore, the Bahrain Directorate of Archaeology and Museums which conducted the excavation from 1985 to 1987 had a policy of complete site clearance and recovery.

The burials comprised three main types:

1 A very small number of jar burials ($n = 25$) containing the remains of subadults. Some of these date early in the sequence.
2 Single graves comprising a limestone and plaster rectangular cist constructed into the slope of the mound and containing the single, generally supine body of the deceased.
3 The same grave structure but containing multiple individuals (average 6.5). While at times it is possible to identify the first individual and the last, most of these remains are commingled which means that, despite the large number of human remains recovered the number of identifiable and complete individuals is significantly smaller.

It has been argued that the jar burials represent single inhumations (Littleton 2000). On this basis there is no significant difference by age and sex in the frequency of those buried in single or multiple tombs, with the sole exception of the 0–1 year age group where less than 2 percent of infants were buried alone compared to approximately 20 percent for the remainder of the cemetery (Table 13.1). This high frequency of multiple graves at the site (c. 80 percent of individuals and 50 percent of all graves) is much greater than in other cemeteries from the same period (Herling 1999). There is no clear evidence for a temporal sequence. At the moment all that can be said is that most mounds contain a mixture of single and multiple graves and that this does not vary across the site which suggests that multiple burial is not a practice restricted to the later period of the site (Littleton 2000).

Methods

The following model builds upon a range of standard bioarchaeological analyses. The indicators recorded, along with age at death, are sufficiently common in the population to be informative. In addition, both porotic hyperostosis and linear enamel hypoplasia (LEH) are persistent lesions formed in childhood during the period of enamel mineralization (LEH) and when hematopoietic marrow occupies the diploe of the skull (Stuart-Macadam 1985). They thus remain as potentially permanent markers of survivorship of childhood insults and any subsequent mortality. In the following, the methods used and summary data are briefly described.

Table 13.1 Distribution of single and multiple graves by age at death

Age (years)	*Single (%)*	*Multiple (%)*	N
0.0–0.9	7.1	92.9	397
1.0–2.9	11.7	88.3	94
3.0–4.9	32.2	67.9	28
5.0–9.9	25.0	75.0	20
10.0–14.9	62.5	37.5	8
15.0–19.9	28.6	71.4	14
Adult (20+)	31.8	68.1	308
Total	**18.4**	**81.6**	**869**

Chi-square = 80.1, $p < 0.001$.

Table 13.2 Age distribution of deaths less than 20 years

Age period	*N*	%
0.0–1.0	405	64.2
1.1–4.9	170	26.9
5.0–9.9	31	4.9
10.0–14.9	25	4.3
15.0–20.0	44	7.5
Total	**631**	**100**

Age at death

The DS3 sample from Bahrain is unusual in the high proportion of subadult remains. Of the 1,051 individuals excavated, 631 were aged to less than 20 years (60 percent). Table 13.2 gives the distribution of deaths less than 20 years of age.

Because such a high proportion of the remains were from commingled burials, frequently dental remains were aged independently of the rest of the skeleton using Ubelaker's modification of Schour and Massler's dental formation and eruption (Ubelaker 1999). In order to age those remains without associated teeth, average long bone length for complete individuals was calculated so that isolated long bones could be assigned to broader age categories. Hence the sample is internally referenced to dental age. When different age categories were used children were apportioned across the possible range of ages accordingly.

Growth status of those who died

The number of subadults (n = 77) sufficiently complete to permit estimates of growth is low when compared with the total number of subadult remains. Hence the database for examining growth among those who died is small. Femoral lengths are presented in Table 13.3.

Table 13.3 Femoral length against dental age

Dental age (years)	*Average (mm)*	*S.D.*	*N*
0.0	67.5	–	3
0.1–0.5	84.1	6.8	10
0.5–1.0	100.5	9.8	26
1–2	122.2	13.3	12
2–3	136.0	2.8	2
3–4	158.6	16.6	8
4–5	173.0	7.5	5
5–6	196.4	2.6	5
6–7	–	–	–
7–8	212.0		1
8–9	230.5	17.7	2
9–10	–	–	–
10–11	280.0		1
11–12	260.0		1
13–14	319.0		1

Sample sizes of femora are small but reasonable up to age six. After that the numbers are simply too small to identify trends. A comparative study by Humphrey (2003) that includes these measurements demonstrates their overall small size compared to other prehistoric skeletal populations with comparable data. The significance of this, however, is hard to evaluate given relatively small adult statures (Littleton 2007).

A persistent difficulty in assessing growth of skeletal populations is the data are cross-sectional records of nonsurvivors and the comparative standards (sometimes longitudinal) are of modern populations (Saunders and Hoppa 1993). One way of dealing with growth, irrespective of the final size is to examine the proportion of average adult femur length attained at each age (AFLA) (Mensforth 1985; Wall 1991). More informatively these can be compared to a living population sample (Maresh 1955) using the study's increments at a given age as a percentage of mean adult values as the standard against which the skeletal populations can be compared (Table 13.4).

Porotic hyperostosis

Porous expansive lesions on the roof of the orbit (cribra orbitalia) and similar lesions on the remainder of the crania (porotic hyperostosis) were recorded across the sample. These lesions were distinguished from periosteal deposits where the appositional new bone has a distinct edge compared to the internal expansion of porotic hyperostosis (Littleton 1998a). There is a possibility of confusion between the two sets of lesions (Wapler et al. 2004) and five instances of periosteal deposits were recorded.

Table 13.4 Percentage of adult femur length attained, standardized to U.S. growth standards

Age (years)	*Bahrain*	N
0.0–0.4	0.6	13
0.5–0.9	−1.5	26
1.0–1.5	−1.9	7
1.5–1.9	−0.8	5
2.0–2.9	−0.5	2
3.0–3.9	−5.3	8
4.0–4.9	−3.2	5
5.0–5.9	−7.0	5

Source: Maresh (1955), derived from Wall, C. 1991 Evidence of Weaning Stress and Catch-up Growth in the Long Bones of a Central Californian Amerindian Sample. Annals of Human Biology 18:9–22.

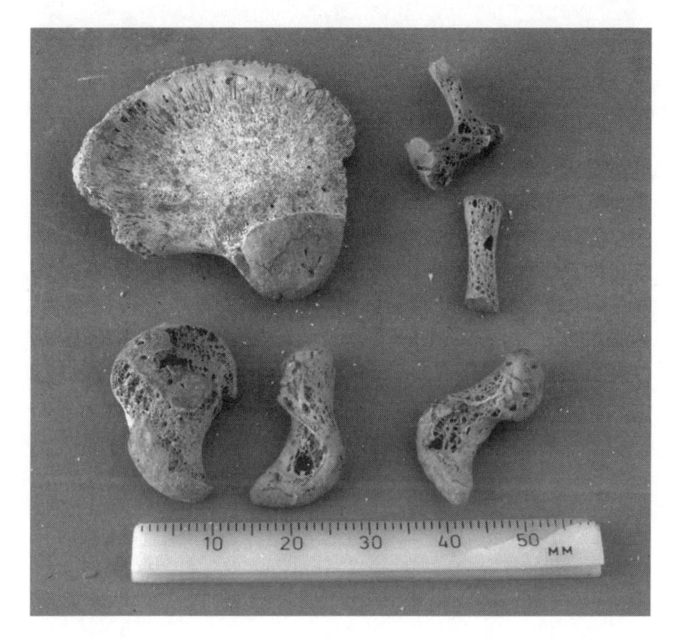

Figure 13.2 Cortical thinning resulting in a lattice-like pattern due to marrow expansion on infant bones (one ilium, two pubi, one ischium, one vertebral arch, and one metacarpal).
Source: Author's photograph

Nathan and Haas' (1966) coding of lesions was followed:

1 porosity – limited expansion of the diploe;
2 cribrotic – larger porosity, slight swelling of the contours;
3 trabecular – clear and marked swelling, confluent porosity; and
4 lattice work cortical thinning (postcranial only) (Figure 13.2).

Remodeled lesions were recognized by the infilling of the porosity (Stuart-Macadam 1989a) (Figure 13.3). Among adults, remodeled lesions were at times accompanied by still marked thickening of the cranium. Apart from bones of the

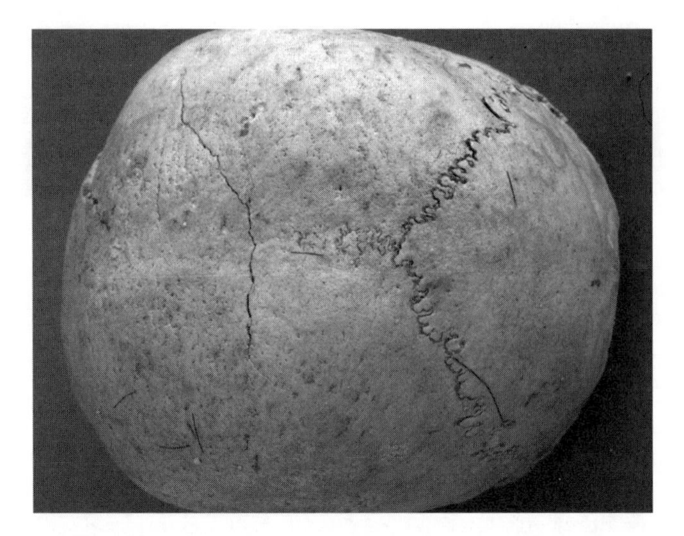

Figure 13.3 Remodeling and cranial thickening in an adult with healed porotic hyperostosis. There is still marked bossing on both parietals with the surface showing infilling of cribrotic and porotic areas.
Source: Author's photograph

Table 13.5 Frequency of cribra orbitalia and lesions associated with porotic hyperostosis

Age	*Cribra orbitalia*		*Porotic hyperostosis*	
	%	*N*	%	*N*
0–0.5	8.7	18	8.3	13
0.5–1	55.9	28	56.5	21
1–1.5	53.3	12	57.1	10
1.5–2	87.5	6	76.9	8
2–3	100	3	71.4	7
3–4	88.9	7	90.4	9
4–5	100	4	71.4	4
5–6	80	5	50	3
6–7	100	2	100	2

calvaria, the facial bones also showed marked changes, including porosity affecting the entire maxilla with marrow hyperplasia of the maxillae observed. At times the sinuses were completely blocked. The frequency of these lesions among children is given in Table 13.5.

Rickets

At least four of the infants from DS3 experienced rickets, most typically caused by vitamin D deficiency (Littleton 1998a). The lesions were recognized by flaring and cupping of the metaphyses of the long and short bones, compression of the vertebral endplates and distortion of the bones including swelling near the distal ends (Brickley and Ives 2008). The overall frequency of rickets cannot be calculated but considering the most frequently observed and affected postcranial bone (the femur), the maximum frequency is estimated to have been 2.4 percent of the subadults who died.

Deciduous enamel hypoplasia

Hypoplasia, enamel defects due to a lack of mineralization during development, was recorded as present or absent for all deciduous teeth. Defects were in the form of both pits and lines in the enamel. As observed by Skinner and Goodman (1992) the canines are most susceptible to these isolated defects. This may well be due, in part, to their susceptibility to traumatic lesions as opposed to more systemic causes such as fever or other stressors (Lukacs et al. 2001; Suckling 1989). The slightly protruding position of the canines and thinner bone over the crypt makes them susceptible to damage from falls. With the exception of the canines (with frequencies over 30 percent) the frequency of deciduous defects ranges between 9–14 percent (Table 13.6).

Permanent dental defects

Linear enamel hypoplasia was recorded for the anterior permanent dentitions: incisors and canines (Figure 13.4). The high frequency of ante-mortem tooth loss (approximately more than 30 percent of individuals) made it impractical to record defects on the posterior dentition. The data have been analyzed in three ways. Firstly the percentage of individuals with defects and the mean number of defects

Table 13.6 Frequency of deciduous enamel hypoplasia

Tooth	*N*	*%*	*Period of crown formation*
Li1	11	9.1	5th month–0.3 year
Li2	11	0	5th month–0.5 year
Lc	23	21.7	6th month–0.75 year
Lm1	23	8.7	5th month–0.5 year
Lm2	23	13	6th month–0.8 year
Ui1	14	14.3	5th month–0.3 year
Ui2	19	10.5	5th month–0.5 year
Uc	22	22.7	6th month–0.75 year
Um1	20	0	5th month–0.5 year
Um2	30	10	6th month–0.8 year

Note: Crown formation times based on Blakey, M. L., and G. J. Armelagos, 1985 Deciduous Enamel Defects in Prehistoric Americans from Dickson Mounds: Prenatal and Postnatal Stress. American Journal of Physical Anthropology 66(4):371–380.

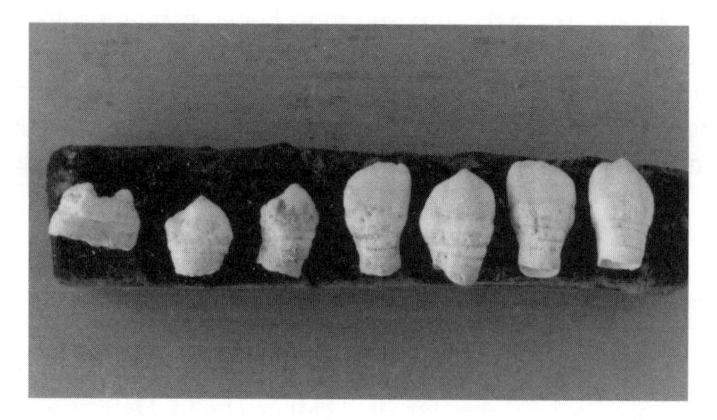

Figure 13.4 Multiple linear enamel defects on unerupted permanent tooth crowns.
Source: Author's photograph

Table 13.7 Percentage of individuals with permanent hypoplasic defects and mean

Age	*% with defect*	*% with >1 defect*	*N*
Subadult	96.7	90.0	30
Adult	96.0	91.9	99
	Mean	*S.D.*	*N*
5–9.9	3.4	1.7	14
10–14.9	3.0	1.8	7
15–19.9	3.8	1.5	6
20–29	2.5	1.5	18
30–44	3.4	1.4	26
45+	3.0	1.2	19
Total	3.1	1.4	132
Total, all units present	3.6	1.4	19
Male	3.1	1.4	44
Female	3.1	1.4	44

per individual were calculated (Table 13.7). Secondly, individual data for the central maxillary incisor and maxillary canine have been analyzed since these are the most hypoplasic teeth. Calculations were undertaken for the entire sample and then, in order to account for tooth wear, solely for those with more than 80 percent of the tooth crown length remaining (Table 13.8). These teeth cover slightly different periods of childhood: the canines 1.7–5.3 years, accounting for hidden enamel (Reid and Dean 2000); the central incisors 1.1–5.0 years. Defects were analyzed in terms of their number on each of these teeth and the distance from the cervicoenamel junction (CEJ), avoiding the difficulties of assuming constant crown formation as in many methods of hypoplasia recording (Goodman and Song 1999 versus Reid and Dean 2000). Finally in order to obtain a chronology of defects, hypoplasias were assigned to an age unit as per Hillson (1996). The scheme used

Table 13.8 Mean defects on the maxillary central incisor and upper canine (only including those with >80% of crown height)

Category	*Upper central incisor*			*Upper canine*		
	Mean	*S.D.*	*N*	*Mean*	*S.D.*	*N*
Subadult	2.83	1.59	12	2.43	1.27	7
Adult	2.09	1.13	33	2.72	1.25	29
Male	2.05	1.13	22	2.67	1.18	15
Female	2.25	1.24	16	2.60	1.58	10

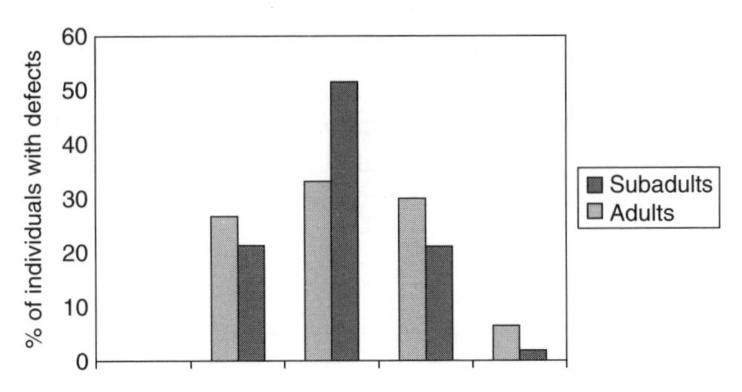

Figure 13.5 Distribution of linear enamel hypoplasia by dental development unit.
Source: Author's drawing

was revised based on experience of recording and maximizing matches between teeth in a sample from central Australia (Littleton 2005). The result is consistent with the pattern of crown formation recorded by Gustafsson and Koch (1974). The age units are not mutually exclusive; they overlap. Attention was paid to achieving the greatest possible number of matches between teeth (Figure 13.5).

Model assumptions

Using these data, a model is drawn linking age at death and lesion distribution with information on the formation of lesions, and the relationship between stressors and response, in the context of the island's particular ecological setting. In skeletal studies, however, analyzing these networks is problematic. While skeletal lesions may be related to the cause of death, they frequently do not reflect acute or accidental causes of death which may be significant in some populations (Ortner 1991). Analysis of age at death, however, might be able to highlight these lacunae by identifying ages at which high numbers of deaths occur in the absence of evident skeletal lesions. Furthermore, aging of adult skeletons in particular, and subadults to a lesser extent, is difficult. In this current analysis the problem of adult aging has been avoided. Subadult aging refers to dental development (Jackes 2000). However, interpretations of what lesions might mean rely on comparison with modern populations aged chronologically. Hence there is an unavoidable reliance ultimately on the relationship between dental and chronological age. In this current study, the most informative ages are 0–6 years which fortunately is also the group of ages which can be aged with the least error (Ubelaker 1999).

The following analysis treats this Bahrain mortuary collection as if they represent a balanced stable population. While it is possible to do a similar analysis allowing for population growth, in the context of limited water supplies and the extent of settlements on Bahrain at that time, there was relatively little potential for significant population growth (Littleton 1998b). In the following analysis the population is considered closed: those who survived into adulthood are assumed to be part of the same population as those who died in childhood. The following analysis is not an analysis of a cross-sectional viewpoint of an agricultural population at a single point in time rather it is a time-averaged model of a cohort or generation passing through childhood. For the ease of calculation 100 is used as the beginning size of the modeled cohort. This has been calculated as probably approximately equivalent to the size of the original population that used the cemetery (Littleton 1998b).

Results

The risks of being born

The majority of children died within the first year of life but there are differences in the frequency of deaths across that year. The lack of complete individuals among the very youngest presents a problem in distinguishing stillbirths from live births, and distinguishing preterm from small for gestational age (or low birth weight) babies (Saunders and Barrans 1999). All that can be pointed to in the Bahrain sample is that all of the infants were found independent of a female skeleton indicative of expulsion of the fetus, but this does not distinguish between live and stillbirths. In addition, identifying the proportion premature versus that proportion having experienced intrauterine growth retardation is important in reconstructing the fetal environment and ultimately the health of mothers (Saunders and Barrans 1999; Storey 1986).

Three children could be aged dentally to between 38–41 weeks gestational age. Their average femur length was 67.5 mm. This is approximately 6.5 mm less than the average femoral length for a range of modern Western populations, for example 73 mm (Olivier and Pineau 1958) and 74.3 mm (Kosa 1989). It suggests that there may be a significant proportion of either preterm births or small for gestational age babies, but which and to what extent?

Using Scheuer's regressions (Scheuer et al. 1980) for femurs less than 85 mm (aged in this sample through individual remains as c. 3 months of age), the resultant distribution indicates that most infants in this early group were between 38.5 and 41.5 weeks gestational age, i.e., full term and normal size (Figure 13.6). Three percent were very small size (30–32 weeks gestation) and it can be assumed that these represent truly premature births that failed to survive. The remaining 30.7

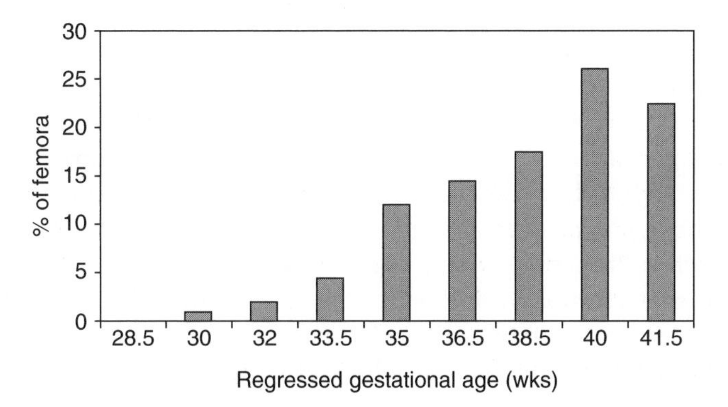

Figure 13.6 Distribution of gestational ages among infants less than three months of age at death, based on regression of femur length.
Source: Scheuer, J., J. Musgrave, and S. Evans 1980 The Estimation of Late Fetal and Perinatal Age from Limb Bone Length by Linear and Logarithmic Regression. Annals of Human Biology 7:257–265

percent of these infants were calculated to be between 33.5 and 36.5 weeks gestational age at death. From 33 to 38 weeks gestational age is the period of the most obvious growth retardation (King and Ulijaszek 1999; Tanner 1978) suggesting that around a third of the infants who died had experienced intrauterine growth retardation. This is indicative only, it cannot be a true percentage and in reality is an upper estimate. Scheuer's regressions are based on modern populations and hence could be assumed to underage rather than overage Bahraini infants, but it does suggest there were a small number of infants whose mothers were challenged during the third trimester of pregnancy. The rate, however, is not extraordinary (cf. Storey 1986). Compared to modern populations, a third of neonatal deaths associated with immaturity, is low to moderate. For example, Puffer and Serrano's (1973) review of subadult mortality in Central and Latin America indicates immaturity as an underlying or associated cause in more than 50 percent of all births. Such differences relate to the underlying structure of the causes of death. Given that intrauterine growth retardation is an important risk factor for neonatal and post-neonatal death, the prevalence of low birth weight among the dead is liable to be higher than in the entire cohort. The frequency suggests that poor maternal health was important for a subset of infants although not the entire population.

The causes of such conditions are numerous: maternal infection, maternal dietary deficiency, and age of the mother are all possible causal factors (Ulijaszek et al. 1998). Nor are such conditions unrelated within themselves, for example anemia during pregnancy can be precipitated by malarial attack and both intrauterine growth retardation and prematurity are higher in populations subject to malaria (Bray and Anderson 1979; Manir and Khalegque 1969; McGregor et al. 1983). In particular, primaparous women experience heavier malaria parasite loads making first births even more at risk of adverse fetal environments (McGregor et al. 1983). Nevertheless the evidence only applies to a third of those children presumably the remainder are deaths due to the combination of other causes of early infant death – birth trauma, congenital defects, droplet infections, even conditions such as neonatal tetanus (Maru et al. 1988) responsible in rural India for 16–72 percent of all deaths. As Oris et al. (2004) point out while it is often assumed that early infant deaths are due to endogenous causes, such conditions are often dependent on the health, well-being and environment of the mother. While the lack of skeletal lesions among these infants might suggest acute versus chronic conditions (Palkovich 1978b), this is also an age at which many of the conditions affecting an infant may not create a visible bony response (Ortner 1991; Stuart-Macadam 1985). Certainly the deciduous defects do suggest (as does the femur lengths) that a small proportion of infants did have difficult gestations.

Surviving the first year

The probability of dying in the first year in this population is estimated at 0.385 (including potential stillbirths). In a cohort of 100 births this represents 38 infants. Of these 38 children who died, 11 could be assumed to be neonatal deaths, deaths within one month of birth. Examining deaths more closely through the remainder of the first dental year the ratio of neonatal to post-neonatal mortality is 29 percent versus 71 percent. This is comparable to Saunders et al.'s (1995) study of the 19th-century St. Thomas' Church, Ontario. The two sets of data are not strictly comparable, the Bahrain figure is elevated by including potential stillbirths, although as argued above this may be as low as 3 percent.

Deaths are not equally distributed across the year (Table 13.9). The fewest deaths occur between 3 to 6 months, while the number of deaths from 6 to 12 months is only slightly lower when apportioned per month than in the first month. Using a base of 100 deaths in the first five years, it can be calculated that there were an average 7.1 deaths in the first month compared to 6.4 deaths per month between 6 to 12 months.

The ratio of early childhood to infant deaths (i.e. Deaths 1–4 years/Deaths 0–1 years) is similarly informative. In modern Western nations proportionately most

Table 13.9 Number of deaths per month of life (based on 100 deaths between birth and 6 years of age)

Age (months)	*Calculated deaths/month*	*N*
0–1	7.1	7.1
1–5	4.7	23.5
6–11	6.5	39
12–17	1.8	10.8
18–23	1.2	7.2
24–35	0.3	3.6
36–47	0.4	4.8
48–59	0.2	2.4
60–72	0.1	1.2

Table 13.10 Weaning ratios (D1-4/D0-1) among selected skeletal and modern populations

Location	*Ratio*	*Source*
DS3, Bahrain	0.42	
Skeletal populations		
Inamgaon Early Jarwe	0.41	(Kennedy 1984)
Inamgaon Lte Jarwe	0.2	(Kennedy 1984)
Late Woodland	1.02	(Goodman et al. 1984)
Middle Misissippian	0.73	(Goodman et al. 1984)
Arroyo Hondo	0.69	(Palkovich 1978b)
Small populations		
Matlab, Bangladesh	0.58	(Chen et al. 1980)
Guyana 1937–46	0.38	(Giglioli 1972)
Zapotec, Mexico	0.99	(Malina and Himes 1978)
Punjab 1886–1900	0.28	(Wyon and Gordon 1971)
Punjab 1957–9	0.14	(Wyon and Gordon 1971)
Rural El Salvador	0.67	(Puffer and Serrano 1973)
Keneba, Gamiba	2.09	(Billewicz and McGregor 1981)
Rural Morrocco 1961–3	0.65	(Cairo Demographic Centre 1982)
National populations		
Egypt 1936–40	0.80	(Cairo Demographic Centre 1982)
Jordan 1972	0.39	(Cairo Demographic Centre 1982)
Jordan 1961	0.67	(Cairo Demographic Centre 1982)
Kuwait 1965	0.3	(Cairo Demographic Centre 1982)

deaths occur in the neonatal period reflecting the intractability of endogenous causes of death and improvements in public health which have reduced post-neonatal death rates (Wills and Waterlow 1958). In lesser developed countries, on the other hand (particularly in places with famines and epidemics), mortality affects somewhat more equally all those less than five years of age at death. In particular, those in poor nutritional circumstances will tend to have higher early child-to-infant death ratios as the effects of under-nutrition impact on the attainment of active immunity to infections (McGregor et al. 1961; Wills and Waterlow 1958). In Bahrain the ratio is 0.42 which sits in the mid range of a series of recent historic populations from the surrounding Middle East and Asian subcontinent reflecting a population with a marked burden of infant and child mortality but most of which is concentrated within the first year of postnatal life (Table 13.10).

Despite a hypothesized number of small for gestational age children, those who died in the first six months were either experiencing catch-up growth or had entirely normal growth. This is a common pattern due to the protection of the quality of breast milk regardless of the nutritional state of the mother (King and Ulijaszek 1999). Fewer deaths occurred in the 3–6 month period attesting to the adequacy of the diet and protection from the effect of infection by breast milk.

During the second half of the first year, however, the number of infants dying increased. Referring to the modeled cohort of 100, 20 infants would have died during this time. These infant deaths are accompanied by skeletal evidence of stressors. Approximately half of those dead had active porotic hyperostosis (Table 13.5), primarily affecting the parietals and orbits. In addition, among the infants who died growth had slowed (Table 13.4), suggesting a synergy of infectious and nutritional problems where infections can directly affect nutritional status (e.g., malaria and hookworm infestation both lead directly to anemia), infections can be affected by nutritional status (e.g., severity of malarial infection is affected by iron status), and both infections and nutritional deficiencies can impair immune function (King and Ulijaszek 1999).

This interaction is seen very directly in the DS3 sample where three children: two 0–6 months of age and one 6–12 months had rickets. There is no evidence of bone deformity in older children suggesting that vitamin D deficiency is restricted to only a small percentage of the population who either died or who healed completely without major deformation. The lesions associated with rickets in this population are more commonly seen in well-nourished children (Stuart-Macadam 1989b). In the Bahrain situation, it is hypothesized the rickets is probably a sequel to childhood infection. This is a common pattern among Middle Eastern countries. Small children experiencing illness become lethargic and are confined indoors. Traditional adaptations to high levels of sunlight and cultural beliefs (such as the evil eye) mean that housing is dark and enclosed, children are swaddled, and exposure to sunlight among those who are not toddling is limited (Costeff and Breslaw 1962; Underwood and Margetts 1987). The interaction of these cultural patterns and preferences with early childhood illness increases the risk of vitamin D deficiency occurring in ill children. In the DS3 population, there is no evidence among older children of healed rickets just these few cases among the very young suggesting that it is a sequel to other illness and its presence among those who died is a reflection of how sick these children were.

The increase in death rates in the second six months is associated with the period when supplementation of breast milk often begins in many populations. While unsupplemented children may become undernourished, children who are supplemented are exposed to greater risk for gastrointestinal infections (King and Ulijaszek 1999). Work in Nigeria shows diarrheal disease peaking between 6–12 months of age and historically gastrointestinal disease is one of the major causes of death among children (Rousham and Humphrey 2002). During this same period, particularly relevant to Bahrain, passive immunity to infectious disease, including malaria, declines, and malarial attacks of children may begin to occur although the prevalence of such attacks will peak slightly later (McGregor et al. 1961).

Untangling the interaction of nutritionally deficient mothers, infectious organisms, and nutrition during this period, however, is impossible. All three conditions will interact. It appears highly likely that a particular sector of the population, possibly including some with nutritionally deficient mothers (also suggested by the frequency of deciduous defects, (Skinner and Newell 2003), became vulnerable to infection and nutritional deficiency, one prompting the other. It is probable that infection may be the major factor here given the nature of rickets which is similar to that among well-nourished children (Salimpour 1975). In particular the pattern of deaths in the first year of life suggest that particular children were at risk: the subset who experienced intrauterine growth retardation, the group subject to early infections (not necessarily mutually exclusive), and possibly a small number with thalassemia (see below).

Becoming a child

A complication and an advantage of the 1–5 year age group is that more skeletal indicators, in particular, enamel hypoplasia of the permanent dentition becomes visible. While mortality declines after the first year, the evidence for morbidity is greater. In the original cohort of 62 survivors who entered their second year, 16 would have died by age five (Table 13.2). Most of these deaths occur in the 1–2 year age period: five in the first six months, four in the second.

Growth remains slow among those who died between one and two years of age. Three-quarters of them (approximately 6–7 of our cohort) had active porotic hyperostosis at the time of death (Table 13.5). This includes a small subset of children (c. 1–2 of the cohort) who had extreme thickening of the crania with hair-on-end appearance and cortical thinning of the long bones leading, in severe cases, to the bones appearing like a lattice work (Figure 13.2). Bones particularly affected were the tubular bones (long and short) (between 1.7–6.7 percent per individual subadult element), especially near the epiphyses, and the flat bones (pelvis bones and scapulae, 3.1–4.4 percent). Ribs and clavicles were less frequently observed (1.6–1.7 percent). Among the complete individual subadults who died, 8 percent had postcranial thinning: one infant 0–6 months, one infant 6–12 months, two children 12–18 months, one child 18 months–2 years, a child 2–3 years, and a child 4 years. In each of these cases porotic hyperostosis was observed except in the youngest child who nevertheless had markedly porous temporal bones. This group with extreme lesions is small, limited to less than 4 years of age. I have previously argued (Littleton 1998b) that these cases are indicative of thalassemia affecting a subset of those who died, primarily because of the extensiveness and extreme severity of the lesions, particularly on the postcrania. Mortality, however, was much higher than could be accounted for by this disease alone. The increase in the prevalence of porotic hyperostosis with a decline in mortality suggests a transition from acute episodes with high mortality to chronic stress with high morbidity.

The distribution of LEH among the survivors supports this interpretation. Between 1.5 and 2.4 years (LEH Unit B) approximately a third of the survivors experience their first LEH episode (Figure 13.7). Hypoplasic defects were extremely frequent in this population. All but one subadult (n = 30) had at least one dental defect, while 95 of 99 adults had a defect (Table 13.7). There is no evidence of any statistically significant differences in defect frequency by age or adult sex and hence no evidence of heterogeneity in the population in relation to hypoplasia defects based on simple presence/absence. Hypoplasia was effectively ubiquitous.

The children who experienced a defect between 1.5–2.4 years of age experienced significantly more disruptions than the rest of the population (4.13 versus 2.95, $t = 4.817, p < 0.001$). This argues for a pattern of repeat infections where an episode

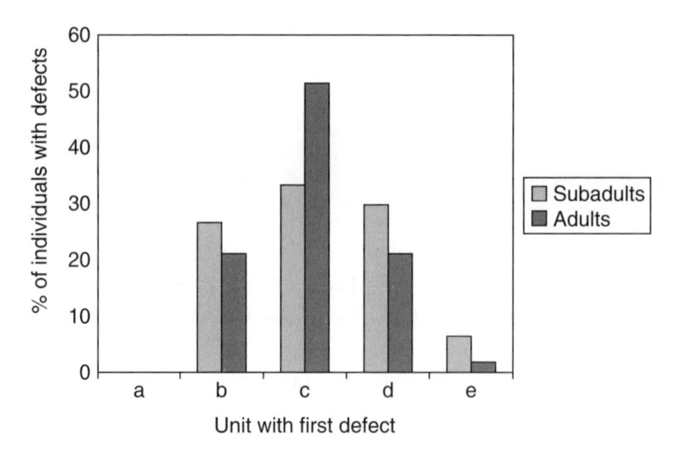

Figure 13.7 Dental development unit in which first linear hypoplasic defect occurs.
Source: Author's drawing

of illness does not protect the child from further illness. There are multiple ways in which this can occur. The consequences of infection may worsen nutritional status making the child more vulnerable or the range of infectious agents may be diverse or it may take repeated attacks to develop full immunity. However, despite the frequency of these stressful episodes, fewer children are dying in this period.

In the following years of childhood, the number of deaths decline even further although the markers of stressors increase among both the dead (as seen in porotic hyperosostis, Table 13.5) and survivors. By c. 2.5–4 years of age (Age Unit C) more than two-thirds of the children have experienced a stressor sufficient to cause a hypoplasic defect (Figure 13.5). These do not appear to reflect a permanent deficit in the child's constitution. This contrasts to other populations (including a broader study of Hellenistic remains from Bahrain, Littleton 2007) where higher numbers of defects or earlier onset of defects have been associated with an earlier average age at death (Goodman et al. 1984b; Stodder 1997). Among the DS3 skeletons, LEH does not effectively distinguish mortality risk. What it indicates is that hypoplasia was ubiquitous, virtually everyone 2.5 years and older was physiologically stressed but survived long enough to develop skeletal modifications.

Having experienced one such period, children go on to experience multiple episodes (most between two to four on each tooth). Stress is repetitive and closely spaced. The mean distance between the earliest two defects on the maxillary incisor is 1.7 mm ($n = 42$) and on the canine 2.0 mm ($n = 21$). This reflects a pattern of repeat insults probably between 6 to 12 months apart (based on the amount of enamel). Later defects are closer together but this does not necessarily reflect a decrease in the amount of time but the geometry of the tooth crown (Reid and Dean 2000). One candidate cause in the Bahrain context is malaria which tends to peak among children between 1 to 4 years of age (McGregor et al. 1961). Children are subject to repeat bouts until effective active immunity develops. However, there are a range of other candidates including acute respiratory and gastrointestinal infections.

Malaria and other forms of parasitism probably account for at least some of the frequency of porotic hyperostosis in this group which remains high (Table 13.5). The route between malaria and anemia is direct due to increased destruction of red blood cells in infected individuals and decreased production of those same cells. Infections with both *Plasmodium vivax* and *falciparum* cause significant anemia (Menendez et al. 2000). Similarly parasitism (including hookworm and schistosomiasis) are associated with blood loss and consequent anemia (Kent 1986).

After 6 years of age healed cases of cribra orbitalia and porotic hyperostosis predominated among the children who died, corresponding with Stuart-Macadam's hypothesized development of porotic lesions in young children only (Stuart-Macadam 1985). Between 6–10 years of age 20 percent of children who died have healed porotic hyperostosis and an equivalent number with active lesions. Assuming that all such lesions initially occurred during the 0–6 year age period, 15 percent of adult survivors and 70 percent of those dying between 1–6 years of life had sufficient anemia to create lesions. This would translate in a rate of anemia among our surviving cohort of around 25 percent (excluding the 8 percent presumed to have some form of thalassemia). This can only be a minimum estimate; not all anemic children would develop skeletal signs. In addition, while malaria is certainly implicated in those repeated stress episodes and in the context of this particular island environment, it would certainly not be the only infection to impact upon the children.

Nutritional deficiencies may (and almost certainly did) play a role in this pattern of dental and skeletal lesions. Both the evidence of rickets and the hypothesized malaria attacks point to two such sets of interactions. On the other hand, the decrease in child mortality after one year of age tends to argue against the chronic effect of under-nutrition but a more cyclical process. In the Bahrain context of multiple protein and vitamin sources, it is hard to argue for generalized under-nutrition although nutrition–infection interactions are highly likely. The depressed child growth which is most acute between 3–5 years of age, suggests that those who died were chronically and repeatedly subjected to stressors, with insufficient ability to withstand them. Saunders and Hoppa (1993) demonstrate how any mortality bias in growth may not be visible in terms of the other errors (sampling, aging) but given the small number who die relative to all the children there is a reason not to extrapolate this slowing across the entire population without further consideration of the other covariates with age at death.

The majority of children at this age were affected by chronic stress and attendant morbidity but relatively little mortality. There is, however, one intrinsically frail group: the small number argued to have had thalassemia. The 4–8 percent of deaths in this 0–6 year age group, tentatively attributed to an inherited anemia, potentially represent the maximum frequency of thalassemia homozygotes in the population. There are, of course, caveats with this – the extreme cortical thinning (8 percent) and hair-on-end thickening of the skull (4 percent) are indeed consistent with thalassemia but it is always possible that severely anemic children could develop such signs making the 8 percent an overestimate (Larsen 1997). Furthermore, the phenotypes of thalassemia major, intermedia and minor do not correspond simply to either homozygosity or heterozygosity (Lagia et al. 2007). This 8 percent of children who died corresponds to 3.2 percent of the entire living population between 0–6 years of age emphasizing as Lagia et al. (2007) point out that in a skeletal population inherited anemias will contribute relatively little to cases of porotic hyperostosis.

This combination of dental, cranial, and postcranial lesions suggests that in the 1–6 year age group, there were three different groups:

1. Children too sick, with no recovery and early mortality (including a small number with thalassemia), with slowed growth, and a high proportion of porotic hyperostosis – c. 25 percent of the cohort.
2. Children with porotic hyperostosis, who experienced repeated episodes of growth arrest, and who died in later childhood (between 6–15 years of age) – 8 percent.
3. Children with repeated periods of growth arrest and recovery but who survive presumably having developed effective immunity, with a lower frequency of porotic hyperostosis (at least 15 percent of this group based on the adults as evidence of childhood rates) – 67 percent of the population.

There is no evidence, with the exception of those with thalassemia that any of these children were experiencing conditions not experienced by the others. Survivors and those who failed to survive both experienced repeated stressors which continued throughout this 1–6 year age group. Those who died may have experienced higher levels of anemia though the adult frequency represents an underestimate since lesions may heal completely (it is uncertain whether this occurs). Nevertheless the ubiquity of these signs does not suggest a group protected from environmental stressors. The population appears to be homogeneous in exposure.

The high frequency of porotic hyperostosis and LEH, plus high mortality in early childhood, demonstrates that this was a stressful environment. Morbidity was widespread. As children grew older their natural immunity developed as did their chances of survival.

Juveniles and adolescents

After 6 years of age, mortality steadily declines. Children develop active immunity by this age to numerous infections. If we consider the projected cohort, of the 46 entering this period, only 3–4 die before 10 years of age. Sample sizes are too small to hypothesize what was happening in terms of childhood growth. Half of the children who died still had visible lesions from porotic hyperostosis but most had experienced healing by the time of death, with the exception of one possible case of thalassemia major.

The situation is even better between 10–15 years of age given the small number of deaths and the lack of active porotic hyperostosis. It suggests that surviving early childhood guaranteed good survivorship. It becomes much harder to define what would be potential causes of death in this group. Chronic stress evidenced by porotic hyperostosis evidently constituted a risk for some children but apart from chronic stress, causes such as accidents or acute infection are invisible.

Discussion

The implications for family and community life

So how does this pattern of mortality and hypothesized morbidity (Table 13.11) correspond to what is known about Bahrain during this period? Historically Bahrain was holoendemic for malaria, i.e., malarial attacks would have been constant and nonepidemic in occurrence, children would develop effective natural immunity as they aged (Daggy 1959). Populations were reliant upon irrigation to maintain agricultural production, standing water would be common, and expansion of settlement is constrained by the supply of water (Larsen 1983). While there are major settlements in the north of the island (Qal'at al-Bahrain), DS3 cemetery is most likely associated with a small agricultural village located adjacent to the burial field (Littleton 1998b). A modern settlement exists in the same location.

The high frequency of children in the Bahrain sample indicates two things. First, this is probably a high fertility population. Buikstra and coworkers have demonstrated that the ratio of deaths over 30 years to deaths over 5 years reflects fertility (Buikstra et al. 1986). In Bahrain this ratio is 0.50, less than very high fertility national populations (e.g., Egypt 1936–40 0.88, Syria 1965–70 0.93, (Cairo Demographic Centre 1982) or skeletal populations (Late Woodland 0.67 (Goodman et al. 1984a), Arroyo Hondo 0.69, (Palkovich 1978a) but still in a high fertility range. Second, in the absence of evidence for marked population growth (Littleton 1998b) this is probably a population that experienced a relatively high mortality. Statements regarding absolute levels of mortality, however, do rely upon the accuracy of adult age estimation.

Traditionally in such agricultural societies the productivity of children is significant within the economic life of the family (Fernea 1995) and household wealth and survival relies upon access to land and labor (Layton 2002). Payne's work in rural Bangladesh indicates that male children are net producers by the age of 12. By 15 years of age they have begun repaying accumulated consumption (Payne 1985). In Egypt children begin contributing to productivity by 5 years of age. Early tasks include shepherding and minding younger siblings (particularly girls). With age such responsibilities increase (Mueller 1976).

In the DS3 sample, the peak period of death is the first year of a child's life. In general, infant deaths are most common among the children of the lowest and highest parity (Oris et al. 2004). In natural fertility populations, one offshoot of frequent infant death is a shorter interbirth interval between births (Wood 1994). This means that in a household cycle the family would be particularly vulnerable during the early phases of establishing the household. Risk would be high during this early stage when the first born children are both too young and experiencing too many episodes of illness to assist with later siblings. The susceptibility of primaparous mothers to malarial attack and more difficult births exacerbates the household's vulnerability. The more active adults in a household and the fewer very young or very elderly people, the higher a household's productive output. Single parent families with young children are particularly vulnerable. These variations in household ability to withstand periods of shortage and illness are potentially the

Table 13.11 Model of survivorship and risk

Age	*Dead*		*Survivors*		*Hypothesised Risks*
	Indicators	*N*	*N*	*Indicators*	
0–3 m	C30% IUGR, 3% premature	–11	88	[10% deciduous defects among those who died between 1.5–6 yr]	Poor maternal health (between 4–30% of mothers) Birth conditions Droplet infections Genetic defects
3–6 m	Normal/catch up growth V. small number with rickets *Small number with thalassemia*	–6	82	"	Protective effect of breast feeding
6–12 m	50% active porotic hyperostosis Slight growth deficit Small number with rickets *Small number with thalassemia*	–20	62	"	Possible beginning of supplementation Increasing gastrointestinal infections Declining passive immunity to malaria
12–24 m	75% active porotic hyperostosis *Small number with thalassemia*	–9	53	30% with first hypoplasic defect	Continuing malarial attack Repeat infections
2–6 yr	50–100% with chronic porotic hyperostosis Depressed child growth	–7	46	Up to 90% with linear enamel hypoplasia Multiple growth arrests 15% with porotic hyperostosis	Continuing chronic stress but development of active immunity
6–10 yr	50% with porotic hyperostosis but half of these healed	–3	43	Healed porotic hyperostosis	Active immunity to infection Limited impact of infectious disease
10–15 yr	None with porotic hyperostosis	–2	41	Healed porotic hyperostosis	Unknown

source of inequality in agricultural villages. Furthermore the presence of children with repeated episodes of illness potentially increases the exposure of infants to an infectious environment.

In this respect then it is not necessary or indeed even possible to attribute the pattern of infant mortality and chronic child morbidity observed at DS3 to malarial attack alone. An agricultural village practicing irrigation means the presence of standing water in storage containers, and irrigation channels which provided a useful breeding place for multiple pathogens not just malaria. Enclosed and dense housing promotes the possible transmission of infectious organisms. Faroughy (1951) writing of the first half of the 20th century describes schistosomiasis, malaria, dysentery, and trachoma as endemic diseases on the island.

I would argue that the pattern of mortality and chronic morbidity reflects selective survivorship. While most children experienced repeat stressors, only some fail to survive infancy. Some were clearly at risk from a very early age because of intrauterine growth retardation, others are not at risk until exposed to new infectious agents probably around the stage when breast milk is supplemented. What this study indicates is that the distribution of skeletal lesions and their relationship to death is dependent upon local ecological and social conditions.

Social implications

It may be that this disproportionate number of deaths in the first year helps explain why among all of the age groups, it is the 0–1-year-old children who are most frequently buried in multiple tombs although such tombs do not exclusively contain infants. Infant death would not be a rare occurrence in this population and it appears as if, at different times, some graves were left accessible exclusively for the repeated interments of young children (but also of adults). However, it cannot be argued that the expectation of these deaths meant that infants were buried with less care than the other age groups. In all age groups multiple burials occur and from one year and above there is no significant difference in the frequency of individual to multiple graves. There is no evidence to argue that infants were not considered members of society.

The impact of morbidity and mortality beyond its possible reflection in burial practices is harder to gauge. Given the fragility of households with this level of fertility, child death and illness, it is likely that communal units were larger than the nuclear household. This is unsurprising given the organization of traditional agricultural societies. Within a household cycle there would have been particular periods of risk, in periods of smaller harvests some households would be better able to survive. Seasonal fluctuations are also likely. Nevertheless, most of the pathogens implicated in the DS3 locale appear to be the result of the external environment so that the whole village experienced a high level of risk.

The vulnerability of such marginal agricultural villages as that associated with the DS3 cemetery, may also help explain the close link seen between settlement and water levels (Larsen 1983). The high rates of mortality and fertility hypothesized for DS3 mean that even a slight decline in mortality or rise in fertility could lead within a short space of time to a burgeoning or rapidly contracting group of people. In this particular instance the long time period the cemetery covers, argues against extinction or some form of environmental crisis but on an island-wide scale this vulnerability may help explain why there may be a close and visible link between water levels and settlement.

The causes of death observed for this Bahrain population may be seen as typical for intensive agriculturalists in malarial areas. Comparable figures of death and disease are observed in modern agricultural villages in malarial areas (e.g., Billewicz and McGregor 1981; Wyon and Gordon 1971). As McGregor et al. describe for Keneba village in the Gambia, "the pattern of infection remains remarkable for its variety and intensity, and communicable diseases are much more often manifest in young children than in older children and adults. Thus the clinical manifestations of malaria, ascariasis, whooping-cough, measles, gastroenteritis, and active trachoma are often seen in the very young but seldom at later ages" (McGregor et al. 1961:1664). I would argue that there are signs of a similarly intense transmission of infectious and parasitic organisms at DS3. Elucidating in more detail the contributions of infection and nutrition to child health and survivorship requires more intensive analysis of both dental remains (analyzing Wilson bands for existence) and of nutritional indicators such as isotopic analysis which will not only potentially assist with identifying the relationship between weaning and age at death but also identify the extent to which marine resources could potentially ameliorate the village's reliance upon its agricultural production.

Conclusion

Achieving a view of the risks and opportunities of childhood from skeletons requires a detailed analysis of not just the frequency of skeletal lesions and ages at death but the interactions between these and the particular ecological circumstances of the society under analysis. In the Bahrain case, there are clearly areas for further analysis. Attempting to disaggregate what is happening among those who die in the first year of life is crucial, associating children with adults and adult health will also help flesh out whether a subset of the population was subject to particular risk. Bahrain also shows, however, that we should not assume that particular stress indicators will have the same impact or associations in different ecological settings. As is so clearly shown in the presence of rickets among a small percentage of dead children, we need to focus upon developing models that take into account "local biologies" (Lock and Kaufert 2001): the local circumstances and understandings that place children at risk or may protect them from the external environment, rather than to assume that a particular indicator reflects the same processes and risks in different ecological settings.

What is missing in this model is the contextualization that can come from more finely dated remains and from the incorporation of data from settlement excavations and written records. This is a first attempt at a reconstruction of more fully realized local biologies, there is further yet to go. However, the aim is not simply to make a "better story." The purpose of the model here is to provide a new perspective, to develop a reconstruction more comparable to the data attained from modern populations and then to test those results in new ways.

The analysis of this particular sample also reflects back to the osteological paradox (Wood et al. 1992). This sample is unusually complete and the pattern of both death and illness it indicates is, contrary to Wood et al.'s (1992) concerns, remarkably similar to historic villages facing similar conditions (e.g., Billewicz and McGregor 1981, Wyon and Gordon 1971). However, it does also show that selective mortality occurs and that any mortality sample is a complex aggregation of frailty, variation in exposure, and variation in the sensitivity of the skeletal indicators. Disentangling this remains a challenge.

REFERENCES

Andersen, S. F. 2007 The Tylos Period Burials in Bahrain, Vol. 1, The Glass and Pottery Vessels. Volume 1. Aarhus: University of Aarhus Press.

Baxter, J. ed. 2005 Children in Action: Perspectives on the Archeology of Children. Archeological Papers of the American Anthropological Association 15(1).

Billewicz, W., and I. McGregor 1981 The Demography of Two West African (Gambian) Villages, 1951–75. Journal of Biosocial Science 13:219–240.

Blakey, M. L., and G. J. Armelagos 1985 Deciduous Enamel Defects in Prehistoric Americans from Dickson Mounds: Prenatal and Postnatal Stress. American Journal of Physical Anthropology 66(4):371–380.

Blau, S. 1998 Finally the Skeleton: An Analysis of Archaeological Human Skeletal Remains the United Arab Emirates. Unpublished Ph.D. Dissertation, University of Sydney.

Blau, S. 1999 The People at Sharm: An Analysis of the Archaeological Human Skeletal Remains. Arabian Archaeology and Epigraphy 7:143–176.

Bray, R., and M. Anderson 1979 Falciparum Malaria and Pregnancy. Transactions of the Royal Society of Tropical Medicine and Hygiene 73(4):427–431.

Brickley, M., and R. Ives 2008 The Bioarchaeology of Metabolic Bone Disease London: Academic Press.

Buikstra, J., L. Konigsberg, and J. Bullington 1986 Fertility and the Development of Agriculture in the Prehistoric Northwest. American Antiquity 51(3):528–546.

Cairo Demographic Centre 1982 Mortality Trends and Differentials in Some African and Asian Countries. Cairo: Cairo Demographic Centre.

Chen, L., M. Rahman, and A. Sarder 1980 Epidemiology and Causes of Death Among Children in a Rural Area of Bangladesh. International Journal of Epidemiology 9(1):25–33.

Cohen, M. N. 1994 The Osteological Paradox Reconsidered. Current Anthropology 35:629–631.

Cohen, M. N., and G. Armelagos eds. 1984 Paleopathology at the Origins of Agriculture. Orlando: Academic Press.

Costeff, H., and Z. Breslaw 1962 Rickets in Southern Israel. Some Epidemiologic Observations. Journal of Pediatrics 61:919–924.

Daggy, R. 1959 Malaria in Oases of Eastern Saudi Arabia. American Journal of Tropical Medicine and Hygiene 8:229–291.

Faroughy, A. 1951 The Bahrein Islands (750–1951). A Contribution to the Study of Power Politics in the Persian Gulf. An Historical, Economic, and Geographical Survey. New York: Very, Fisher and Co.

Fernea, E. W. 1995 Childhood in the Muslim Middle East. *In* Children in the Muslim Middle East. E. W. Fernea, ed. Pp. 3–16. Austin: University of Texas Press.

Fitzgerald, C., S. Saunders, L. Bondioli, and R. Macchiarelli 2006 Health of Infants in an Imperial Roman Skeletal Sample: Perspective from Dental Microstructure. American Journal of Physical Anthropology 130(2):179–189.

Giglioli, G. 1972 Changes in the Pattern of Mortality Following the Eradication of Hyperendemic Malaria from a Highly Susceptible Community. Bulletin W.H.O. 46:181–202.

Goodman, A. 1993 On the Interpretation of Health from Skeletal Remains. Current Anthropology 34(3):281–288.

Goodman, A., and R.-J. Song 1999 Sources of Variation in Estimated Ages at Formation of Linear Enamel Hypoplasias. *In* Human Growth in the Past: Studies from Bone and Teeth. R. Hoppa, and C. Fitzgerald, eds. Pp. 210–240. Cambridge: Cambridge University Press.

Goodman, A., G. J. Armelagos, and J. C. Rose 1984 The Chronological Distribution of Enamel Hypoplasias from Prehistoric Dickson Mounds Populations. American Journal of Physical Anthropology 65(3):259–266.

Gustafsson, G., and G. Koch 1974 Age Estimation Up to 16 Years of Age Based on Dental Development. Odontologisk Revy 25:297–306.

Hansen, H. 1968 Investigations in a Shi'ite Village in Bahrain. Copenhagen: National Museum of Denmark.

Herling, A. 1999 Necropoli and Burial Customs in the Tylos Era. *In* Bahrain: The Civilisation of the Two Seas. P. Lombard, ed. Pp. 156–159. Paris: Institute du Monde Arabe.

Hillson, S. 1996 Dental Anthropology. Cambridge: Cambridge University Press.

Højgaard, K. 1980 Dentition on Umm an-Nar (Trucial Oman), c 2500 B.C. Scandinavian Journal of Dental Research 88: 355–364.

Humphrey L. T. 2003. Linear Growth Variation in the Archaeological Record. *In* Patterns of Growth Variation in the Genus Homo. J. L. Thompson, G. E. Krovitz, and A. J. Nelson eds. Pp. 144–169. Cambridge: Cambridge University Press.

Insoll, T., and E. Hutchins 2005 The Archaeology of Disease: Molluscs as Potential Disease Indicators in Bahrain. World Archaeology 37(4):579–588.

Jackes, M. K. 2000 Building the Bases for Paleodemographic Analysis: Adult Age Determination. *In* Biological Anthropology of the Human Skeleton. S. Saunders, and M. Katzenberg, eds. Pp. 417–466. New York: Wiley-Liss.

Kennedy, K. 1984 Growth, Nutrition and Demography in Changing Paleodemographic Settings in South Asia. *In* Paleopathology at the Origins of Agriculture. M. Cohen, and G. Armelagos, eds. New York: Academic Press.

Kent, S. 1986 The Influence of Sedentism and Aggregation on Porotic Hyperostosis and Anaemia: A Case Study. Man 21:605–636.

Kieswetter, H. 2006. Analyses of the Human Remains from the Neolithic Cemetery at al-Buhais 18 (Excavations 1996–2000). *In* Funeral Monuments and Human Remains from Jebel al-Buhais. P. Uerpmann, M. Uerpmann, and S. Abboud Jasim, eds. Pp. 103–265. Sharjah: Department of Culture and Information.

King, S., and S. Ulijaszek 1999 Invisible Insults During Growth and Development: Contemporary Theories and Past Populations. *In* Human Growth in the Past: Studies from Bone and Teeth. R. Hoppa, and C. Fitzgerald, eds. Pp. 161–182. Cambridge: Cambridge University Press.

Kosa, F. 1989 Age Estimation from the Fetal Skeleton. *In* Age Markers in the Human Skeleton. M. Iscan, ed. Pp. 21–54. Springfield, IL: C.C.Thomas.

Kunter, M. 1983 Chronologische und Regionale Unterschiede bei Pathologischen Zahribefunden auf der Arabischen Halbinsel. Archäologisches Korrespondenzblatt 13: 339–343.

Kunter, M. 1991 Die Menschlichen Skelettreste aus den Gräbern von Umm an-Nar, Abu Dhabi, U.A.E. (3. Jt. v. Chr.). *In* The Island of Umm an-Nar. Vol 1: Third

Millennium Graves. K. Frifelt, ed. Pp. 163–179. Aarhus: Jutland Archaeological Society Publications.

Kunter, M. 2001 Individuelle Skelettdiagnose aller Ausgegrabenen Skelette. *In* Die Graberfellder in Samad al Shan (Sultanat Oman). P. Yule, ed. Pp. 477–480. Rahden: Marie Leidorf.

Lagia, A., C. Eliopoulos, and S. Manolis 2007 Thalassemia: Macroscopic and Radiological Study of a Case. International Journal of Osteoarchaeology 17(3):269–285.

Larsen, C. S. 1997 Bioarchaeology: Interpreting Behavior from the Human Skeleton. New York: Cambridge University Press.

Larsen, C. S. 1983 Holocene Land Use on the Bahrain Islands. Chicago: Chicago University Press.

Layton, R. 2002 Population, Community and Society in Peasant Societies. *In* Human Population Dynamics: Cross-Disciplinary Perspectives. H. MacBeth, and P. Collinson, eds. Pp. 65–82. Cambridge: Cambridge University Press.

Lewis, M. E. 2004 Endocranial Lesions in Non-adult Skeletons: Understanding Their Aetiology. International Journal of Osteoarchaeology 14:82–97.

Lewis, M. E. 2007 The Bioarchaeology of Children. Cambridge: Cambridge University Press.

Littleton, J. 1998a Middle Eastern Paradox: Rickets in Skeletons from Bahrain. Journal of Palaeopathology 10:13–30.

Littleton, J. 1998b Skeletons and Social Composition: Bahrain 250 BC–250 AD. Oxford: Tempus Reparatum.

Littleton, J. 2000 Excavation of a Tylos Cemetery in Bahrain: A Preliminary Report. Journal of Oman Studies 11:121–131.

Littleton, J. 2005 Invisible Impacts but Long-term Consequences. American Journal of Physical Anthropology 126:295–304.

Littleton, J. 2007 The Political Ecology of Health on Bahrain. *In* Ancient Health: Skeletal Indicators of Agricultural and Economic Intensification. M. Cohen, and G. Crane-Kramer, eds. Pp. 176–189. Gainesville: University Press of Florida.

Lock, M., and P. Kaufert 2001 Menopause, Local Biologies, and Cultures of Aging. American Journal of Human Biology 13(4):494–504.

Lukacs, J. R., S. R. Walimbe, and B. Floyd 2001 Epidemiology of Enamel Hypoplasia in Deciduous Teeth: Explaining Variation in Prevalence in Western India. American Journal of Human Biology 13(6):788–807.

Macchiarelli, R. 1989 Prehistoric "Fish-eaters" Along the Eastern Arabian Coasts: Dental Variation, Morphology, and Oral Health in the Ra's al-Hamra Community (Qurum, Sultanate of Oman, 5th-4th Millennia BC). American Journal of Physical Anthropology 78:575–594.

Malina, R., and J. Himes 1978 Patterns of Childhood Mortality and Growth Status in a Rural Zapoteec Community. Annals of Human Biology 5(6):517–531.

Manir, S., and K. Khalegque 1969 Anaemia in Pregnancy in East Pakistan. Transactions of the Royal Society of Tropical Medicine and Hygiene 63:120–124.

Maresh, M. 1955 Linear Growth of Long Bones of Extremities from Infancy Through Adolescence. American Journal of Disease in Childhood 89:725–742.

Maru, M., A. Getahun, and S. Hosana 1988 A House-to-House Survey of Neonatal Tetanus in Urban and Rural Areas in the Gondar Region, Ethiopia. Tropical and Geographical Medicine 40:233–235.

McGregor, I., M. Wilson, and W. Billewicz 1983 Malaria Infection of the Placenta in The Gambia, West Africa; Its Incidence and Relationship to Still Birth, Birthweight and Placental Weight. Transactions of the Royal Society of Tropical Medicine and Hygiene 77(2):232–244.

McGregor, I., W. Billewicz, and A. Thomson 1961 Growth and Mortality in Children in an African Village. British Medical Journal 5268:1661–1666.

Menendez, C., A. Fleming, and P. Alonso 2000 Malaria-related Anaemia. Parasitology Today 16:469–476.

Mensforth, R. 1985 Relative Tibia Long Bone Growth in the Libben and Bt-5 Prehistoric Skeletal Populations. American Journal of Physical Anthropology 68:247–262.

Mueller, E. 1976 The Economic Value of Children in Peasant Agriculture. *In* Population and Development: The Search for Selective Interventions. R. Rielker, ed. Pp. 98–153. Baltimore: Johns Hopkins University.

Nathan, H., and N. Haas 1966 "Cribra Orbitalia." A Bone Condition of the Orbit of Unknown Nature. Israel. Journal of Medical Science 2:171–191.

Olivier, G., and H. Pineau 1958 Determination de L'Age de Foetus et de L'Embryon. Archives d'Anatmie Pathologique 6:21–28.

Oris, M., R. Derosas, and M. Breschi 2004 Infant and Child Mortality. *In* Life Under Pressure. Mortality and Living Standards in Europe and Asia, 1700–1900. T. Bengtsson, C. Campbell, and J. Lee, eds. Pp. 359–398. Cambridge, MA: MIT Press.

Ortner D. 1991 Theoretical and Methodological Issues in Paleopathology. *In* Human Paleopathology: Current Syntheses and Future Options. D. Ortner, and A. Aufderheide, eds. Pp. 5–11. Washington, DC: Smithsonian Institution.

Palkovich, A. 1978a A Model of the Dimensions of Mortality and its Application to Paleodemography. Ph.D. Dissertation, Northwestern.

Palkovich, A. 1978b Interpreting Prehistoric Morbidity Incidence and Mortality Risk: Nutritional Stress at Arroyo Hondo Peublo, New Mexico. *In* Health and Disease in the Prehistoric Southwest. C. Merbs, C. Miller, and P. Alcauskas, eds. Pp. 128–138. Tempe: Arizona State University Press.

Payne, P. 1985 Nutritional Adaptation in Man: Social Adjustments and Their Nutritional Implications. *In* Nutritional Adaptation in Man. K. Blaxter, and J. Waterlow, eds. Pp. 71–88. London: John Libbey.

Puffer, R., and C. Serrano 1973 Patterns of Mortality in Chidlhood. Washington, DC: Pan American Health Organization.

Rautman, A. 2000 Reading the Body: Representations and Remains in the Archaeological Record. Philadelphia: University of Pennsylvania Press.

Reid, D. J., and M. C. Dean 2000 Brief Communication: The Timing of Linear Hypoplasias on Human Anterior Teeth. American Journal of Physical Anthropology 113(1):135–139.

Rousham, E., and L. T. Humphrey 2002 The Dynamics of Child Survival. *In* Human Population Dynamics: Cross-Disciplinary Perspectives. H. MacBeth, and P. Collinson, eds. Pp. 124–140. Cambridge: Cambridge University Press.

Rumney, G. 1968 Climatology and the World's Climates. New York: Macmillan.

Salimpour, R. 1975 Rickets in Tehran. Archives of Diseases in Childhood 50:63–66.

Salles, J.-F. 1999 Bahrain, from Alexander the Great to the Sassanians. *In* Bahrain: The Civilisation of the Two Seas. P. Lombard, ed. Pp. 150–155. Paris: Institute du Monde Arabe.

Saunders, S., and L. Barrans 1999 What Can Be Done About the Infant Category in Skeletal Samples. *In* Human Growth in the Past: Studies from Bone and Teeth. R. Hoppa, and C. Fitzgerald, eds. Pp. 183–209. Cambridge: Cambridge University Press.

Saunders, S. R., and R. D. Hoppa 1993 Growth Deficit in Survivors and Nonsurvivors – Biological Mortality Bias in Subadult Skeletal Samples. Yearbook of Physical Anthropology 36:127–151.

Saunders, S. R., D. A. Herring, and G. Boyce 1995 Can Skeletal Samples Accurately Represent the Living Population They Come From? The St Thomas' Cemetery Site,

Belleville, Ontario. *In* Bodies of Evidence. Reconstructing History Through Skeletal Analysis. A. Grauer, ed. Pp. 69–90. New York: Wiley-Liss.

Scheuer, J., J. Musgrave, and S. Evans 1980 The Estimation of Late Fetal and Perinatal Age from Limb Bone Length by Linear and Logarithmic Regression. Annals of Human Biology 7:257–265.

Schutkowski, H., and B. Herrmann 1987 Anthropological Report on Human Remains from the Cemetery at Shimal. *In* Shimal 1985/6 Excavation of the German Archaeological Mission in Ras al-Khaimah, U.A.E.: A Preliminary Report. B. Vogt, and U. Franke-Vogt eds. Pp.55–65. Berlin: Berliner Beiträge zum Vorderen Orient 8.

Skinner, M., and A. Goodman 1992 Anthropological Uses of Developmental Defects of Enamel. *In* Skeletal Biology of Past Peoples: Research Methods. S. Saunders, and M. Katzenberg, eds. Pp. 153–173. New York: Wiley-Liss.

Skinner, M., and E. Newell 2003 Localized Hypoplasia of the Primary Canine in Bonobos, Orangutans, and Gibbons. American Journal of Physical Anthropology 120:61–72.

Stodder, A. L. 1997 Subadult Stress, Morbidity, and Longevity in Late Period Populations on Guam, Mariana Islands. American Journal of Physical Anthropology 104(3):363–380.

Storey, R. 1986 Perinatal Mortality at Pre-Columbian Teotihuacan. American Journal of Physical Anthropology 69:541–548.

Stuart-Macadam, P. 1985 Porotic Hyperostosis: Representative of a Childhood Condition. American Journal of Physical Anthropology 66:391–398.

Stuart-Macadam, P. 1989a Porotic Hyperostosis: Relationship Between Orbital and Vault Lesions. American Journal of Physical Anthropology 80:187–193.

Stuart-Macadam, P. 1989b Nutritional Deficiency Diseases: A Survey of Scurvy, Rickets, and Iron-Deficiency Anemia. *In* Reconstruction of Life from the Skeleton. M. Iscan, and K. Kennedy, eds. Pp. 201–222. New York: Liss.

Suckling, G. 1989 Developmental Defects of Enamel – Historical and Present-day Perspectives of Their Pathogenesis. Advances in Dental Research 3(2):87–94.

Tanner, J. 1978 Foetus into Man. London: Open Books.

Ubelaker, D. 1999 Excavating Human Skeletal Remains. Chicago: Aldine.

Ulijaszek, S., F. Johnston, and M. Preece 1998 Cambridge Encyclopaedia of Growth and Development. Cambridge: Cambridge University Press.

Underwood, P., and B. Margetts 1987 High Levels of Childhood Rickets in Rural North Yemen. Social Science and Medicine 24:37–41.

Vine, P. 1986 Pearls in Arabian Waters. London: Immel.

Wall, C. 1991 Evidence of Weaning Stress and Catch-up Growth in the Long Bones of a Central Californian Amerindian Sample. Annals of Human Biology 18:9–22.

Wapler, U., E. Crubezy, and M. Schultz 2004 Is Cribra Orbitalia Synonymous with Anemia? Analysis and Interpretation of Cranial Pathology in Sudan. American Journal of Physical Anthropology 123(4):333–339.

Wills, V., and J. Waterlow 1958 The Death-Rate in the Age-Group 1-4 Years as an Index of Malnutrition. Journal of Tropical Pediatrics (March):167–170.

Wood, J. 1994 Dynamics of Human Reproduction: Biology, Biometry, Demography. New York: Aldine de Gruyter.

Wood, J., G. Milner, H. Harpending, and K. Weiss 1992 The Osteological Paradox. Current Anthropology 33:343–370.

Wright, L., and C. Yoder 2003 Recent Progress in Bioarchaeology: Approaches to the Osteological Paradox. Journal of Archaeological Research 11:43–70.

Wyon, J., and J. Gordon 1971 The Khanna Study. Cambridge, MA: Harvard University Press.

14

Skeletal Injury Across the Life Course

Towards Understanding Social Agency

Bonnie A. Glencross

Introduction

Skeletal injury involves deformation, dislocation, wounding, crushing or fracture of living bones and/or joints under unfavorable loading or environmental conditions. Evidence of injury can become ingrained in the skeleton leaving behind an indelible record of life's events long after the bones are no longer living. Skeletal injury also has remarkable antiquity. Spanning millennia, evidence for skeletal injury predates the origin of our hominid ancestors while remaining a leading cause of disability and mortality in many modern populations (Woolf and Pfleger 2003). Bone fracture, the most frequently reported class of skeletal injury today (Cheng and Shen 1993; Thanni and Kehinde 2006), is also confirmed "ubiquitous" in the archaeological record (Ortner and Pustschar 1981:55).

The conceptual frameworks used by anthropologists and bioarchaeologists to investigate skeletal injury vary, as do the research questions, how data are collected and handled, and the types of inferences drawn. Initially, paleopathology and specifically skeletal injury garnered only occasional interest by archaeologists and biological anthropologists. With their focus trained on skeletal variation and typology, anomalies like traumatic lesions were relegated to an appendix in an archaeological report or occasionally described in a general treatise on anthropometry (Ubelaker 1982). It wasn't until work conducted by American anthropologist, Earnest Hooton in 1930 that new methodology and the basis for an epidemiological

Social Bioarchaeology. Edited by Sabrina C. Agarwal and Bonnie A. Glencross
© 2011 Blackwell Publishing Ltd

approach in paleopathology was heralded. Hooton is known for his innovative statistical applications for the time and the description of disease and skeletal injury in prehistoric populations in the context of the ecological triad (Armelagos and Van Gerven 2003). More recently, skeletal biologists and bioarchaeologists have developed a broader biocultural approach to the analysis of human skeletal remains (Buikstra and Cook 1980). Under this new regime, health status and skeletal injury are discussed from a cross-cultural, adaptive, and evolutionary perspective (Armelagos and Van Gerven 2003; Ubelaker 1982; see also Zuckerman and Armelagos this volume). Today, biological anthropologists and bioarchaeologists continue to introduce new elements to the mix that emphasize the political, social, and economic contexts of their specialized research foci. Studies of skeletal injury in particular address issues of social inequality, gender, and violence (see for example Jurmain and Kilgore 1998; Walker 2001). However, missing from this diverse body of literature are detailed examinations of skeletal injury from a life course perspective and particularly with an eye to exploring cultural age and social agency.

In this chapter, I review the conceptual framework of the life course and examine applications of the life course perspective to theory and research in bioarchaeology. Further, this essay outlines a life course model for the analysis of skeletal injury in past peoples that defines the role of biological age in constructing personal and social identities from skeletal injuries as well as highlighting the importance of the cumulative nature of traumatic lesions for exploring the relationship between biological outcome and social process in individual lives, generations, and extended time. Finally with an example from bioarchaeology, I examine the role of skeletal injury and the value of the life course model for studies of cultural age and social agency in past peoples.

The Life Course Perspective

Now considered the pre-eminent theoretical orientation in studies of the complex and dynamic lives of individuals and groups (Bengston and Allen 1993; Elder et al. 2003), life course analysis combines concepts and methods from multiple disciplines (e. g. biology, psychology, sociology, history, economics, and demography), involves multiple levels (from macro social and economic structures and institutions to micro individual experiences), and draws upon both quantitative and qualitative data.

The life course perspective offers both a developmental and historical framework. Life course theory is applied to the study of peoples' lives with an emphasis on the powerful connections between lives and the changing social and historical context in which they unfold as well as the social meaning of the events or transitions experienced across the life course. Several tenets are identified as forming the basis of this theoretical orientation and these same principles can also be used to strengthen current bioarchaeological interpretations by providing a more contextualized portrayal of past lives. First and foremost is the framing of individual development where the individual is an active agent both exerting and being influenced by social context and structures (Bengston and Allen 1993). Equally important is the idea that lives are lived interdependently and the influence of shared relationships is also profound on the life course (Elder et al. 2003). These aspects relating to development are then placed within the interconnected multiple temporal contexts extending from the micro-level unfolding of individual biographies and mid-range telling of stories shared by generations, to macro-level dimensions of time that stretch across era (Bengston and Allen 1993).

Bioarchaeology and particularly studies of skeletal injury are currently poised to advance more informed pictures of life course in the past. It has been argued that the current biocultural approach continues to evolve becoming more socially, culturally and politically informed and thus positioning bioarchaeology as a link between the biological and sociocultural subdisciplines of anthropology (Knudson and Stojanowski 2008; see also Zuckerman and Armelagos this volume), as well as promoting cross-disciplinary research with other fields within the social and biological sciences. Through the life course approach the biocultural model in bioarchaeology can be further extended so that bony changes are no longer viewed simply as adaptations to environmental change but rather holistic analyses of the unfolding of people's lives and their connections to changing social and historical contexts. For studies of skeletal injury in the past this means that aspects of age, gender, inequality, and agency can be explored anew as social phenomena in the lives of individuals, communities and across generations.

As with all theoretical approaches, researchers continually strive to better integrate knowledge and improve the methods that they use. Key to skeletal-based investigations of the life course and identity in the past is the notion of age or age structure. Bioarchaeologists and paleopathologists, in dealing with biological materials, have necessarily sought to refine methods for estimating biological or physiological age as this forms the basis for interpreting biological maturation. However, bioarchaeologists have recently begun to advocate the use of social age categories in explorations of individual identity and life course research (Halcrow and Tayles 2008; Perry 2005). This approach relies on the construction of cultural age and social identity from patterns evident in material culture that are ultimately organized around biological determinants of age and sex. Assumed is a direct link between the two despite known variation in biological and social development between individuals and groups through time (Halcrow and Tayles 2008; Sofaer 2006; see also Sofaer this volume). I would argue that skeletal injury, as with other age-related pathology (e.g., dental caries, osteoarthritis), has the potential to bridge this gap by providing evidence of accumulated biological indicators of behavior and attitudes that speak to the life course as a process, and that can be further strengthened when woven into the broader contextual fabric constructed from biological, material culture, and ethnographic data.

The Life Course Perspective in Bioarchaeology

The life course approach offers a conceptual framework only recently adopted in bioarchaeological analyses (Knudson and Stojanowski 2008). Work in this area has placed an emphasis on the need to no longer view corporeal remains as lifeless and static snapshots of the past but rather animate and fluid evidence of lived experiences across historical time. Also crucial to the developmental context is an understanding of age structuring, recognition of differences in physiological, chronological and/or cultural age, and the need to employ social age analyses in bioarchaeology (see for example Halcrow and Tayles 2008; Kamp 2001; Perry 2005; Knudson and Stojanowski 2008). This approach necessarily requires mixed methodology that actively weaves together as many contextual threads as available to the researcher. New insights on the social meaning of the passage of biological time are currently being gained from analyses of idealized figurative and written representations, data on treatment of the dead body, and skeletal evidence of habitual activities.

Examples include work by Stoodley (2000) and Halsall (1996) who use mortuary data to investigate social life during the Middle Ages. Stoodley (2000) uses mortuary objects associated with the skeletal remains to construct aspects of social identity and cultural age in Early Anglo-Saxon England. He suggests that biological maturation provides a "template" for social identity and that changing identities across the female life course are more closely linked to biological events than for males. Similarly Halsall (1996) in his analysis uses grave goods and historical documents to reconstruct female status and power in Merovingian, Austrasia. He explores shifting gender identities and social relations across the female life course in the context of historical change during the sixth century. He too suggests that burial rites were loosely structured around biological categories of age and sex, with particular emphasis on women of child-bearing age. For sixth-century northern Gaul women, social importance was secured in roles as wife and mother, and the value of these roles in maintaining family alliances.

Research by Sofaer Derevenski (2000) examines metal work as an expression of gendered identities across the life course and changing age–gender constructions through the Copper Age in the Carpathian Basin. Sofaer Derevenski (2000) notes that in the early Copper Age, social identities expressed in metal artifacts are strongly divided along gendered categories and multiple ages. In contrast, at the end of the Copper Age, the number of distinguishable age grades declines indicative of a restructuring of age and gender relations. Meskell (2000) explores the life course from conception through death for villagers from Deir el Medina in Ancient Egypt. She notes that "intimate knowledge of lifecycles and individual life experience" can only be gained from the synthesis of multiple sources of evidence (Meskell 2000:423). By combining evidence from mortuary data, art, historical documents, and personal letters Meskell (2000) suggests the concept of life cycle and the transition through biologically defined life stages best models religious ideology and life experiences at Deir el Medina.

Paleopathology and particularly skeletal injury provide another source of biological evidence that when combined with other contextual information has the ability to make significant contributions to explorations of social identity and life course in the past. For example, Harke (1990) draws on a variety of paleopathology data, including stature, indicators of biological stress and even skeletal injury data, to examine the "Anglo-Saxon weapon burial rite." Harke (1989) tests the underlying assumption that weapons found in Anglo-Saxon burials are the personal property and weaponry of the deceased through an examination of artifact patterns and associations with skeletal pathology. He found that weapons do not correlate with intensity of warfare, functionality, or an individual's ability to participate in fighting but rather the deceased family's wealth and status. Here I argue that skeletal injury and specifically skeletal fracture data have unique dimensions making them important additions to bioarchaeological analyses of the life course. Epidemiological data demonstrate patterns of skeletal injury and fracture that correlate with skeletal growth and development across the life span, and that have been linked with specific behavior and lifestyles. Further, skeletal fractures are one of few pathological conditions that have the potential to accumulate over the life course while remaining visible in the archaeological record.

The Life Course Perspective in Bioarchaeological Studies of Skeletal Injury

In recent decades, paleopathology research on bone fractures has moved from being primarily descriptive in nature to concentrating on the biocultural contextualization of fracture patterns in both the individual and population. This approach emphasizes the use of historical (when available), archaeological, and clinical evidence in support of bioarchaeological interpretations. The vast majority of bioarchaeological studies are concerned with the exploration of the frequency of unintentional injury and fracture in relation to specific physical activities or life styles (Domett and Tayles 2006; Djurić et al. 2006; Jiménez-Brobeil et al. 2007; Mitchell 2006), intentional injury and fracture whether self-inflicted, interpersonal or collective (Buckley 2000; Buzon and Richman 2007; Judd 2004, 2006; Jurmain 2001; Standen and Arriaza 2000; Torres-Rouff and Costa Junqueira 2006; Weber and Czarnetzki 2001), and fragility fracture resulting from underlying pathology and poor health that weakens the skeleton making it more vulnerable to injury (Brickley 2002, 2006; Mays 2006; Mays et al. 2006). None of these studies have explicitly examined skeletal fractures from within the conceptual framework of the life course. However, work by Lovejoy and Heiple (1981) does provide detailed fracture frequencies and examines fracture risk in relation to developmental age across the life span. This work is often cited as exemplary in its emphasis on the strong association between biological age and increasing frequencies of fractures in skeletal samples, and their quantitative assessment of fracture risk and cause.

To further strengthen our understanding of skeletal injury in the past, bioarchaeologists might benefit by borrowing and incorporating a life course approach that is already well established in experimental and epidemiological research (see Fausto-Sterling 2005; Kuh and Ben Shlomo 1997). Recent research in both these areas has shown the developmental origins of bone fragility and the impact of cumulative effects experienced across the life course on fracture risk. Until recently, bioarchaeologists have necessarily hesitated to contextualize fracture patterns in relation to age distributions (exception Lovejoy and Heiple 1981) and thus the life course due to a number of perceived hurdles. The fact that most skeletal collections are composed of individuals showing healed fractures that have accumulated over a lifetime and with no indication of when the injury occurred is often considered an insurmountable methodological challenge. Variability in the healing process within and between individuals makes it difficult to know exactly at what age a healed fracture was sustained. Only those fractures that happen at or around the time of death without the chance to heal are able to indicate a precise age-at-injury. Further, because bone is a dynamic living tissue any evidence of injury sustained at an early age has the potential to be lost during growth and remodeling. So how do we identify at what age an injury was sustained?

Modern clinical and epidemiological studies demonstrate a strong relationship between fracture patterns, growth, development, and chronological age (see for example, Alffram and Bauer 1962; Buhr and Cooke 1959; Cheng and Shen 1993; Fife and Barancik 1985; Garraway et al. 1979; Johansen et al. 1997; Jones 1994; Kowal-Vern et al. 1992; Peclet et al. 1990; Reed 1977; Singer et al. 1998). As a person grows and matures structural and chemical changes in bone tissue lend to bone fragility and an increased susceptibility to bone fracture. Definite patterns of specific fracture types, fracture locations, and peak frequencies are correlated with biological age so that each age cohort has its own unique pattern of skeletal trauma (Jones 1994). Modern clinical data also show that the risk of specific mechanisms of injury change with age-appropriate behavior (Johansen et al. 1997; Peclet et al. 1990; Tibbs et al. 1998). Further, fractures sustained at a young age have been shown to leave residual deformities despite the influences of growth and remodeling. Growth and remodeling are highly variable themselves with a number of influential factors now identified as having little or no effect on erasing evidence of fracture sites including time left to grow, location of injury, and degree of apposition (Gasco de Pablos 1997; Jones 1994; Odgen 1982). This knowledge forms the basis for identifying age-related skeletal injuries incurred over a lifetime in archaeological remains.

Such an analytical strategy, as well as diagnostic methods for the identification of some specific childhood fracture types, has recently been applied to healed fractures witnessed in adult archaeological skeletal remains (Glencross 2003; Glencross and Stuart-Macadam 2000). A paper by Stuart-Macadam et al. (1998) identifies two probable cases of traumatic bowing deformities in adult ulnae from the 15th-century Milton Ossuary and the 17th-century Glen Williams Ossuary, Ontario. Acute plastic bowing deformation (APBD) describes permanent bowing of tubular bones due to either acute or chronic compression experienced during a childhood fall typically onto outstretched hands (Borden 1974). Differences in bone porosity and surface anatomy are believed to make children's bones more susceptible to permanent bone deformation and the highest prevalence of traumatic bowing deformities as opposed to overt fractures experienced by adults (Jones 1994). In another study of the same adult skeletal materials, childhood trauma causing injury about the elbow was quantitatively assessed on x-rays (Glencross and Stuart-Macadam 2001). Supracondylar fractures at the distal humerus often result in subtle non-displaced or mildly displaced fractures from a simple fall onto extended arms, or less often flexed elbows in children and adolescents. This anatomical location is highly susceptible to injury in children and adolescents due to structural changes during growth (Jones 1994). These studies demonstrate that careful noting of the bones involved, intra-skeletal locations of fractures, and the types of fractures, allow the identification and interpretation of childhood lesions even in adult archaeological remains (Glencross 2003; Glencross and Stuart-Macadam 2000, 2001; Stuart-Macadam et al. 1998).

The observation of skeletal fractures as *accumulated* pathology gives us the potential to understand skeletal injury in the context of lifelong processes. This contrasts with, our current static cross-sectional view of skeletal injury in the past where injury is viewed as a singular event. The fact that skeletal factures accumulate over a lifetime can be used to evaluate variable risk across the life span of an individual (Glencross 2003; Lovejoy and Heiple 1981), as well as address heterogeneity in risk among individuals of a group, and through time (see for example Glencross 2003; Judd 2002). However, in examining and comparing health risks between groups Djurić et al. (2006) caution that it is important to be able to distinguish between behavioral and cultural causal factors and results confounded by the cumulative nature of age-related pathology and varied lengths of exposure. Recent methodological developments that involve fracture prevalence measured over years of exposure, negates this concern by taking into account the age structure of the sample and the impact of age structure on prevalence data (Glencross and Sawchuk 2003; Lovejoy and Heiple 1981). Sofaer Deverenski (2000) also notes that we need not only to conduct micro-individual level analyses in life course research but we also need to address macro-level variation given that social change occurs over generations and even longer spans of time. This too has recently been addressed with the application of the Poison model when testing the random distribution of fractures in samples known to have accumulated over lengthy periods of time (Glencross 2003). As an example, I will present preliminary evidence from a prehistoric archaeological site that necessarily involves the combined evidence from skeletal dimensions of age, sex and skeletal fracture as well as information derived from mortuary data and ethnographic analogy.

Indian Knoll: An Example

Indian Knoll is one of the best-known Late Archaic shell midden sites (5,500–1,500 B.C.) located in the Middle Green River region of Ohio County, west-central Kentucky (Herrmann and Konigsberg 2002; Winters 1974). First excavated in 1915 by C. B. Moore, Indian Knoll was the subject of a later more thorough investigation by W. S. Webb in 1939–1940 (Jefferies 1988). Webb's large project is responsible for the recovery of over 880 human burials and 55,000 artifacts. The large sample and excellent preservation of skeletal remains has made it one of the most investigated and important sources of data on prehistoric complex hunter-gatherers in the New World (see for example, Blakey 1971; Brothwell and Burleigh 1975; Cassidy 1972, 1980, 1984; Finkenstaedt 1984; Glencross 2003; Herrmann and Konisgsberg 2002; Howells 1960; Johnston 1961, 1962, 1968; Johnston and Snow 1961; Kelley 1980, 1982; Mensforth 2007; Nagy 2000; Perzigian 1971, 1973, 1977; Pierce 1987; Snow 1948; Stewart 1974; Sundick 1972, 1978; Webb 1946; Wilczak 1998). This diverse body of literature when combined with ethnographic data provides the necessary context for testing new hypotheses that are life course driven and particularly concerned with the expression of cultural age and attendant social roles. Here I present preliminary evidence from Indian Knoll on artifact associations with different physiological ages and sexes, along with age-centered patterns of skeletal injury that together highlight the powerful connections between people's lives and their changing social roles.

Cultural age and mortuary analyses

The grave goods from Indian Knoll are the subject of several past studies that reflect the theoretical and methodological concerns of the time and now cover cultural-historic classification to cultural age. Beginning with Webb (1946), the grave goods were thoroughly described and used to compile a "trait list" characterizing Indian Knoll culture for placement within the larger Midwestern Taxonomic System. Details concerning the human burials from Indian Knoll; their biological profiles, burial form, position, placement and artifacts included, are detailed elsewhere (see, for example, Webb 1946). It is important to note here that Webb (1946) reports a relatively small number of Indian Knoll burials with grave goods (31.2 percent) which he considered typical of most shell mound sites, and an even smaller proportion with artifacts other than beads (21.5 percent). By the 1960s research focuses turn to questions of settlement and economics. Winters (1968, 1969, 1974) reviews the grave goods from a functional perspective for the purpose of determining the relative importance of specific artifacts and associated activities. He notes that marine shells, Marginella and Olivella, are found exclusively with infants and children (Winters 1968) and this is also confirmed by Watson (2005).

Work by Rothchild (1979) was the first to address status and social organization at Indian Knoll as expressed in the burials and associated materials. Concerned with addressing the "egalitarian–non-egalitarian axis," multivariate techniques showed sex to have a weak tendency for discriminating clusters composed mainly of adult males and children buried with multiple artifacts and age a weak discriminator of clusters of adults buried with technomic artifacts. Watson (2005) cautions that these results should be considered tentative due to error inherent in original field assessments but also notes that Rothchild's basic conclusion that mortuary evidence does not support the idea of ascribed status does seem to hold. More recently, Marquardt (1985) interprets the technomic artifacts as evidence for broad inclusion of men, women, and children in subsistence practices and ritual life while the inclusion of exotic items such as cooper and marine shell with some individuals as symbolic of their membership in a "corporate group of traders/ travelers."

Nagy (2001) also examines social complexity through her analyses of activity markers in relation to artifact categories and quantities. She suggests that different types of habitual activities are evident at Indian Knoll and, while some appear associated with gender, variability exists with both males and females participating. In addition, significant patterns with health and mortuary components may reflect differences in status, occupational specialization, and/or temporal variation (Nagy 2001). All of these analyses, however, come to a similar general conclusion: that Indian Knoll probably had some type of social variability. Missing from this body of research is consideration of the mortuary evidence from a life course perspective. Here I provide some preliminary observations based exclusively on the grave goods. The data presented below must be considered a partial analysis as it currently ignores certain elements such as body position and deposition as well as burial location. Despite these shortcomings some subtle age-related patterns are evident (Table 14.1).

Table 14.1 Indian Knoll mortuary analysis – number of individuals having each grave good type by age cohort

Age cohort	*1*	*2*	*3*	*4*	*5*	*6*	*7*	*8*
Jewelry								
Beads	13	52	17	15	5	61	3	0
Pendant	1	9	2	4	1	7	0	0
Ring	0	0	1	1	0	0	0	0
Drilled animal teeth	0	1	1	2	1	4	0	0
Pin	0	2	4	1	1	8	1	0
Hair pin	0	0	2	1	0	8	0	0
Equipment								
Atlatl	1	7	3	3	3	18	2	0
Antler hook	0	6	3	3	3	15	2	0
Point	1	4	6	3	2	21	1	0
Knife	0	1	0	1	0	2	0	0
Axe	0	1	1	1	0	0	0	0
Fish hook	0	0	0	0	1	2	0	0
Plummet	0	0	0	0	0	1	0	0
Hammer	0	0	0	0	0	1	0	0
Drift (billet)	0	0	1	0	0	4	1	0
Reamer (used to make holes)	0	1	0	0	0	0	0	0
Drill	0	0	0	0	1	2	1	0
Graver	0	3	0	0	0	2	0	0
Awl	1	6	5	1	0	6	0	0
Scraper	1	0	1	2	0	7	0	0
Whet stone	0	1	0	0	0	0	0	0
Pestle	0	1	4	1	0	11	1	0
Spatula	0	1	0	0	0	0	0	0
Furnishings								
Rattle	1	5	4	1	1	11	0	0
Dog burial	0	2	2	0	0	6	0	0
Copper	0	2	0	0	0	0	0	0
Shell	2	2	2	1	0	9	0	0
Animal bone	1	9	7	4	1	17	0	0
Tine	0	2	3	0	1	6	0	0
Worked bone/antler	0	2	5	1	0	8	0	0
Flint/Stone	0	0	3	1	0	4	1	0
Total (847)	55	209	75	52	29	390	34	3

1 = newborn; 2 = infant (0–3 years); 3 = child (4–12 years); 4 = adolescent (13–17 years); 5 = subadult (18–20 years); 6 = young adult (21–35 years); 7 = mature adult (36–55 years), 8 = old adult (56+).

Infant underrepresentation is often reported for archaeological skeletal samples and attributed to either the result of poor preservation or differential burial practices (Saunders 1992; Sutton 1988). Webb (1946) reports that newborns and infants (0–3 years of age at death) make up 30 percent of the burials at Indian Knoll. When the proportion of neonatal remains (birth–1 month equals 33 percent) are compared to the proportion of post-neonatal skeletons (birth–1 year equals 67 percent) it becomes apparent that newborns and infants are not underrepresented and even suggests that the youngest individuals are recognized and accepted as part of the larger social collective. If grave goods are considered a measure of status, the number of newborns and infants with grave goods appears low (39 percent). However, given that the number of individuals in the full sample with grave goods stands at 30 percent it can be argued that newborns and infants were not necessarily accorded a low status in society. In fact, shell, stone, and bone beads, the most common grave good, are most frequently encountered among infants, children, and adolescents with marine shells, Marginella and Olivella, found exclusively with infants and children.

Other grave goods also show similar age-related patterning. Childhood is marked by peak frequencies in turtle rattle and awl inclusions, while animal bone and points suddenly increase during childhood only to decline in frequency with age. Adolescence, often considered a crucial age marking the transition to maturity, is reflected in the significant number of pendants and the inclusion of drilled animal teeth almost to the exclusion of any other age grade. Atlatls and antler hooks were placed with individuals of any age, however, peak frequencies occur among young adults. Older adults are seldom accompanied by grave goods and none were adorned by beads. All of these sources demonstrate a tendency to differentiate between individuals categorized as children, adolescents, young, and old adults. Biological sex, on the other hand, does not appear to have as much influence on the construction of the life course with no specific grave goods found with males or females to the exclusion of the other. As noted earlier only a weak association was found between males and multiple grave inclusions (Rothchild 1979).

Cultural age and skeletal trauma

Skeletal trauma at Indian Knoll has also received attention, however, only work by Glencross (2003) considers the skeletal fractures in the context of the life course. Beginning with the earliest osteological analysis, Snow (1948) reports total fractures for most of the major long bones while also providing detailed descriptions of eight of the "more striking examples of … healed fractures." Following Snow's work fracture data are included in two analyses with the intent of characterizing and comparing health and disease under contrasting dietary regime, lifestyles, and environments. Cassidy (1972) found no difference in the number of males and females with fractures and in the number of Indian Knoll hunter-gatherers and later agriculturalists with fractures leading her to conclude similar activity levels across all groups. Kelley (1980) provides a more detailed analysis of fractures by bone, age, and sex. In contrast, he concludes that male frequencies are nearly two times that in females, and that the Indian Knoll hunter-gatherers have the highest number of bone fractures compared to two other groups. He suggests that the high rate of skeletal trauma at Indian Knoll indicates a physically arduous life, particularly for the males. These contradictory findings are in part the result of methodological shortcomings and the need to address age at injury, the effect of age on fracture rates, and age-related risk.

Table 14.2 Indian Knoll prevalence of long bone fractures by bone and sex

	Female	*Male*	*Unknown*	*Total*
Humerus	1	7	1	9
Radius	11	17	1	29
Ulna	7	10	0	17
Femur	1	4	1	6
Tibia	3	3	0	6
Fibula	4	4	0	8
Total	27	45	3	75

Glencross' (2003) analysis of Indian Knoll long bone fractures highlights the importance of age when exploring the relationship between the biological outcome of bone fracture and social process among the Archaic hunter-gatherers. To address age at injury, this work necessarily relies on clinical data that demonstrates a link between physiological age, chronological age, and specific patterns of long bone fractures and related behavior. This work also models individual risk of skeletal injury based on the random distribution of long bone fractures with implications for generational differences and issues of time scale. Most importantly, the analysis recognizes the cumulative nature of skeletal injuries in archaeological samples and the need to account for length of exposure when weighing the effects of social experience across the life course.

Amongst 748 individuals (composed of over 2,200 complete subadult long bone diaphyses and 3,400 adult long bones) analyzed, 75 long bone fractures are identified (Table 14.2). The majority of fractures are located in the forearm (28 radii and 17 ulnae), and all are well healed. The nature and patterning of the long bone fractures are also indicative of unintentional injuries. Determinations of age at injury are drawn from modern clinical and epidemiological studies demonstrating a strong relationship between fracture patterns, growth, development, and age (see, for example, Buhr and Cooke 1959; Jones 1994). Six fractures are found in individuals less than 17 years of age at death, although an estimated 45 percent (34) of total fractures observed in remains of all ages are thought to have occurred during childhood and youth (Figure 14.1). Of the remaining fractures, a number of fragility fractures are observed in the wrist and upper arm of aged females and males.

Nonfatal skeletal trauma accumulates over the life course serving as a physical record (albeit incomplete) of an individual's exposure to events that lead to a fracture outcome. This information is used to model individual risk of skeletal injury. A subsample of 183 individuals, 53 with fracture and 130 controls each with six complete long bones from the same side are used to estimate the probability of fracture and to test for the random distribution of fractures. Fracture events appear infrequent, random and unrelated among the prehistoric foragers of Indian Knoll suggesting no single individual or group demonstrating a significant tendency toward skeletal injury or across time. Frequency data for the population when manipulated with age controls reveal peak fracture frequencies in relation to specific age cohorts and reflect risks that vary with different phases of the life course. Results

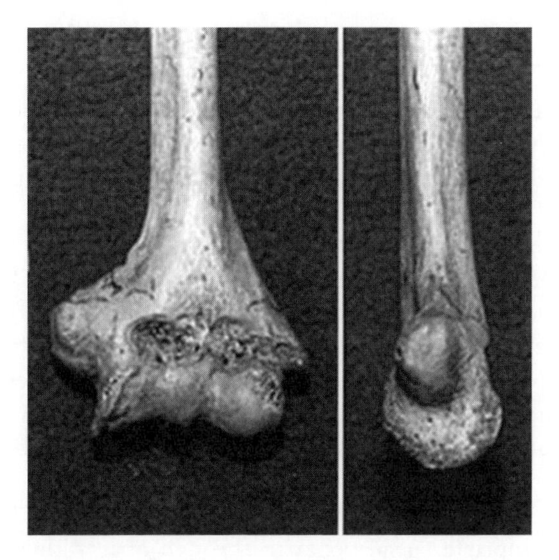

Figure 14.1 A healed childhood supracondylar fracture of distal humerus from Indian Knoll. Note the subtle posterior displacement of the articular region when viewed from the lateral side.
Source: Author's photograph

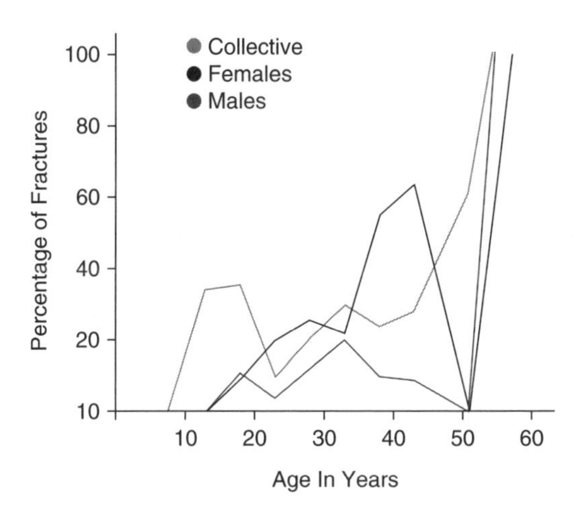

Figure 14.2 Graph showing Indian Knoll fracture distributions across the life course – collective, males and females separately.
Source: Author's drawing

for the collective group show peak frequencies at adolescence, in middle age, and in old adults (Figure 14.2). When males and females are separated each shows a distinct pattern. The male pattern mirrors that of the collective with peak frequencies in adolescents, middle aged and old adults. In contrast the female pattern shows a gradual rise that begins in adolescence, drops off slightly and quickly rises in old age.

Social agency

The concept of the circle is a fundamental theme common to most Aboriginal cultures in North America (Figure 14.3). The continuous nature of the circle symbolizes the interrelationships and interdependency of individuals, families, and communities (Red Horse 1997) where identity is to a large extent a collective experience (King et al. 2009). Further, the circle symbolizes the inclusion of all community members from all stages of life with each person bringing a unique contribution or role to the community. Within the circle, the stages of life: infant and child, youth, adult, elder are marked by status, responsibility, and respect. It is important to note that while Western models of development follow a linear progression with independence increasing with age, in traditional Aboriginal culture age and independence are negatively correlated with those growing older expected to assume more kinship responsibilities (Red Horse 1997).

Most scholars agree that children are highly regarded in traditional communities and are taught appropriate social behavior through observation and emulation of older children and adult role models (Garrett 1993; Red Horse 1997). As children and youths mature, their spiritual and community responsibilities increase and passage from one life stage to the next is marked by ritual events. The greatest responsibilities, however, are carried by elders, those not necessarily old but who have shown wisdom and leadership in cultural and spiritual matters within their communities and who are considered the keepers of community knowledge (Garrett 1993). So what might the combined mortuary, skeletal injury, and ethnographic

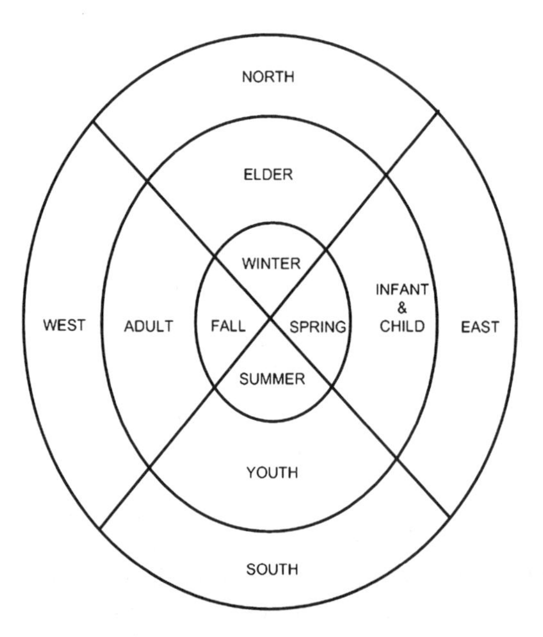

Figure 14.3 Graphic representation of the traditional life cycle.
Source: Author's drawing

data tell us about the life course and social agency among the ancient hunter-gatherers of Indian Knoll?

While we have no clear indication of whether or not birth was surrounded by particular social events, mortuary evidence reinforces the idea that infants were recognized as social beings and considered an integral part of the larger collective body. It is interesting to note that infant grave goods consist almost exclusively of shell adornment and, while it is difficult to know whether the decorative beads were worn in life or simply constructed for the grave, that the beads are included suggests recognition and social importance. Skeletal injury is absent among infants and this speaks to a lack of social risks such as neglect or physical abuse within this vulnerable cohort and further highlights the importance of family responsibility and the social value of children.

Youth appears to be a period of continued social significance that is infused with personal growth both physically and publicly. Grave goods associated with youths demonstrate a tendency to recognize this life stage as a distinct social entity. The large number of awls and sudden increase in points, artifacts considered technomic suggest a newly achieved level of responsibility that is only realized at this point in the life cycle. If we can assume that the types of grave artifacts found also reflect the types of activities engaged in, it seems highly probably that youths began to prepare for their roles as adults by emulating adult behaviors at least in relation to some aspects of subsistence. Evidence of skeletal injury in subadult remains and evidence of injury in older individuals that can be attributed to youth also suggest a level of risk not previously experienced. However, due to the accidental nature of the injuries it can be argued that this might be expected given individuals are exploring and negotiating new-found social identities.

The transition from youth to adult is often considered an important milestone in most cultures. Again, while we do not know if this transition was marked with particular rites of passage by the hunter-gatherers of Indian Knoll, the significant number of specific artifacts that are associated with this group of individuals almost to the exclusion of any other age grade supports the idea of a distinctive social identity. The artifacts, drilled animal teeth, and shell pendants are sociotechnic marking the transition to adulthood and increasing family and community responsibilities. Not surprisingly skeletal injury increases for both males and females during this period although long bone fracture events remain random and unrelated with no single individual or group apparently exhibiting a tendency towards high risk-taking behavior or activities. The long bone fractures are characterized as unintentional injuries that may relate to individual experience and behavior, increased participation in varied aspects of subsistence (e.g., hazards associated with tool use, interactions with animals), heightened interactions with the physical environment (e.g., rough terrain, seasonal conditions) as well as the social environment (e.g., interpersonal relations).

Adulthood also appears to have social variability at Indian Knoll. Mortuary evidence in the form of atlatls and antler hooks are found most often with young adults of both sex, despite skeletal evidence for habitual atlatl use mainly by males at Indian Knoll (Nagy 2000). This suggests that the presence of atlatls and antler hooks in graves were not necessarily used by the deceased and that perhaps these artifacts articulate a form of social identity that transcends gender lines and is in keeping with traditional values. In contrast, skeletal long bone fracture patterns show distinct differences between males and females at Indian Knoll. Males show a slight decline in fracture frequency following adolescence only to see rates increase through adult life before peaking and gradually declining through the later years. This distribution has been identified in contemporary populations and is largely associated with occupation and activity-related behavior (see Buhr and Cooke 1959). Females, on the other hand, show a gradual rise in fracture rates from youth through young adulthood only to plateau briefly around age 35 years and begin to sharply rise again. This pattern is more difficult to explain, however, it is important to note here that the suite of hormones involved in both pregnancy and lactation play a significant role in bone maintenance in young adult females. Still the occurrence of atlatls and antler hooks can be interpreted as symbolic of self-actualization and realization of an individual's capabilities and responsibilities within the community with the fracture patterns speaking more directly to social roles of husbands and wives.

Another important distinction to be made is between "elders" and the elderly although both are key in the structure of traditional societies. Both Marquardt (1985) and Nagy (2000) have identified a small number of individuals at Indian Knoll that they say display unusual mortuary patterns and that they believe are symbolic of the social roles of ritual healers and/or occupational specialists such as traders. These individuals would then in fact also fit the role of "elders," individuals that hold community knowledge in support of the collective. In contrast aged adults, both males and females, have very few if any grave inclusions. Patterns of skeletal injury reflect skeletal maturation, deterioration, and underlying fragility that occurs with physiological aging and that has the potential to limit the behavior and physical activities of aged individuals. This, however, should not in any way suggest less social importance if we consider this in the context of traditional behavioral attributes of responsibility, respect, and inclusion (Red Horse 1997). Within traditional family and community systems, aged individuals carry responsibilities that act to bind generations as respected teachers, mentors of, and role models for youth.

Conclusions

In sum, the life course approach has emerged as a powerful analytical tool in the social sciences. Certainly many benefits can be gained by the adoption of the life course approach by bioarchaeologists. Paleopathology and particularly skeletal injury are proven a unique source of biological data that, when combined with other contextual information, has the ability to make significant contributions to the exploration of social identity, cultural age, and social agency. A strong relationship between growth, development, chronological age, fracture patterns, and associated behaviors forms the basis for identifying age-related patterns of skeletal injury. The added dimension of skeletal fractures as visible *accumulated* pathology also underlies our ability to understand skeletal injury in the context of lifelong processes. The combined evidence from mortuary data and age-centered patterns of skeletal injury at Indian Knoll, when considered in the context of traditional value systems highlights how communities shape and guide individual behavior in social relations and responsibilities across the life course.

REFERENCES

Alffram, P., and G. C. H. Bauer 1962 Epidemiology of Fractures of the Forearm. Journal of Bone and Joint Surgery 44A:105–114.

Armelagos, G. J., and D. P. Van Gerven 2003 A Century of Skeletal Biology and Palaeopathology: Contrasts, Contradictions, and Conflicts. American Anthropologist 105(1):53–64.

Bengston, V. L. and K. R. Allen 1993 The Life Course Perspective Applied to Families Over Time. *In* Sourcebook of Families, Theories, Methods: A Contextual Approach. P. G. Boss, W. J. Doherty, R. LaRossa, W. R. Scham, and S. K. Steinmetz eds. Pp. 469–498. New York: Plenum Press.

Blakey, R. L. 1971 Comparison of the Mortality Profiles of Archaic, Middle Woodland, and Middle Mississippian Skeletal Populations. American Journal of Physical Anthropology 34:43–54.

Borden, S. 1974 Traumatic Bowing of the Forearm in Children. Journal of Bone and Joint Surgery 56A:611–616.

Brickley, M. 2002 An Investigation of Historical and Archaeological Evidence for Age-related Bone Loss and Osteoporosis. International Journal of Osteoarchaeology 12:364–371.

Brickley, M. 2006 Rib Fractures in the Archaeological Record: A Useful Source of Sociocultural Information? International Journal of Osteoarchaeology 16:61–75.

Brothwell, D., and R. Burleigh 1975 Radiocarbon Dates and the History of Treponematoses in Man. Journal of Archaeological Science 2:393–396.

Buckley, H. R. 2000 A Possible Fatal Wounding in the Prehistoric Pacific Islands. International Journal of Osteoarchaeology 10:135–141.

Buikstra, J. E., and D. C. Cook 1980 Paleopathology: An American Account. Annual Review of Anthropology 9:433–470.

Buhr, A. J., and A. M. Cooke 1959 Fracture Patterns. Lancet 1:531–536.

Buzon, M. R., and R. Richman 2007 Traumatic Injuries and Imperialism: The Effects of Colonial Strategies at Tombos in Upper Nubia. American Journal of Physical Anthropology 133:783–791.

Cassidy, C. M. 1972 Comparison of Nutrition and Health in Preagricultural and Agricultural Ameridian Skeletal Populations. Ph.D. Dissertation, University of Wisconsin.

Cassidy, C. M. 1980 Nutrition and Health in Agriculturalists and Hunter-gatherers: A Case Study of Two Prehistoric Populations. *In* Nutritional Anthropology: Contemporary Approaches to Diet and Culture. N. W. Jerome, R. F. Kandel, and H. Pelto eds. Pp. 117–145. New York: Redgrave.

Cassidy, C. M. 1984 Skeletal Evidence for Prehistoric Subsistence Adaptation in the Central Ohio River Valley. *In* Paleopathology at the Origins of Agriculture. M. Cohen, and G. J. Armelagos eds. Pp. 307–338. New York: Academic Press.

Cheng, J. C. Y., and W. Y. Shen 1993 Limb Fracture Pattern in Different Pediatric Age Groups: A Study of 3350 Children. Journal of Orthopaedic Trauma 7:15–22.

Domett, K. M., and N. Tayles 2006 Adult Fracture Patterns in Prehistoric Thailand: A Biocultural Interpretation. International Journal of Osteoarchaeology 16:185–199.

Djurić, M. P., C. A. Roberts, Z. B. Rakocevics, D. D. Djonic, and A. R. Lesic, 2006 Fractures in Late Medieval Skeletal Populations from Serbia. American Journal of Physical Anthropology 130:167–178.

Elder, G., M. K. Johnson, and R. Crosnoe 2003 The Emergence and Development of Life Course Theory. *In* Handbook of the Life Course. B. J. Lee, J. T. Mortimer, and M. J. Shanahan eds. Pp. 3–19. New York: Plenum Press.

Fausto-Sterling, A. 2005 The Bare Bones of Sex: Part 1 – Sex and Gender Signs. Journal of Women in Culture and Society 30:1491–1527.

Fife, D., and J. I. Barancik 1985 Northeastern Ohio Trauma Study III: Incidence of Fractures. Annals of Emergency Medicine 14:244–248.

Finkenstaedt, E. 1984 Age at First Pregnancy Among Females at the Indian Knoll Oh-2 Site. Transactions of the Kentucky Academy of Science 45:51–54.

Garraway, W. M., R. N. Stauffer, L. T. Kurland, and W. M. O'Fallon 1979 Limb Fractures in a Defined Population. I. Frequency and Distribution. Mayo Clinic Proceedings 54: 701–707.

Garrett, J. T. 1993 Understanding Indian Children, Learning from Indian Elders. Children Today 22(4):18.

Gasco, J., and J. de Pablos 1997 Bone Remodelling in Malunited Fractures in Children. Is it Reliable? Journal of Pediatric Orthopaedics Part B6:126–132.

Glencross, B. A. 2003 An Approach to the Epidemiology of Bone Fractures: Methods and Techniques Applied to Long Bones from the Indian Knoll Skeletal Sample, Kentucky. Ph. D. Dissertation, University of Toronto.

Glencross, B. A., and L. Sawchuk 2003 The Person-years Construct: Ageing and the Prevalence of Health Related Phenomena from Skeletal Samples. International Journal of Osteoarchaeology 13:369–374.

Glencross, B. A., and P. Stuart-Macadam 2000 Childhood Trauma in the Archaeological Record. International Journal of Osteoarchaeology 10:198–209.

Glencross, B. A., and P. Stuart-Macadam 2001 Radiographic Clues to Fractures of Distal Humerus in Archaeological Remains. International Journal of Osteoarchaeology 11:298–310.

Halcrow, S. E., and N. Tayles 2008 The Bioarchaeological Investigation of Childhood and Social Age: Problems and Prospects. Journal of Archaeological Method and Theory 15: 190–215.

Halsall, G. 1996 Female Status and Power in Early Merovingian Central Austrasia: The Burial Evidence. Early Medieval Europe 5:1–24.

Harke, H. 1989 Knives in Early Saxon Burials: Blade Length and Age at Death. Medieval Archaeology 33:144–148.

Harke, H. 1990 Warrior Graves? The Background of the Anglo-Saxon Weapon Burial Rite. Past and Present 126:22–43.

Herrmann, N. P., and L. W. Konigsber 2002 A Re-examination of the Age-at-Death Distribution of Indian Knoll. *In* Paleodemography: Age Distributions from Skeletal Samples. R. D. Hoppa, and J. W. Vaupel, eds. Pp. 243–257. Cambridge: Cambridge University Press.

Hooton, E. A. 1930 The Indians of Pecos Pueblo: A Study of Their Skeletal Remains. New Haven, CT: Yale University Press.

Howells, W. W. 1960 Estimating Population Numbers Through Archaeological and Skeletal Remains. *In* The Application of Quantitative Methods in Archaeology. R. F. Heizer, and S. F. Cook eds. Pp. 158–176. Viking Fund Publications in Anthropology No. 28. New York.

Jefferies, R. W. 1988 The Archaic in Kentucky: New Deal Archaeological Investigations. *In* New Deal Archaeology and Current Research in Kentucky. D. Pollack, and M. L. Powell eds. Pp. 14–25. Frankfort, KY: Kentucky Heritage Council.

Jiménez-Brobeil, S. A., I. Al Oumaoui, and P. H. Du Souich 2007 Childhood Trauma in Several Populations from the Iberian Peninsula. International Journal of Osteoarchaeology 17:189–198.

Johansen, A., R. J. Evans, M. D. Stone, P. W. Richmond, Su Vui Lo, and K. W. Woodhouse 1997 Fracture Incidence in England and Wales: A Study Based on the Population of Cardiff. Injury 28:655–660.

Johnston, F. E. 1962 Growth of the Long Bones of Infants and Young Children at Indian Knoll. American Journal of Physical Anthropology 20:249–254.

Johnston, F. E. 1968 Growth of the Skeleton in Earlier Peoples. *In* The Skeletal Biology of Earlier Human Populations. D. R. Brothwell, ed. Pp. 57–66. Oxford: Pergamon Press.

Johnston, F. E., and C. E. Snow 1961 The Reassessment of the Age and Sex of the Indian Knoll Skeletal Population: Demographic and Methodological Aspects. American Journal of Physical Anthropology 19:237–244.

Jones, E. 1994 Skeletal Growth and Development as Related to Trauma. *In* Skeletal Trauma in Children. N. E. Green, and M. F. Swiontkowski, eds. Pp. 1–14. Philadelphia: W. B. Saunders.

Judd, M. A. 2002 Ancient Injury Recidivism: An Example for the Kerma Period of Ancient Nubia. International Journal of Osteoarchaeology 12(2):9–106.

Judd, M. A. 2004 Trauma in the City of Kerma: Ancient Versus Modern Injury Patterns. International Journal of Osteoarchaeology 14:34–51.

Judd, M. A. 2006 Continuity of Interpersonal Violence Between Nubian Communities. American Journal of Physical Anthropology 131:324–333.

Jurmain, R. 2001 Paleoepidemiological Patterns of Trauma in a Prehistoric Population from Central California. American Journal of Physical Anthropology 115:13–23.

Jurmain, R., and L. Kilgore 1998 Sex-related Patterns of Trauma in Humans and African Apes. *In* Sex and Gender in Paleopathological Perspective. A. L. Grauer, and P. L. Stuart-Macadam eds. Pp. 11–26. Cambridge: Cambridge University Press.

Kamp, K. A. 2001 Where Have All the Children Gone? The Archaeology of Childhood. Journal of Archaeological Method and Theory 8:1–34.

Kelley, M. A. 1980 Disease and Environment: A Comparative Analysis of Three Early American Indian Skeletal Collections. Ph.D. Dissertation, Case Western Reserve University.

Kelley, M. A. 1982 Intervertebral Osteochondrosis in Ancient and Modern Populations. American Journal of Physical Anthropology 59:271–279.

King, M., A. Smith, and M. Gracey 2009 Indigenous Health Part 2: The Underlying Causes of the Health Gap. Lancet 374:76–85.

Knudson, K. J., and C. M. Stojanowski 2008 New Directions in Bioarchaeology: Recent Contributions to the Study of Human Social Identities. Journal of Archaeological Research 16(4):397–432.

Kowal-Vern, A., T. P. Paxton, S. P. Ros, H. Lietz, M. Fitzgerald, and R. L. Gamelli 1992 Fractures in the Under-3-Year-Old Age Cohort. Clinical Pediatrics 31:653–659.

Kuh, D., and Y. Ben-Shlomo 1997 A Life Course Approach To Chronic Disease Epidemiology. Oxford: Oxford Medical Publications.

Lovejoy, C. O., and K. G. Heiple 1981 The Analysis of Fractures in Skeletal Populations With an Example From the Libben Site, Ottowa County, Ohio. American Journal of Physical Anthropology 55:529–541.

Marquardt, W. H. 1985 Complexity and Scale in the Study of Fisher-gatherer-hunters: An Example from the Eastern United States. *In* The Emergence of Cultural Complexity. D. Price, and J. A. Brown, eds. Pp. 59–98. Orlando: Academic Press.

Mays, S. A. 2006 A Palaeopathological Study of Colles' Fracture. International Journal of Osteoarchaeology 16:415–428.

Mays, S., M. Brickley, and R. Ives 2006 Skeletal Manifestations of Rickets in Infants and Young Children in a Historic Population from England. American Journal of Physical Anthropology 129:362–374.

Meskell, L. 2000 Cycles of Life and Death: Narrative Homology and Archaeological Realities. World Archaeology 31(3):423–441.

Mensforth, R. P. 2007 Human Trophy Taking in Eastern North America During the Archaic Period: The Relationship to Warfare and Social Complexity. *In* The Taking and Displaying of Human Body Parts as Trophies by Ameridians. R. J. Chacon, and D. H. Dye, eds. Pp. 222–277. New York: Springer.

Mitchell, P. D. 2006 Trauma in the Crusader Period City of Caesarea: A Major Port in the Medieval Eastern Mediterranean. International Journal of Osteoarchaeology 16:493–505.

Nagy, B. L. B. 2000 The Life Left in Bones: Evidence of Habitual Activity Patterns in Two Prehistoric Kentucky Populations. Ph.D. Dissertation, Arizona State University.

Odgen, J. A. 1982 Skeletal Injury in the Child. Philadelphia: Lea and Febiger.

Ortner, D. J., and W. G. J. Pustschar 1981 Identification of Pathological Conditions in Human Skeletal Remains. Washington, DC: Smithsonian Institution Press.

Peclet, M. H., K. D. Newman, M. R. Eichelberger, C. S. Gotschall, P. C. Guzzetta, K. D. Anderson et al. 1990 Patterns of Injury in Children. Journal of Pediatric Surgery 25:85–91.

Perry, M. 2005 Redefining Childhood Through Bioarchaeology: Toward an Archaeological and Biological Understanding of Children in Antiquity. Archeological Papers of the American Anthropological Associations 15(1):89–111.

Perzigian, A. J. 1971 Gerontal Osteoporotic Bone Loss in Two Prehistoric Indian Populations. Ph.D. Dissertation, Indiana University.

Perzigian, A. J. 1973 Osteoporotic Bone Loss in Two Prehistoric Indian Populations. American Journal of Physical Anthropology 39:87–96.

Perzigian, A. J. 1977 Fluctuating Dental Asymmetry: Variation Among Skeletal Populations. American Journal of Physical Anthropology 47:81–88.

Pierce, L. K. C. 1987 A Comparison of the Pattern of Involvement of Degenerative Joint Disease Between an Agricultural and Non-agricultural Skeletal Series (Averbuch, Indian Knoll). Ph.D. Dissertation, University of Tennessee.

Red Horse, J. 1997 Traditional American Indian Family Systems. Families, Systems and Health 15(3):243–250.

Reed, M. H. 1977 Fracture and Dislocations of the Extremities in Children. Journal of Trauma 17:351–354.

Rothchild, N. A. 1979 Mortuary Behavior and Social Organization at Indian Knoll and Dickson Mounds. American Antiquity 44(4):658–675.

Saunders, S. R. 1992 Subadult Skeletons and Growth Related Studies. *In* Skeletal Biology of Past Peoples: Research Methods. S. R. Saunders, and M. A. Katzenberg, eds. Pp. 1–20. New York: Wiley.

Singer, B. R., G. J. McLauchlan, C. M. Robinson, and J. Christie 1998 Epidemiology of Fractures in 15,000 Adults: The Influence of Age and Gender. Journal of Bone and Joint Surgery 80B:243–248.

Snow, C. E. 1948 Indian Knoll Skeletons of Site Oh2, Ohio County, Kentucky. University of Kentucky Reports in Anthropology IV(3): Part II. Lexington, KY: University of Kentucky Press.

Sofaer, J. R. 2006 The Body as Material Culture: A Theoretical Osteoarchaeology. Cambridge: Cambridge University Press.

Sofaer Derevenski, J. 2000 Rings of Life: The Royal of Early Metal Work in Mediating the Gendered Life Course. World Archaeology 31(3):389–406.

Standen, V. G., and B. T. Arriaza 2000 Trauma in the Preceramic Costal Populations of Northern Chile: Violence or Occupational Hazards? American Journal of Physical Anthropology 112:239–249.

Stewart, T. D. 1974 Nonunion of Fractures in Antiquity, with Descriptions of Five Cases from the New World Involving the Forearm. Bulletin of the New York Academy of Medicine 50:675–891.

Stoodley, N. 2000 From Cradle to Grave: Age Organization and Early Anglo-Saxon Burial Rite. World Archaeology 31:456–472.

Stuart-Macadam, P., B. A. Glencross, and M. Kricun 1998 Traumatic Bowing Deformities in Tubular Bones. International Journal of Osteoarchaeology 8:252–262.

Sundick, R. I. 1972 Human Skeletal Growth and Dental Development as Observed in the Indian Knoll Population. Ph.D. Dissertation, University of Toronto.

Sundick, R. I. 1978 Human Skeletal Growth and Age Determination. Homo 29:228–249.

Sutton, R. E. 1988. Paleodemography and Late Iroquoian Ossuary Samples. Ontario Archaeology 48:42–50.

Thanni, L. O., and O. A. Kehinde 2006 Trauma at a Nigerian Teaching Hospital: Pattern and Documentation of Presentation. African Health Sciences 6(2):104–107.

Tibbs, R. E., D. E. Haines, and A. D. Parent 1998 The Child as a Projectile. Anatomical Record 253:167–175.

Torres-Rouff, C., and M. A. Costa Junqueira 2006 Interpersonal Violence in Prehistoric San Pedro de Atacama, Chile: Behavioral Implications of Environmental Stress. American Journal of Physical Anthropology 130:60–70.

Ubelaker, D. H. 1982 The Development of American Paleopathology. *In* A History of American Physical Anthropology 1930–1980. F. Spencer, ed. Pp. 337–356. New York: Academic Press.

Walker, P. L. 2001 A Bioarchaeological Perspective on the History of Violence. Annual Review of Anthropology 30:573–596.

Watson, P. J. 2005 WPA Excavations in the Middle Green River Area: A Comparative Account. *In* Archaeology of the Middle Green River Region, Kentucky. W. H. Marquardt, and P. J. Watson, eds. Pp. 515–628. Mongraph 5, Institute of Archaeology and Paleoenvironmental Studies. Gainesville: University of Florida.

Webb, W. S. 1946 Indian Knoll, Site Oh 2, Ohio County, Kentucky. University of Kentucky, Department of Anthropology, Reports in Anthropology and Archaeology IV(3): Part I. Lexington, KY: University of Kentucky Press.

Weber, J., and A. Czarnetzki 2001 Brief Communication: Neurotraumatological Aspects of Head Injuries Resulting from Sharp and Blunt Force in the Early Medieval Period of Southwestern Germany. American Journal of Physical Anthropology 114:352–356.

Wilczak, C. A. 1998 A New Method for Quantifying Musculoskeletal Stress Markers (MSM): A Test of the Relationship Between Enthesis Size and Habitual Activity in Archaeological Populations. Ph.D. Dissertation, Cornell University.

Winters, H. D. 1968 Value Systems and Trade Cycles of the Late Archaic in the Midwest. *In* New perspectives in Archaeology. S. R. Binford, and L. R. Binford, eds. Pp. 175–221. Chicago: Adeline.

Winters, H. D. 1969 The Riverton Culture: A Second Millennium Occupation in the Central Wabash Valley. Reports of Investigations 13. Springfield: Illinois State Museum.

Winters, H. D. 1974 Introduction to the New Edition. *In* Indian Knoll. W. S. Webb, Pp. v–xxvii. Knoxville: University of Tennessee Press.

Woolf, A. D., and B. Pfleger 2003 Burden of Major Musculoskeletal Conditions. Bulletin of the World Health Organization 81(9):646–656.

15

Diet and Dental Health through the Life Course in Roman Italy

Tracy L. Prowse

Introduction

The life course perspective emphasizes connections between events throughout the lifetime of an individual, and situates these events within larger historical contexts (Elder 1985; Giele and Elder 1998). Alternative trajectories are possible and individual choice may help determine the unique direction of an individual's life course, but these choices are made within constraints defined by social context (Elder 1998; Giele and Elder 1998). Within these larger trajectories are short-term transitions (e.g., weaning, puberty, marriage) that can alter the path of an individual's life course (Elder 1998). As individuals are linked by shared experiences and social expectations, their individual trajectories are also linked to larger cultural and economic conditions (Giele and Elder 1998). Elder (1994, 1998) identifies the four central themes of the life course perspective as: historical time and place; human agency; timing of lives; and linked lives. This life course perspective has clear connections with the biocultural approach, which emphasizes the study of biological variation within the context of social relations and recognizes the importance of history and historical contingency in studying human health (Goodman and Leatherman 1998).

Longitudinal studies or historical cohort studies are preferred methods to track individual progress throughout the life course, and studies on modern populations have shown that early developmental stress can have a significant impact on later morbidity and mortality (Humphrey and King 2000; Kuh and Ben-Schlomo 2004). When investigating the lives of past human populations through the analysis of their skeletal remains, we are dealing with a cross-sectional sample of the original popula-

Social Bioarchaeology. Edited by Sabrina C. Agarwal and Bonnie A. Glencross
© 2011 Blackwell Publishing Ltd

tion (Sofaer 2006; Wood et al. 1992). The life course perspective is nonetheless a useful organizational framework for examining the human condition in the past, because it emphasizes the interrelationship between biological and social processes and the connections between different stages of the life span (Giele and Elder 1998; Kuh and Ben-Schlomo 2004). We can therefore investigate the trajectory of the life course, marked by both biological and social transitions, through the analysis of the physical remains.

Our bodies need food to survive, but access to food and what is ultimately consumed vary through the life course and what we eat is influenced by social, political, and economic factors. This chapter explores diet and dental health throughout the life course in the Roman Imperial (1st–3rd centuries A.D.) skeletal sample from Isola Sacra, Italy. This approach is interdisciplinary, integrating the analysis of stable isotopes and dental pathology data situated within the historical context of the life course in Roman Italy. Isotopic analysis of infants and children provides evidence on patterns of infant feeding and weaning, and deciduous dental health data can tell us about texture and consistency of the post-weaning diet. Comparable analyses of adult bones and teeth provide evidence of diet and dental health in the later stages of the life course.

Much of what is known about the stages of the life course in the Roman world is derived from textual and iconographic evidence (e.g., Brandt 2002; Cohen and Rutter 2007; Cokayne 2003; Fraschetti 1998; Harlow and Laurence 2002; Parkin 2003; Rawson 1999; Revell 2005). The information derived from these sources can be used to partially reconstruct the social and historical context that shaped the life course of each individual. Life course approaches to aging in the ancient world recognize that the experience of aging is affected by an individual's status in society, and that aging is both a biological and social process (e.g., Cokayne 2003; Harlow and Laurence 2002; Parkin 2003). In addition, the socially defined age stages overlap with gender categories, resulting in different life course trajectories for males and females in Roman society (Revell 2005).

Historians acknowledge that the surviving written evidence typically focuses on literate, elite, male members of society, while other members (e.g., women, children, and slaves) are underrepresented in the literature (Dixon 2001; Harlow and Laurence 2002). Further, some ancient authors relied heavily on earlier sources for their information and were often describing circumstances or details from earlier centuries. Thus, the information that has survived from the Roman period has passed through a succession of unknown "filters" that may have significantly altered these texts. Their usefulness here is in providing a framework and historical context for interpreting the evidence provided by the human remains. This study integrates the skeletal, archaeological, and historical evidence to better understand the historical and social context within which these people lived.

The stages of the life course in antiquity were described by various ancient writers as the "ages of man," and each age division was distinguished based on physical, mental, and behavioral characteristics (Cokayne 2003; Sharpe 1964). Infancy (*infantia*) lasted from birth to 7 years and it was during this time that the first important transition in the life course occurred, the process of transitional feeding and weaning. The transition to the second stage of the life course, childhood (*pueritia*) was typically marked by the loss of the deciduous dentition and was a time when formal education began (for those who could afford it) (Dixon 1988; Rawson 2003). Childhood ended around 14 years of age with the transition to adolescence (*adulescentia*), a stage that lasted to the 30th year, but was a period during which individuals were not considered fully "adult." Roman writers divided the remaining adult years into three stages: youth (*juventus*) from 30 to 49 years; adulthood (*gravitas*) from 50 to 69 years; and old age (*senectus*) beyond the age of 70 years. Much of the historical evidence focuses on the male experience of aging and these transitions were marked by a male's changing role in military or public life (e.g., the Senate) (Cokayne 2003). Transitions in the adult female life course were associated with changes in reproductive status and the process of aging was associated with a decline in fertility and the initiation of menopause (Fraschetti 1998; Harlow 2007).

The Isola Sacra Skeletal Sample

The necropolis of Isola Sacra is located on an artificial island approximately 23 km southwest of Rome and is associated with the harbor town of *Portus Romae*, a key trading center for the city of Rome (Figure 15.1). The necropolis extends approximately 1.5 km along the road between Ostia and *Portus Romae* and was used by the inhabitants of *Portus Romae* from the first to third centuries A.D. The inhabitants of the town were likely middle-class administrators, traders, and merchants, although there is no reference to a local aristocracy in the inscriptional evidence, unlike other Roman towns from the Imperial period (Garnsey 1998). Commercial activity in the port started to decline by the end of the fourth century A. D., but it remained a harbor for the Roman fleet during the fifth and sixth centuries A.D. After that time the port and the cemetery fell into disuse.

Various sections of the cemetery have been excavated since the 1920s, mainly focusing on the monumental tomb structures, but the most recent work in the late 1980s focused on the areas around the monumental tombs where over 600 additional burials were excavated (Baldassare 1990). It is estimated that approximately 2,000 individuals have been recovered from the Isola Sacra cemetery, many of which are commingled remains from the monumental tombs (Sperduti 1995). Nearly 1,000 individuals have been individually catalogued and analyzed and are now curated at the L. Pigorini Museum in Rome. Most of the individuals included in this study came from burials found around the monumental tombs.

Research on the Isola Sacra skeletal collection has examined various aspects of this ancient Roman population, including: evidence of early stress in the life course using enamel hypoplasias, dental asymmetry, and dental microstructure (Bondioli and Macchiarelli 1999; FitzGerald et al. 2006; Hoover et al. 2005; Manzi et al. 1989); skeletal evidence for rickets (Wood 2004); histological analyses (Cho and Stout 2003; Geusa et al. 1999; Savorè 1996); behavior-induced auditory exostoses (Manzi and Passarello 1991); cranial and dental morphology (Argenti and Manzi 1988; Manzi and Sperduti 1988; Manzi et al. 1997); DNA evidence for thalassemia (Yang 1998); and isotopic evidence for diet, weaning, and migration (Prowse et al. 2004, 2005, 2007, 2008).

Figure 15.1 Map showing Portus Romae (Porto) comprising both Porto di Claudio and Porto di Traiano.

Stable Isotopes and Diet

Carbon ($\delta^{13}C$) and nitrogen ($\delta^{15}N$) isotopes are now routinely used in bioarchaeological analyses of past human diets based on the observation that isotopes in body tissues reflect the isotopic composition of the diet (see reviews by Katzenberg 2008; Katzenberg and Harrison 1997; Pate 1994; Schwarcz and Schoeninger 1991; Sealy 2001). The most commonly used material in the isotopic analysis of past human diet is bone, which is made up of both organic (collagen) and inorganic (apatite) components. Controlled feeding experiments demonstrate that collagen principally reflects dietary protein, whereas apatite reflects total diet (proteins, carbohydrates, and lipids) (Ambrose and Norr 1993; Jim et al. 2004; Tiezen and Fagre 1993), although diets low in protein may derive a greater proportion of the carbon from non-protein sources (Coltrain and Leavitt 2002; Schwarcz 2000). This is not the case for $\delta^{15}N$ isotopes, as dietary protein is the only significant source of nitrogen.

Carbon

Carbon from the atmosphere enters food webs through the process of photosynthesis in terrestrial and marine plants, and plants using different photosynthetic pathways will have characteristic $\delta^{13}C$ values. C_3 plants (e.g., shrubs, trees, wheat, barley, most terrestrial fruits and vegetables) have more negative $\delta^{13}C$ values ranging from −20 to −35‰ (per million).[1] C_4 plants (e.g., millet, sorghum, maize) are enriched in ^{13}C, so they have more positive $\delta^{13}C$ values , ranging from −9 to −14‰ (Katzenberg 2008).[2]

The distinctive photosynthetic pathways of terrestrial C_3 and C_4 plants produce $\delta^{13}C$ ranges that do not overlap, which is useful in paleodietary analysis when both types of plants are present in the food web. Carbon isotopes can also be used to investigate the relative contribution of terrestrial versus marine resources in the diet of coastal populations, because marine foods have more positive $\delta^{13}C$ values (see Katzenberg 2008 for a review). The analysis of trophic level positions in animals has demonstrated a slight enrichment (0–2‰) in the $\delta^{13}C$ values of predators over their prey (Bocherens and Drucker 2003; DeNiro and Epstein 1978, 1981; Roth and Hobson 2000). The slight differences in $\delta^{13}C$ values between successive trophic levels are often smaller than the overall range of $\delta^{13}C$ in potential food sources, so they are not as useful as differences in $\delta^{15}N$ levels for distinguishing trophic level relationships (Ambrose and DeNiro 1986; Chisholm et al. 1982; Schoeninger and DeNiro 1984). However, a slight enrichment in $\delta^{13}C$ values (~1‰) has been observed in breastfeeding human infants (Fuller et al. 2006).

Humans obtaining their protein exclusively from terrestrial C_3 sources will have $\delta^{13}C$ collagen values around −20‰. Those consuming exclusively terrestrial C_4 sources will have $\delta^{13}C$ values of approximately −8‰, while those getting most of their protein from marine resources will have values around −12‰ (Choy and Richards 2009; Richards and Hedges 1999). Humans do not typically consume a 100% C_3-, C_4-, or marine-based diet, so $\delta^{13}C$ values are typically between these extremes; consequently we need to examine the distribution of $\delta^{15}N$ values in order to distinguish different types of diets.

Nitrogen

The range of nitrogen isotope values in food webs is related to the manner in which terrestrial and marine plants obtain their nitrogen, either through direct N_2-fixation (e.g., legumes), or indirectly through bacterial decomposition of organic matter. Nitrogen levels are also related to a consumer's position in the food web; $\delta^{15}N$ levels are higher in the tissue of consumers by approximately 3–5‰ relative to the diet for each successive step in the food chain, known as the "trophic level effect"

(Bocherens and Drucker 2003; Minagawa and Wada 1984; Schoeninger and DeNiro 1984). The magnitude of the trophic level effect was previously assumed to be constant for all consumers (i.e., 3‰), however, recent research has demonstrated that there is variability in the size of this shift, so a range of 3–5‰ is now accepted (Bocherens and Drucker 2003; Hedges and Reynard 2007; Vanderklift and Ponsard 2003).

Bioarchaeological studies have used $\delta^{15}N$ in bone collagen to determine the relative proportion of plant versus animal protein in terrestrial diets and to detect the contribution of marine foods in past human diets (e.g., Choy and Richards 2009; Keenleyside 2006; Prowse et al. 2004; Richards and Hedges 1999; Schoeninger et al. 1983; Schwarcz et al. 1985; Walker and DeNiro 1986). Higher $\delta^{15}N$ levels in marine organisms are due to the more complex food chains in marine environments, with further enrichment of ^{15}N in each successive trophic level (Schoeninger and DeNiro 1984).

Nitrogen isotopes have also been widely used to investigate the timing and pattern of weaning in past human populations based on this observed trophic level effect (e.g., Clayton et al. 2005; Dupras et al. 2001; Dupras and Tocheri 2007; Jay et al. 2008; Katzenberg and Pfeiffer 1995; Katzenberg et al. 1996; Prowse et al. 2008; Richards et al. 2002, 2006; Schurr 1997, 1998; Schurr and Powell 2005).[3] This is because breastfeeding infants are enriched by approximately 2–3‰ relative to maternal levels, since the infants are consuming their mothers' tissues in the form of breast milk (Fogel et al. 1997; Fuller et al. 2006). Weaning is defined here as the termination of breastfeeding and "transitional feeding" as the period of the life course during which both breast milk and complementary foods are provided (after Sellen 2007).[4] Once transitional feeding has started there is a decrease in $\delta^{15}N$ due to the removal of breast milk, ultimately reaching levels representative of childhood diet once weaning has occurred. Since we do not know the rate of bone deposition and turnover in infants and young children, it is not possible to accurately determine the lag time between complete cessation of breastfeeding and the associated decrease in $\delta^{15}N$ levels in bone collagen (Herring et al. 1998; Richards et al. 2002).

Nondietary variables can affect $\delta^{15}N$ and $\delta^{13}C$ levels in humans including nutritional stress (Fuller et al. 2005), disease (Katzenberg and Lovell 1999; White and Armelagos 1997), and environmental conditions such as extreme aridity (Schwarcz et al. 1999) or the "canopy effect" (van der Merwe and Medina 1991). However, among adults there is no physiological basis for age-related differences with respect to the incorporation of carbon and nitrogen isotopes into the body's tissues (Lovell et al. 1986). Isotopic studies of animals and modern humans have shown age-related shifts in isotopic values, but the differences are explained by variation in food choices between younger and older individuals and not physiological differences associated with aging (e.g., Wilkinson et al. 2007; Witt and Ayliffe 2001). Consequently, if age-related variations in $\delta^{13}C$ and $\delta^{15}N$ levels exist within a sample, it can be inferred that the differences are due to variability in diet. Controlled feeding studies also demonstrated that differences in isotopic values between adult males and females consuming the same diet are not significant, so any sex-related differences in isotopic values will reflect actual variability in diet (DeNiro and Schoeninger 1983).

Diet and Dental Pathology

Another method to explore diet in past human populations is through the analysis of dental pathology in the deciduous and permanent dentition. Once the teeth have erupted in the mouth, the crowns are exposed to the oral environment and the foods consumed, and the type and texture of the foods consumed have a significant impact on dental health. A diet composed largely of hard, abrasive foods can rapidly wear away dental enamel and dentine, whereas a diet consisting of soft, sticky foods will produce little wear, but can promote the formation of carious lesions. The overall dental health of an individual is the result of the interaction between diet, tooth morphology, food preparation techniques, and conditions in the oral environment (Littleton and Frohlich 1993; Powell 1985).

Dental caries is a disease process caused by the activity of bacteria in the mouth. When food is consumed, small particles remain in the mouth and bacteria ingest these particles. The metabolic activity of bacteria produce acids that demineralizes tooth enamel and it is the acid that produces a carious lesion (or cavity). Foods high in sugar and soft, sticky foods promote bacterial activity, and both can lead to the formation of cavities. In contrast, abrasive foods help reduce the number of cavities, because the grit in the food naturally cleans the teeth, removing both food and bacteria. A diet composed of coarse, abrasive foods will also lead to attrition of enamel and remove pits and fissures where bacteria (and food) accumulate, so there is usually an inverse relationship between the degree of tooth wear and the prevalence of caries (Larsen 1997).

Lukacs and Thompson's (2008) meta-analysis of caries in prehistoric skeletal samples concluded that females have consistently higher rates of caries due to life history variables including: earlier eruption of teeth; variation in dietary patterns (e.g., food preparation, frequency of consumption); and hormonal changes associated with pregnancy and menopause. These variables affect the biochemical composition and flow rate of saliva, both of which have an impact on the formation of cavities; thus, all other things being equal, females should have more cavities. Explanations for sex-based differences in caries prevalence must therefore consider both biological and behavioral explanations.

The pattern and degree of tooth wear reflect a combination of variables including the consistency and texture of the food consumed, methods of food preparation, and nondietary uses of teeth (Molnar 1972). Tooth wear is a normal process caused by contact between teeth and materials introduced into the mouth (e.g., food, grit) and results in loss of enamel with increasing age (Hillson 2000). In addition, food preparation techniques can affect the degree of occlusal wear. High rates of occlusal wear in permanent teeth have been attributed to the use of coarse stone grinders for the processing of plants, especially grains (Molleson and Jones 1991; Powell 1985).

The formation of dental calculus (mineralized plaque) is related to multiple factors in the mouth including saliva flow, activity of microorganisms, and an alkaline pH in the oral environment that promotes the formation of calculus (Lieverse 1999). There tends to be an inverse relationship between caries and calculus, but this relationship is not consistent and it is possible to have both calculus and cavities on the same tooth (Hillson 2000). There are a number of dietary and nondietary factors that may have a significant influence on the formation of calculus, such as mineral content of water, food preparation methods, oral hygiene, culturally specific behaviors, and individual susceptibility (Lieverse 1999).

Abscesses occur when the pulp chamber of a tooth is exposed to the oral environment through the development of carious lesions, trauma to the tooth, or heavy wear. If this occurs, microorganisms can enter the pulp cavity and infect the surrounding tissue. A drainage channel can develop in the surrounding bone to allow accumulated pus to drain from the area of infection. Abscesses usually occur on the thinner buccal surface of the alveolus and their prevalence increases with age due to the interaction with dental caries and tooth wear (Hillson 1996; Jurmain 1990).

During life, a tooth will fall out if there is loss or destruction of the alveolar bone surrounding the tooth. Ante-mortem tooth loss (AMTL) can be caused by a number of pathological processes, and it is often difficult to determine the cause, or causes, of AMTL, particularly when the alveolus has remodeled (Hillson 1996). Severe caries, extreme tooth wear leading to pulp exposure and necrosis, continuous eruption, periapical abscesses, alveolar resorption, and heavy calculus are all factors that can cause AMTL (Clark and Hirsch 1991; Littleton and Frohlich 1993; Lukacs 1989; Scott and Turner 1988).

Poor dental health not only affects the teeth and jaws, but can also have consequences for the overall health of an individual through the life course. AMTL, abscesses, or inflammation of the gum tissues may reduce the ability to chew certain foods effectively restricting diet (Powell 1985). This may lead to both malnutrition and under-nutrition, with potentially serious consequences for health. An individual under nutritional stress may be less resistant to infectious diseases and the ability to recover from disease may be compromised by poor nutritional status. There is growing evidence from modern clinical studies of an association between periodontal disease and cardiovascular disease; although a direct causal relationship is still debated (Demmer and Desvarieux 2006). In addition, the unchecked development of dental caries may lead to infection of the tooth and the surrounding tissues and bone, promoting the formation of abscesses.

Infant and Childhood Diet

Isotopic evidence

Prowse and colleagues (2008) investigated the pattern of breastfeeding and weaning in the Isola Sacra skeletal sample through isotopic analysis of 37 rib samples from individuals ranging in age between birth and 13 years of age.[5] Collagen preparation and analysis followed standard procedures outlined in Longin (1971) and Chisholm et al. (1982) (see Prowse et al. 2008). Delta ^{15}N values range from 8.7 to 16.1‰ (mean = 12.5 ± 1.9‰) and δ^{13}C values range from −17.8 to −19.8‰ (mean = −18.7 ± 0.5‰). Figure 15.2 shows the distribution of the δ^{13}C and δ^{15}N data in comparison to the adult female means for each isotope (indicated by the horizontal dashed and solid lines, respectively).

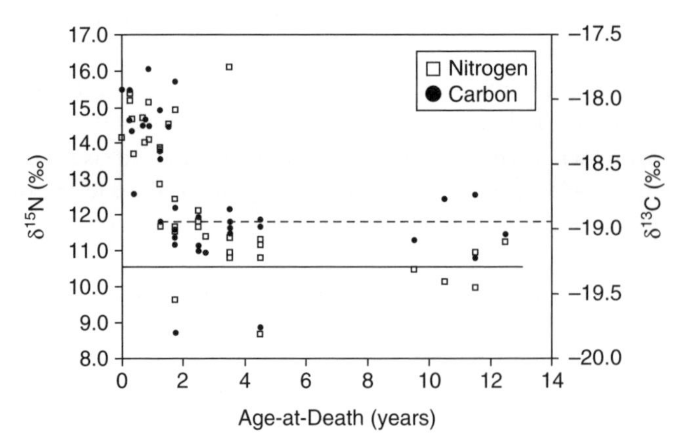

Figure 15.2 Scatter plot of the Isola Sacra rib data (n = 37), showing $\delta^{15}N$ and $\delta^{13}C$ versus estimated age-at-death. Solid horizontal line represents the adult female $\delta^{15}N$ mean; dashed horizontal line represents the adult female $\delta^{13}C$ mean.
Source: Author's drawing

The $\delta^{15}N$ values are highest among individuals under one year of age and there is little variability in these values (±1‰). The maximum offset from the adult female mean (10.6 ± 1.1‰) is approximately 4‰, which falls within the expected range of 3–5‰, and is an indication of the trophic level effect associated with breastfeeding. Delta ^{15}N values are lower in the 1–2 year age categories, indicating the cessation of breast feeding and the introduction of complementary foods (Prowse et al. 2008). After 2.5 years of age, the remaining $\delta^{15}N$ values approximate the female adult mean (Figure 15.2). This suggests that that transitional feeding began between 1–2 years of age and that weaning was complete by approximately 2.5 years of age.

The $\delta^{13}C$ data show a similar pattern, with slightly higher (~1‰) $\delta^{13}C$ values among infants under 1 year of age that is indicative of the smaller carbon trophic level effect (Figure 15.2). There is considerable variability in the isotopic signatures of individuals between 1 and 2.5 years of age, but there is an overall decrease in $\delta^{13}C$ values among individuals in this age range. After 2.5 years of age, the majority of the $\delta^{13}C$ values are below the adult female mean (−18.9 ± 0.3‰). This suggests that the complementary foods provided to infants and children were isotopically more negative than the foods consumed by the adult females in the sample, suggesting a predominantly terrestrial diet (Prowse et al. 2008).

Dental pathology evidence

The composition and texture of complementary foods and the post-weaning diet can be investigated through the analysis of dental pathology in the deciduous dentition. Data on caries, calculus, tooth wear, AMTL, and abscesses were collected on 78 individuals between 1 to 12 years of age (n = 556 observable teeth) according to the standards outlined in Buikstra and Ubelaker (1994) (Prowse et al. 2008).

Of the 37 individuals included in the isotopic analysis, 19 had erupted dentitions that could be used in this part of the study.

The prevalence of AMTL and abscesses are low in the deciduous dentition (0.4% and 1.3%, respectively), but an interesting pattern emerges when calculus, caries, and tooth wear are examined (Prowse et al. 2008). Nearly half of the individuals studied (42.3%) have some calculus present on the deciduous teeth, almost exclusively on the molars, and calculus is present on the teeth of individuals as young as 2.5 years of age (Prowse et al. 2008). Similarly, carious lesions (only on the molar teeth) are present in individuals aged 2.5 years and older, and some degree of tooth wear is visible on nearly all of the individuals in the sample (74/78 = 94.9%, or 438/554 teeth = 79.1%) (Prowse et al. 2008). Most notably, individuals as young as 1.5 years of age show evidence of tooth wear on the deciduous incisors and first molars (Prowse et al. 2008).

The dental evidence provides us with information on the timing of the introduction of foods other than breast milk, or the possible use of feeding implements and/or teething aids. The presence of wear on children under 2 years of age suggests that transitional feeding began early in this stage of the life course. Descriptions of infant and childhood feeding practices refer to the use of ceramic feeding vessels during the Roman period (Fildes 1986), but there is no information from the Roman period on the use of teething aids. The presence of calculus and cavities on the molar teeth of children over the age of 2.5 years suggests that complementary foods consisted of soft carbohydrates that adhered to the posterior teeth and led to the development of calculus and caries (Prowse et al. 2008).

The historical evidence from the Greco-Roman period relating to weaning consists mainly of medical texts (e.g., Soranus, Galen, and Oribasius) as well as wet-nursing contracts from Roman Egypt. Some of the recommendations by medical writers were based on the physical development of the infant, such as Galen's recommendation that the introduction of more solid food could occur when the anterior teeth started to emerge (normally between 6–12 months of age). Similarly, Soranus recognized that an infant might not be physically ready to consume more solid food prior to the age of 6 months (Temkin 1956). The written evidence from these medical sources, although equivocal, indicate that transitional feeding should start after 6 months of age and that weaning should occur around 2 to 3 years of age, but this timetable was likely very flexible.[6] According to some of these ancient sources, the first weaning foods consisted of bread softened with milk, honey, and wine, followed later by porridge, eggs, and the shoots of figs (Garnsey 1999; Fildes 1986).

Both the $\delta^{13}C$ and $\delta^{15}N$ data indicate a pattern of breastfeeding and weaning that is roughly consistent with historical sources from the Roman period. We do not know the precise amount of time required for the change in diet (i.e., removal of breast milk and introduction of complementary foods) to be registered in the bones of these infants, but if there is a pattern of decreasing isotopic values starting around 1 year of age, this suggests that the process began earlier than that. The process of transitional feeding and weaning is a period of high risk in the life of an infant or young child, so we would expect to see evidence of stress during this time period. The analysis of Wilson bands in 274 deciduous teeth from the Isola Sacra sample reveals peaks in the prevalence of Wilson bands between the ages of 2 to 5 months and 6 to 9 months (FitzGerald et al. 2006). This suggests that infants were experiencing (and surviving) stress events during this time period, and this may be attributed, in part, to the process of removing breast milk and introducing complementary foods (Prowse et al. in press). However, Wilson bands only indicate that some kind of stressor occurred, so we cannot exclude the possibility that some other morbidity event occurred, but the fact that the histological data indicate periods of stress prior to 1 year of age is consistent with both the isotopic and historical data.

The deciduous dental data also correlate well with the isotopic and histological evidence; those individuals with declining $\delta^{15}N$ and $\delta^{13}C$ values (indicating a trophic level shift) also present evidence of tooth wear, calculus, and caries at relatively young ages. The lower $\delta^{13}C$ values among children older than 2.5 years of age suggest that they were consuming a largely terrestrial diet composed of C_3 plants and animal products (Prowse et al. 2008). Both the isotopic and dental data can be understood in terms of the social relationships and status of children in Roman society, although little has been written about specific dietary practices relating to children after they were weaned. Children were expected to eat a more frugal type of food in keeping with their inferior status in the household, although specific foods were not mentioned (Bradley 1998). Children, like women, were perceived as weak and had similar humors (Wiedemann 1989). Although the information is scant, the similarity of women and children in terms of their status in Roman society and their perceived characteristics suggests that the subadults at Isola Sacra were probably consuming foods similar in isotopic composition to those consumed by the females. Unfortunately, we cannot examine sex-related differences in children and prepubescent adolescents because of the lack of sexually dimorphic characteristics.

Diet into Adulthood and Through the Aging Process

Isotopic Evidence

To analyze variation among individuals within the Isola Sacra sample beyond infancy and early childhood, 105 (male = 48; female = 32; unknown = 25) femoral bone collagen samples were analyzed to explore diet between the sexes and through the remaining stages of the life course (ages 5 to 45+ years) (Prowse et al. 2005).[7] Isotopic analyses suggest that the overall diet of the people buried at Isola Sacra was composed of a mixture of C_3 terrestrial marine resources (Prowse et al. 2004). The results of the intra-sample analysis reveal that females have slightly more negative average $\delta^{13}C$ values (–18.9‰) than males (–18.7‰), and females also have lower average $\delta^{15}N$ values (10.7‰) than males (11.0‰) (Prowse et al. 2005). Although the differences between the male and female means is small, a t-test indicates that the female $\delta^{13}C$ values are significantly more negative than those of the males ($t = 2.789$; $p = .007$) (Prowse et al. 2005). This suggests that females ate more terrestrial (C_3) resources than males, and that males consumed a greater proportion of high trophic level marine foods in their diets.

When the data are compared by both age and sex, a clear pattern emerges in all stages of the life course. The mean $\delta^{15}N$ and $\delta^{13}C$ values are consistently higher among males in all age categories (Figure 15.3a and b). There is no known physi-

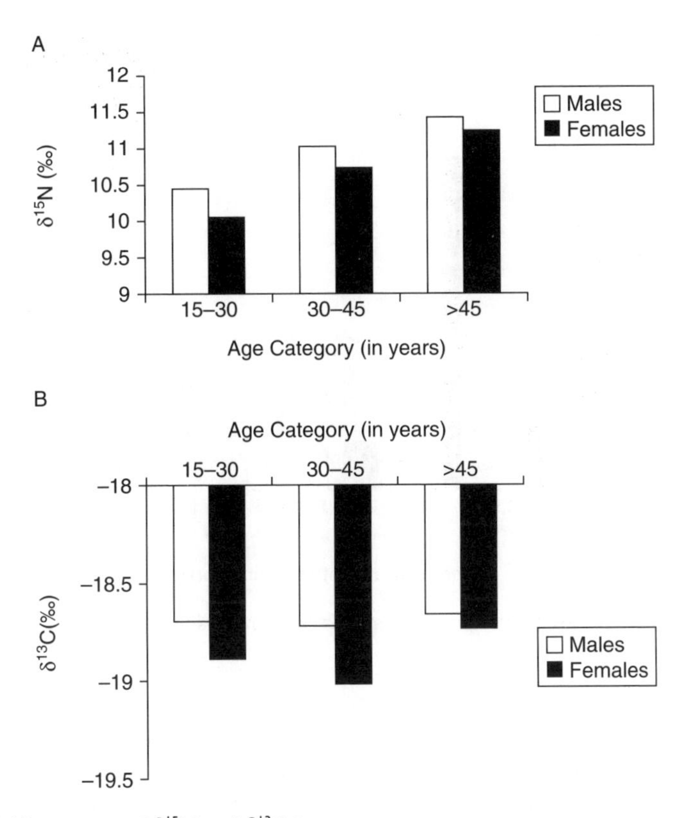

Figure 15.3 Histograms of $\delta^{15}N$ and $\delta^{13}C$ by age category.
Source: Author's drawing

ological explanation for higher isotopic values among males (or lower ones among females), so these differences are clearly due to sex-based dietary differences. Female $\delta^{15}N$ values are higher with increased age-at-death, whereas there is only a moderate increase in relation to age among males (Prowse et al. 2005). The $\delta^{13}C$ values for both sexes only show a slight increase with increased age-at-death, although female $\delta^{13}C$ values remain consistently lower than males in all age categories, indicating that there were sex-based dietary preferences (or prohibitions) that were consistent throughout the life course (Prowse et al. 2005).

Dental pathology evidence

An earlier study by Manzi and coworkers (1999) analyzed dental pathology in a subsample ($n = 65$) of the Isola Sacra skeletal collection. The researchers estimated sex based on cranial morphology alone, in contrast to the present study that uses sex estimates derived from a combination of morphological characteristics from the both the cranial and postcranial remains (after Sperduti 1995). The sample used by Manzi et al. (1999) also consisted of a larger proportion of males ($n = 42$) versus

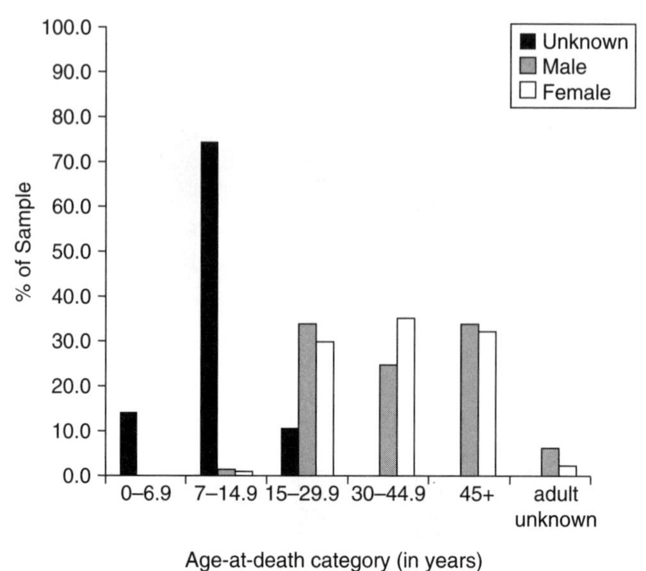

Figure 15.4 Age and sex distribution of the Isola Sacra permanent dentition sample (n = 325). Source: Author's drawing

females (n = 21) and their analysis did not include any permanent teeth from individuals less than 20 years of age, thus underrepresenting the adolescent segment of this skeletal sample.

To investigate dental health in the permanent dentition, data were collected for AMTL, caries, calculus, abscesses, and tooth wear on 325 individuals (5,549 teeth and 7,474 observable sockets) following the standards outlined in Buikstra and Ubelaker (1994), unless noted otherwise. The age and sex distribution of the sample is shown in Figure 15.4. The age categories are divided into 15-year intervals in order to approximate the stages of the life course in ancient Rome. Due to limitations in the ability to accurately age older individuals based on skeletal features (i.e., >50 years of age), the final age category consists of all individuals over the age of 45 years.

Ante-mortem tooth loss (AMTL)

Females have a slightly higher overall rate of AMTL than males (7.9 percent vs. 6.5 percent, respectively), however, a Mann-Whitney U test indicates no significant difference in the overall percentage of AMTL between the sexes (Table 15.1) ($p \leq 0.05$). When the data are analyzed by age categories, individuals younger than 15 years of age show no evidence of AMTL, but there is a consistent increase in percentage AMTL in the older age categories among both sexes (Figure 15.5). However, it is only within the oldest age group (45+) that the difference between males and females are statistically significant, with females having significantly more AMTL (Mann-Whitney U test, $p = .04$).

Table 15.1 Percentage of teeth lost ante-mortem

Age category (years)	*Total # of sockets*	*Teeth lost AM*	*% AMTL*
Females			
7–14.9	14	0	0
15–29.9	1020	11	1.1
30–44.9	1128	60	5.3
45+	1079	187	17.3
Adult	32	0	0
Total	**3273**	**258**	**7.9**
Males			
7–14.9	28	0	0
15–29.9	1123	9	0.8
30–44.9	871	55	6.3
45+	1115	140	12.6
Adult	179	10	5.6
Total	**3316**	**214**	**6.5**

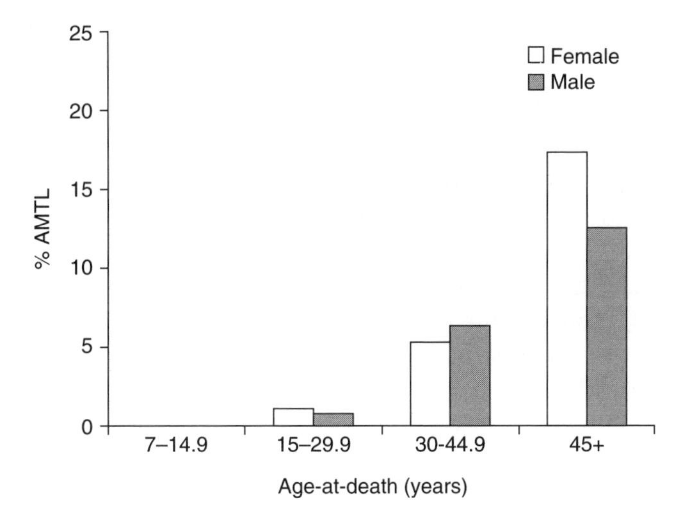

Figure 15.5 Percent AMTL by age and sex category.
Source: Author's drawing

Calculus

Calculus is present in 90 percent of the individuals with permanent teeth, although this figure only tells us that most of the people in the Isola Sacra sample had *some* degree of calculus on their teeth, but not how many teeth were affected or the amount of calculus present. To examine variability within the sample, the

Table 15.2 Calculus rate by age and sex category

Age category (years)	*Females*	*Males*
7–14.9	14.00*	8.57**
15–29.9	21.24	15.07
30–44.9	27.02	30.42
45+	44.13	32.31
Adult	30.99	28.82

*one individual; **two individuals.

presence of calculus was scored for each tooth surface (buccal, lingual, interproximal) on all teeth, as well as on the occlusal surface of molar teeth, combining the ranked scoring method of Brothwell (1981) and a modified version of the Simplified Oral Hygiene Index used by oral hygienists (Greene and Vermilion 1964). This method takes into account the potential variability in the amount of calculus on different surfaces of one tooth, and permits the calculation of a "Modified Calculus Rate" for each tooth, tooth type, and individual (Prowse 2001).[8] Using this method to quantify calculus for each individual permits a more detailed analysis of differences between sexes and age categories rather than simply reporting "mild, moderate, or severe" calculus in a sample (see Greene et al. 2005 for a similar method).

Once calculus rates were determined for each tooth and individual, averages were calculated for each age and sex category (summarized in Table 15.2). The mean calculus rate is 25.4 for males and 30.8 for females, indicating that females have more calculus on their teeth. A Mann-Whitney U test indicates that the difference between the sexes is statistically significant ($p = .010$). The results in Table 15.2 demonstrate that (with the exception of the 30–44.9 age category) females have consistently more calculus in each age group, although there is only one identified female and two males in the 7–14.9 age category. These data suggest that there was a consistent difference in the diet between males and females throughout the life course. The amount of calculus present on the permanent teeth also increases with age-at-death.

Abscesses

There are a total of 56 abscesses out of 7,426 observable sockets in the sample (0.8 percent), with the same prevalence of abscesses among males (30/3,316 = 0.9 percent) and females (25/3,287 = 0.8 percent). When abscesses are examined by number of individuals affected (vs. teeth), the results are the same, with equal proportions of males (15 percent) and females (15 percent) affected by one or more abscesses. There is also no difference in the pattern of abscesses, that is, females and males have similar numbers of single and multiple abscesses. Like the other dental health indicators examined, there is an age-related increase in the prevalence of abscesses within the sample (Table 15.3).

Table 15.3 Prevalence of abscesses by age and sex category

Sex and age category	*# of individuals*	*# with abscesses*
Female		
7–14.9	1	0
15–29.9	38	4
30–44.9	44	6
45+	41	9
Adult	3	0
	127	**19**
Male		
7–14.9	2	0
15–29.9	47	2
30–44.9	35	6
45+	48	13
Adult	9	0
	141	**21**

Caries

The prevalence of caries in the permanent dentition is quantified using the Caries Rate calculation and the Diseased Missing Index, or DMI (after Moore and Corbett 1971).[9] Both of these methods permit analysis of caries prevalence within the sample and in relation to age and sex. The Caries Rate calculation may underestimate the actual prevalence of caries, because it does not consider AMTL that may have been caused by caries, although the DMI may actually overestimate this prevalence as it assumes that all AMTL is due to caries (Hillson 2000). Thus, both the Caries Rate and DMI are presented here and the actual values are likely somewhere between the two extremes.

The female mean Caries Rate (6.6) is lower than that for males (7.8), while the DMI values are the same for both sexes (15.3). A Mann-Whitney U test confirms that the differences between the sexes are not statistically significant for either Caries Rate or DMI. When the prevalence of caries is examined by age-at-death there is a pattern of increased caries prevalence with increased age (Table 15.4). This is to be expected as caries is an age-progressive disease, so we expect to see more carious lesions in older members of the sample. The higher DMI values, particularly in the older age categories, reflect higher levels of AMTL. When the data are examined by both age and sex (Figure 15.6a), males have higher Caries Rates in all age categories (except the 7–14.9 age category, where there is only one individual). The DMI values are higher among males in the younger age categories, but females have a higher mean DMI among older adults (45+ years) (Figure 15.6b). When the prevalence of AMTL by sex is examined (Table 15.1), females in the 45+ age category have a higher prevalence of AMTL (17.3 percent) than males (12.6 percent), which explains the shift to higher DMI levels among older females; however, a Mann-Whitney U test indicates that the differences in DMI values is not statistically significant ($p \leq 0.05$).

Table 15.4 Prevalence of Caries Rate and DMI by age and sex category

Age category (years)	*Female*	*Male*
Caries Rate		
0–7	*	*
7–14.9	7.1*	0.0
15–29.9	3.4	4.2
30–44.9	6.5	9.3
45+	10.1	12.0
Adult	0.0	1.2
Average	6.6	7.8
DMI		
0–7	*	*
7–14.9	7.1*	0.0
15–29.9	4.3	5.0
30–44.9	11.5	14.9
45+	30.8	27.7
Adult	0.0	10.3
Average	15.3	15.3

*one individual.

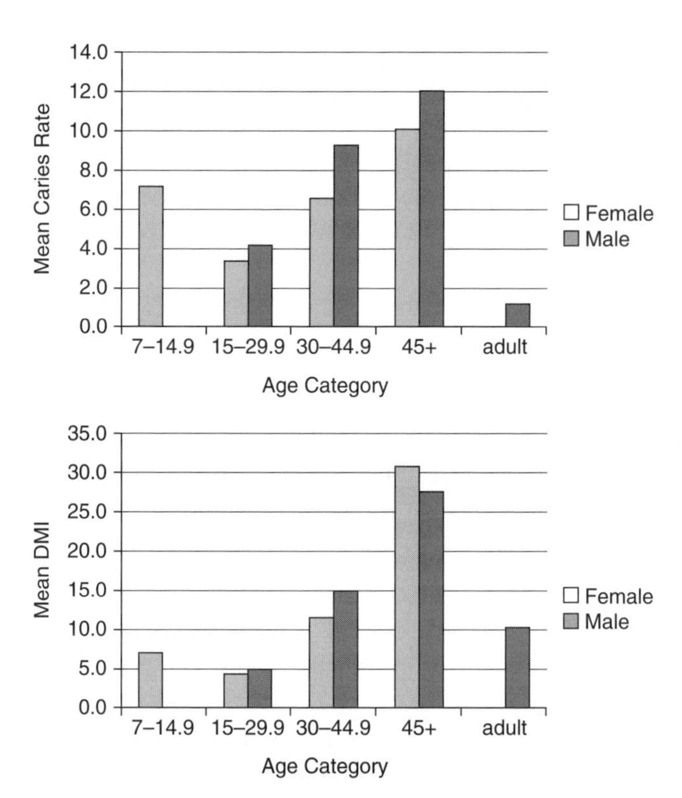

Figure 15.6 Caries Rate and DMI (Diseased Missing Index) by age and sex category.
Source: Author's drawing

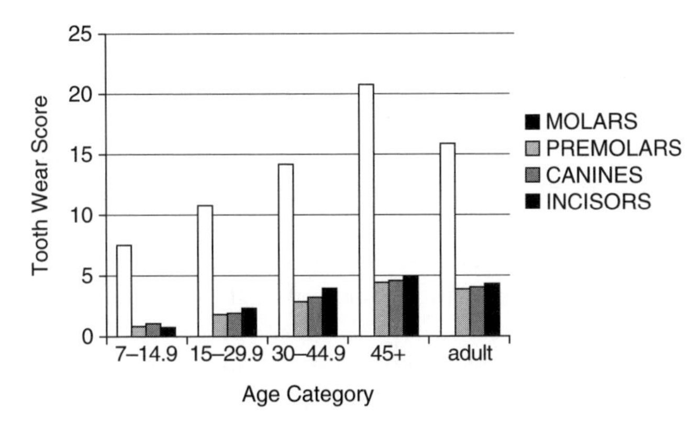

Figure 15.7 Tooth wear scores by age category.
Source: Author's drawing

Tooth wear

Tooth wear scores were recorded for all permanent teeth, using the Scott (1979) scoring system for molars and the Smith (1984) system for all other teeth. In order to investigate variability within the sample based on age and sex, tooth wear scores were calculated for each individual by adding up the scores of each tooth type (e.g., molars, premolars) in relation to the number of observable teeth present (see Prowse 2001).[10] When tooth wear is analyzed by sex, males had higher tooth wear scores for all tooth types, but these differences were not statistically significant (Mann-Whitney U test) ($p \leq 0.05$). When the data are examined by age category (sexes combined), once again a clear age-related pattern emerges. For all tooth types, there is an increase in the amount of tooth wear with age, which is not unexpected as tooth wear is an age-progressive process (Figure 15.7).

The combined dental pathology and isotopic evidence can be examined in relation to the food choices available to males and females throughout the later stages of the life course. In general, the adult Roman diet was characterized by the "Mediterranean triad" of cereals, wine, and olives, as well as dry legumes (Garnsey 1999). Literary and archaeological evidence indicates that pork was the most commonly consumed meat, although goats and sheep (and their byproducts) may have made up a significant portion of the diet (Brothwell 1988). Living in a port city on the western coast of Italy, the people buried at Isola Sacra would have had access to a wide variety of marine resources. Fish and seafood were considered expensive food items, so it is possible that only certain members of society would have regular access to them, although small fish were consumed in the form of salsamenta and garum (Curtis 1991; Frayn 1993).

Access to food is affected by social relationships between people and the relative status of an individual within the household and in society (Garnsey 1999). Historical evidence from the Roman period suggests that there was considerable control over the bodies and behaviors of women, particularly during their reproductive years. Medical writers from the first and second centuries A.D.

described appropriate health regimes for adolescent girls, primarily intended to restrict sexual development until an "appropriate" age for marriage (Garnsey 1999). It was said that young females should be "moderate" in the amount of food consumed, restrict their intake of meat and wine, and engage in work and exercise (Garnsey 1999). Ancient authors made connections between diet, exercise, and chastity, with the idea that a woman should not gain too much weight, eat rich food, or engage in strenuous physical activity lest it encourage immoral thoughts or behaviors (Alberici and Harlow 2007). Greek and Roman writers recognized the need for pregnant women to eat well and exercise regularly, although "eating well" was not clearly defined, other than the recommendation to eat "hot and dry" instead of "cold and wet" foods, particularly fish and fatty meats in order to maintain the proper balance of body humors (Garnsey 1999; Pomeroy 1995). The isotopic evidence suggests that women consumed a more terrestrial diet throughout the life course, which is consistent with the evidence from the historical records.

In contrast to the relatively rigorous and controlled lifestyle of females in their reproductive years, there do not seem to have been as many restrictions on diet and behavior as those for young males. Males possessed higher status both in the household and in Roman society; therefore they likely had greater access to certain types of food or at least to greater varieties of food. The higher $\delta^{15}N$ values of males in all age categories suggest that males were consuming higher trophic level foods, such as fish. Fishing was an important part of the economy in the ports of Rome, and fish were generally considered high status food items (Frayn 1993). As fishermen or merchants involved in the production and trade of fish and fish products (e.g., garum), males would have had greater access to marine resources. In addition, higher levels of tooth wear have been attributed to an increased reliance on seafood and the presence of sand and grit in foods (Sealy et al. 1992; Walker 1978). A higher contribution of marine foods in the diet of males may be partially responsible for the greater degree of wear among males in the sample, even though these differences are not statistically significant. There are higher levels of tooth wear among individuals with higher $\delta^{15}N$ and $\delta^{13}C$ values, which suggests that individuals consuming more marine foods in their diet were experiencing greater tooth wear.

The age-related isotopic trends can be explained by a pattern of increased consumption of meat and fish among older individuals, or those individuals who were consuming a diet rich in terrestrial and marine protein were the ones who survived into old age and ended up in the 45+ age category (Prowse et al. 2004). The overall increase in $\delta^{15}N$ values among females with increased age-at-death suggests that females in the Isola Sacra sample had greater access to higher trophic level foods at older ages. Once again, there is no physiological cause (e.g. menopause) that would explain higher $\delta^{15}N$ values among older women, so this shift represents actual differences in diet. Perhaps this relates to the changing social roles of females later in the life course and a loosening of dietary restrictions in old age. The ancient authors Galen, Plutarch, and Cicero emphasized a lifestyle that promoted continued mental and physical health in old age, consisting of a frugal diet and moderate exercise (Cokayne 2007). It was thought that the process of aging for both men and women was related to a decrease in the body's heat making older individuals dryer and harder (van Tilburg 2009). To counterbalance this, more "wet" foods (e.g., fish) may have been consumed. Unfortunately, there is very little in the ancient medical sources concerning women after menopause, although both Hippocrates and Aristotle wrote that a woman's body aged faster than a man's because of its inherently weak nature (Cokayne 2003).

The dental pathology data indicate that females tended to have more calculus present on their teeth and higher levels of AMTL, the latter becoming more pronounced among older individuals (45+ years). Without more precise age estimates for the older individuals, it is difficult to determine if the reason for the higher levels of AMTL among the older females is because they survived longer than males (i.e., had longer life expectancies) and lost more teeth over time. The higher prevalence of calculus may have provided the females some protection against the development of carious lesions, which are more prevalent among males. The alternative explanation is that males were consuming a more cariogenic diet.

The hypothesis by Lukacs and Thompson (2008) that females tend to have higher rates of caries due to life history variables (i.e., early tooth eruption, variation in dietary patterns, and hormonal changes associated with pregnancy and menopause) is not supported by the data presented here. This suggests that the prevalence of caries is affected by the changing dietary patterns of women throughout the life course. The combined higher prevalence of tooth wear and caries among the males, although not statistically significant, suggests that males were consuming a coarser and more cariogenic diet throughout the life course. The fact that males had more carious lesions even with higher levels of tooth wear suggests that they were consistently consuming food that facilitated the formation of carious lesions.

Conclusions

The combined information derived from the isotopic and dental pathology data provide us with the opportunity to explore diet throughout the life course in this Roman period sample. The integration of literary sources gives us insight into some of the potential reasons for variability between the sexes and in different stages of the life course, which can be explained within the framework of social relations and status in Roman society.

Isotopic evidence of the pattern of breastfeeding, weaning, and transitional feeding practices in this sample is further supplemented by the written evidence from the Roman period, providing a comprehensive picture of diet during the early stages of the life course. The $\delta^{13}C$ and $\delta^{15}N$ data from the infants and children reveal a pattern of weaning that is consistent with written records from the Roman period, and indicate that transitional feeding began soon after 6 months of age, and weaning was complete by 2 to 2.5 years of age. The dental pathology data demonstrate that caries, calculus, and tooth wear began early in the life course and are likely related to the complementary foods consumed by children during transitional feeding and weaning.

The transitions through different stages of the life course are marked by both biological and social changes, to which can be added the differences associated with gender (Revell 2005). The expression of these differences is manifested in the patterns of isotopic and dental pathology seen in the Isola Sacra sample. The isotopic and dental evidence quantify the relative importance of certain dietary components in the people from Isola Sacra, and also contribute to a greater understanding of dietary patterns among women and children in the sample, who are often not well represented in the ancient literature. It is clear that men and women were eating different diets and this was likely tied to the disparate status of males and females in Roman society. The various sources of information do not necessarily blend together seamlessly, but the combination of historical, archaeological, and skeletal evidence provide us with a richer, more nuanced understanding of variation in diet throughout the life course.

ACKNOWLEDGEMENTS

I would like to thank Dr. Luca Bondioli, Dr. Roberto Macchiarelli, and the Soprintendenza al Museo Nazionale Preistorico Etnografico "L. Pigorini" for permission to study the Isola Sacra skeletal sample and their continued support of this research. I would also like to thank A. Gallina Zevi, at the Superintendency of Ostia. My sincere gratitude to Dr. Shelley Saunders and Dr. Henry Schwarcz.

NOTES

1 The data are presented in the δ-notation: $\delta R = \{(R_x/R_s) - 1\} \times 1000$, where $R = {}^{15}N/{}^{14}N$ or ${}^{13}C/{}^{12}C$ (written as $\delta^{15}N$ and $\delta^{13}C$, respectively); x = sample; s = standard. The standards are PDB (Peedee Belemnite) for $\delta^{13}C$ and atmospheric N_2 (AIR) for $\delta^{15}N$. The precision of analysis is ±0.1‰ (per million) for $\delta^{13}C$ and ±0.2‰ for $\delta^{15}N$.

2 CAM (Crassulacean Acid Metabolism) plants (e.g., cacti and other succulents) are not considered here, because they are not a major component of prehistoric diets.

3 See Waters-Rist and Katzenberg (2009) for a summary of all nitrogen isotope studies on infant feeding practices.

4 Complementary foods are items that are processed specifically for consumption by infants over the age of six months (Sellen 2007).

5 Age-at-death was determined by the development and eruption of the dentition, development of the temporal and occipital bones, development and fusion of the epiphyses, and maximum diaphyseal length by Sperduti (1995).

6 See Prowse et al. (2008) for a detailed discussion of the historical sources on infant breastfeeding and weaning.

7 For a discussion of bone carbonate analysis, see Prowse et al. 2005.

8 $\text{Modified calculus rate} = \sum \frac{\text{Calculus scores for all tooth surfaces}}{\text{Number of observable tooth surfaces}}$

Tooth scores: 0–absent; 0.5–small, sporadic patches; 1–small amount (less than one-third of tooth covered); 2–moderate amount (not more than two-thirds of tooth covered); 3–large amount (more that two-thirds of tooth covered) (modified from Greene and Vermillion 1964, and Brothwell 1981).

9 Caries rate by individual $= \frac{\text{\# of carious teeth}}{\text{\# of teeth present}} \times 100$

Diseased missing index $= \frac{(\text{\# of carious teeth} + \text{\# of teeth lost ante-mortem})}{(\text{\# of observable teeth} + \text{\# of teeth lost ante-mortem})}$

10 Tooth wear score $= \sum \frac{\text{Scores for each tooth type}}{\text{\# of observable teeth for each tooth type}}$

This controls for the fact that an individual with more teeth would have a higher tooth wear score than one with fewer teeth.

REFERENCES

Alberici, L. A., and M. Harlow 2007 Age and Innocence: Female Transitions to Adulthood in Late Antiquity. *In* Constructions of Childhood in Ancient Greece and Italy. Hisperia Supplement No. 41. A. Choen, and J. B. Rutter, eds. Pp. 193–203. Princeton: The American School of Classical Studies at Athens.

Ambrose, S. H., and M. J. DeNiro 1986 The Isotopic Ecology of East African Mammals. Oecologia 69:395–406.

Ambrose, S. H., and L. Norr 1993 Experimental Evidence for the Relationship of the Carbon Isotope Ratios of Whole Diet and Dietary Protein to Those of Bone Collagen and Carbonate. *In* Prehistoric Human Bone: Archaeology at the Molecular Level. J. B. Lambert, and G. Grupe, eds. Pp. 1–36. Berlin: Springer-Verlag.

Argenti, M., and G. Manzi 1988 Morfometria Cranica delle Popolazioni Romane di Eta Imperiale: Isola Sacra e Lucus Feroniae. Rivista di Antropologia 66:179–200.

Baldassare, I. 1990 Nuove Richerche nella Necropoli dell'Isola Sacra. Quaderni di Archeologia Etrusco-Italica 19:164–172.

Bocherens, H., and D. Drucker 2003 Trophic Level Isotopic Enrichment of Carbon and Nitrogen in Bone Collagen: Case Studies from Recent and Ancient Terrestrial Ecosystems. International Journal of Osteoarchaeology 13:46–53.

Bondioli, L., and R. Macchiarelli eds. 1999 Osteodental Biology of the People of Portus Romae (Necropolis of Isola Sacra, 2nd–3rd century A.D.). Enamel Microstructure and Developmental Defects of the Primary Dentition. Rome: National Prehistoric Ethnographic "'L. Pigorini'" Museum, CD-ROM.

Bradley, K. R. 1998 The Roman Family at Dinner. *In* Meals in a Social Context. I. Nielsen and H. S. Nielsen, eds. Pp. 36–55. Aarhus: Aarhus University Press.

Brandt, H. 2002 Wird Auch Silbern Mein Haar: Eine Geschichte des Alters in der Antike. Munich: C. H. Beck.

Brothwell, D. R. 1981 Digging Up Bones. Ithaca, NY: Cornell University Press.

Brothwell, D. R. 1988 Foodstuffs, Cooking, and Drugs. *In* Civilization of the Ancient Mediterranean: Greece and Rome. M. Grant, and R. Kitzinger, eds. Pp. 247–261. New York: Charles Scribner.

Buikstra, J. E., and D. H. Ubelaker 1994 Standards for Data Collection from Human Skeletal Remains. Fayetteville, Arkansas: Arkansas Archaeological Survey.

Chisholm, B. S., D. E. Nelson, N. Tuross, and H. P. Schwarcz 1982 Stable-carbon Isotope Ratios as a Measure of Marine Versus Terrestrial Protein in Ancient Diets. Science 216:1131–1132.

Cho, H., and S. D. Stout 2003 Bone Remodeling and Age-associated Bone Loss in the Past: A Histomorphometric Analysis of the Imperial Roman Skeletal Population of Isola Sacra. *In* Bone Loss and Osteoporosis: An Anthropological Perspective. S. C. Agarwal, and S. D. Stout eds. Pp. 207–228. New York: Kluwer Academic/Plenum Publishers.

Choy, K., and M. P. Richards 2009 Stable Isotope Evidence of Human Diet at the Nukdo Shell Midden Site, South Korea. Journal of Archaeological Science 36:1312–1318.

Clarke, N. G., and R. S. Hirsch 1991 Physiological, Pulpal, and Periodontal Factors Influencing Alveolar Bone. *In* Advances in Dental Anthropology. M. A. Kelley, and C. S. Larsen, eds. Pp. 241–266. New York: Wiley-Liss.

Clayton, F., J. Sealy, and S. Pfeiffer 2005 Weaning Age Among Foragers at Matjes River Rock Shelter, South Africa, from Stable Nitrogen and Carbon Isotope Analyses. American Journal of Physical Anthropology 129:311–317.

Cohen, A., and J. B. Rutter eds. 2007 Constructions of Childhood in Ancient Greece and Italy. Hisperia Supplement, 41. Princeton: The American School of Classical Studies at Athens.

Cokayne, K. 2003 Experiencing Old Age in Ancient Rome. London: Routledge.

Cokayne, K. 2007 Age and Aristocratic Self-identity: Activities for the Elderly. *In* Age and Ageing in the Roman Empire. Journal of Roman Archaeology, Supplement No. 65. M. Harlow, and R. Laurence, eds. Pp. 209–220. Portsmouth, RI: Journal of Roman Archaeology.

Coltrain, J. B., and S. W. Leavitt 2002 Climate and Diet in Fremont Prehistory: Economic Variability and Abandonment of Maize Agriculture in the Great Salt Lake Basin. American Antiquity 67(3):453–485.

Curtis, R. I. 1991 Garum and Salsamenta. New York: E. J. Brill.

Demmer, R. T., and M. Desvarieux 2006 Periodontal Infections and Cardiovascular Disease: The Heart of the Matter. Journal of the American Dental Association 137:14S–20S.

DeNiro, M. J., and S. Epstein 1978 Influence of Diet on the Distribution of Carbon Isotopes in Animals. Geochimica et Cosmochimica Acta 42:495–506.

DeNiro, M. J., and S. Epstein 1981 Influence of Diet on the Distribution of Nitrogen Isotopes in Animals. Geochimica et Cosmochimica Acta 45:341–351.

DeNiro, M. J., and M. J. Schoeninger 1983 Stable Carbon and Nitrogen Isotope Ratios of Bone Collagen: Variations Within Individuals, Between Sexes, and Within Populations Raised on Monotonous Diets. Journal of Archaeological Science 10:199–203.

Dixon, S. 1988 The Roman Mother. London: Croom Helm.

Dixon, S. ed. 2001 Childhood, Class, and Kin in the Roman World. London: Routledge.

Dupras, T. L., H. P. Schwarcz, and S. I. Fairgrieve 2001 Infant Feeding and Weaning Practices in Roman Egypt. American Journal of Physical Anthropology 115:204–211.

Dupras, T. L, and M. W. Tocheri 2007 Reconstructing Infant Weaning Histories at Roman Period Kellis, Egypt Using Stable Isotope Analysis of Dentition. American Journal of Physical Anthropology 134:63–74.

Elder, G. H. Jr. ed. 1985 Life Course Dynamics: Trajectories and Transitions, 1968–1980. Ithaca: Cornell University Press.

Elder, G. H. Jr. ed. 1994 Time, Human Agency, and Social Change: Perspectives on the Life Course. Social Psychology Quarterly 57(1):4–15.

Elder, G. H. Jr. 1998 The Life Course as Developmental Theory. Child Development 69(1):1–12.

Fildes, V. 1986 Breast, Bottles, and Babies: A History of Infant Feeding. Edinburgh: Edinburgh University Press.

FitzGerald, C. M., S. R. Saunders, R. Macchiarelli, and L. Bondioli 2006 Health of Infants in an Imperial Roman Skeletal Sample: Perspective from Dental Microstructure. American Journal of Physical Anthropology 130:179–189.

Fogel, M. L., N. Tuross, B. J. Johnson, and G. H. Miller 1997 Biogeochemical Record of Ancient Humans. Organic Geochemistry 27:275–287.

Fraschetti, A. 1998 Roman Youth. *In* History of Young People in the West. Volume 1: Ancient and Medieval Rights of Passage. G. Levi, and J.-C. Schmitt, eds. Pp. 51–82. Cambridge, MA: Belknap Press.

Frayn, J. M. 1993 Markets and Fairs in Roman Italy. Oxford: Clarendon Press.

Fuller, B. T., J. L. Fuller, D. A. Harris, and R. E. M. Hedges 2006 Detection of Breastfeeding and Weaning in Modern Human Infants with Carbon and Nitrogen Stable Isotope Ratios. American Journal of Physical Anthropology 129:279–293.

Fuller, B. T., J. L. Fuller, N. E. Sage, D. A. Harris, T. C. O'Connell, and R. E. M. Hedges 2005 Nitrogen Balance and δ^{15}N: Why You're Not What You Eat During Nutritional Stress. Rapid Communications in Mass Spectrometry 19:2497–2506.

Garnsey, P. 1998 The People of Isola Sacra. *In* Enamel Microstructure and Developmental Defects of the Primary Dentition. Osteodental Biology of the People of Portus Romae (Necropolis of Isola Sacra, 2nd–3rd Cent. AD. I. P. F. Rossi, L. Bondioli, G. Geusa, and R. Macchiarelli, eds. Pp. 1–4. Rome: Soprintendenza Speciale al Museo Nazionale Preistorico Etnografico "L. Pigorini', CD-ROM.

Garnsey, P. 1999 Food and Society in Classical Antiquity. Cambridge: Cambridge University Press.

Geusa, G., L. Bondioli, E. Capucci, A. Cipriano, G. Grupe, C. Savorè, and R. Machiarelli 1999 Osteodental Biology of the People of Portus Romae (Necropolis of Isola Sacra, 2n –3rd Cent. A.D.). Vol. 2, Dental Cementum Annulations and Age at Death Estimates. Rome: Soprintendenza Speciale al Museo Nazionale Preistorico Etnografico, CD-ROM.

Giele, J. Z., and G. H. Elder Jr. 1998 Methods of Life Course Research: Qualitative and Quantitative Approaches. London: Sage.

Goodman, A. H., and T. L. Leatherman 1998 Building a New Biocultural Synthesis: Political-Economic Perspectives on Human Biology. Ann Arbor: University of Michigan Press.

Greene, J. C., and J. R. Vermillion 1964 The Simplified Oral Hygiene Index. Journal of the American Dental Association 68:7–13.

Greene, T. R., C. L. Kuba, and J. T. Irish 2005 Quantifying Calculus: A Suggested New Approach for Recording an Important Indicator of Diet and Dental Health. HOMO 56:119–132.

Harlow, M. 2007 Blurred Visions: Male Perceptions of the Female Life Course – The Case of Aemilia Pudentilla. *In* Age and Ageing in the Roman Empire. Journal of Roman Archaeology, Supplement No. 65. M. Harlow, and R. Laurence, eds. Pp.195–208. Portsmouth, RI: Journal of Roman Archaeology.

Harlow, M., and R. Laurence 2002 Growing Up and Growing Old in Ancient Rome: A Life Course Approach. London: Routledge.

Hedges, R. E. M., and L. M. Reynard 2007 Nitrogen Isotopes and the Trophic Level of Humans in Archaeology. Journal of Archaeological Science 34:1240–1251.

Herring, D. A., S. R. Saunders, and M. A. Katzenberg 1998 Investigating the Weaning Process in Past Populations. American Journal of Physical Anthropology 105:425–439.

Hillson, S. 1996 Dental Anthropology. Cambridge: Cambridge University Press.

Hillson, S. 2000 Dental Pathology. *In* Biological Anthropology of the Human Skeleton. M. A. Katzenberg, and S. R. Saunders, eds. Pp. 249–286. New York: Wiley-Liss.

Hoover, K. C., R. S. Corruccini, L. Bondioli, and R. Macchiarelli 2005 Exploring the Relationship Between Hypoplasia and Odontometric Asymmetry in Isola Sacra, an Imperial Roman Necropolis. American Journal of Human Biology 17(6):752–764.

Humphrey, L. T., and T. King 2000 Childhood Stress: A Lifetime Legacy. Anthropologie XXXVII/1:33–49.

Jay, M., B. T. Fuller, M. P. Richards, C. J. Knüsel, and S. S. King 2008 Iron Age Breastfeeding Practices in Britain: Isotopic Evidence from Wetwang Slack, East Yorkshire. American Journal of Physical Anthropology 136:327–337.

Jim, S., S. H. Ambrose, and R. P. Evershed 2004 Stable Carbon Isotopic Evidence for Differences in the Dietary Origin of Bone Cholesterol, Collagen, and Apatite: Implications for Their Use in Palaeodietary Reconstruction. Geochimica et Cosmochimica Acta 68(1): 61–72.

Jurmain, R. 1990 Paleoepidemiology of a Central California Prehistoric Population from CA-Ala-329: Dental Disease. American Journal of Physical Anthropology 81(3): 333–342.

Katzenberg, M. A. 2008 Stable Isotope Analysis: A Tool for Studying Past Diet, Demography, and Life History. *In* Biological Anthropology of the Human Skeleton, 2nd edition. M. A. Katzenberg, and S. R. Saunders, eds. Pp. 413–442. Hoboken, NJ: John Wiley & Sons, Inc.

Katzenberg, M. A., and R. G. Harrison 1997 What's in a Bone? Recent Advances in Archaeological Bone Chemistry. Journal of Archaeological Research 5:265–293.

Katzenberg, M. A., and S. Pfeiffer 1995 Nitrogen Isotope Evidence for Weaning. *In* Bodies of Evidence: Reconstructing History Through Skeletal Analysis. A. L. Grauer, ed. Pp. 221–235. New York: Wiley-Liss.

Katzenberg, M. A., A. Herring, and S. R. Saunders 1996 Weaning and Infant Mortality: Evaluating the Skeletal Evidence. Yearbook of Physical Anthropology 39:177–199.

Katzenberg, M. A., and N. C. Lovell 1999 Stable Isotope Variation in Pathological Bone. International Journal of Osteoarchaeology 9:316–324.

Keenleyside, A. 2006 Stable Isotopic Evidence of Diet in a Greek Colonial Population from the Black Sea. Journal of Archaeological Science 33(9):1205–1215.

Kuh, D., and Y. Ben-Schlomo eds. 2004 A Life Course Approach to Chronic Disease Epidemiology. 2nd Edition. Oxford: Oxford University Press.

Larsen, C. S. 1997 Bioarchaeology: Interpreting Behavior from the Human Skeleton. Cambridge: Cambridge University Press.

Lieverse, A. R. 1999 Diet and the Aetiology of Dental Calculus. International Journal of Osteoarchaeology 9:219–232.

Littleton, J., and B. Frohlich 1993 Fish-eaters and Farmers: Dental Pathology in the Arabian Gulf. American Journal of Physical Anthropology 92:427–447.

Longin, R. 1971 New Method for Collagen Extraction for Radiocarbon Dating. Nature 230:241–242.

Lovell, N., D. E. Nelson, and H. P. Schwarcz 1986 Carbon Isotopes in Paleodiet: Lack of Age or Sex Effect. Archaeometry 28(1):51–55.

Lukacs, J. R. 1989 Dental Paleopathology: Methods for Reconstructing Dietary Patterns. *In* Reconstruction of Life from the Skeleton. M. Y. Iscan, and K. A. R. Kennedy, eds. Pp. 261–286. New York: Liss.

Lukacs, J. R., and L. M. Thompson 2008 Dental Caries Prevalence by Sex in Prehistory: Magnitude and Meaning. *In* Technique and Application in Dental Anthropology. J. D. Irish, and G. C. Nelson, eds. Pp. 136–177. Cambridge: Cambridge University Press.

Manzi, G., L. Censi, A. Sperduti, and P. Passarello 1989 Linee di Harris e Ipoplasia dello Smalto nei Resti Scheletrici delle Popolazioni Umane di Isola Sacra e Lucus Feroniae (Roma, I–III sec. d.C.). Rivista di Antropologia 67:129–148.

Manzi, G., and P. Passarello 1991 Behavior-induced Auditory Exostoses in Imperial Roman Society: Evidence from Coieval Urban and Rural Communities near Rome. American Journal of Physical Anthropology 85(3):253–260.

Manzi, G., L. Salvadei, A. Vienna, and P. Passarello 1999 Discontinuity of Life Conditions at the Transition from the Roman Imperial Age to the Early Middle Ages: Example from

Central Italy Evaluated by Pathological Dento-alveolar Lesions. American Journal of Human Biology 11(3):327–341.

Manzi, G., E. Santandrea, and P. Passarello 1997 Dental Size and Shape in the Roman Imperial Age: Two Examples from the Area of Rome. American Journal of Physical Anthropology 102(4):469–479.

Manzi, G., and A. Sperduti 1988 Variabilita Morfologica nei Campioni Cranici di Isola Sacra e Lucus Feroniae (Roma, I-III Secolo d.C.). Rivista di Antropologia 66:201–216.

Minagawa, M., and E. Wada 1984 Stepwise Enrichment of ^{15}N Along Food Chains: Further Evidence and the Relation Between δ^{15}N and Animal Age. Geochimica et Cosmochimica Acta 48:1135–1140.

Molleson, T., and K. Jones 1991 Dental Evidence for Dietary Change at Abu Hureyra. Journal of Archaeological Science 18:525–539.

Molnar, S. 1972 Tooth Wear and Culture: A Survey of Tooth Functions Among Some Prehistoric Populations. Current Anthropology 13:511–525.

Moore W. J., and M. E. Corbett 1971 Distribution of Dental Caries in Ancient British Populations: I Anglo-Saxon Period. Caries Research 5:151–168.

Parkin, T. G. 2003 Old Age in the Roman World: A Cultural and Social History. Baltimore: Johns Hopkins University Press.

Pate, F. D. 1994 Bone Chemistry and Paleodiet. Journal of Archaeological Method and Theory 1:161–209.

Pomeroy, S. B. 1995 Goddesses, Whores, Wives, and Slaves: Women in Classical Antiquity. New York: Schocken Books.

Powell, M. L. 1985 The Analysis of Dental Wear and Caries for Dietary Reconstruction: *In* The Analysis of Prehistoric Diets. R. I. Gilbert Jr., and J. H. Mielke, eds. Pp. 307–388. Orlando: Academic Press.

Prowse, T. L. 2001 Isotopic and Dental Evidence for Diet from the Necropolis of Isola Sacra (1st–3rd Centuries AD), Italy. Unpublished Ph.D. Dissertation. McMaster University, Hamilton, Ontario.

Prowse, T. L., S. R. Saunders, C. FitzGerald, L. Bondioli, and R. Macchiarelli In press Growth, Morbidity, and Mortality in Antiquity: A Case Study from Imperial Rome. *In* Human Diet and Nutrition in Biocultural Perspective. T. Moffat, and T. L. Prowse, eds. Studies of the Biosocial Society. New York: Berghahn Press.

Prowse, T. L., S. R. Saunders, H. P. Schwarz, P. Garnsey, R. Macchiarelli, and L. Bondioli 2008 Isotopic and Dental Evidence for Infant and Young Child Feeding Practices in an Imperial Roman Skeletal Sample. American Journal of Physical Anthropology 137(3):294–308.

Prowse, T. L., H. P. Schwarcz, P. Garnsey, L. Bondioli, R. Macchiarelli, and M. Knyf 2007 Isotopic Evidence for Age-related Immigration to Imperial Rome. American Journal of Physical Anthropology 132(4):510–519.

Prowse, T. L., H. P. Schwarcz, S. R. Saunders, L. Bondioli, and R. Macchiarelli 2005 Isotopic Evidence for Age-related Variation in Diet from Isola Sacra, Italy. American Journal of Physical Anthropology 128(1):2–13.

Prowse, T. L., H. P. Schwarcz, S. R. Saunders, R. Macchiarelli, and L. Bondioli 2004 Isotopic Paleodiet Studies of Skeletons from the Imperial Roman-age Cemetery of Isola Sacra, Italy. Journal of Archaeological Science 31(3):259–272.

Rawson, B. 1999 The Iconography of Roman Childhood. *In* The Roman Family in Italy: Status, Sentiment, Space. B. Rawson, and P. Weaver, eds. Pp. 205–238. Oxford: Oxford University Press.

Rawson, B. 2003 Children and Childhood in Roman Italy. Oxford: Oxford University Press.

Revell, L. 2005 The Roman Life Course: A View from the Inscriptions. European Journal of Archaeology 8(1):43–63.

Richards M. P., B. T. Fuller, and T. I. Molleson 2006 Stable Isotope Palaeodietary Study of Humans and Fauna from the Multi-period (Iron Age, Viking and Late Medieval) Site of Newark Bay, Orkney. Journal of Archaeological Science 33(1):122–131.

Richards, M. P., and R. E. M. Hedges 1999 Stable Isotope Evidence for Similarities in the Types of Marine Foods Used by Late Mesolithic Humans at Sites Along the Atlantic Coast of Europe. Journal of Archaeological Science 26:717–722.

Richards, M. P., S. Mays, and B. T. Fuller 2002 Stable Carbon and Nitrogen Isotope Values of Bone and Teeth Reflect Weaning Age at the Medieval Wharram Percy Site, Yorkshire, UK. American Journal of Physical Anthropology 119:205–210.

Roth, J. D., and K. A. Hobson 2000 Stable Carbon and Nitrogen Isotopic Fractionation Between Diet and Tissue of Captive Red Fox: Implications for Dietary Reconstruction. Canadian Journal of Zoology 78:848–852.

Savorè, C. 1996 I Rest Scheletrici Umani di Isola Sacra: Stimà dell'Eta alla Morte con Metodo Istomorfometrico ed Analisi Paleoistologica di un Campione. Master's Thesis. Università di Pavia, Pavia, Italy.

Schoeninger, M. J., and M. J. DeNiro 1984 Nitrogen and Carbon Isotopic Composition of Bone Collagen from Marine and Terrestrial Animals. Geochimica et Cosmochimica Acta 48:625–639.

Schoeninger M. J., M. J. DeNiro, and H. Tauber 1983 Stable Nitrogen Isotope Ratios in Bone Collagen Reflect Marine and Terrestrial Components of Prehistoric Human Diet. Science 220:1381–1383.

Schurr, M. R. 1997 Stable Isotopes as Evidence for Weaning at the Angel Site: A Comparison of Isotopic and Demographic Measures of Weaning Age. Journal of Archaeological Science 24:919–927.

Schurr, M. R. 1998 Using Stable Nitrogen Isotopes to Study Weaning Behavior in Past Populations. World Archaeology 30:327–342.

Schurr, M. R., and M. L. Powell 2005 The Role of Changing Childhood Diets in the Prehistoric Evolution of Food Production: An Isotopic Assessment. American Journal of Physical Anthropology 126:278–294.

Schwarcz, H. P. 2000 Some Biochemical Aspects of Carbon Isotopic Paleodiet Studies. *In* Biogeochemical Approaches to Paleodietary Analysis. S. H. Ambrose, and M. A. Katzenberg, eds. Pp. 189–210. New York: Kluwer Academic/Plenum Publishers.

Schwarcz, H. P., T. L. Dupras, and S. I. Fairgrieve 1999 ^{15}N-enrichment in the Sahara: In Search of a Global Relationship. Journal of Archaeological Science 26:629–636.

Schwarcz , H. P., J. Melbye, M. A. Katzenberg, and M. Knyf 1985 Stable Isotopes in Human Skeletons of Southern Ontario: Reconstructing Paleodiet. Journal of Archaeological Science 12:187–206.

Schwarcz, H. P., and M. J. Schoeninger 1991 Stable Isotope Analyses in Human Nutritional Ecology. Yearbook of Physical Anthropology 34:283–321.

Scott, E. C. 1979 Dental Wear Scoring Technique. American Journal of Physical Anthropology 51:213–218.

Scott, G. R., and C. G. Turner II 1988 Dental Anthropology. Annual Review of Anthropology 17:99–126.

Sealy, J. 2001 Body Tissue Chemistry and Palaeodiet. *In* Handbook of Archaeological Science. D. R. Brothwell, and A. M. Pollard, eds. Pp. 269–279. Chichester: Wiley.

Sealy, J. C., N. J. van der Merwe, J. A. Lee Thorp, and J. L. Lanham 1987 Nitrogen Isotopic Ecology in Southern Africa: Implications for Environmental and Dietary Tracing. Geochimica et Cosmochimica Acta 51:2707–2717.

Sellen, D. W. 2007 Evolution of Infant and Young Child Feeding: Implications for Contemporary Public Health. Annual Review of Nutrition 27:123–148.

Sharpe, W. D. 1964 Isidore of Seville: The Medical Writings. An English Translation with an Introduction and Commentary. Transactions of the American Philosophical Society 54(2):1–75.

Smith, B. H. 1984 Patterns of Molar Wear in Hunter-gatherers and Agriculturalists. American Journal of Physical Anthropology 63:39–56.

Sofaer, J. R. 2006 The Body as Material Culture: A Theoretical Osteoarchaeology. Cambridge: Cambridge University Press.

Sperduti, A. 1995 I Resti Scheletrici Umani Della Necropoli Di Età Romano Imperiale Di Isola Sacra (I-III Sec DC): Analisi Paleodemografica. Ph.D. Dissertation. Rome, Università di Roma La Sapienza, Rome, Italy.

Temkin, O. (translator) 1956 Soranus of Ephesus. Gynecology. Baltimore: Johns Hopkins Press.

Tiezen, L. L., and T. Fagre 1993 Effect of Diet Quality and Composition on the Isotope Composition of Respiratory CO2, Bone Collagen, Bioapatite and Soft Tissues. *In* Prehistoric Human Bone: Archaeology at the Molecular Level. J. B. Lambert, and G. Grupe, eds. Pp. 121–155. Berlin: Springer-Verlag.

van der Merwe, N. J., and E. Medina 1991 The Canopy Effect, Carbon Isotope Ratios and Foodwebs in Amazonia. Journal of Archaeological Science 18:249–259.

Vanderklift, M. A., and S. Ponsard 2003 Sources of Variation in Consumer-diet δ^{15}N Enrichment: A Meta-analysis. Stable Isotope Ecology 136:169–182.

van Tilburg, M. 2009 Tracing Sexual Identities in "Old Age": Gender and Seniority in Advice Literature of the Early-modern and Modern Periods. Journal of Family History 34(4):369–386.

Walker, P. L. 1978 A Quantitative Analysis of Dental Attrition Rates in the Santa Barbara Channel Area. American Journal of Physical Anthropology 48:101–106.

Walker, P. L., and M. J. DeNiro 1986 Stable Nitrogen and Carbon Isotope Ratio in Bone Collagen as Indices of Prehistoric Dietary Dependence on Marine and Terrestrial Resources in Southern California. American Journal of Physical Anthropology 71:51–61.

Waters-Rist, A. L., and M. A. Katzenberg 2009 The Effect of Growth on Stable Nitrogen Isotope Ratios in Subadult Bone Collagen. International Journal of Osteoarchaeology. Early View. DOI: 10.1002/oa.1017.

White, C. D., and G. J. Armelagos 1997 Osteopenia and Stable Isotope Ratios in Bone Collagen of Nubian Female Mummies. American Journal of Physical Anthropology 103:185–199.

Wiedemann, T. 1989 Adults and Children in the Roman Empire. London: Routledge.

Wilkinson, M. J., Y. Youlin, and D. M. O'Brien 2007 Age-related Variation in Red Blood Cell Stable Isotope Ratios (δ^{13}C and δ^{15}N) from Two Yupik Villages in Southwest Alaska: A Pilot Study. International Journal of Circumpolar Health 66:31–41.

Witt, G. B., and L. K. Ayliffe 2001 Carbon Isotope Variability in the Bone Collagen of Red Kangaroos (*Macropus rufus*) is Age Dependent: Implications for Palaeodietary Studies. Journal of Archaeological Science 28:247–252.

Wood, C. 2004 An Investigation of the Prevalence of Rickets among Subadults from the Roman Necropolis of Isola Sacra (1st to 3rd centuries AD), Italy. Master's Thesis, McMaster University.

Wood, J. W., G. R. Milner, H. C. Harpending, and K. M. Weiss 1992 The Osteological Paradox: Problems in Inferring Prehistoric Health from Skeletal Samples. Current Anthropology 33:343–358.

Yang, D. 1998 DNA Diagnosis of Thalassemia from Ancient Italian Skeletons. Ph.D. Dissertation, McMaster University, Hamilton, Ontario.

Index

Social Bioarchaeology. Edited by Sabrina C. Agarwal and Bonnie A. Glencross

© 2011 Blackwell Publishing Ltd